T0212483

Lecture Notes in Computer Science 10041

Commenced Publication in 1973
Founding and Former Series Editors:
Gerhard Goos, Juris Hartmanis, and Jan van Leeuwen

More information about this series at http://www.springer.com/series/7409

Wojciech Cellary · Mohamed F. Mokbel
Jianmin Wang · Hua Wang
Rui Zhou · Yanchun Zhang (Eds.)

Web Information Systems Engineering – WISE 2016

17th International Conference
Shanghai, China, November 8–10, 2016
Proceedings, Part I

 Springer

Editors

Wojciech Cellary
Poznań University of Economics
Poznan
Poland

Hua Wang
Victoria University
Melbourne, VIC
Australia

Mohamed F. Mokbel
University of Minnesota
Minneapolis, MN
USA

Rui Zhou
Victoria University
Melbourne, VIC
Australia

Jianmin Wang
Tsinghua University
Beijing
China

Yanchun Zhang
Victoria University
Melbourne, VIC
Australia

ISSN 0302-9743   ISSN 1611-3349   (electronic)
Lecture Notes in Computer Science
ISBN 978-3-319-48739-7   ISBN 978-3-319-48740-3   (eBook)
DOI 10.1007/978-3-319-48740-3

Library of Congress Control Number: 2016955509

LNCS Sublibrary: SL3 – Information Systems and Applications, incl. Internet/Web, and HCI

This Springer imprint is published by Springer Nature
The registered company is Springer International Publishing AG
The registered company address is: Gewerbestrasse 11, 6330 Cham, Switzerland

Preface

Welcome to the proceedings of the 17th International Conference on Web Information Systems Engineering (WISE 2016), held in Shanghai, China, during November 8–10, 2016. The series of WISE conferences aims to provide an international forum for researchers, professionals, and industrial practitioners to share their knowledge in the rapidly growing area of Web technologies, methodologies, and applications. The first WISE event took place in Hong Kong, China (2000). Then the trip continued to Kyoto, Japan (2001); Singapore (2002); Rome, Italy (2003); Brisbane, Australia (2004); New York, USA (2005); Wuhan, China (2006); Nancy, France (2007); Auckland, New Zealand (2008); Poznan, Poland (2009); Hong Kong, China (2010); Sydney, Australia (2011); Paphos, Cyprus (2012); Nanjing, China (2013); Thessaloniki, Greece (2014); Miami, USA (2015); and this year, WISE 2016 was held in Shanghai, China, supported by Fudan University, China.

A total of 233 research papers were submitted to the conference for consideration, and each paper was reviewed by at least three reviewers. Finally, 39 submissions were selected as full papers (with an acceptance rate of 16.7 % approximately), plus 31 as short papers. The research papers cover the areas of social network data analysis, recommender systems, topic modeling, data diversity, data similarity, context-aware recommendation, prediction, big data processing, cloud computing, event detection, data mining, sentiment analysis, ranking in social networks, microblog data analysis, query processing, spatial and temporal data, graph theory and non traditional environments.

In addition to regular and short papers, the WISE 2016 program also featured a special session on Data Quality and Trust in Big Data (QUAT-16) and a medical big data forum.

QUAT is a forum for presenting and discussing novel ideas and solutions related to the problems of exploring, assessing, monitoring, improving, and maintaining the quality of data and trust for big data. It aims to provide researchers in the areas of web technology, e-services, social networking, big data, data processing, trust, and information systems and GIS with a forum for discussing and exchanging their recent research findings and achievements. This year, the QUAT 2016 program featured eight accepted papers on data cleansing, data quality analytics, reliability assessment, and quality of service for domain applications. As the organizers of QUAT 2016, Prof. Deren Chen, Prof. William Song, Dr. Xiaolin Zheng, Dr. Johan Håansson, and Prof. Shaozhong Zhang, we would like to thank all the authors for their enthusiastic high-quality submissions, the reviewers (Program Committee members) for their careful and timely reviews, and the Organizing Committee, Dr. Roger Nyberg, Dr. Zukun Yu, Dr. Xiaofeng Du, and Dr. Xiaoyun Zhao, for their excellent publicity.

The medical big data forum aims to promote the analysis and application of big data in healthcare. Experts and companies related to the domain of big data in healthcare were invited to present their reports in this forum. Many hot research points of big data in healthcare were discussed, including analysis and mining, application and value exploration, interoperability standards, security and privacy protection. The objective of this forum is to provide forward-looking ideas and views for research and application of big data in healthcare, which will promote the development of big data in healthcare, accelerate practical research, and facilitate the innovation and industrial development of mobile healthcare. The forum was organized by Prof. Yan Jia, Prof. Weihong Han, and Prof. Hua Wang.

We also wish to take this opportunity to thank the honorary conference chair, Prof. Maria Orlowska; the general co-chairs, Prof. Hong Mei, Prof. Marek Rusinkiewicz, and Prof. Yanchun Zhang; the program co-chairs, Prof. Wojciech Cellary, Prof. Mohamed F. Mokbel, and Prof. Jianmin Wang; the special area chairs, Prof. Xueqi Cheng, Prof. Yan Jia, and Prof. Jianhua Ma; the workshop co-chairs, Prof. Zhiguo Gong and Prof. Yong Tang; the tutorial and panel chair, Prof. Xuemin Lin; the publication co-chairs, Prof. Hua Wang and Dr. Rui Zhou; the publicity co-chairs, Dr. Jing Yang and Dr. Quan Bai; the website chair, Dr. Rui Zhou; the local arrangements chair, Prof. Shangfei Zhu; the finance co-chairs, Ms. Lanying Zhang and Ms. Irena Dzuteska; the sponsor chair, Dr. Tao Li; and the WISE society representative, Prof. Xiaofang Zhou. The editors and chairs are grateful to Ms. Sudha Subramani and Mr. Sarathkumar Rangarajan for their help in preparing the proceedings and updating the conference website.

We would like to sincerely thank our keynote and invited speakers:

- Professor Maria Orlowska, Fellow of the Australian Academy of Sciences, Vice-President of the Polish-Japanese Institute of Information Technology, former Secretary of State in the Ministry of Science and Higher Education, Poland
- Professor Binxing Fang, academician of CAE (Chinese Academy of Engineering) and the former president of BUPT (Beijing University of Posts and Telecommunications), China
- Dr. Phil Neches, Advisor, Member of National Academy of Engineering, Chairman of Foundation Ventures LLC, founder of Teradata Corporation, USA
- Professor Ramamohanarao (Rao) Kotagiri, Fellow of the Institute of Engineers Australia, Fellow of the Australian Academy Technological Sciences and Engineering, and Fellow of Australian Academy of Science, The University of Melbourne, Australia.

In addition, special thanks are due to the members of the international Program Committee and the external reviewers for the rigorous and robust reviewing process. We are also grateful to Fudan University, China, Victoria University, Australia, and the International WISE Society for supporting this conference. The WISE Organizing Committee is also grateful to the QUAT special session organizers and medical big data forum organizers for their great efforts to help promote web information system research to broader domains.

We expect that the ideas that emerged at WISE 2016 will result in the development of further innovations for the benefit of scientific, industrial, and social communities.

November 2016
Wojciech Cellary
Mohamed F. Mokbel
Jianmin Wang
Hua Wang
Rui Zhou
Yanchun Zhang

Organization

Honorary Conference Chair

Maria Orlowska — Polish-Japanese Institute of Information Technology, Poland

General Co-chairs

Hong Mei — Shanghai Jiao Tong University, China
Marek Rusinkiewicz — New Jersey Institute of Technology, USA
Yanchun Zhang — Victoria University, Australia and Fudan University, China

Program Co-chairs

Wojciech Cellary — Poznań University of Economics, Poland
Mohamed F. Mokbel — University of Minnesota, USA
Jianmin Wang — Tsinghua University, China

Special Area Chairs

Big Data Area Chair

Xueqi Cheng — Chinese Academy of Sciences, China

Medical Big Data Analysis Area Chair

Yan Jia — National University of Defense Technology, China

Transparent Computing and Service Area Chair

Jianhua Ma — Hosei University, Japan

Tutorial and Panel Chair

Xuemin Lin — The University of New South Wales, Australia and East China Normal University, China

Workshop Co-chairs

Zhiguo Gong — University of Macau, Macau, China
Yong Tang — South China Normal University, China

Publication Co-chairs

Hua Wang	Victoria University, Australia
Rui Zhou	Victoria University, Australia

Publicity Co-chairs

Jing Yang	Chinese Academy of Sciences, China
Quan Bai	Auckland University of Technology, New Zealand

Conference Website Chair

Rui Zhou	Victoria University, Australia

Conference Finance Co-chairs

Lanying Zhang	Fudan University, China
Irena Dzuteska	Victoria University, Australia

Local Arrangements Chair

Shangfeng Zhu	Fudan University, China

Sponsorship Chair

Tao Li	Florida International University, USA

Wise Society Representative

Xiaofang Zhou	The University of Queensland, Australia and Soochow University, China

Program Committee

Karl Aberer	EPFL, Switzerland
Imad Afyouni	GIS Technology Innovation Center, Saudi Arabia
Marco Aiello	University of Groningen, The Netherlands
Mohammed Eunus Ali	Bangladesh University of Engineering and Technology, Bangladesh
Toshiyuki Amagasa	University of Tsukuba, Japan
Farnoush Banaei-Kashani	University of Colorado Denver, USA
Jie Bao	Microsoft Research Asia, China
Denilson Barbosa	University of Alberta, Canada
Boualem Benatallah	University of New South Wales, Australia
Azer Bestavros	Boston University, USA

Antonis Bikakis	University College London, UK
Bin Cao	Zhejiang University of Technology, China
Barbara Catania	University of Genoa, Italy
Richard Chbeir	LIUPPA Laboratory, France
Cindy Chen	University of Massachusetts Lowell, USA
Jinchuan Chen	Renmin University of China, China
Jacek Chmielewski	Poznań University of Economics, Poland
Alex Delis	University of Athens, Greece
Schahram Dustar	Vienna University of Technology, Austria
Islam Elgedawy	Middle East Technical University, Turkey
Hicham Elmongui	Alexandria University, Egypt
Marie-Christine Fauvet	Université Grenoble Alpes, France
Yunjun Gao	Zhejiang University, China
Thanaa Ghanem	Metropolitan State University, USA
Claude Godart	Université de Lorraine, France
Daniela Grigori	Laboratoire LAMSADE, Université Paris Dauphine, France
Venkata Gunturi	IIIT-Delhi, India
Hakim Hacid	Bell Labs, USA
Armin Haller	Australian National University, Australia
Mohammad Hammoud	CMU Qatar, Qatar
Tanzima Hashem	Bangladesh University of Engineering and Technology, Bangladesh
Rafiul Hassan	King Fahd University of Petroleum and Minerals, Saudi Arabia
Xiaofeng He	East China Normal University, China
Yuh-Jong Hu	National Chengchi University, Taiwan
Peizhao Hu	Rochester Institute of Technology, USA
Jianbin Huang	Xidian University, China
Marta Indulska	University of Queensland, Australia
Yoshiharu Ishikawa	Nagoya University, Japan
Adam Jatowt	Kyoto University, Japan
Yan Jia	National University of Defense Technology, China
Lili Jiang	Max Planck Institute for Informatics, Germany
Wei Jiang	Missouri University of Science and Technology, USA
Peiquan Jin	University of Science and Technology of China, China
Byeong Ho Kang	University of Tasmania, Australia
Raymond Lau	City University of Hong Kong, Hong Kong, SAR China
Dan Lin	Missouri University of Science and Technology, USA
Shuai Ma	Beihang University, China
Murali Mani	University of Michigan-Flint, USA
Natwar Modani	Adobe Research, India
Mikolaj Morzy	Poznań University of Technology, Poland
Wilfred Ng	Hong Kong University of Science and Technology, Hong Kong, SAR China

Kjetil Nørvåg	Norwegian University of Science and Technology, Norway
Mitsunori Ogihara	University of Miami, USA
George Pallis	University of Cyprus, Cyprus
Wen-Chih Peng	National Chiao Tung University, Taiwan
Olivier Pivert	ENSSAT, France
Tieyun Qian	Wuhan University, China
Jarogniew Rykowski	Poznań University of Economics, Poland
Yucel Saygin	Sabanci University, Turkey
Wei Shen	Nankai University, China
John Shepherd	University of New South Wales, Australia
Lawrence Si	University of Macau, Macau, SAR China
Dandan Song	Beijing Institute of Technology, China
Shaoxu Song	Tsinghua University, China
Reima Suomi	University of Turku, Finland
Stefan Tai	Karlsruhe Institute of Technology, Germany
Dimitri Theodoratos	New Jersey Institute of Technology, USA
Yicheng Tu	University of South Florida, USA
Xiaojun Wan	Peking University, China
Hua Wang	Victoria University, Australia
Junhu Wang	Griffith University, Australia
De Wang	Google, USA
Ingmar Weber	Qatar Computing Research Institute, Qatar
Adam Wojtowicz	Poznań University of Economics, Poland
Jei-Zheng Wu	Soochow University, Taiwan
Takehiro Yamamoto	Kyoto University, Japan
Hayato Yamana	Waseda University, Japan
Yanfang Ye	West Virginia University, USA
Hongzhi Yin	The University of Queensland, Australia
Tetsuya Yoshida	Nara Women's University, Japan
Ge Yu	Northeastern University, China
Jeffrey Xu Yu	Chinese University of Hong Kong, Hong Kong, SAR China
Qi Zhang	Fudan University, China
Xiangmin Zhou	RMIT University, Australia
Xingquan Zhu	Florida Atlantic University, USA

QUAT General Co-chairs

| Deren Chen | Zhejiang University, China |
| William Song | Dalarna University, Sweden |

QUAT Program Committee Co-chairs

Xiaolin Zheng   Zhejiang University, China
Johan Håkansson   Dalarna University, Sweden
Shaozhong Zhang   Zhejiang Wanli University, China

QUAT Organizing Committee Co-chairs

Roger G. Nyberg   Dalarna University, Sweden
Zukun Yu   Britich Telecom, UK
Xiaofeng Du   British Telecom, UK
Xiaoyun Zhao   Dalarna University, Sweden

QUAT Program Committee

Adriana Marotta   Universidad de la República, Uruguay
Anders Avdic   Dalarna University, Sweden
Fei Chiang   McMaster University, Canada
Hasan Fleyeh   Dalarna University, Sweden
Jacky Keung   City University of Hong Kong, Hong Kong, SAR China
Jun Hu   Nanchang University, China
Preben Hansen   Stockholm University, Sweden
Rajeev Agrawal   North Carolina A&T State University, USA
Sheng Zhang   Nanchang Hangkong University, China
Yuansheng Zhong   Jiangxi University of Finance and Economics, China
Yuhao Wang   Nanchang University, China

QUAT Sponsors

Complex Systems & Microdata Analysis, Dalarna University, Sweden
E-Service Research Center, Zhejiang University, China

Contents – Part I

Social Network Data Analysis

Attribute-Based Influence Maximization in Social Networks 3
Jiuxin Cao, Tao Zhou, Dan Dong, Shuai Xu, Ziqing Zhu, Zhuo Ma,
and Bo Liu

Twitter Normalization via 1-to-N Recovering 19
Yafeng Ren, Jiayuan Deng, and Donghong Ji

A Data Cleaning Method for CiteSeer Dataset 35
Yan Wang, Hao Zhang, Yaxin Li, Deyun Wang, Yanlin Ma, Tong Zhou,
and Jianguo Lu

Towards Understanding URL Resources in Recent Sina Weibo 50
Yifang Wan, Peng Li, Rui Li, Meilin Zhou, Yongjun Ye, and Bin Wang

Recommender Systems

Nonparametric Bayesian Probabilistic Latent Factor Model for Group
Recommender Systems 61
Nipa Chowdhury and Xiongcai Cai

Joint User Knowledge and Matrix Factorization for Recommender Systems 77
Yonghong Yu, Yang Gao, Hao Wang, and Ruili Wang

GEMRec: A Graph-Based Emotion-Aware Music Recommendation
Approach 92
Dongjing Wang, Shuiguang Deng, and Guandong Xu

Topic Modeling

Domain Dictionary-Based Topic Modeling for Social Text 109
Bo Jiang, Jiguang Liang, Ying Sha, Rui Li, and Lihong Wang

Towards an Impact-Driven Quality Control Model for Imbalanced
Crowdsourcing Tasks 124
Kinda El Maarry and Wolf-Tilo Balke

Modeling and Analyzing Engagement in Social Network Challenges 140
Marco Brambilla, Stefano Ceri, Chiara Leonardi, Andrea Mauri,
and Riccardo Volonterio

Data Diversity

Select, Link and Rank: Diversified Query Expansion and Entity
Ranking Using Wikipedia   157
  Adit Krishnan, Deepak Padmanabhan, Sayan Ranu, and Sameep Mehta

Multi-dimension Diversification in Legal Information Retrieval.   174
  Marios Koniaris, Ioannis Anagnostopoulos, and Yannis Vassiliou

Generating Multiple Diverse Summaries   190
  Natwar Modani, Balaji Vasan Srinivasan, and Harsh Jhamtani

Diversifying the Results of Keyword Queries on Linked Data   199
  Ananya Dass, Cem Aksoy, Aggeliki Dimitriou, Dimitri Theodoratos,
  and Xiaoying Wu

Data Similarity

Semantic Similarity of Workflow Traces with Various Granularities   211
  Qing Liu, Quan Bai, and Yi Yang

Intermediate Semantics Based Distance Metric Learning for Video
Annotation and Similarity Measurements   227
  Wen Qu, Xiangmin Zhou, Daling Wang, Shi Feng, Yifei Zhang,
  and Ge Yu

A Community Detection Algorithm Considering Edge Betweenness
and Vertex Similarity.   243
  Hongwei Lu, Chang Liu, and Zaobin Gan

Measuring and Ensuring Similarity of User Interfaces: The Impact
of Web Layout.   252
  Sebastian Heil, Maxim Bakaev, and Martin Gaedke

Context-Aware Recommendation

Semantic Context-Aware Recommendation via Topic Models Leveraging
Linked Open Data.   263
  Mehdi Allahyari and Krys Kochut

Optimizing Factorization Machines for Top-N Context-Aware
Recommendations   278
  Fajie Yuan, Guibing Guo, Joemon M. Jose, Long Chen, Haitao Yu,
  and Weinan Zhang

Taxonomy Tree Based Similarity Measurement of Textual Attributes of
Items for Recommender Systems   294
    Longquan Tao, Fei Liu, and Jinli Cao

A Personalized Recommendation Algorithm for User-Preference Similarity
Through the Semantic Analysis.   302
    Haolin Zhang and Feiyue Ye

Prediction

Learning-Based SPARQL Query Performance Prediction   313
    Wei Emma Zhang, Quan Z. Sheng, Kerry Taylor, Yongrui Qin,
    and Lina Yao

Can Online Emotions Predict the Stock Market in China?   328
    Zhenkun Zhou, Jichang Zhao, and Ke Xu

Predicting Replacement of Smartphones with Mobile App Usage   343
    Dun Yang, Zhiang Wu, Xiaopeng Wang, Jie Cao, and Guandong Xu

Real Time Prediction on Revisitation Behaviors of Short-Term
Type Commodities   352
    Xiangzhen Xu, Jinghua Fu, Yuliang Shi, Shijun Liu, and Lizhen Cui

Big Data Processing

Parallel Materialization of Datalog Programs with Spark
for Scalable Reasoning   363
    Haijiang Wu, Jie Liu, Tao Wang, Dan Ye, Jun Wei, and Hua Zhong

A Data Type-Driven Property Alignment Framework for Product Duplicate
Detection on the Web   380
    Gijs van Rooij, Ravi Sewnarain, Martin Skogholt, Tim van der Zaan,
    Flavius Frasincar, and Kim Schouten

A Semantic Data Parallel Query Method Based on Hadoop   396
    Liu Yang, Liu Yang, Jiangbo Niu, Zhigang Hu, Jun Long,
    and Meiguang Zheng

A Strategy to Improve Accuracy of Multi-dimensional Feature Forecasting
in Big Data Stream Computing Environments.   405
    Dawei Sun, Hao Tang, Shang Gao, and Fengyun Li

Cloud Computing

Automatic Creation and Analysis of a Linked Data Cloud Diagram   417
  Alexander Arturo Mera Caraballo, Bernardo Pereira Nunes,
  Giseli Rabello Lopes, Luiz André Portes Paes Leme,
  and Marco Antonio Casanova

Fast Multi-keywords Search over Encrypted Cloud Data   433
  Cheng Hong, Yifu Li, Min Zhang, and Dengguo Feng

A Scalable Parallel Semantic Reasoning Algorithm-Based on RDFS
Rules on Hadoop   447
  Liu Yang, Xiao Wen, Zhigang Hu, Chang Liu, Jun Long,
  and Meiguang Zheng

Cloud Resource Allocation from the User Perspective: A Bare-Bones
Reinforcement Learning Approach   457
  Alexandros Kontarinis, Verena Kantere, and Nectarios Koziris

Event Detection

Event Phase Extraction and Summarization   473
  Chengyu Wang, Rong Zhang, Xiaofeng He, Guomin Zhou,
  and Aoying Zhou

ESAP: A Novel Approach for Cross-Platform Event Dissemination
Trend Analysis Between Social Network and Search Engine   489
  Yan Tang, Pengju Ma, Boyuan Kong, Wenqian Ji, Xiaofeng Gao,
  and Xuezheng Peng

Learning Event Profile for Improving First Story Detection
in Twitter Stream   505
  Yongqin Qiu, Rui Li, Lihong Wang, and Bin Wang

A Novel Approach of Discovering Local Community Using Node
Vector Model   513
  Jinglian Liu, Daling Wang, Shi Feng, Yifei Zhang, and Weiji Zhao

Data Mining

Labeled Phrase Latent Dirichlet Allocation   525
  Yi-Kun Tang, Xian-Ling Mao, and Heyan Huang

WTEN: An Advanced Coupled Tensor Factorization Strategy for Learning
from Imbalanced Data   537
  Quan Do, Thanh Pham, Wei Liu, and Kotagiri Ramamohanarao

Mining Actionable Knowledge Using Reordering Based Diversified
Actionable Decision Trees   553
    Sudha Subramani, Hua Wang, Sathiyabhama Balasubramaniam,
    Rui Zhou, Jiangang Ma, Yanchun Zhang, Frank Whittaker, Yueai Zhao,
    and Sarathkumar Rangarajan

Improving Distant Supervision of Relation Extraction
with Unsupervised Methods   561
    Min Peng, Jimin Huang, Zhaoyu Sun, Shizhong Wang, Hua Wang,
    Guangping Zhuo, and Gang Tian

Author Index   569

Contents – Part II

Sentiment Analysis

Dynamic Topic-Based Sentiment Analysis of Large-Scale Online News.   3
  Peng Liu, Jon Atle Gulla, and Lemei Zhang

Improving Object and Event Monitoring on Twitter Through Lexical
Analysis and User Profiling   19
  Yihong Zhang, Claudia Szabo, and Quan Z. Sheng

Aspect-Based Sentiment Analysis Using Lexico-Semantic Patterns   35
  Kim Schouten, Frederique Baas, Olivier Bus, Alexander Osinga,
  Nikki van de Ven, Steffie van Loenhout, Lisanne Vrolijk,
  and Flavius Frasincar

Multilevel Browsing of Folksonomy-Based Digital Collections   43
  Joaquín Gayoso-Cabada, Daniel Rodríguez-Cerezo,
  and José-Luis Sierra

Ranking in Social Networks

Faderank: An Incremental Algorithm for Ranking Twitter Users   55
  Massimo Bartoletti, Stefano Lande, and Alessandro Massa

Personalized Re-ranking of Tweets   70
  Yukun Zhao, Shangsong Liang, and Jun Ma

Ranking Microblog Users via URL Biased Posts.   85
  Yongjun Ye, Peng Li, Rui Li, Meilin Zhou, Yifang Wan, and Bin Wang

Identifying Implicit Enterprise Users from the Imbalanced Social Data.   94
  Zhenni You, Tieyun Qian, Baochao Zhang, and Shi Ying

Microblog Data Analysis

Understanding Factors That Affect Web Traffic via Twitter   105
  Chunjing Xiao, Zhiguang Qin, Xucheng Luo,
  and Aleksandar Kuzmanovic

Analysis of Teens' Chronic Stress on Micro-blog   121
  Yuanyuan Xue, Qi Li, Liang Zhao, Jia Jia, Ling Feng, Feng Yu,
  and David A. Clifton

Large-Scale Stylistic Analysis of Formality in Academia and Social Media. . .   137
Thin Nguyen, Svetha Venkatesh, and Dinh Phung

Discriminative Cues for Different Stages of Smoking Cessation
in Online Community   146
Thin Nguyen, Ron Borland, John Yearwood, Hua-Hie Yong,
Svetha Venkatesh, and Dinh Phung

Query Processing

POL: A Pattern Oriented Load-Shedding for Semantic Data
Stream Processing.   157
Fethi Belghaouti, Amel Bouzeghoub, Zakia Kazi-Aoul, and Raja Chiky

Unsupervised Blocking of Imbalanced Datasets for Record Matching   172
Chenxiao Dou, Daniel Sun, and Raymond K. Wong

Partially Decompressing Binary Interpolative Coding for Fast
Query Processing   187
Xi Fu, Peng Li, Rui Li, and Bin Wang

Using Changesets for Incremental Maintenance of Linkset Views   196
Vânia M.P. Vidal, Marco A. Casanova, Elisa S. Menendez,
Narciso Arruda, Valeria M. Pequeno, and Luiz A. Paes Leme

Spatial and Temporal Data

Graph-Based Metric Embedding for Next POI Recommendation.   207
Min Xie, Hongzhi Yin, Fanjiang Xu, Hao Wang, and Xiaofang Zhou

Temporal Pattern Based QoS Prediction.   223
Liang Chen, Haochao Ying, Qibo Qiu, Jian Wu, Hai Dong,
and Athman Bouguettaya

Searching for Data Sources for the Semantic Enrichment of Trajectories   238
Luiz André P. Paes Leme, Chiara Renso, Bernardo P. Nunes,
Giseli Rabello Lopes, Marco A. Casanova, and Vânia P. Vidal

On Impact of Weather on Human Mobility in Cities   247
Jun Pang, Polina Zablotskaia, and Yang Zhang

Graph Theory

Minimum Spanning Tree on Uncertain Graphs   259
Anzhen Zhang, Zhaonian Zou, Jianzhong Li, and Hong Gao

A Block-Based Edge Partitioning for Random Walks Algorithms
over Large Social Graphs.   275
    Yifan Li, Camelia Constantin, and Cedric du Mouza

Differentially Private Network Data Release via Stochastic Kronecker
Graph   290
    Dai Li, Wei Zhang, and Yunfang Chen

An Executable Specification for SPARQL   298
    Mihaela Bornea, Julian Dolby, Achille Fokoue,
    Anastasios Kementsietsidis, Kavitha Srinivas, and Mandana Vaziri

Non-traditional Environments

Towards a Scalable Framework for Artifact-Centric Business Process
Management Systems   309
    Jiankun Lei, Rufan Bai, Lipeng Guo, and Liang Zhang

Bridging Semantic Gap Between App Names: Collective Matrix
Factorization for Similar Mobile App Recommendation   324
    Ning Bu, Shuzi Niu, Lei Yu, Wenjing Ma, and Guoping Long

Summarizing Multimedia Content   340
    Natwar Modani, Pranav Maneriker, Gaurush Hiranandani,
    Atanu R. Sinha, Utpal, Vaishnavi Subramanian, and Shivani Gupta

Bridging Enterprise and Software Engineering Through an User-Centered
Design Perspective   349
    Pedro Valente, Thiago Silva, Marco Winckler, and Nuno Nunes

Special Session on Data Quality and Trust in Big Data

Region Profile Based Geo-Spatial Analytic Search   361
    Xiaofeng Du and Zhan Cui

Segmentation and Enhancement of Low Quality Fingerprint Images   371
    Hasan Fleyeh

Feature Selection and Bleach Time Modelling of Paper Pulp Using
Tree Based Learners   385
    Karl Hansson, Hasan Fleyeh, and Siril Yella

Trust Model of Wireless Sensor Networks Based on Shannon Entropy   397
    Jun Hu and Chun Guan

Assessing the Quality and Reliability of Visual Estimates in Determining
Plant Cover on Railway Embankments   404
   Siril Yella and Roger G. Nyberg

Community-Based Message Transmission with Energy Efficient in
Opportunistic Networks   411
   Sheng Zhang, Xin Wang, Minghui Yao, and William Wei Song

A Multi-Semantic Classification Model of Reviews Based on Directed
Weighted Graph   424
   Shaozhong Zhang, William Wei Song, Minjie Ding, and Ping Hu

Data Warehouse Quality Assessment Using Contexts.   436
   Flavia Serra and Adriana Marotta

Author Index   449

Social Network Data Analysis

Attribute-Based Influence Maximization in Social Networks

Jiuxin Cao$^{(\boxtimes)}$, Tao Zhou, Dan Dong, Shuai Xu, Ziqing Zhu,
Zhuo Ma, and Bo Liu

School of Computer Science and Engineering, Southeast University, Nanjing, China
{jx.cao,zhoutao,dongdan,xushuai7,zzqxztc,mazhuo,bliu}@seu.edu.cn

Abstract. As traditional advertising model exposes its weakness of ignoring consumer interests, the concept of narrow advertising draws increasingly more attention which considers the feature of each user. Under this specific environment, effective viral marketing has to select a set of initial users to maximize their influence on the targeted customers. This paper aims at the integration of viral marketing and narrow advertising, by proposing a novel problem called attribute-based influence maximization. Firstly, the problem definition is presented with the consideration of user features. Then the influence probability between two nodes is modeled and two heuristic algorithms, Sum of Probability Covered Algorithm (SoPCA) and Community-based Algorithm (CBA), are designed. Finally, experiments on six datasets are conducted to verify the effectiveness of proposed algorithms.

Keywords: Influence maximization · Influence probability · Social networks · User attribute

1 Introduction

In recent years, with the development of the Internet and the success of online social network services, such as Facebook, Twitter and Weibo, large-scale social data has been generated, which has promoted the related researches in information science. In the field of information dissemination, viral marketing is a well-known problem. A specific need for viral marketing in social networks is to identify the most influential users as information source from which the information could reach the most of the network through the effect of word-of-mouth, which is the prototype of influence maximization.

Current viral marketing is based on the traditional adverting model. However, the model does not consider the difference between customers, which has led to the weakness in practical application. The seller could not benefit too much, the customer would be bothered by information overflow and the social media would suffer the deluge of advertisement. To improve the effect of advertisements, a new model called narrow advertising or Narrow-ad was proposed. It pushes ads to the customers who have real interests in the product by identifying user preferences, habits, location historical access information and other user attributes.

© Springer International Publishing AG 2016
W. Cellary et al. (Eds.): WISE 2016, Part I, LNCS 10041, pp. 3–18, 2016.
DOI: 10.1007/978-3-319-48740-3_1

The traditional influence maximization problem lacks the concern on user attributes and does not differentiate the targeted customer. However, the result in [17] claimed that different customers have different attributes which would cause the difference in accepting information on certain topics. So different topics would have different propagation path. Meanwhile, as indicated by narrow advertising, the targeted users of the influence maximization under this environment should be specific as we prefer the customers with real interest. To sum up, user attributes should be regarded as important content when we try to integrate viral marketing and narrow advertising.

In this paper, we will focus on the influence maximization with the consideration of narrow advertising based on the real social network data. The definition of attribute-based influence maximization problem will be given. To solve the problem, we will present the calculation of the influence probability between two nodes according to their attributes and then two heuristic algorithms will be proposed, namely Sum of Probability Covered Algorithm (SoPCA) and Community-based Algorithm (CBA). Experimental results prove the algorithms effective in solving the attribute-based influence maximization problem.

The rest of the paper is structured as follows. In Sect. 2, the related work is presented. The attribute-based influence maximization is defined and the method of calculating the influence probability is described in Sect. 3. In Sect. 4, two heuristic algorithms are proposed. Section 5 shows the experimental results. Finally, Sect. 6 concludes this paper.

2   Related Work

The pioneer work of influence maximization is carried out by Domingos and Richardson [6]. Then, Kempe, Kleinberg and Tardos [11] are the first to formulate the problem as a discrete optimization problem and prove it as a NP-hard problem. They also provide a Greedy algorithm with approximation guarantees.

To improve the efficiency of the Greedy algorithm, Leskovec et al. [13] present a "lazy-forward" optimization in selecting new seeds which greatly reduces the running time. However, it still takes too much time to find 100 seed nodes in a network with tens of thousands of nodes. Then, Chen Wei et al. [4] provide two algorithms to improve the Greedy algorithm, namely NewGreedy and Mixed-Greedy. In Independent Cascade Model, NewGreedy deletes edges in the network with certain probability to obtain a smaller network and then finds seed nodes in the new generated network. MixedGreedy applies NewGreedy algorithm in the first round and then CELF is used to select seed nodes. The experimental result shows that the performance of MixedGreedy is better than NewGreedy. The improved greedy algorithms are still not scalable for large-scale networks.

Many other works [1,2,4,10] focus on the efficient heuristic algorithms. Chen Wei et al. [4] propose the DegreeDiscount algorithm based on MaxDegree algorithm which assumes that the influence spread increases with the node degree. PageRank [1] is a method to rank web pages. IRIE [10] is the state-of-the-art heuristic algorithm both in running time and memory usage. CCA is proposed in our previous work [2] and is based on coreness and covered distance.

Moreover, Galstyan et al. [8] are the first to apply the community structure to deal with the problem. Wang Yu et al. [18] propose a Community-based Greedy Algorithm which is composed of two steps: dividing the whole network into some communities, and selecting seed nodes from these communities by dynamic programming algorithm. Cao Tianyu et al. [3] propose OASNET algorithm based on the community and transfer the influence maximization problem into optimal resource dynamic allocation problem which is proved as an NP-hard problem.

Compared with the original problem, attribute related influence maximization is rarely studied. Fa-Hsien Li et al. [14] propose the labeled influence maximization problem based on the idea that different products get different profits on different users and propose two kinds of algorithms: one is to consider the total profits produced by seed nodes rather than the number of nodes to be activated when selecting seed nodes, and the other is a novel algorithm whose main idea is to offline compute the pairwise proximities of nodes and online find the seed node set. Siyuan Liu et al. [15] propose the categorical influence maximization problem and present the probability distribution based search method which is suitable for the networks where every vertex has very small degree.

3 Problem Description and Probability Modeling

3.1 Attribute-Based Influence Maximization

Based on the analysis above, we propose a new problem called attribute-based influence maximization, by introducing the attributes into traditional influence maximization. Note that we will use "targeted individual" to refer to the consumer that has the exact attributes matched with the product. So given a product, the targeted individual could be determined according to the product attributes. In other words, the targeted individual is the ideal consumer defined by the seller and could represent the feature of the product.

The definition of the new problem is as follows: Given (a) a social network $G(V, E, \Lambda)$, in which V is the vertex set, E is the edge set, Λ is the attribute category set and $\Lambda = \{A_1, ..., A_m\}$, $dom(A_j)$ is the range of $A_j (1 \leqslant j \leqslant m)$, and each vertex $v_i \in V$ is associated with an attribute vector $AttributeVector(i) = \left[a_1^i, ..., a_m^i\right]$, in which a_j^i represents the value of jth attribute of v_i, $a_j^i \in dom(A_j)$, (b) targeted attribute vector $AV = [a_1, ..., a_m]$, in which a_j is the value of jth attribute of targeted individual and for each vertex $v_i \in V$, we could compute the similarity between the vertex and targeted individual, denoted by $similarity_i^{AV}$, based on the attribute vectors, (c) the number of selected seed nodes k. The attribute-based influence maximization problem is under the premise of considering the impact of attributes on influence probability between two nodes to find a set of seed nodes $S \subseteq V$, $|S| = k$ such that the total influence spread (i.e., similarity) caused by S, denoted by $\sum_{i \in S'} similarity_i^{AV}$, is maximized, where S' is the final influenced node set.

3.2 Influence Probability Modeling

Influence diffusion model illustrates the way information spreads across the network, such as Independent Cascade Model [9,19] and Thread Linear Model [20]. For Independent Cascade Model, each edge (u, v) is associated with an influence probability p_{uv}, which represents the probability that node u influences node v when u is active. p_{uv} is mostly set as a constant value or random value. However, in most real-world cases, the influence probability varies with the relationship of two individuals and even influence probability between the same two individuals could be different for different information. So we will present the method of modeling the influence probability in this part.

We consider two factors influencing the probability between two individuals. One is the similarity between the targeted individual and the individual who receives the information. Whether a user accepts the particular information depends on the user interests to a large extent. The other is the interaction strength between these two individuals. We assume that the closer they are, the greater the influence probability is. Next, we will introduce how to assemble the factors.

Similarity between Individual and Targeted Individual. Datasets in this paper are crawled from Sina Weibo[1], which contain the relatively integrated networks, statuses, comments, user tags and user profiles. The tags are irregular phrases (i.e. music, liking food) and we need to classify them into attribute categories before utilization. For those users who have no tags, we attach tags to them according to the keywords in their statuses. Sogou Lexicon[2] is applied for classification. At last, we get eighteen most common attribute categories to represent the interests of individuals. In addition, gender is set as the 19th attribute category. So every individual is associated with a 19-dimensional attribute vector with binary values. A specific product would have targeted individual who has certain attributes, so targeted individual could be described with a 19-dimensional attribute vector. Thus we transfer the similarity between individual and targeted individual to the similarity between two vectors.

For each individual in the network, the similarity with the targeted individual is represented by the Jaccard Coefficient, denoted by J. For example, X and Y are two vectors and the similarity between them can be calculated with Eq. 1,

$$J(X, Y) = \frac{f_{11}}{f_{01} + f_{10} + f_{11}} \tag{1}$$

where f_{11} is the number of elements whose value is 1 both in X and Y, f_{01} is the number of elements whose value is 0 in X and 1 in Y and f_{10} is the opposite.

Interaction Strength between Individuals. In order to measure the relationship strength of two individuals, we utilize the status information and comment information crawled from Sina Weibo. We assume that the number of u's

---

[1] http://open.weibo.com.
[2] http://pinyin.sogou.com/dict/cell.php?id=11640.

statuses which are forwarded or commented by v is positively correlated to u's influence to v. The interaction strength between v and u's statuses, denoted by p_{uv}^r, can be calculated with Eq. 2,

$$p_{uv}^r = \frac{\frac{\#v\ forwards\ u's\ statuses}{\#u's\ statuses} + \frac{\#v\ comments\ u's\ statuses}{\#u's\ statuses}}{2} \quad (2)$$

where $\#v\ forwards\ u's\ statuses$ is the number of u's statuses forwarded by v, $\#v\ comments\ u's\ statuses$ is the number of u's statuses commented by v and $\#u's\ statuses$ is the total number of u's statuses. To make the influence probability between 0 and 1, the parameter is normalized, i.e. divided by 2.

Calculation of Influence Probability. Given a directed edge (u, v) in the network and a specific product (i.e. targeted individual), we can obtain the similarity between node v and targeted individual, denoted by s_v and the interaction strength between v and u, denoted by p_{uv}^r. We then calculate the influence probability through Eq. 3.

$$p_{uv} = p_{uv}^r * s_v \quad (3)$$

Note that p_{uv}^r is relatively fixed between two users, but s_v should be changed along with the targeted individual (i.e. product), so p_{uv} is various for different products. To avoid the occasion of s_v being zero, we add a very small offset ζ to s_v and the value of ζ is decided by the magnitude of s_v.

4 Heuristic Algorithms

4.1 Sum of Probability Covered Algorithm

In the previous works, many metrics have been proposed to measure the importance of nodes in social network. The most common ones are degree centrality, betweenness centrality [7] and coreness [12]. These metrics only consider the position of nodes in topology. Although they are useful in simple independent cascade model because influence probabilities between any two nodes are the same, they might not be suitable for the problem with inconstant probability. So we present a new metric called Sum of Probability, denoted by SoP in Eq. 4.

$$SoP(u) = \sum_{v \in N(u)} p_{uv} \quad (4)$$

where $N(u)$ is the node set of u's neighbors.

According to Eq. 4, the new metric is not only relevant to the degree of u but also related to the influence probability between u and its neighbors. In order to avoid influence overlapping, we further introduce covered distance d so that every two nodes in selected node set maintain a certain distance. Then we propose Sum of Probability Covered Algorithm (SoPCA) described as follows: (a) for each node in the network, there are two states (i.e. uncovered and covered) and it could be in only one state at a time, and (b) nodes which are in the state of covered cannot be chosen as seed nodes, and (c) at first, every node in the network is in the state of uncovered, and (d) every round we select one node with the largest SoP as a seed node and cover all the nodes whose distance from the seed node is less than or equal to d.

Algorithm 1. Sum of Probability Covered Algorithm

Input: A social network $G(V, E)$, size of seeds k, covered distance d
Output: The set of seed nodes S
 1: initialize $S = \varnothing$;
 2: **for** each vertex $v \in V$ **do**
 3:     $ComputeSumOfProbability(v)$;
 4:     $CO_v = false$;
 5: **end for**
 6: **for** $i = 1$ to k **do**
 7:     $w = argmax_u \{SoP(u)|u \in V \backslash S, CO_u = false\}$;
 8:     $S = S \cup \{w\}$;
 9:     **for** each vertex v in $\{v|d_{u,v} \le d, v \in V\}$ **do**
10:         $CO_v = true$;
11:     **end for**
12: **end for**
13: **return** S;

SoPCA is summarized in Algorithm 1. $SoP(v)$ represents the sum of probability of node v. CO_v is a boolen variable representing whether the node v has been covered. CO_v is $true$ if v is covered, otherwise CO_v is $false$. $d_{u,v}$ is the shortest distance between u and v. Assume that the number of vertex in social network $G(V, E)$ is m and the number of seed nodes is k, then the total time complexity of the algorithm is $O(km)$.

4.2 Community-Based Algorithm

Because of the existence of community structure in social networks and the homophily [16] between neighboring nodes, some nodes with similar attributes could form a community. Then, we can find seed nodes from different communities. As existing methods of detecting communities [8, 18] are not applicable for our special needs, we present a new community-based algorithm. Given a product (i.e. targeted individual), the main idea is to detect the communities in which all the nodes are similar with the targeted individual and are connected (including direct connection and indirect connection) on the original topology. The influence probability between two indirectly connected individuals is the product of all probabilities in the shortest path. For example, in Fig. 1, the left part is the original network in which different shapes represent different attributes. We need to detect communities which are composed of the circular nodes. The right part could be obtained using our community-based algorithm. At last, influential nodes will be selected from the right two communities.

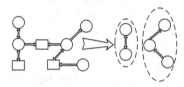

Fig. 1. The example of detecting communities

Algorithm 2. Community Detection Algorithm (CDA)

Input: A social network $G(V, E)$, number of degrees t, the set of nodes V_s whose similarity with targeted individual is not equal to zero
Output: The network G' which consists of V_s

 1: **for** each vertex $v_s \in V_s$ **do**
 2:     Queue q;
 3:     q.INQUEUE(v_s);
 4:     q.INQUEUE(#);
 5:     Hash_map hm;
 6:     hm.INSERT(v_s,1.0);
 7:     **while** q != ∅ and $t > 0$ **do**
 8:       u = q.DEQUEUE();
 9:       **if** u equals to # **then**
10:         q.INQUEUE(#);
11:         $t--$;
12:       **else**
13:         basePro = hm[u];
14:         **for** $w \in Adj(u)$ **do**
15:           **if** $t > 1$ or $w \in V_s$ **then**
16:             q.INQUEUE(w);
17:             totalP = basePro*$p(u, w)$;
18:             **if** hm.Contains(w) && totalP $>$ hm[w] **then**
19:               hm[w] = totalP;
20:             **else**
21:               hm.INSERT(w, totalP);
22:             **end if**
23:           **end if**
24:         **end for**
25:       **end if**
26:     **end while**
27:     **for** each vertex $v \in V_s \setminus v_s$ **do**
28:       **if** hm.Contains(v) **then**
29:         $G' \leftarrow (v_s, v, \text{hm}[v])$;
30:       **end if**
31:     **end for**
32: **end for**
33: **return** G';

However, the time complexity of detecting communities using the method directly is quite high because we must determine whether each two nodes are directly or indirectly connected. Although we could use the idea of Dijskstra algorithm to make an optimization, the time complexity is still unacceptable for a large-scale network. Moreover, if two nodes are indirectly connected through too many mid-nodes such as 10 or more, they basically cannot influence each other and further it may bring negative effect on finding the influential nodes.

Therefore we make an improvement on detecting communities based on the theory of three degrees of influence [5]. It means every two nodes within three hops are strongly connected and could influence each other, and then the influence gradually disappears when the distance exceeds three. In this paper, the number of degrees within which people can influence others is set as a variable, denoted by t, which is an integer from 1 to 3. Two nodes are considered connected only when there are at most two mid-nodes on their shortest path.

The improved community detection algorithm is shown in Algorithm 2. Here, q is FIFO queue. # is just a mark. Hm is a hash mapping table composed by (key, value) in which the key is the ID number of a node and the value is the influence probability of node v_s to the ID node. $p(u, w)$ represents the probability of node u to node w. $Adj(u)$ is a set of nodes to which node u points.

Algorithm 3. Community-Based Algorithm(CBA)

Input: A social network $G(V, E)$, size of seeds k, number of degrees t, the set of targeted nodes V_s

Output: The set of initial seed nodes S

1: initialize $S = \varnothing$;
2: $G' = CommunityDetection(V_s, t)$; // Algorithm 2
3: **for** each vertex $v_s \in V_s$ **do**
4:    $ComputeSumOfProbability(G', v_s)$;
5:    $CO_{vs} = false$;
6: **end for**
7: **for** $i = 1$ to k **do**
8:    $w = argmax_u \{SoP(u)|u \in V_s \backslash S, CO_u = false\}$;
9:    $S = S \cup \{w\}$;
10:   **for** each vertex v in $\{v|d_{u,v} \leq t, v \in V_s\}$ **do**
11:       $CO_v = true$;
12:   **end for**
13: **end for**
14: **return** S;

In Algorithm 2, Lines (7–26) acquire the nodes that node v_s can influence in t degrees and calculate the influence probability. Lines (27–31) extract the targeted nodes[3] that node v_s can influence in t degrees and add the directed

---

[3] Targeted nodes in this paper all represent the nodes whose similarity with targeted individual is not equal to 0.

edge (v_s, v) and the probability in the edge to the new network G'. Assume that the number of targeted nodes is $|V_s|$ and the number of nodes that node v_s can influence in t degrees is w. The complexity of CDA is $O((w + |V_s|)|V_s|)$.

By applying community detection, the candidate set of seed nodes could be shrunk and the following selection process could take the indirectly connected nodes into consideration. After detecting communities, we can find the top-k influential nodes in the new network G' which is composed of one or more connected graphs (i.e. communities). The selected seeds are nodes with the largest SoP in G'.

The Community-Based Algorithm is described in Algorithm 3. Here lines (3–6) calculate the sum of probability for every node in V_s. The total complexity of Algorithm 3 is $O(|V_s|(w + |V_s| + k))$.

5 Experiments

5.1 Experiment Setup

Environment and Method. The experiments are conducted in the PC machine with Intel Core i5 processor, 4G memory and 64 bit Windows 7 operating system. The program is coded in C++. Monte-Carlo simulation is applied to estimate of the influence spread with a given set of seed nodes. We run the simulation 10000 times for each set of seed nodes and take the average influence spread. Note that influence spread in this paper is the sum of activated nodes' similarity with targeted individual. Influence diffusion process is based on the Weighted Cascade (WC) Model [11].

Datasets. The datasets in the experiments are crawled from Sina Weibo. The original dataset includes 1,935,391 users and 137,284,538 relationships with the information of user profiles, statuses, comments and tags.

Two connected networks are obtained from the original one by excluding the inactive nodes (i.e. the nodes which rarely publish statuses). One contains 68,243 users and 1,013,164 directed relationships which refer to the following relationships. The other one includes 44,514 users and 162,398 undirected relationships which refer to the relationships of following each other.

Given a product (i.e. targeted individual), we can calculate the influence probabilities, which means a dataset we can directly use is generated. We select three kinds of products to verify our algorithms in multiple datasets with different number of targeted nodes. These three products are cars, health sport equipments and fashion beauty products. The targeted individuals of cars are those who have attribute of liking cars. The health sport equipments are sold to those who have attributes of taking attention to health and liking sports. The audiences of fashion beauty products are usually post-80s girls who take much time on beauty. From the analysis above, we can get three targeted attribute vectors correspondingly. At last, we get six datasets by combining two topologies and three targeted attribute vectors. The details are shown in Table 1.

Table 1. Dataset list

Dataset	Constituents
DataSet 1	Undirected networks and cars
DataSet 2	Undirected networks and health sport equipments
DataSet 3	Undirected networks and fashion beauty products
DataSet 4	Directed networks and cars
DataSet 5	Directed networks and health sport equipments
DataSet 6	Directed networks and fashion beauty products

Algorithms. Algorithms in the experiments include Degree [4], DegreeDiscount [4], PageRank [1], IRIE [10] (proved as the best heuristic algorithm in terms of time complexity and influence spread), CCA(d) [2] (d is an integer and is set in the range of 0 to 3), SoPCA(d) (d is covered distance and is set in the range of 0 to 3) and CBA(t) (t is degrees in the algorithm and is set in the range of 1 to 3). The greedy algorithm is not included in the comparison due to the high time complexity in large-scale social networks.

5.2   Experiment Results and Analysis

In SoPCA, we introduce the covered distance d. To expound the effectiveness of bringing in the covered distance d, we compare the influence spread of the algorithm under different covered distance. In CBA, we introduce the degree t and in this section, we will observe the influence spread of the algorithm under different degree t and choose the best degree value. At last, the two algorithms with best covered distance or degree are compared with other heuristic algorithms in terms of influence spread and running time.

Every algorithm is run on the six datasets under WC model to obtain influence spread. The seed set size k ranges from 1 to 100. To facilitate visualization, in all influence spread figures of this section, the legend ranks the algorithms top-down in the same order as the influence spreads of the algorithms when k equals to 100. The average percentage difference of the algorithms is also calculated.

Effectiveness of introducing covered distance d in SoPCA. To evaluate the effectiveness of bringing in covered distance d into SoPCA, we conduct experiments on SoPCA with d's value being in the range of 0 to 3 on the six real-world datasets. The result is shown in Fig. 2. It is obvious in Fig. 2 that (1) the SoPCA with covered distance is better than SoPCA without covered distance (i.e. the distance d equals to 0) except for DataSet 6 and this is because that on DataSet 6, most of the seeds nodes selected are closely connected but cannot influence each other since the influence probabilities on the edges are relatively small, and (2) the best covered distances for every dataset are different because of the different topologies and the different influence probabilities on the edges, and (3) in Fig. 2(d) and (e), the influence spreads with different covered distances go up alternately, which is reasonable since the propagation probability for each edge in the network is different and the best covered distance for each node is also different, and (4) when $d = 3$, in most cases the performance of SoPCA is relatively the worst since the influence scope of a node is limited and if d is too large, many influential nodes will be covered by accident which will influence the quality of seed nodes.

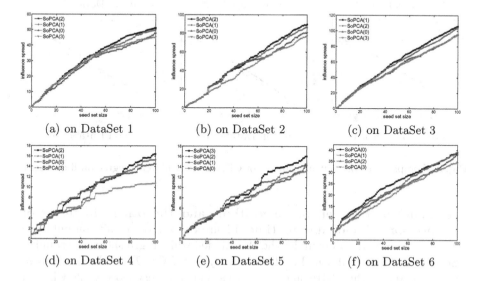

(a) on DataSet 1   (b) on DataSet 2   (c) on DataSet 3

(d) on DataSet 4   (e) on DataSet 5   (f) on DataSet 6

Fig. 2. Influence spread of SoPCA under different covered distance on 6 datasets

In summary, it is effective to consider covered distance when selecting seed nodes using SoPCA and the best distance is related to the network topology and propagation probabilities on the edges. In general, the best distance is limited since influence scope is finite for the nodes. For a certain network, the best distance does not exist since different nodes have different influence scopes.

Observation of influence spread of CBA under different degrees. To observe when CBA performs best for different degrees, we conduct experiments on CBA with t's value being in the range of 1 to 3 on the six real-world datasets. The result is shown in Fig. 3. In Fig. 3(a), we can easily find that the influence spreads of CBA with three different degrees are almost the same. Further we find that seed nodes selected in different t are almost the same. The reason for the phenomenon may be that the targeted nodes are very sparse in this dataset and their influences are almost the same when different degrees are selected. In other datasets, there are obvious differences for different degrees. However, the best degree for each dataset is different because of different topologies and propagation probabilities. Similar to the result of SoPCA, when $t = 3$, in most cases the influence spread of CBA is relatively the worst.

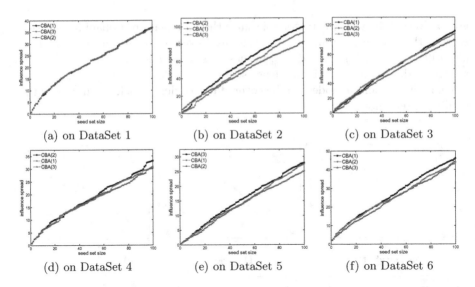

Fig. 3. Comparisons of influence spread of CBA under different degree on 6 datasets

A comparison between our algorithms and the others in terms of influence spread and running time. Figure 4 shows the results on influence spreads for the six datasets under the WC model, while Fig. 5 shows the running time results. The results in Fig. 4(a) show that SoPCA(2) produces the best influence spread, while IRIE and DegreeDiscount are very close to it. CBA(1) does not perform so well because the targeted nodes are very sparse on this dataset. SoPCA(2) is very close to DegreeDiscount and Degree in running time and is faster than the rest algorithms.

In Fig. 4(b), we can find that CBA(2) has the best influence spread when the seed set size is 100. However, when the set of seed nodes is small, PageRank performs better than CBA(2) and the possible reason is that on this dataset, the influence probabilities on the edges coincidently are accordance with the similarities of nodes that will be activated. This coincidence just shows the instability of PageRank. For other datasets, PageRank always behaves much worse. SoPCA(2) is 23.4 % worse than CBA(2) while it is 6.9 % and 4.5 % better compared with DegreeDiscount and IRIE respectively. The advantage of SoPCA in influence spread is more obvious when the seed set size is large. The running times for all algorithms are similar with those in DataSet 1.

Figure 4(c) presents the results on DataSet 3. CBA(1) and SoPCA(1) perform better than other heuristic algorithms. From the figure, we cannot clearly see the difference between four algorithms-CBA(1), SoPCA(1), IRIE and DegreeDiscount, so we calculate the difference between them. CBA(1) is 3.1 %, 4.2 % and 6.8 % better compared with SoPCA(1), IRIE and DegreeDiscount respectively. Comparing with the results in DataSet 1 and DataSet 2, we can find the advantage of our algorithms is much smaller. The reason for the phenomenon is that

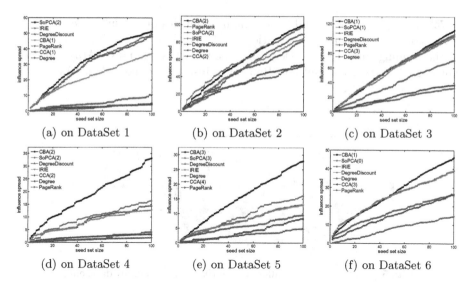

Fig. 4. Comparisons between seven algorithms in terms of influence spread on 6 datasets

the number of targeted nodes in DataSet 3 is very large (reaching 70%). When the targeted nodes size becomes larger and larger, the attribute-based influence maximization will gradually degenerate into the original influence maximization. IRIE and DegreeDiscount are good algorithms for the original problem and our algorithms are designed for the attribute-based influence maximization. So the phenomenon is also reasonable. The running times of all algorithms for this dataset are similar with those on two previous datasets.

On DataSet 4, CBA(2) outperforms SoPCA(2) with a large margin (95.1%) and SoPCA(2) outperforms other heuristic algorithms (4.5%, 8.8%, 75.8%, 83.0%, and 89.5% better than DegreeDiscount, IRIE, Degree, CCA(2) and PageRank, respectively). For the running time, SoPCA(2), DegreeDiscount, Degree and CCA(2) take less than 10 s. IRIE and CBA(2) are just a little more than 10 s, but they are still acceptable.

On DataSet 5, CBA(3) again outperforms SoPCA(3) with a large margin (59.2%). SoPCA(3) outperforms other heuristics (7.6%, 8.5%, 31.8%, 62.2%, and 74.2% better than DegreeDiscount, IRIE, Degree, CCA(4) and PageRank, respectively). For the running time, Degree, DegreeDiscount and SoPCA(3) take less than 10 s. IRIE and PageRank take less than 100 s. CCA(4) and CBA(3) take much more time than those in DataSet 4 because the covered distance d and the degree t in the algorithms are larger than those in CCA(2) and CBA(2) on DataSet 4. Once the covered distance or the degree becomes larger, the time complexity of covering nodes becomes higher.

In Fig. 4(f), CBA(1) outperforms SoPCA(0) with a relatively smaller margin (3.9%). SoPCA(0), DegreeDiscount and IRIE almost have the same influence spread because in this dataset, the number of targeted nodes is very large (reaching 75%). Degree, CCA(3) and PageRank have much worse results because they do not consider influence probabilities on the edges when finding seed nodes. For the running time, Degree, DegreeDiscount SoPCA(0) and CCA(3) take less than 10 s. CBA(1), IRIE and PageRank take just a little more than 10 s.

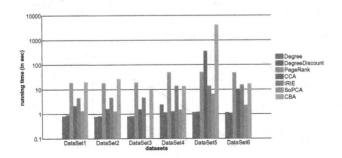

Fig. 5. Running time of different algorithms for different datasets

Overall, in the majority of the datasets, CBA performs the best in terms of influence spread and is followed by SoPCA. The less the number of targeted nodes in the social network is, the more obvious the advantages of our algorithms are. Although the running time of SoPCA and CBA is not the shortest, it is still acceptable in large-scale social networks.

6   Conclusion and Future Work

This paper introduces node attributes into the influence maximization and proposes the attribute-based influence maximization. To solve the new problem, we firstly calculate the influence probabilities of the edges based on the node attributes and the interaction between the two nodes. Then we propose two algorithms to find seed nodes in the social networks on WC model. Finally, the experiment results on six datasets prove that in most cases, the proposed algorithms perform best in the influence spread and they are scalable for large-scale social networks. In the future, we will further explore the relationship between covered distance d and features of the network.

Acknowledgments. This work is supported by National Natural Science Foundation of China (61272531, 61202449, 61272054, 61370207, 61370208, 61300024, 61320106007 and 61472081), China high technology 863 program (2013AA013503), Jiangsu Technology Planning Program (SBY2014021039-10), Jiangsu Provincial Key Laboratory of Network and Information Security under Grant No. BM2003201 and Key Laboratory of Computer Network and Information Integration of Ministry of Education of China under Grant No. 93k-9.

References

1. Brin, S., Page, L.: Reprint of: the anatomy of a large-scale hypertextual web search engine. Comput. Netw. **56**(18), 3825–3833 (2012)
2. Cao, J.-X., Dong, D., Xu, S., Zheng, X., Liu, B., Luo, J.-Z.: A k-core based algorithm for influence maximization in social networks. Chinese J. Comput. **38**(2), 238–248 (2015). (in Chinese)
3. Cao, T., Wu, X., Wang, S., Hu, X.: Oasnet: an optimal allocation approach to influence maximization in modular social networks. In: 2010 ACM Symposium on Applied Computing, pp. 1088–1094. ACM (2010)
4. Chen, W., Wang, Y., Yang, S.: Efficient influence maximization in social networks. In: 15th ACM SIGKDD International Conference on Knowledge Discovery and Data Mining, pp. 199–208. ACM (2009)
5. Christakis, N.A., Fowler, J.H.: Connected: The surprising power of our social networks and how they shape our lives. hachette digital (2009)
6. Domingos, P., Richardson, M.: Mining the network value of customers. In: Seventh ACM SIGKDD International Conference on Knowledge Discovery and Data Mining, pp. 57–66. ACM (2001)
7. Freeman, L.C.: Centrality in social networks conceptual clarification. Soc. Netw. **1**(3), 215–239 (1978)
8. Galstyan, A., Musoyan, V., Cohen, P.: Maximizing influence propagation in networks with community structure. Phys. Rev. E. **79**(5), 056102 (2009)
9. Goldenberg, J., Libai, B., Muller, E.: Talk of the network: a complex systems look at the underlying process of word-of-mouth. Mark. Lett. **12**(3), 211–223 (2001)
10. Jung, K., Heo, W., Chen, W.: Irie: scalable and robust influence maximization in social networks. In: 2012 IEEE 12th International Conference on Data Mining, pp. 918–923. IEEE (2012)
11. Kempe, D., Kleinberg, J., Tardos, É.: Maximizing the spread of influence through a social network. In: Ninth ACM SIGKDD International Conference on Knowledge Discovery and Data Mining, pp. 137–146. ACM (2003)
12. Kitsak, M., Gallos, L.K., Havlin, S., Liljeros, F., Muchnik, L., Stanley, H.E., Makse, H.A.: Identification of influential spreaders in complex networks. Nature Phys. **6**(11), 888–893 (2010)
13. Leskovec, J., Krause, A., Guestrin, C., Faloutsos, C., VanBriesen, J., Glance, N.: Cost-effective outbreak detection in networks. In: 13th ACM SIGKDD International Conference on Knowledge Discovery and Data Mining, pp. 420–429. ACM (2007)
14. Li, F.-H., Li, C.-T., Shan, M.-K.: Labeled influence maximization in social networks for target marketing. In: 2011 IEEE Third International Conference on Privacy, Security, Risk and Trust and 2011 IEEE Third Inernational Conference on Social Computing, pp. 560–563. IEEE (2011)
15. Liu, S., Chen, L., Ni, L.M., Fan, J.: Cim: categorical influence maximization. In: 5th International Conference on Ubiquitous Information Management and Communication, p. 124. ACM (2011)
16. McPherson, M., Smith-Lovin, L., Cook, J.M.: Birds of a feather: Homophily in social networks. Annu. Rev. Sociol. **27**, 415–444 (2001)
17. Tang, J., Sun, J., Wang, C., Yang, Z.: Social influence analysis in large-scale networks. In: 15th ACM SIGKDD International Conference on Knowledge Discovery and Data Mining, pp. 807–816. ACM (2009)

18. Wang, Y., Cong, G., Song, G., Xie, K.: Community-based greedy algorithm for mining top-k influential nodes in mobile social networks. In: 16th ACM SIGKDD International Conference on Knowledge Discovery and Data Mining, pp. 1039–1048. ACM (2010)
19. Watts, D.J.: A simple model of global cascades on random networks. Proc. National Acad. Sci. $99(9)$, 5766–5771 (2002)
20. Young, H.P.: The diffusion of innovations in social networks. The economy as an evolving complex system III: Current perspectives and future directions. 267 (2006)

Twitter Normalization via 1-to-N Recovering

Yafeng Ren[✉], Jiayuan Deng, and Donghong Ji

Computer School, Wuhan University, Wuhan, China
renyafeng@whu.edu.cn

Abstract. Twitter messages are written in an informal style, which hinders many information retrieval and natural language processing applications. Existing normalization systems have two major drawbacks. The first is that these methods largely require large-scale annotated training data. The second is that these systems assume that a nonstandard token is recovered to one standard word. However, there are many nonstandard tokens that should be recovered to two or more standard words, so the problem remains to be highly challenging. To address the above issues, we propose an unsupervised normalization system based on the context similarity. The proposed system does not require any annotated data. Meanwhile, a nonstandard token will be recovered to one or more standard words. Results show that the proposed approach achieves state-of-the-art performance.

Keywords: Twitter normalization · Forward search · Random walk · Spell checker

1 Introduction

User-generated contents have drastically increased in the past few years, driven by the development of microblogs. These user-generated contents, which contains rich information, become a heated research topic in natural language processing and text mining [1,25,26]. Twitter is one among these microblogging services that count about one billion of active users and 500 million of daily messages[1]. However, due to the nature of posts, twitter messages contain many nonstandard tokens, which are created both intentionally and unintentionally by people. For examples, substituting numbers for letters such as *2gether* (*together*) and *2morrow* (*tomorrow*), repeating letters for emphasizing the expression such as *coollllll* (*cool*) and *birthdayyyyyy* (*birthday*), eliminating vowels such as *ppl* (*people*), and substituting phonetically similar letters such as *fon* (*phone*). Besides, there are another type of nonstandard tokens in reality, this type of nonstandard tokens mainly contains the omission of the spaces, punctuation and letters among successive multiple standard words, these nonstandard tokens also should be converted into their standard forms. e.g. *theres* (*there is*), *thatthe* (*that the*), *untiltheend* (*until the end*), and *ndyou* (*and you*). We randomly count 1000

---

[1] http://expandedrambling.com/.

© Springer International Publishing AG 2016
W. Cellary et al. (Eds.): WISE 2016, Part I, LNCS 10041, pp. 19–34, 2016.
DOI: 10.1007/978-3-319-48740-3_2

tweets, and find that there are 1326 nonstandard tokens in which 294 tokens should be recovered to two or more standard words.

Twitter has become a very valuable information source for many natural language processing (NLP) applications, such as information extraction [29], summarization [17], sarcasm detection [34], sentiment analysis [27,28] and event discovery [3]. However, the nonstandard tokens limit the performance of standard NLP tools [10,29]. Previous work reported that the Stanford named entity recognizer (NER) experienced a performance drop from 90.8 % to 45.8 % on tweets [19], and the part-of-speech (POS) tagger and dependency parser degraded 12.2 % and 20.65 % on tweets, respectively. It is therefore of great importance to normalize twitter message before applying standard NLP techniques.

In recent years, some attempts have been made to normalize nonstandard tokens to their standard forms [10–12,15,18,33]. But these approaches all assume that a nonstandard token is normalized to one standard word. For example, *hiiiiii* is converted into *hi*. Note that *theyre* is converted into *they* and *itmay* is deployed as *it*. It is obvious that previous systems limit the performance of this task.

In this paper, we focus on the task of the normalization of English Twitter messages, which can be regarded as a pre-processing step for NLP applications. This is a challenging task based on two reasons. First, for a nonstandard token, it should be recovered to one or more standard words? Second, text normalization as a preprocessing step should have high precision and recall to have a good impact on various NLP applications.

In this paper, we propose an unsupervised normalization system to address the above challenges. Firstly, according to the characteristics of nonstandard tokens, a nonstandard token will be divided into the possible multiple words (standard word or noisy word) using forward search and backward search. Then the normalization candidates are generated for each noisy word by integrating random walk and spell checker, the essence of these two step is transform 1-to-N recovering into 1-to-1 recovering. Finally, the best normalization candidate is selected based on a n-gram language model.

The main contributions of this paper are as follows:

- In our system, a nonstandard token can be recovered to one or more standard words. To our knowledge, 1-to-N recovering in twitter normalization is proposed for the first time.
- Results show that our system achieves the best performance (a 10 % absolute increment compared to state-of-the-art). The proposed approach can be deployed as a preprocessing step for various NLP application to handle twitter message.

2   Related Work

With the rapid development of social media, text normalization system has drawn increasing attention in recent years. There are many systems proposed for tackling text normalization. Previous methods are mainly divided into two directions: NC-based model (Noisy Channel) and SMT-based model (Statistical Machine Translation).

Suppose the ill-formed text is T and its corresponding standard form is S. NC-based model [31] aims to find $argmax\ P(S|T)$ by computing $argmax\ P(T|S)P(S)$, in which $P(S)$ is usually a language model and $P(T|S)$ is an error model. Brill and Moore (2000) characterise the error model by computing the product of operation probabilities on slice-by-slice string edits [4]. Toutanova and Moore (2002) improve the model by incorporating the pronunciation information [32]. Choudhury et al. (2007) model the word-level text generation process for SMS messages [5], by considering graphemic/phonetic abbreviations and unintentional typos. Cook and Stevenson (2009) expand the error model by introducing the inferences from different erroneous formation processes [7]. However, these models make the strong assumption that a token $t_i \in T$ only depends on $s_i \in S$, ignoring the context around a token, which could be utilized to help in resolving ambiguity.

SMT-based model, which has been widely proposed as a means of context-sensitive text normalization, treats the ill-formed text as the source language, and the standard form as the target language. For example, Aw et al. (2006) propose a phrase-level SMT SMS normalization method with bootstrapped phrase alignments [2]. However, SMT method tends to suffer from a critical lack of training data. It is labor intensive to construct an annotated corpus to sufficiently cover ill-formed words and context-appropriate corrections.

Recent work focuses on normalizing the twitter message, which map a noisy form to a normalized form. Han and Baldwin (2011) develop a classifier for detecting the ill-formed words, and generate corrections based on the morphophonemic similarity [10]. Gouws et al. (2011) use a normalization lexicon based on string and distribution similarity to detect noisy words. Han et al. (2012) introduce a similar approach by generating a normalization lexicon based on distributional similarity and string similarity [11].

More recently, Hassan and Menezes (2013) first find the possible candidates based on bipartite graph [12], then they construct a lattice from possible normalization candidates, and determine the best normalization sequence according to a n-gram language model. Wang and Ng (2013) propose a beam-search decoder to effectively integrate various normalization operations [33], and apply their normalization to machine translation tasks for both Chinese and English. Li and Liu (2014) propose a re-ranking strategy to combine the results from different systems [15]. Besides, Li and Liu (2015) propose a joint Viterbi decoding process to determine each token's POS tag and non-standard token's correct form at the same time [16]. Cotelo et al. (2015) and Schulz et al. (2016) explore the text normalization on Spanish and Dutch, respectively [8,30].

However, the above methods have two major problems. The first is these methods largely require large-scale annotated training data, limiting their adaptability to new domains and languages. The second is these method assume that the relationship between nonstandard token and standard word is 1-to-1 recovering. But there are large amounts of nonstandard tokens (e.g. *howyou*, *havent*), which these nonstandard tokens should be converted into two or more standard words. Some key information may lost if we only use 1-to-1 recovering (e.g. *howyou* will be converted into *how*, and *havent* will be converted into *have*). In this paper, we propose an unsupervised normalization system, where a nonstandard token is converted into its best possible candidate with 1-to-N recovering so that the proposed system can satisfy various NLP applications in reality.

3   Text Normalization System

The main objective of this paper is convert the noisy tweet text as the source input into the normalized tweet as the target output. The overall framework of the proposed system is shown in Fig. 1. For a tweet, a standard dictionary first is used to determine whether it is need to be normalized. Then, a nonstandard token is considered 1-to-1 or 1-to-N recovering based on the characteristics of nonstandard tokens. For 1-to-N recovering, the nonstandard token will be

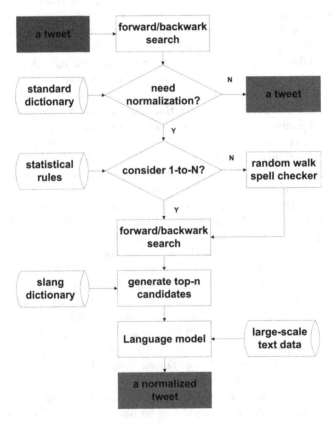

Fig. 1. The overall framework of the proposed normalization system. The blue area represents the input, and the red area represents the output. (Color figure online)

divided into multiple possible words using forward search and backward search. Then, noisy words among multiple possible words and the overall token (the nonstandard token itself) are generated some normalization candidates by integrating random walks and spell checker. Finally, we get the best normalized tweet by taking all candidates into consideration of n-gram language model. In the following section, we will introduce the proposed system in details.

3.1   Pre-processing and Standard Dictionary Generation

The first step of the proposed system is determine whether a tweet is needed to normalize. According to the nature of social media text, we need some pre-processing before a tweet is considered to normalize. The following tokens are excluded for normalization:

- The tokens begun with the symbol # or @, e.g. #kingJames, @michael;
- The tokens constructed completely by numbers, e.g. 2014;
- The emoticons or URLs;

In twitter, there are many named entities which should not be normalized. Our standard dictionary must enough broad so that it can include these named entities. To get a standard dictionary, we first calculate the frequency of all different tokens based on a large-scale clean corpus (LDC2011T07), and delete the tokens that its predefined threshold is lower than 5. Then, we filter out some tokens using the GUN Aspell dictionary (v0.60.6)[2]. The remaining tokens construct our standard vocabulary, which includes nearly 68,000 words. Naturally, a nonstandard token is defined as a token that does not exist in the standard vocabulary.

3.2   Statistical Rules

In the above section, we discuss which token need to be normalized. Next, the key challenge is that a nonstandard token should be normalized to one or more standard words. We conclude some characteristics by observing and analyzing large amounts of nonstandard tokens. For 1-to-1 recovering, there are following rules:

- The nonstandard token that its length is lower than 4 usually should be recovered to one standard word, e.g. *u* (*you*) and *r* (*are*);
- The nonstandard token that contains letters and numbers is usually should be recovered to one standard word, e.g. *b4* (*before*) and *2day* (*today*);

In our system, the nonstandard tokens that meets the one of these two rules are normalized one standard word. Conversely, the nonstandard token is considered to normalize one or more standard words. These rules can not capture all situations, for example, *ur* may be converted into *you are*, but *ur* may be converted into *your* or *our*, this type of nonstandard token is ambiguous, its best

---

[2] http://aspell.net/.

normalization depends on the specific context, but it belongs to 1-to-1 recovering normalization in most situations.

We also analyze the nonstandard tokens that should be recovered to two or more standard words. These tokens mainly include the following three classes:

- **Type 1:** The nonstandard tokens are constructed by two or more standard word, the space among these words are omitted by people, intentionally and unintentionally. e.g. *rememberwith* (*remember with*) and *Iloveyousomuch* (*I love you so much*);
- **Type 2:** The nonstandard tokens are constructed by one or more standard and a nonstandard word. e.g. *wasnt* (*was not*) and *ndyou* (*and you*);
- **Type 3:** The nonstandard tokens are constructed by the acronym of two or more standard words. e.g. *ur* (*you are*) and *sm* (*so much*);

Type 3 is very limited, and this type of nonstandard tokens are much ambiguity. In our system, for this type of nonstandard token, its possible normalization candidates are added into the slang dictionary[3] created by the web users. Finally, we find the best candidate according to its context by using language model.

3.3 Forward and Backward Search

For a nonstandard token, we must find the possible multiple words if this token need to be normalized two or more standard words. According to statistical nature of the nonstandard token. We propose to generate the possible multiple words by using forward search and backward search. Forward search can solve two types of nonstandard tokens, the first type is composed of one standard word and the nonstandard word from front to back, e.g. *theyre* (*they are*) and *wouldnt* (*would not*). The second type is composed of two or more standard word, e.g. *aboutthem* (*about them*) and *Iloveyou* (*I love you*). Forward search algorithm is shown as Algorithm 1:

In Algorithm 1, the trie is a tree that is constructed using the standard vocabulary. Based on the trie, we can quickly find the standard word for a nonstandard token.

Similar to forward search, backward search can also solve two types of nonstandard tokens, the first type is composed of one standard word and the nonstandard word from back to front, e.g. *anyou* (*and you*) and *looveyou* (*love you*). The second type is totally composed of two or more standard words. The specific algorithm of backward search is similar to forward search, we only change the search order for a nonstandard token.

After forward search and backward search, the nonstandard token is divided into multiple words, which include standard word and noisy word, e.g., *ndyou* is divided into a noisy word *nd* and a standard word *you*. Meanwhile, the nonstandard token itself (*ndyou*) will be considered as a noisy word, we must find the possible normalization candidates for all noisy words. In the following section, we will discuss how to find the possible candidates for a noisy word.

---

[3] http://www.noslang.com/dictionary.

Algorithm 1. ForwardSearch(token)

Input: a nonstandard token; trie;
Output: SegmentTable;

---

length=token.length();
for(i=1; i ≤ length;)
  {word1=token.substring(0, i);
    if (word1 ∈ trie)
      {Add word1 to SegmentTable; word2=token.substring(i+1, length);
      if (word2 ∈ trie ‖ word2<4)
        {Add word2 to SegmentTable; i++; continue; }
      else
        ForwardSearch(word2);
      }
    else
      { if (i==length)
        Add token in SegmentTable;
      else
         i++;
      }
  }

3.4   Random Walk and Spell Checker

The noisy word includes many types, such as lengthening, letter substitution, letter-number substitution and phonetic substitution. The natural idea is spell checker to recover the noisy word. But spell checker mainly can recover the noisy words that are created unintentionally by users. In order to recover the noisy words created intentionally by user, we use random walk based on bipartite graph to find the proper candidate. In this paper, a noisy word generates some normalization candidates by integrating random walk and spell checker.

For a noisy word, to get its normalization candidates using random walk based on bipartite graph, there are two hypothesis for tweets as follows:

- **Hypothesis 1:** The words share the same or similar contexts should be the same word or semantically related words;
- **Hypothesis 2:** The users usually have different writing styles. For a twitter, some users may use all standard words to express, and other user may use one or more noisy words.

Hypothesis 1 tells us the normalization equivalence between a noisy word and a standard word by sharing the same or similar contexts. Hypothesis 2 tells us there are a large number of such contexts in twitter. Based on two hypothesis, we can generate the candidates for a noisy word by using contextual similarity. For instance, assume 5-gram sequences of words, two words may be the same word if their contexts share the same two words on the left and the same two words on the right, e.g. for the word *be4* and *before*, they share the same contexts *the day * the day* and *dress yourself * contract me*, so we think that these two words are the same or semantically related, and *before* is considered to the best normalization for the noisy word *be4* by using the string similarity and language model. To use the random walk, we firstly introduce the bipartite graph representation.

Bipartite Graph Representation. Contextual similarity can be represented as a bipartite graph where the first partite represents the words, and the second partite represents the contexts that are shared by words. A word node can be either noisy word or standard word. Figure 2 shows a sample of the bipartite graph $G(W, C, E)$, where standard words are shown as blue nodes.

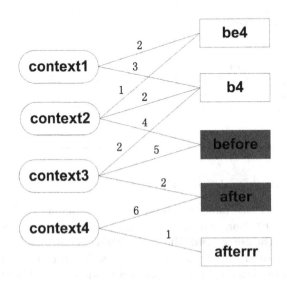

Fig. 2. Bipartite graph representation. Left nodes represent the contexts, right nodes with blue colour represent the standard words, and other nodes represent the noisy words. The weight of the edge is the co-occurrence of a word and its context. (Color figure online)

The bipartite graph, $G(W, C, E)$, is composed of W which includes the nodes representing noisy words and standard words, C includes the nodes representing shared context, and E represents the edges of the bipartite graph connecting word nodes and context nodes. The weight of the edge is the number of occurrences of a given word in a context. The bipartite graph is constructed using Algorithm 2.

Candidates Generation Using Random Walk. Random walk based on bipartite graph is defined [23] and then used in many NLP applications. For example, Hughes and Ramage (2007) used random walk on Wordnet graph to measure lexical semantic relatedness between words [13]. Das and Petrov (2011) used graph-based label propagation for cross-lingual knowledge to induce POS tags between two languages [9]. Minkov and Cohen (2012) introduced a path

---

Algorithm 2. ConstructBipartiteGraph(text)

Input: Twitter corpus;

Output: G(W,C,E);

Extract all 5-gram sequences from twitter corpus;

Store all sequences into NgramTable;

for each sequence $\in$ NgramTable

{

   $W \leftarrow$ Add(CenterWord);

   $C \leftarrow$ Add(Context);

   $E \leftarrow$ Add(Context, Word, count);

}

---

constrained graph walk algorithm given a small number of labeled examples to assess nodes relatedness in the graph [22]. Hassan and Menezes (2013) used Markov random walk to consider normalization equivalences based on twitter and LDC data [12]. In this paper, we use graph-based random walk to find the normalization candidates for a noisy word by only using the twitter text.

For a noisy word, the random walk algorithm repeats independent random walk for 4 steps where the walks traverse the graph randomly according to roulette rule. Each walk starts from a noisy word node and ends at a standard word node, or consumes the maximum number of steps (4 steps in our algorithm) without hitting a standard word node. In our random walk, for a noisy word, the process of random walk algorithm will be iterated 100 times.

Consider a random walk on the bipartite graph $G(W, C, E)$ starting a noisy word and ending at a standard word. For the bipartite graph in Fig. 2, assume a random walk staring at the node representing the noisy word *be4* then moves to the context node **context2** then to the node representing the standard word *before*. This random walk will associate *be4* with *before*. Meanwhile, the noisy word *be4* can first find the noisy word *b4* by sharing **context1**, then noisy word *b4* can find the standard word *after* by sharing **context3**, this random walk will associate *be4* with *after*. For noisy word *be4*, which word is better candidate for normalization?

For a noisy word, we rank its all candidates based on the confidence $Conf(N, S)$ as the result of random walk, N represents the noisy word and S represents the standard word. $Conf(N, S)$ is calculated as:

$$Conf(N, S) = \alpha F(S) + \beta Sim(N, S) \qquad (1)$$

Where, α and β are weights, we use uniform interpolation, both $\alpha = \beta = 1$. $F(S)$ is calculated as follows:

$$F(S) = S\_frequency/total\_frequency \qquad (2)$$

$S\_frequency$ is the times of standard word S in all random walks, and $total\_frequency$ is total numbers of all standard words in 100 iteration. $Sim(N, S)$ is a similarity function based on Longest Common Subsequence Ratio

(LCSR) [6, 20]. This function is defined as the ratio of LCSR and Edit distance between two string as follows:

$$Sim(N, S) = \frac{LCS(N, S)/MaxLength(N, S)}{ED(N, S)} \tag{3}$$

$LCS(N, S)$ represents the length of Longest Common Subsequence between word N and S. Edit distance calculation $ED(N, S)$ is modified to be more adequate for social media text according to previous work [12];

For a noisy word N, we can construct a lexicon that includes all normalization candidates by using random walk. We can prune the lexicon to take top-5 according to $ConValue(N, S)$. The advantage of random walk is that it can find the proper candidates for a nonstandard word. But random walk can not find candidates for all noisy words due to the limitation of our twitter data. So we must use another method to generate the normalization candidates for a noisy word. In the paper, we use spell checker to generate some candidates as a supplement for random walk.

Spell Checker. Spell checker is a simple and effective method in normalizing misspellings. In this paper, we use the Jazzy spell checker [14] that integrates the DoubleMetaphone phonetic matching algorithm and the Levenshtein distance using the near-miss strategy, which enables the interchange of two adjacent letters, and the replacing/deleting/adding of letters. In this paper, the edit distance is set to 2 for generating candidates.

Generating Top-N Candidates. For a nonstandard token, some candidates are generated by using random walk and spell checker. We need to rank for these candidates. In addition to considering the string similarity, we use the semantic similarity. The final score between a nonstandard token N and standard word S is:

$$S(N, S) = Sim(N, S) + C(vec(N), vec(S)) \tag{4}$$

$Sim(N, S)$ is defined in the above section, where the function $C(vec(N), vec(S))$ represents the cosine similarity between $vec(N)$ and $vec(S)$. In order to compute the semantic similarity of words, we use the tool word2vec[4] to implement it based on 8 million twitter message by using feed-forward neutral network language model [21], the vector dimension is set to 200. According to the value of $S(N, S)$, we select the top-n candidates for a nonstandard token N ($n = 1, 3$ and 5).

3.5   Normalizing Twitter Using Language Model

After generate top-n candidates for a nonstandard token, we need determine the best candidate to convert a noisy tweet into a normalized tweet. We must take consideration of the context information. Based on large amount of text data, we score the best Viterbi path with 3-gram and 4-gram language model. We use three different types of texts, which include twitter text, clean text (newspaper data) and mixed text (twitter text and newspaper data).

---

[4] http://code.google.com/p/word2vec/.

4 Experiments

4.1 Experimental Setup

The following datasets are used in our experiments. Dataset (1) and (2) are used for word-level evaluation, and dataset (3) is used for both word- and message-level evaluation. Note that dataset (1) and (2) only contain 1-to-1 recovering.

- **Dataset 1:** 3,802 nonstandard tokens along with their human-annotated normalized word forms. The nonstandard tokens are collected from a corpus with about 6,150 tweets between 2009 and 2010 [18].
- **Dataset 2:** 2,333 unique pairs of nonstandard tokens and standard words, collected from 2,577 Twitter messages [15,24].
- **Dataset 3:** 1,000 tweets are annotated by three native human. This dataset includes 850 noisy tweets and 150 clean tweets. The dataset contains 1,345 nonstandard tokens in which 297 tokens need to recovered to two or more standard words.
- **Dataset 4:** 8M tweets, which are collected from October to December 2014 using the Twitter Streaming APIs[5], these tweets are constructed to bipartite graph for random walk.
- **Dataset 5:** clean data from English LDC Gigaword corpus[6]. This dataset is used to construct a standard vocabulary.

The goal of word-level normalization is to convert the nonstandard tokens into standard words. For each nonstandard token, the system is considered correct if any of the corresponding standard words among the top-n candidates from the system. We adopt this word-level top-n accuracy to make our results comparable to the state-of-the-art systems. On the message-level, we evaluate the top-1, top-3 and top-5 system output using precision, recall, and F-score, calculated respectively to the nonstandard tokens based on language model.

4.2 Word-Level Results

The word-level results are presented in Table 1, evaluated on dataset (1) and (2) respectively. We present the top-n accuracy (n = 1, 3, 5) of the proposed approach.

The spell checker gives only 40 % to 60 % accuracy on dataset (1) and (2), indicating that the vast amount of intentionally created nonstandard tokens can hardly be tackled by a system relies solely on lexical similarity. The random walk performs surprisingly well, and shows robust performance across two datasets, random walk achieves 65 % to 80 % accuracy on all datasets. Compared to spell checker, random walk is effective for normalizing intentionally created tokens. Finally, the mixed system of spell checker and random walk achieves 90 % to 94 % accuracy in top-5 on dataset (1) and (2), showing the effectiveness of our proposed system.

---

[5] http://dev.twitter.com/docs/streaming-apis.
[6] http://www.ldc.upenn.edu/Catalog/LDC2011T07.

Table 1. Word-level results.

Method	Accuracy (%)					
	Dataset (1)			Dataset (2)		
	top-1	top-3	top-5	top-1	top-3	top-5
Spell checker	47.2	56.9	58.3	39.9	46.5	47.1
Random walk	67.2	73.7	79.6	64.7	71.4	76.4
Mixed system	76.8	87.3	**94.1**	75.2	85.1	**90.3**
Hassan and Menezes, 2013	74.4	85.3	92.8	74.1	83.0	88.2

Compared with previous work [12], our approach achieves better performance in accuracy. The main reason is that the normalization candidates by integrating random walk and spell checker have better coverage for nonstandard tokens created both unintentionally and intentionally.

4.3   Message-Level Results

The goal of message-level normalization is to replace a nonstandard token with the candidate that best fits the context. Our system is evaluated on dataset (3), and the results are shown in Table 2.

In Table 2, the *"w/o Context"* results are generated by replacing each nonstandard token using top-1 word-level candidate. Although the replacement process is static, it results in 84.3 % F-score due to the high performance of the word-level system. We explore three different data as language models (LM) for the Viterbi decoding process. For the twitter data as language model, it is clear that increasing the amount of candidate gives better precision and recall. When n is fixed to 5, we can get 86.4 % in F-score. As Table 2 shows, Mixed LM and

Table 2. Message-level results.

Dataset (3)	Language model	Message-level		
		Precision	Recall	F-score
Word-level (top-1)	*w/o Context*	78.2	91.3	84.3
w/Context (top-3)	Twitter LM	79.8	93.1	86.0
	LDC LM	77.6	90.6	83.6
	Mixed LM	79.6	92.9	85.7
w/Context (top-5)	Twitter LM	**80.2**	**93.6**	**86.4**
	LDC LM	77.2	90.1	83.1
	Mixed LM	79.7	93.0	85.8
Hassan and Menezes, 2013	Twitter LM	74.4	71.3	72.8
Li and Liu, 2014	Twitter LM	76.9	72.5	74.6

Twitter LM achieve better performance than the previous best results, demonstrating the effectiveness of language model. Results show that our proposed system have better performance in F-score to serve as a reliable preprocessing step for standard NLP applications.

For the LDC data as language model, the proposed system experiences a performance drop with the increasing of the amount of candidates. The main reason is twitter text is informal, there are some grammatical errors even if the nonstandard tokens are converted into the standard words. It is not difficult to explain the reason that twitter text is better than mixed text as language model for twitter normalization.

Compared with previous systems [12,15], our model outperforms the current systems, which reach a 10 % absolute increment. The main reason is our proposed method exploits 1-to-N recovering, but previous work limits 1-to-1 recovering in this task. The proposed system can be well applied in reality.

4.4   Output Analysis

Table 3 lists three examples from dataset (3) and their normalizations using previous methods and our proposed system. At the first example, the nonstandard tokens can both be recovered to their standard words. For example 2 and 3, previous methods can not get the proper normalization because they only use 1-to-1 recovering in which some information may be lost. Our proposed 1-to-N recovering can get the best normalization. Our proposed system is desirable as a preprocessing step for various NLP applications.

Table 3. Twitter normalization examples, **S** represents the source tweet, **B** represents the current approach (Li and Liu, 2014) and **O** is our proposed method.

S: *u* just follow the wrong *ppl* but that is ok !
B: *you* just follow the wrong *people* but that is ok !
O: *you* just follow the wrong *people* but that is ok !
S: *hiii*! I hope *youre* doing *welland* having a nice day
B: *hi*! I hope *you* doing *well* having a nice day
O: *hi*! I hope *you are* doing *well and* having a nice day
S: *youre* such *anamazing* human *beingwho* deserves so *muchlove*
B: *you* such *amazing* human *being* deserves so *much*
O: *you are* such *an amazing* human *being who* deserves so *much love*

5   Conclusion

In this paper, we proposed an unsupervised normalization system in which a noisy tweet can be converted into a normalized tweet. For a nonstandard token, our proposed system uses 1-to-N recovering to get its standard form, which contains one or more standard words. Experimental results show that the proposed system significantly outperforms the state-of-the-art systems in F-score on the dataset. The proposed approach can be deployed as a preprocessing step for various NLP applications to handle the tweet text.

Acknowledgments. This work is supported by the State Key Program of National Natural Science Foundation of China (Grant No. 61133012), the National Natural Science Foundation of China (Grant Nos. 61173062, 61373108) and the National Philosophy Social Science Major Bidding Project of China (Grant No. 11&ZD189).

References

1. Almeida, T.A., Silva, T.P., Santos, I., Hidalgo, J.M.G.: Text normalization and semantic indexing to enhance instant messaging and sms spam filtering. Knowl. Based Syst. **108**, 25–32 (2016)
2. Aw, A., Zhang, M., Xiao, J., Su, J.: A phrase-based statistical model for sms text normalization. In: Proceedings of the Joint Conference on Annual Meeting of the Association for Computational Linguistics and International Conference on Computational Linguistics, pp. 33–40 (2006)
3. Benson, E., Haghighi, A., Barzilay, R.: Event discovery in social media feeds. In: Proceedings of the 49th Annual Meeting of the Association for Computational Linguistics, pp. 389–398 (2011)
4. Brill, E., Moore, R.C.: An improved error model for noisy channel spelling correction. In: Proceedings of the 38th Annual Meeting on Association for Computational Linguistics, pp. 286–293 (2000)
5. Choudhury, M., Saraf, R., Jain, V., Mukherjee, A., Sarkar, S., Basu, A.: Investigation and modeling of the structure of texting language. Int. J. Doc. Anal. Recogn. **10**(3–4), 157–174 (2007)
6. Contractor, D., Faruquie, T.A., Subramaniam, L.V.: Unsupervised cleansing of noisy text. In: Proceedings of the 23rd International Conference on Computational Linguistics, pp. 189–196 (2010)
7. Cook, P., Stevenson, S.: An unsupervised model for text message normalization. In: Proceedings of the Workshop on Computational Approaches to Linguistic Creativity, pp. 71–78 (2009)
8. Cotelo, J.M., Cruz, F.L., Troyano, J., Ortega, F.J.: A modular approach for lexical normalization applied to spanish tweets. Expert Syst. Appl. **42**(10), 4743–4754 (2015)
9. Das, D., Petrov, S.: Unsupervised part-of-speech tagging with bilingual graph-based projections. In: Proceedings of the 49th Annual Meeting of the Association for Computational Linguistics, pp. 600–609 (2011)
10. Han, B., Baldwin, T.: Lexical normalisation of short text messages: makn sens a# twitter. In: Proceedings of the 49th Annual Meeting of the Association for Computational Linguistics, pp. 368–378 (2011)
11. Han, B., Cook, P., Baldwin, T.: Automatically constructing a normalisation dictionary for microblogs. In: Proceedings of the 2012 Joint Conference on Empirical Methods in Natural Language Processing and Computational Natural Language Learning, pp. 421–432 (2012)
12. Hassan, H., Menezes, A.: Social text normalization using contextual graph random walks. In: Proceedings of the 51st Annual Meeting of the Association for Computational Linguistics, pp. 1577–1586 (2013)

13. Hughes, T., Ramage, D.: Lexical semantic relatedness with random graph walks. In: Proceedings of the 2007 Joint Conference on Empirical Methods in Natural Language Processing and Computational Natural Language Learning, pp. 581–589 (2007)

14. Idzelis, M.: Jazzy: the java open source spell checker (2005)

15. Li, C., Liu, Y.: Improving text normalization via unsupervised model and discriminative reranking. In: Proceedings of the 52nd Annual Meeting of the Association for Computational Linguistics, pp. 86–93 (2014)

16. Li, C., Liu, Y.: Joint pos tagging and text normalization for informal text. In: Proceedings of the 24th International Conference on Artificial Intelligence, pp. 1263–1269 (2015)

17. Liu, F., Liu, Y., Weng, F.: Why is sxsw trending? exploring multiple text sources for twitter topic summarization. In: Proceedings of the Workshop on Languages in Social Media, pp. 66–75 (2011)

18. Liu, F., Weng, F., Wang, B., Liu, Y.: Insertion, deletion, or substitution? normalizing text messages without pre-categorization nor supervision. In: Proceedings of the 49th Annual Meeting of the Association for Computational Linguistics, pp. 71–76 (2011)

19. Liu, X., Zhang, S., Wei, F., Zhou, M.: Recognizing named entities in tweets. In: Proceedings of the 49th Annual Meeting of the Association for Computational Linguistics, pp. 359–367 (2011)

20. Melamed, I.D.: Bitext maps and alignment via pattern recognition. Comput. Linguist. **25**(1), 107–130 (1999)

21. Mikolov, T., Chen, K., Corrado, G., Dean, J.: Efficient estimation of word representations in vector space. arXiv preprint arXiv:1301.3781 (2013)

22. Minkov, E., Cohen, W.W.: Graph based similarity measures for synonym extraction from parsed text. In: Proceedings of the Workshop on Graph-based Methods for Natural Language Processing, pp. 20–24 (2012)

23. Norris, J.R.: Markov Chains. Cambridge University Press, New York (1998)

24. Pennell, D., Liu, Y.: A character-level machine translation approach for normalization of sms abbreviations. In: Proceedings of the 5th International Joint Conference on Natural Language Processing, pp. 974–982 (2011)

25. Ren, Y., Ji, D., Yin, L., Zhang, H.: Finding deceptive opinion spam by correcting the mislabeled instances. Chin. J. Electron. **24**(1), 52–57 (2015)

26. Ren, Y., Ji, D., Zhang, H.: Positive unlabeled learning for deceptive reviews detection. In: Proceedings of the 2014 Joint Conference on Empirical Methods in Natural Language Processing, pp. 488–498 (2014)

27. Ren, Y., Zhang, Y., Zhang, M., Ji, D.: Context-sensitive twitter sentiment classification using neural network. In: Proceedings of the 30th AAAI Conference on Artifical Intelligence, pp. 215–221 (2016)

28. Ren, Y., Zhang, Y., Zhang, M., Ji, D.: Improving twitter sentiment classification using topic-enriched multi-prototype word embeddings. In: Proceedings of the 30th Conference on Artificial Intelligence, pp. 3038–3044 (2016)

29. Ritter, A., Clark, S., Etzioni, O., et al.: Named entity recognition in tweets: an experimental study. In: Proceedings of the 2011 Conference on Empirical Methods in Natural Language Processing, pp. 1524–1534 (2011)

30. Schulz, S., De Pauw, G., De Clercq, O., Desmet, B., Hoste, V., Daelemans, W., Macken, L.: Multi-modular text normalization of dutch user-generated content. ACM Trans. Intell. Syst. Technol. **7**(4), 1–22 (2016)

31. Shannon, C.E.: A mathematical theory of communication. ACM SIGMOBILE Mob. Comput. Commun. Rev. **5**(1), 3–55 (2001)

32. Toutanova, K., Moore, R.C.: Pronunciation modeling for improved spelling correction. In: Proceedings of the 40th Annual Meeting on Association for Computational Linguistics, pp. 144–151 (2002)
33. Wang, P., Ng, H.T.: A beam-search decoder for normalization of social media text with application to machine translation. In: Proceedings of the 2013 Conference of the North American Chapter of the Association for Computational Linguistics, pp. 471–481 (2013)
34. Wang, Z., Wu, Z., Wang, R., Ren, Y.: Twitter sarcasm detection exploiting a context-based model. In: Proceedings of the International Conference on Web Information Systems Engineering, pp. 77–91 (2015)

A Data Cleaning Method for CiteSeer Dataset

Yan Wang[1(✉)], Hao Zhang[1], Yaxin Li[1], Deyun Wang[1], Yanlin Ma[1],
Tong Zhou[2], and Jianguo Lu[2]

[1] School of Information, Central University of Finance and Economics, Beijing, China
`dayanking@gmail.com`, {`zhhao1991,petalli,dywang,mmylcu`}`@163.com`
[2] School of Computer Science,
University of Windsor, Windsor, ON N9B 3P4, Canada
{`zhou142,jlu`}`@uwindsor.ca`

Abstract. CiteSeer is considered as the first academic search engine
that have been serving data for almost twenty years. Recently, CiteSeer
graciously makes all the data public, including raw PDF files, text transformed from PDF, and metadata extracted from the text. Numerous
efforts have been tried to improve the accuracy of the metadata extraction. The problem is inherently challenging and errors are abundant. In
this paper, we propose an innovative record-linkage-based method for
data cleaning, which use two new matching algorithms to significantly
improve the cleaning performance for the CiteSeer dataset. One is an
enhanced matching algorithm for local datasets, the other is developed
for online datasets. Experimental results show that 48.1 % wrong metadata entries can be corrected by our method in total and the improvement is more than 539 % compared to existing state-of-the-art data cleaning methods.

Keywords: Scholarly data · Record linkage · Data cleaning ·
Identification

1 Introduction

CiteSeer indexes around 2 million publications for searching and also offers its
data for academic purpose, which includes the metadata information and the full
text of publications. The CiteSeer dataset has been proven as a powerful resource
in many applications such as text classification [1,2], collective classification [3],
document and citation recommendation [4–6].

On the other side, severe errors observed occur in the metadata information
(such as title, authors' name, page number) of the publications of the dataset
since it is extracted automatically [7]. In this paper, 1000 metadata entries are
randomly selected from 1,926,882 entries of the CiteSeer dataset (provided by
CiteSeer website in Jan 2015 [8]), 439 titles are identified with errors. The error
rate is up to 43.9 % and this result is similar to the observation reported in [9].

Such many errors could result in inaccurate application and research conclusions based on it. For instance, these errors will lead to low quality scholarly

© Springer International Publishing AG 2016
W. Cellary et al. (Eds.): WISE 2016, Part I, LNCS 10041, pp. 35–49, 2016.
DOI: 10.1007/978-3-319-48740-3_3

networks (i.e., co-author and citation network) and their corresponding properties could be far from the truth. However, ordinary users of CiteSeer usually
do not recognize such high error rate. One reasonable explanation is that many
publications wrongly extracted are not from "popular" journals or conferences
and usually have special writing formats, which have much less probabilities to
be searched by users. Another proper reason is that the errors of many metadata entries is "minor" and usually there are few of wrong terms (or symbols).
Sometimes users are hard to detect it.

In order to correct the metadata entries with errors, CiteSeer provided two
distinct methods [7,10]. One method is user correction, which is a kind of manual
correction, they allows users to correct the metadata information of the related
papers as long as they login their CiteSeer account. [10] reports that the user
correction can achieve 100 % and 94.55 % accuracy for authors' names and titles
respectively. The other method is based on record linkage [7,11]. In [7], the
information from the DBLP dataset is used to automatically correct errors by
matching the metadata entries of the two datasets. The DBLP dataset provides
manually created metadata information from publications in computer science
and related fields so that the information in DBLP has a much higher accuracy.

The two data cleaning methods work well but with obvious constraints such
that they cannot satisfy the following requirements.

- Automation requirement: although the first method can achieve good accuracy, it cannot handle with the large amount of data collected from the Web.
  In fact, more than 300,000 scholarly documents from web crawlers are added
  into the CiteSeer dataset monthly and it is out of the capacity of manual
  correction.
- High-accuracy requirement: the method of [7] is automated but its correction
  accuracy is far from satisfactory since the method cannot tackle the errors
  that are irreverent to the actual ones.
- High-coverage requirement: the percentage of corrected entries (*correction
  coverage*) by the method of [7] is low due mainly to only a fairly small portion
  of the CiteSeer dataset covered by the DBLP dataset.

Based on the above reasons, we propose a record-linkage-based method that
contains two phases: correction based on local and online datasets respectively, as
shown in Fig. 1. Local datasets refer to all correct scholarly datasets downloaded
in local machines, such as the DBLP dataset used [7]. Online datasets include
not only the data of standard scholarly search engines (e.g., Goolge Scholar,
Microsoft Academic Search and Aminer.com [12]) but also publishers' websites
(e.g., www.link.springer.com), which usually only can be accessed by queries [14].
In our method, we first correct the metadata entries with errors by matching
with local datasets (Phase 1), and then the rest of wrong unmatching entries
will be further corrected by using online scholarly datasets (Phase 2).

In Phase 1, we match each metadata entry in a local scholarly dataset with
the entries in the CiteSeer dataset, then the matched entry will be replaced by the
corresponding one in the local dataset. Unlike [7], our local matching algorithm
utilizes first-several-line text of each publication in the CiteSeer dataset to do

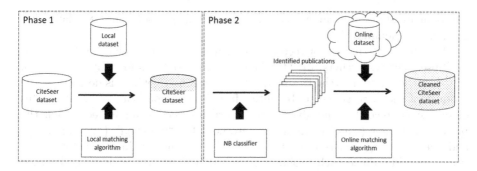

Fig. 1. The framework of our method for correcting the CiteSeer dataset (the shadow represents the corrected metadata entries in the CiteSeer data).

match such that it can correct a significant portion of wrong entries that cannot be handled by [7] and the correction accuracy is greatly improved.

In Phase 2, for the rest of unmatched metadata entries in the CiteSeer dataset, we need to identify wrong ones first since there is a huge cost to match each rest of entries with the metadata entries inside an online dataset by queries [14]. The identification process is implemented by a Naive Bayes (NB) classifier with an automated labeling algorithm and two feature selection methods. Then, our online matching algorithm attempts to find the matched one for each identified entry in an online dataset. Since more datasets are not allowed to be downloaded but open for queries, the online matching algorithm plus error identification significantly increase the correction coverage.

The performance of our method is tested based on 1000 randomly selected publications of the CiteSeer dataset containing 1,926,882 publications. Compared to existing cleaning methods [7,10], our contributions are shown as follows:

- Our method is program-based and can handle large mount of data without manual work;
- Our method outperforms the method of [7] at the correction accuracy around 130 % merely based on the local DBLP dataset;
- Our method significantly improves the correction coverage by using two new matching algorithms, the improvement is up to 539 % compared to the method of [7];
- In total, 48.1 % metadata entries with errors can be corrected by our method.

2   Related Work

Our data cleaning method is based on record linkage. Record linkage refers to recognizing (matching) records in two different data sources, which represent identical entity [11]. In the last decades, record-linkage-based methods are extensively used in many applications, such as duplicate detection [15,16], approximate string matching [17], reference or entity resolution [18], data integration and cleaning [7,19–21].

A probabilistic model for record linkage was firstly proposed in [11] adopted by subsequent researchers. In [11], there are three options for matching status, i.e., "matched", "unmatched" and "probably matched", which is decided by the product of the conditional probabilities from the comparisons of all attributes of two given records. The given records are matched (unmatched) when the product is bigger (smaller) than the threshold for "matched" ("unmatched"), and the matching status is probably matched if the product is between the two thresholds. Since each probably matched record pair needs to be checked manually, there are only two options (matched and unmatched) are used in our method.

In general, the applications to record linkage contain index step that is to reduce the large number of comparisons [22]. Intuitively, if two datasets (A and B) are to be matched, it is not necessary to compare each record from A with all records in B since most records in B correspond to non-matches and it is time-consuming (the number of comparison increases quadratically as datasets are getting larger). Thus, it is better to only compare potential record pairs. The traditional index techniques include blocking approaches [11,16], canopy clustering approaches [18,21], string-mapping based approaches [23,24] and so on. The underlying idea of these approaches is to shrink the comparison range by comparing a given record with "similar" records. Usually, as an output of index step, a similarity value is calculated for each potential pair based on different matching functions (Jaro-Winkler, Jaccard, edit-distance and longest common sub-string) [17]. As [7], our index step is based on inverted index, i.e., we utilize Boolean and Phrase queries to find "similar" metadata entries for comparison in Phase 1 and 2 respectively. Meanwhile, Jaccard function is used for matching.

The work of [7] is most closely related to ours. The authors firstly pointed out that all wrongly extracted entries can be separated by two categories: *irrelevant* or *relevant* to the actual one. For example, the wrong titles that contain (i) tokens that do not occur at all in the actual title or (ii) only stop words (such as "and" or "the") belong to the irrelevant category. The wrong titles that (i) are incomplete or (ii) contain other tokens besides title tokens (such as author, venue, or year) are in the relevant category. The method in [7] works well on wrong titles with relevant errors, however, it cannot handle with irrelevant errors. The basic steps of the method is that (1) each title in the CiteSeer dataset are divided into trigram word(s) as queries [13]; (2) the titles of all returned entries from the indexed DBLP dataset are matched with the original title based on Jaccard similarity; (3) the matched metadata entry will replace the original entry.

3   Problem Description and Measurement

Our goal is to use external linked scholarly datasets (local or online) to reduce the number of the wrong metadata entries of the CiteSeer dataset as many as possible, which is defined as follows:

$$\Delta N_w = N_w - N_w' = \Delta N_w^1 + \Delta N_w^2, \tag{1}$$

N_w and N_w' are the numbers of wrong metadata entries in the CiteSeer dataset before and after running a data cleaning method respectively. ΔN_w^1 and ΔN_w^2 are the changes of the numbers of entries with errors in Phase 1 and Phase 2 respectively. Note that ΔN_w can be negative and it means that there are more wrong entries in the CiteSeer dataset after data cleaning. Actually, if the matching and identification processes work poorly and many correct entries are wrongly "corrected", ΔN_w could be negative.

In order to facilitate the comparison, ΔN_w is normalized and we obtain the definition of correction coverage shown as follows:

$$CC = \frac{\Delta N_w}{N_w}. \tag{2}$$

CC is the percentage of corrected entries. A higher CC means better performance for a data cleaning method. Note that CC is sensitive to linked datasets and usually more/larger linked datasets lead to higher CC. To remove the effect of different datasets, the correction accuracy is used to measure the performance of data cleaning methods and defined in Eq. 3:

$$CA = \frac{\Delta N_w^D}{N_w^D} \tag{3}$$

where ΔN_w^D and N_w^D are the total number of corrected and wrong entries in the overlap between the CiteSeer dataset and linked datasets. The superscript D represents all linked datasets.

Here we also select the *Recall* (R), *Precision* (P) and *F measure* $(F_1$ value) as the metrics for identification and matching results, which are the standard measurements in Information Retrieval and Machine Learning domains.

4   Phase 1

The title and the authors' names of a publication usually appear in its first-several-line text. Since the CiteSeer dataset provides the full text of each publication, these metadata information can be located by searching its first-several-line text. Based on this idea, we propose our matching algorithm with a local dataset. For the title and the authors' names of each metadata entry in a local dataset, they are used to search the matched publication in the CiteSeer dataset, which contains them in the first-several-line text. Thus, our algorithm can address both of relevant and irrelevant errors since the title and the authors' names of the matched publication are from the text (but not from extracted results by the SVMHeaderParse). Certainly, the search process needs to index the CiteSeer and local datasets and build a local search engine, which is implemented by Lucene [25].

The details of our local matching algorithm for Phase 1 is shown in Algorithm 1. In line 1, the CiteSeer index includes the following fields: title, author, publishing year and venue, page number and the first-40-line text of each publication (usually the metadata information of a publication is included in its first-40-lines text). The fields of the local dataset index are identical to the ones of the CiteSeer index except the first-40-lines text. Note that each field could be empty if its content does not successfully extracted from the datasets. In line 2, it is to initialize the set of the corrected metadata entries of the Cite-Seer dataset. Line 3 is to retrieve all metadata entries of the local dataset. In line 4, it is to get the title of each entry. In line 5, all terms of each title are used to formulate a conjunctive Boolean query send to the CiteSeer index on the first-40-line-text field, it can return the publications containing all the terms. In line 6–8, for a returned publication, it will be considered as the matched one if (1) its first-40-line text contains a similar title ($matchTitle(d, T_e) \geq \theta$ where $matchTitle(d, T_e)$ is to calculating a Jaccard similarity and θ is a predefined threshold) and at least one author's name ($matchAuthors(c, e) \geq 1$); (2) it has the maximum number of identical items ($matchAll(c, e) = MAX$), e.g., title and all authors' names. Finally, the matched one in the CiteSeer dataset are replaced by the one in the local dataset.

Algorithm 1. Our matching algorithm for a local dataset

Input: $D \leftarrow$ all metadata entries of a linked local dataset, $C \leftarrow$ all
    CiteSeer metadata entries with first-40-line text, θ is a predefined
    threshold.
Output: A set of corrected CiteSeer metadata entries C^D using D.
1. Index the local and the CiteSeer datasets into inverted indexes;
2. $C^D = \phi$;
3. **for** all entries $e \in D$ **do**{
4. $T_e \leftarrow getTitle(e)$;
5. $C_e \leftarrow retrieve(T_e)$;
6.   **for** all $c \in C_e$ **do**{
7. $d \leftarrow getText(c)$;
8.   **if** $matchTitle(d, T_e) \geq \theta$ & $matchAuthors(c, e) \geq 1$ &
        $matchAll(c, e) = MAX$ **then**{
9. $m \leftarrow match(c, e)$;
10. $C^D = C^D \bigcup m$;
11.   }
12.   }
13. }
14. **return** C^D

5   Phase 2

5.1   Identifying Incorrect Entries

Before correcting the metadata entries remained from Phase 1, we first need to identify the entries with errors due to the communication cost for network transmission. Our identification method is based on the observation: the vast majority of metadata entries with errors have wrong titles such that we can use identified wrong titles to find wrong metadata entries.

Intuitively, words in the titles of the entries belonging to the classes c_w and c_c (represent the sets of wrong and correct entries respectively) could have different conditional occurrence probabilities, e.g., the word 'abstract' has higher probability in wrong titles than in correct ones. Our NB classifier first calculates the probabilities of each word for c_w and c_c based on labelled entries, and then the probabilities are used to identify wrong entries. Here each title is represented in unigram (single word) and bigram, e.g., for the title "introduction to algorithms", its bigram words are "introduction to" and "to algorithm".

Algorithm 2 shows how to calculate the conditional probability of each word. In this algorithm, the input is the set of n labelled metadata entries S from the CiteSeer dataset. The output is the training data Tr and each element $tr \in Tr$ is a 3-tuple including a term, its conditional probabilities for the classes c_w and c_c respectively. In line 2, V is a vocabulary and each (unigram or bigram) word in V is from all titles of S. In line 4 and 5, the conditional probabilities are calculated by the following formula $\frac{T_{ct}+1}{\sum_{t' \in V}(T_{ct'}+1)}$ $(c \in \{c_c, c_w\})$, in which T_{ct} and $\sum_{t' \in V}(T_{ct'}+1)$ are the numbers of the occurrence of word t and all words in V respectively. Adding one is a simple smoothing strategy.

Algorithm 2. Train NB classifier

Input: The set of n labelled metadata entries of CiteSeer with labelled
     class $(c_c$ or $c_w)$ S.
Output: Conditional occurrence probabilities of each term t for c_c and
     c_w.

1. $Tr = \phi$;   //initialization
2. $V = getVocabulary(S)$; //get all unigram or bigram words
3. **for** each word $t \in V$ **do**{
4. $P(c_w, t) = getProb(t, S, c_w)$;
5. $P(c_c, t) = getProb(t, S, c_c)$;
6. $Tr = Tr \cup (t, P(c_w, t), P(c_c, t))$;
7. }
8. **return** Tr;

Algorithm 3 demonstrates that how to identify a given metadata entry by using the NB classifier. In line 1, not all words in Tr are used to predict. Here the *feature selection* methods based on *Mutual Information* and χ^2 [13] are applied to learning a set of discriminative words that are good at distinguishing c_c and c_w. In line 2, only the words of the given title in the selected words V can be left for prediction. In line 3 and 4, n_w and n_c are the numbers of the entries with and without errors in the training dataset S ($|S| = n_w + n_c$). $\frac{n_w}{|S|}$ and $\frac{n_c}{|S|}$ are the prior probabilities for c_w and c_c respectively.

Algorithm 3. NB classifier prediction

Input: The training data Tr, the title of a given entry T.
Output: The predicted class $c^* \in \{c_c, c_w\}$ of T.
1. $V = featureSelect(Tr)$;   //get selected words by MI or χ^2 methods
2. $I = extractTokens(T, V)$;   // tokens in V and T remain
3. $s_w = \frac{n_w}{|S|}$;   //initial score for c_w
4. $s_c = \frac{n_c}{|S|}$;   //initial score for c_c
5. **for** each word $t \in I$ **do**{
6. $s_w += logP(c_w, t)$;   //multiple conditional probabilities
7. $s_c += logP(c_c, t)$;
8. }
9. **if** $s_w \geq s_c$ **then** $\{c^* = c_w\}$
10. **else** $\{c^* = c_c\}$;
11. **return** c^*;

5.2   Matching Process

After all identified metadata entries of the CiteSeer dataset are selected, we find their correct matched entries from online datasets. Unlike local datasets, online datasets only can be accessed by queries and thus a new matching algorithm is presented in Algorithm 4. Here we assume that each online dataset provides the services of full-text and phrase-query search, i.e., a phrase is considered as a query by using a double quotes syntax.

Given an identified metadata entry, first of all, two random strings (in unigram model) are obtained from its first-40-line text. Then they are sent as phrase queries to the search interface of an online dataset to retrieve returned metadata entries. For a returned entry, it is the matched one if it satisfies the three criteria:

- it must be returned by the two phrase queries simultaneously;
- its more than one authors' surnames can be found in its first-40-line text of the identified entry, i.e., $matchAuthors(d_{40}, A_c) \geq 1$;
- its title can be found in the first-40-line text of the identified entry by using Jaccard similarity coefficient, i.e., $matchTitle(d_{40}, T_c) \geq \theta$.

Note that (1) if any phrase query returns empty, another random string will be extracted and sent as another query; (2) the length of each phrase query is empirically set to 7 (words) since a too long or short phrase query will result in empty or too many metadata entries returned; (3) usually returned entries by a query are ranked by some ranking criteria, here only the top 20 returned metadata entries are retrieved as candidates.

Algorithm 4. Our matching algorithm for an online dataset

Input: $C \leftarrow$ identified metadata entries of CiteSeer with its full text, I $\leftarrow$ the search interface of an online dataset and θ is a predefined threshold.

Output: A set of corrected CiteSeer metadata entries C^I using I.

1. $C^I = \phi$;   //initializing C^I
2. **for** each entry $e \in C$ **do**{
3. $d \leftarrow getText(e)$;   // obtain the full text
4.   **while** $q_1 \leftarrow getRandString(d_{40})$ && $q_2 \leftarrow getRandString(d_{40})$ **do**{
5. $C_{q_1} \leftarrow retrieve(q_1, I)$;   // obtain returned entries
6. $C_{q_2} \leftarrow retrieve(q_2, I)$;
7.   **for** each entry $c \in (C_{q_1} \cap C_{q_2})$ **do**{
8. $T_c \leftarrow getTitle(c)$;
9. $A_c \leftarrow getAuthors(c)$;
10.   **if** $matchTitle(d_{40}, T_c) \geq \theta$ & $matchAuthors(d_{40}, A_c) \geq 1$ {
11. $m \leftarrow match(e, c)$;
12. $C^I = C^I \bigcup m$;
13.   }
14.   }
15. }
16. **return** C^L

---

6   Experimental Results

6.1   Data

Our CiteSeer dataset was downloaded from Amazon Web services provided by the CiteSeer website in Jan 2015, which is for academic purpose and contains around 2 million publication information. For each publication, a text file (.txt) stores its full text and a XML file keeps its metadata and citation information.

To carry out the experiments for testing the performance of Algorithm 1, the DBLP dataset is used as the local dataset as [7], which was downloaded in February 2015 from the DBLP official website. It contains the metadata information of around 2.7 million publications without full texts. For matching with an online dataset, the mirror website of Google Scholar (xs.glgoo.com) is selected as the online dataset for testing. Like Google Scholar, it also supports phrase query and full-text search, meanwhile, currently it is estimated that there are more

Table 1. The statistics information of the experimental datasets.

Dataset	Type	#docs	Size(G)	Full text	Metadata
CiteSeer	Local	1,926,882	4.6	Yes	DOI, title, authors' names, authors' emails, publishing year and venue, page number, full text
DBLP	Local	2,731,554	0.6	No	Title, authors' names, publishing venue, publishing date, pages number
Google Scholar	Online	≈100M	Unknown	Yes	Title, authors' names, publishing venue, publishing date, pages number

than 100 million publications stored in the background dataset [26]. To automatically send search queries and extract metadata entries from resulting pages, we develop a program-based crawler communicating with the search interface. Table 1 shows the statistics information of the experimental datasets.

6.2   Matching Results for Local Dataset

To show the performance of Algorithm 1, 1000 metadata publications are randomly selected from the CiteSeer dataset and manually labelled as 'in DBLP' or 'not in DBLP'. Then 308 documents are found in DBLP dataset and there are 97 entries with wrong titles ($N_w^D = 97$). Meanwhile, we implement the method of [7] (Caragea's method) described in Sect. 2. Table 2 shows the matching results of two methods and we find that the number of corrected entries is small although the best F_1 value of Caragea's method can reach 0.88, it results in relatively low CA and CC. Especially, when the threshold is 0.9 or 1.0, the number of wrong entries just decrease 3. The reason for the results is that Caragea's method cannot find the matched ones for the metadata entries with irrelevant errors in the DBLP dataset and it is proven by the low recalls. Our method outperforms the method at CA and CC around 130 %, and 710 random entries are left for Phase 2 (18 false-negative entries + 692 unmatched entries).

6.3   Identification Results

As mentioned before, the NB classifier with two feature selection methods is used to identify metadata entries with errors. Thus, the size of training data plays an important role in the performance of identification. For 1000 random metadata

Table 2. The matching results of our method and the method in [7] between the DBLP and the CiteSeer datasets (#macthed: the number of matched entries; θ is a threshold for Jaccard similarity coefficient; CC: $\frac{\Delta N_w^1}{N_w}$, $N_w = 439$; CA: $\frac{\Delta N_w^1}{N_w^D}$, $N_w^D = 97$).

Method	θ	Prec.(P_m)	Recall(R_m)	F_1 value	#matched	ΔN_w^1	CC	CA
Ours	0.5	0.90	0.97	0.93	292	60	0.14	0.65
	0.6	0.94	0.96	0.95	289	69	0.16	0.74
	0.7	0.96	0.95	0.95	285	73	0.17	0.78
	0.8	0.96	0.95	0.96	285	75	0.17	0.81
	0.9	0.97	0.95	0.96	284	76	0.17	0.82
	1.0	0.97	0.95	0.96	284	76	0.17	0.82
Caragea's	0.5	0.93	0.82	0.88	270	33	0.08	0.35
	0.6	0.96	0.79	0.87	252	32	0.07	0.34
	0.7	0.97	0.73	0.84	231	20	0.05	0.22
	0.8	0.98	0.70	0.82	221	14	0.03	0.15
	0.9	0.99	0.66	0.79	204	3	0.007	0.03
	1.0	0.99	0.66	0.79	204	3	0.007	0.03

entries, if we only use 900 ones as training data and 100 entries are used as testing data (10-fold cross-validation), F_1 value merely reaches 0.39 on average, which is far from satisfactory. Obviously, much more random labelled entries are required for training. However, the preparation of a big enough training data is an expensive, cumbersome and time-consuming process.

In order to reduce the level of human intervention, Algorithm 1 is used to automatically label metadata entries as follows: suppose a metadata entry c in the CiteSeer dataset is matched with the entry d in the DBLP dataset by using Algorithm 1, c is labelled as c_c (correct class) if the extracted title of c is identical to the title of d, otherwise, c_w (wrong class). Then, we have two collections that contains 132398 and 339340 entries labelled as c_w and c_c respectively.

Figure 2 shows the identification results based on the two collections with 10-fold cross-validation. In each subgraph, x axis is sample size (e.g. the 4000-entry sample contain 2000 correct and wrong metadata entries respectively and 90 % training data and 10 % testing data), y axis is feature number (e.g., 10000 features mean that the top 10000 unigram (or bigram) words are selected by MI or χ^2 method for prediction) and z axis is the F_1 value. Note that, when the feature size of y axis is beyond the total number of words in each training data, the rest F_1 values are equal to the one from the maximum feature number. From Fig. 2, we can see that (1) F_1 value grow quickly with the increase of sample (or feature) size when feature (sample) number is large; (2) F_1 value do not change with the increase of feature (sample) size when sample (feature) size is small; (3) the performance of using unigram or bigram with MI or χ^2 are close to each other. In a

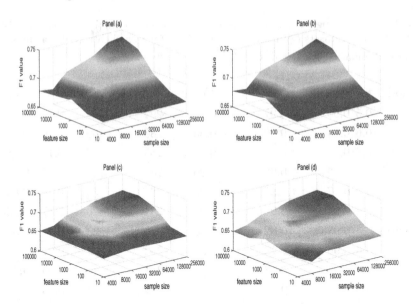

Fig. 2. The identification results based on 471,738 metadata entries labelled by Algorithm 1 with 10-fold cross-validation (Panel (a): implemented by bigram words with MI method; Panel (b): implemented by bigram words with χ^2 method; Panel (c): implemented by unigram words with MI method; Panel (d): implemented by unigram words with χ^2 method).

word, the NB classifier for the 471,738-entry collection can reach acceptable performance especially with large sample and feature sizes.

With the same training data and the NB classifier, the rest 710 random entries remained from Phase 1 is used as testing dataset and Fig. 3 shows the average identification result. At the first glance, the results is counter-intuitive and the best F_1 is better than the best one in Fig. 2. Compared with testing data in Fig. 2, the percentage of short titles in the 710 entries is much higher and most of short titles are wrong ones, meanwhile, short titles are easier to be judged as wrong ones by our method and thus the better F_1 can be reached.

6.4   Matching Results for Online Dataset

After the identification process, all identified metadata entries will be matched with Google Scholar online dataset. First of all, we manually check each random entry in the 1000-entry sample and find that 90.6 % entries are in Google Scholar dataset. To show the performance Algorithm 4, we use the one of identified results (whose F_1 value is 75.5 %) based on 230,400 training dataset with unigram words and χ^2 method to do matching experiment.

Table 3 shows the matching results and we find that (1) the best CA and CC are 0.48 and 0.31 when $\theta = 0.6$; (2) CC is much larger than the ones shown in Table 2 due to the huge size of Google Scholar dataset; (3) CA is between the

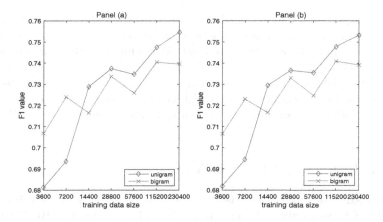

Fig. 3. The average identification results for the 710 random entries remained from Phase 1 based on the training data used in Fig. 2 (Panel (a): feature selection is based on χ^2 method; Panel (b): feature selection is based on MI method; Feature size: all unigram/bigram words of each sample are selected as features).

ones from our local matching algorithm and Caragea's method since Algorithm 4 can deal with irrelevant errors but random phrase queries from text could reduce the precision and the recall.

Table 3. The matching result between Google Scholar and the CiteSeer datasets based on 455 identified entries (#macthing: the number of matched entries; θ is a threshold for Jaccard similarity coeffient; CC: $\frac{\Delta N_w^2}{N_w}$, $N_w = 439$; CA: $\frac{\Delta N_w^2}{N_w^D}$, $N_w^D = 280$).

θ	Prec.(P_m)	Recall(R_m)	F_1 value	#matched	ΔN_w^2	CC	CA
0.5	0.81	0.89	0.85	272	128	0.29	0.46
0.6	0.85	0.87	0.86	265	135	0.31	0.48
0.7	0.86	0.83	0.85	254	130	0.30	0.46
0.8	0.88	0.82	0.85	250	132	0.30	0.47
0.9	0.89	0.79	0.84	243	132	0.30	0.47
1.0	0.90	0.76	0.82	233	129	0.29	0.46

6.5   Overall Performance

Totally, for the 1000 random entries, after the cleaning of Phase 1 and 2, the best performance is $\Delta N_w = \Delta N_w^1 + \Delta N_w^2 = 76 + 135 = 211$ and thus the overall corrected coverage and accuracy can reach 48.1 %. Compared to Caragea's method, the overall improvements on CC and CA are 539 % and 65 % respectively.

All the experiments were carried out on a Xeon E3 3.2 GHz server with 32G memory and all inverted indexes are builded by Lucene v3.2.1. In Phase 1, except the random sample dataset, we also apply Algorithm 1 to the whole CiteSeer dataset and total 471,738 metadata entries are matched with DBLP dataset, and it takes 2,867,079 seconds to the matching process. In Phase 2, each identified entry requires 4 times communication with Google Scholar on average and thus around 2.7 million queries needs to be issued for all rest entries of the CiteSeer dataset.

7   Conclusions and Future Work

We presented a record-linkage-based method to clean the metadata information of the CiteSeer dataset by matching with a local dataset (DBLP) and an online dataset (Google Scholar) respectively. It outperforms existing data cleaning methods for the CiteSeer dataset. The success of our method is from (1) it can address irrelevant errors by utilizing publication text and (2) its new online matching algorithm plus automated identification process creates a feasible way to link web data sources, which significantly improve the performance of data cleaning.

In fact, for Algorithms 3 and 4, there is still much room for improvement. Especially, the random phrase queries of online matching algorithm result in undesirable recall and precision. In the future, we attempt to send keywords with high tf-idf values from full text and random phrase queries together to address this problem.

Acknowledgements. This work has been partially supported by National Key Research Program of China (2016YFB1001101), NSFC (No.61440020, No.61272398 and No.61309030), NSERC Discovery grant (RGPIN-2014-04463) and Programs for Innovation Research in CUFE.

References

1. Peng, F., Schuurmans, D.: Combining naive bayes and n-gram language models for text classification. In: Sebastiani, F. (ed.) ECIR 2003. LNCS, vol. 2633, pp. 335–350. Springer, Heidelberg (2003). doi:10.1007/3-540-36618-0_24
2. Caragea, C., Silvescu, A., Kataria, S., Caragea, D., Mitra, P.: Classifying scientific publications using abstract features. In: SARA (2011)
3. Sen, P., Namata, G., Bilgic, M., Getoor, L., Gallagher, B., Eliassi-Rad, T.: Collective classification in network data. AI Mag. **29**(3), 93–106 (2008)
4. Caragea, C., Silvescu, A., Mitra, P., Giles, C.: Can't see the forest for the trees? a citation recommendation system. In: JCDL, pp. 111–114 (2013)
5. Carage, C., Wu, J., Williams, K., Das, S., Khabsa, M., Teregowda, P., Giles, C.L.: Automatic identification of research articles from crawled documents. In: WSDM-WSCBD (2014)
6. Huang, W., Kataria, S., Caragea, C., Mitra, P., Giles, C., Rokach, L.: Recommending citations: translating papers into references. In: CIKM, pp. 1910–1914 (2012)

7. Caragea, C., Wu, J., Ciobanu, A., Williams, K. ndez Ram rez, J.F., Chen, H., Wu, Z., Giles, L.: Citeseerx: a scholarly big dataset. In: Advances in InformationRetrieval, pp. 311–322 (2014)
8. CiteSeerX. http://csxstatic.ist.psu.edu/about/data
9. Lipinski, M., Yao, K., Breitinger, C., Beel, J., Gipp, B.: Evaluation of header metadata extraction approaches and tools for scientific pdf documents. In: Proceedings of JCDL, pp. 385–386 (2013)
10. Wu, J., Williams, K., Khabsa, M., Giles, C.L.: The impact of user corrections on a crawl-based digital library: a citeseerx perspective. In: Collaborative Computing: Networking, Applications and Worksharing (CollaborateCom) (2014)
11. Fellegi, I.P., Sunter, A.B.: A theory for record linkage. J. Am. Stat. Assoc. **64**(328), 1183–1210 (1969)
12. Tang, J.: https://aminer.org/
13. Manning, C.D., Raghavan, P., Schutze, H.: Introduction to Information retrieval. Cambridge University Press, Cambridge (2008)
14. Wang, Y., Lu, J., Chen, J.: TS-IDS algorithm for query selection in the deep web crawling. In: Chen, L., Jia, Y., Sellis, T., Liu, G. (eds.) APWeb 2014. LNCS, vol. 8709, pp. 189–200. Springer, Heidelberg (2014). doi:10.1007/978-3-319-11116-2_17
15. Manku, G., Jain, A., Sarma, S.A.: Detecting near-duplicates for web crawling. In: WWW, pp. 141–150 (2007)
16. Wu, J., William, K., Chen, H., Khabsa, M., Caragea, C., Tuarob, S., Ororbia, A.G., Jordan, D., Mitra, P., Lee Giles, C.: Citeseerx: AI in a digital library search engine. AI Mag. **36**(3), 35–49 (2015)
17. Cohen, W., Ravikumar, P., Fienberg, S.: A comparison of string metrics for matching names and records. In: KDD Workshop on Data Cleaning and Object Consolidation, vol. 3, pp. 73–78 (2003)
18. McCallum, A., Nigam, K., Ungar, L.H.: Efficient clustering of high-dimensional data sets with application to reference matching. In: Proceedings of SIGKDD, pp. 169–178 (2000)
19. Rahm, E., Do, H.: Data cleaning: problems and current approaches. IEEE Data Eng. Bull. **23**, 3–13 (2000)
20. Tejada, S., Knoblock, C., Minton, S.: Learning object identification rules for information integration. J. Inf. Syst. **26**(3), 607–633 (2001)
21. Cohen, W.W., Richman, J.: Learning to match and cluster large high-dimensional data sets for data integration. In: Proceedings of SIGKDD, pp. 475–480 (2002)
22. Christen, P.: A survey of indexing techniques for scalable record linkage and deduplication. IEEE Trans. Knowl. Data Eng. **24**(9), 1537–1555 (2012)
23. Chakrabarti, S.: Mining the web: discovering knowledge from hypertext data. Morgan-Kauffman (2002)
24. Jin, L., Li, C., Mehrotra, S.: Efficient record linkage in large data sets. In: Proceedings of DASFAA, pp. 137–146 (2003)
25. Hatcher, E., Gospodnetic, O.: Lucene in Action. Manning Publications (2004)
26. Khabsa, M., Giles, C.L.: The number of scholarly documents on the public web. PLoS One 9(5) (2014)

Towards Understanding URL Resources in Recent Sina Weibo

Yifang Wan[1,2], Peng Li[1,2(✉)], Rui Li[1,2], Meilin Zhou[1,2], Yongjun Ye[1,2], and Bin Wang[1,2]

[1] Institute of Informaiton Engieering Chinese Academy of Sciences, Beijing, China
{wanyifang,lipeng,lirui,zhoumeilin,yeyongjun,wangbin}@iie.ac.cn
[2] University of Chinese Academy of Sciences, Beijing, China

Abstract. With the rapid development of Internet, micro-blog service has become the fastest growing Internet application, where URLs play an important role in the social network. However, the studies on analyzing the URL resources especially for Chinese micro-blog system are extremely scarce. In this paper, we construct a corpus which contains the dissemination and classification information about URLs in Sina Weibo. Then we focus on the typical questions who publishes the URLs, what the URLs point to and how the URLs are disseminated and answer all the questions above by analyzing a recent Sina Weibo corpus. We find that verified users tend to publish about twice the amount of URLs as non-verfied users; Video URLs are more easily to disseminate in Sina Weibo. Our findings provide insights on downstream IR applications such as search engine and recommender systems effectively.

Keywords: URLs · Sina Weibo · Online social networks

1 Introduction

Nowadays, micro-blog service has become an important media for information sharing, however, due to the 140 character limits, people have to use external links (e.g., URLs) to enrich the published information. According to previous study [1], around 20 %–30 % of tweets have URLs and the number of URLs per message are still increasing [2]. Despite the rapid growth, we see very limited studies on analyzing URL resources [3,4], specifically for Chinese micro-blog system.

On analyzing the URL resources, we organized our studies in 3 questions: who publish these URLs, what these URLs point to and how these URLs are disseminated. The highlights of our work can be summarized as follows:

- We construct a corpus that can be used for the research of URLs contained in Sina Weibo, which include URLs' dissemination and classification information.
- We find that the average num of URLs that per verified user published is far greater than it of the non-verified users.

© Springer International Publishing AG 2016
W. Cellary et al. (Eds.): WISE 2016, Part I, LNCS 10041, pp. 50–57, 2016.
DOI: 10.1007/978-3-319-48740-3_4

- Video URLs are more easily to disseminate in Sina Weibo, which can help the marketers to develop strategies to gain more benefits.
- Average sharing and commenting counts of arts, reference and video are greater relatively high, which indicates that reference, art, video resources in Sina Weibo are good resources for recommendation.

We believe that our findings provide new insights for better using the URL resources in Sina Weibo and help the marketers to design excellent marketing plan with the URLs.

2   Related Work

Recently, some researches focused on the short URLs in the web and Sina Weibo.

In 2009, Kandylas et al. [4] studied bitly URLs contained in Twitter posts and examined their properties. Their results indicated that unlike frequently book-marked URLs, which were generally of high quality, frequently tweeted URLs tended to fall into two kinds of extreme situations: they are either in high quality, or they are spam. In 2010, Antoniades et al. [3] provided the first analysis on the usage of short URLs. They conducted a series of experiments to describe the web of short URLs, specifically, they examined the content short URLs point to, how they are published, their popularity and activity over time, as well as their potential impact on the performance of the web. Their analysis shed light on the great value of URLs and how to best use these URLs. And in 2013, Wang et al. [5] analyzed URL links within users' contents, and attempted to understand user behaviors using the statistics. By observing the contents and publishing time of the contents, they classified users with commercial purposes into several types, and thus provided an efficient way to immediately and accurately identify the advertisers and Water Army in Sina Weibo.

Compared with the works mentioned above, our paper focus on the URLs in Sina Weibo, and presents a comprehensive and systematic analysis of the resources from different points of view.

3   Data Collection and Defination

3.1   Data Collection

To the best of our knowledge, there is no corpus of URL resources in Sina Weibo, due to the restrictions of Sina API such as the frequency and the account limitation, we try our best to construct the corpus:

(1) At first, 200,000 random users are selected from a public corpus[1], which was crawled According to the forwarding relationship. (2) Next, we collect the 200 thousand users' posts during April and September 2015, finally we got 1,871,484 posts.

The process of data preprocessing is as follows:

---

[1] https://aminer.org/billboard/Influencelocality.

(1) Pick out the posts which contain URLs based on the regular expression. (2) Remove the duplicate URLs, and transfer the short URLs to the long. (3) Send requests to Sina API[2] and ODP system[3] to obtain the file as well as content type of URLs. (4) Label the URLs not included in ODP [6] manually, check Chap. 6.1 for details.

Finally, we have 127,416 posts which contain a URL, 109,238 unique URLs, 91,697 URLs which have a content type and 72,552 URLs which have a file type. We believe that our corpus will contribute significantly to the study of URLs in Sina Weibo.

3.2   Definition

This section is mainly to explain some of the proper nouns in the paper.

URL resources: URL Resources in Sina Weibo refer to the links contained in the Weibo posts.

URLs' sharing/commenting counts: The short links' sharing/commenting counts in Sina Weibo, and a higher sharing/commenting counts represents the URL is more popular.

4   Who Published These URLs?

In this section, the first typical question is answered by analyzing the user features associated with the URLs. It naturally goes to analysis of the impact of whether a user is verified on the popularity of the URLs, as can be seen in Table 1, one conclusion we can draw is the average amount of URLs per verified user published is far greater than the non-verified users, which is consistent with our intuition.

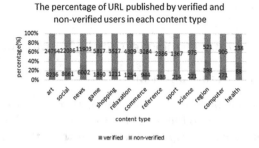

The percentage of URL published by verified and non-verified users in each content type

Table 1. Verified user vs not verified user

	Verified	Non − V
Nums (Pct.)	30,006 (17.8 %)	168,246 (82.2 %)
The nums of posts with URL (Pct.)	29,190 (26.3 %)	81,894 (73.7 %)
The average nums of posts with URL per user	0.97	0.43

Fig. 1. The percentage of URL published by verified and non-verified users in each content type

---

[2] Sina API. http://open.weibo.com/.

[3] http://www.dmoz.org/.

Now we concentrate on the proportion of verified and non-verified users on the each content category based on the ODP and file category, which will be explained in detail in Sect. 6. Figure 1 show that except the categories only have a small number of URLs, but in the dimension of news, verified users publish much more URLs, almost reach 35 %, which indicates users who like to publish URLs of news are generally more likely to certify themselves.

5   What the URLs Point To?

The first question has been discussed in the last section, now we focus on the second question: What the URLs point to? This section will be introduced from three aspects: content type of the URL, file type of the URL.

5.1   Content Type

The data was preprocessed as this: (1) Send the URLs to the dmoz.org site to get the standard answer; (2) Label these URLs according to the key words in domain name [7] manually, for example, if a URL contains the word 'music', we divide it into art; (3) Cluster URLs in accordance with the domain name, for the domain name appearing frequently, we manually label them according to experience, eventually, we get the 91,697 classified URLs. One thing that can't be ignored is that the ODP classification system divide video, music, photos URLs into art category, and the division of region is also not accurate. So in this part, art and region are not involved in the discussion.

Figure 2 shows the number of URLs for each category. There are many art, social, news, and game URLs, the URLs ranking top four accounted for about 80 % of the total. We can infer that URL resource distribution in Sina Weibo is unbalanced, and the user's interest is concentrated on several particular categories.

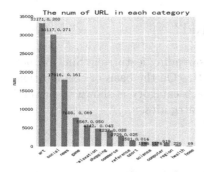

Fig. 2. The num of URL in each category

Fig. 3. Tag cloud of music

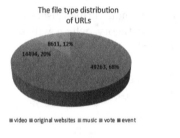

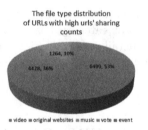

Fig. 4. The file type distribution of URLs

Fig. 5. The file type distribution of URLs with high sharing counts

5.2   File Type

The classification on file type is also studied to understand what the URL point to from another aspect. We use the Sina API to obtain the file type of URLs, the processsing is the same as the content type. After all, we got a total of 72,552 according to the file type of URLs. We investigate the proportion of each file type on the whole URLs and the URLs have a large share counts. Combined Figs. 4 and 5, for the URLs with higher number of sharing counts, 500 in this paper, the proportion of video type has a significant improvement, from 20 % to 36 %, we can draw the conclusion that video URLs are more likely to disseminate in Sina Weibo, which can help marketers to develop strategies to gain more benefit. Now we focus on the content of each category by employing tag cloud. As an example, Fig. 3 shows the word cloud of music, and these words can be divided into two categories according to their pos tag: nouns, verbs. Most of the nouns are the names of music player in China, from which we can find, music.163.com, lizhi.fm, echo.com etc. have become the mainstream music players in Sina Weibo; Verbs related to the words "like", "enjoy", "tune", indicates that Weibo is a platform for sharing, users posts the music they like to others, which shows that the URL in Weibo is a good resource for recommendation.

6   How the URLs Are Disseminated?

In this section, we want to understand URLs' spreading features in Sina Weibo, as well as the regulation hidden behind the URLs. On the basis of classification on content and file type, the distribution of each category is shown to present the differences and relationship between these categories.

6.1   Distribution of Each Category

Content Type: Due to the restrictions of sina Weibo API, we can not get all the posts containing the same URLs, so in this paper, we focus on the study of the spread range of URL instead analyzing the propagation characteristics of URLs in accordance with the time.

At first, we use Figs. 6 and 7 to present the distribution [8] of features have high discrimination on the content type of URL. Significantly differs from what we expect, the distinguishing factors are not the commenting, sharing counts of the URLs, but the users' tweet counts and followings instead. We think these features will help the classification of URLs to some extent.

Table 2 presents the correlation coefficient between the parameter performed by Pearson correlation analysis. In Eq. 1, n is sample size, X,Y are features above, x_i and y_i are samples of X,Y respectively. Compared with other features, the correlation coefficient between the sharing counts and commenting counts of URL is relatively large, which means there is a clear dependency relation between them. As the Table illustrates that correlation coefficient between the update frequency and following num is also large, which means the dependence between these two characteristics is not strong.

$$r_{XY} = \frac{n\sum_{i=1}^{n} x_i y_i - \sum_{i=1}^{n} x_i \cdot \sum_{n=1}^{n} y_i}{\sqrt{n\sum_{i=1}^{n} x_i^2 - (\sum_{i=1}^{n} x_i)^2} \cdot \sqrt{n\sum_{i=1}^{n} y_i^2 - (\sum_{i=1}^{n} y_i)^2}} \tag{1}$$

Figure 8 shows the each category's distribution of URLs' spread features based on the content type. Health and Family are eventually abandoned because the counts of these two type are really small. The blue lines represent the overall distribution, the update frequency of URLs [3] is how many new URLs appear in every day. The first picture of Fig. 8 suggests that between 2015-04-01 and 2015-09-31, the update frequency has a significant upward trend. The second and third pictures of Fig. 8 depict the sharing and commenting counts distribution of the short URLs in Sina Weibo, and we present it with the corresponding Cumulative Distribution Function (CDF). We can see that though URLs concerned with reference accounts for a small proportion (1.4 %) but they always appear at the bottom, which indicates that they are reposted more frequently.

File Type: Figure 9 shows the each category's distribution of URLs' spread features based on the file type. As illustrated in the second an third pictures of

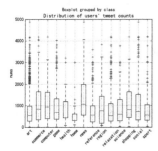

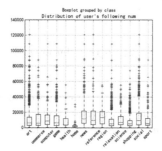

Fig. 6. Distribution of update frequency

Fig. 7. Distribution of users' following num

Table 2. The correlation coefficient between the features

name	symbol		1	2	3	4	5	6	7
following	1	1	1.000	0.421	-0.018	0.005	-0.008	-0.010	-0.001
tweetcounts	2	2	0.421	1.000	0.0256	0.032	0.005	-0.052	-0.033
follower	3	3	-0.018	0.025	1.000	0.170	0.338	-0.007	-0.004
forwardnum	4	4	0.005	0.032	0.170	1.000	0.538	-0.003	0.000
commentnum	5	5	-0.008	0.005	0.338	0.538	1.000	-0.004	-0.001
urlsharenums	6	6	-0.010	-0.052	-0.007	-0.003	-0.004	1.000	0.652
urlcommentnums	7	7	-0.001	-0.03	-0.004	-0.000	-0.001	0.652	1.000

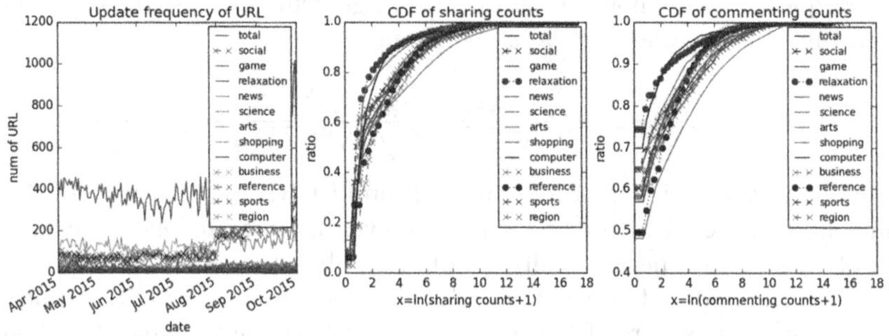

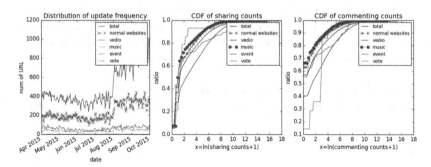

Fig. 8. Distribution of URL features in different content types (Color figure online)

Fig. 9. Distribution of URL features in different file types

Fig. 9, the average sharing and commenting counts of video [9] are greater than that of music and original websites, which indicates that video resource in Sina Weibo is a good resource for video recommendation.

7   Conclusion

With the popularity of social networks, many researchers have found the huge value of the URLs in Sina Weibo, they have expanded some work based on it, including identifying the garbage URLs [10], the threats they will cause to the social security [11] and so on. In this paper, we mainly concern about who publish them, what they point to, how they are disseminated. By properly analyzing the URLs in Sina Weibo, we find some interesting phenomenons as mentioned before. Furthermore, we should concentrate on: (1) extracting high-quality URLs from Sina Weibo; (2) establishing search engine on the high quality URLs; (3) building a classifier to identify the content and file type of these URLs.

Acknowledgments. This work is supported by the National Natural Science Foundation of China (grant Nos. 61402466 and 61572494) and the Strategic Priority Research Program of the Chinese Academy of Sciences (grant No. XDA06030200).

References

1. Suh, B., Hong, L., Pirolli, P., et al.: Want to be retweeted? large scale analytics on factors impacting retweet in twitter network. In: 2010 IEEE Second International Conference on Social computing (socialcom), pp. 177–184. IEEE (2010)
2. Liu, Y., Kliman-Silver, C., Mislove, A.: The tweets they are a-changin: evolution of twitter users and behavior. In: International AAAI Conference on Weblogs and Social Media (ICWSM), vol. 13, p. 55 (2014)
3. Antoniades, D., Polakis, I., Kontaxis, G., et al.: we. b: The web of short URLs. In: Proceedings of the 20th International Conference on World Wide Web, pp. 715–724. ACM (2011)
4. Kandylas, V., Dasdan, A.: The utility of tweeted URLs for web search. In: Proceedings of the 19th International Conference on World Wide Web, pp. 1127–1128. ACM (2010)
5. Wang, Y., Tao, H., Cao, J., Wu, Z.: Understanding user behavior through URL analysis in sina tweets. In: Huang, Z., Liu, C., He, J., Huang, G. (eds.) WISE 2013. LNCS, vol. 8182, pp. 98–108. Springer, Heidelberg (2014). doi:10.1007/978-3-642-54370-8_9
6. Odijk, D., White, R.W., Hassan Awadallah, A., et al.: Struggling and success in web search. In: Proceedings of the 24th ACM International on Conference on Information and Knowledge Management, pp. 1551–1560. ACM (2015)
7. Ritter, A., Etzioni, O., Clark, S.: Open domain event extraction from twitter. In: Proceedings of the 18th ACM SIGKDD International Conference on Knowledge Discovery and Data Mining, pp. 1104–1112. ACM (2012)
8. Spina, D., Gonzalo, J., Amigó, E.: Discovering filter keywords for company name disambiguation in twitter. Expert Syst. Appl. **40**(12), 4986–5003 (2013)
9. Cheng, X., Dale, C., Liu, J.: Statistics and social network of youtube videos. In: 16th International Workshop on Quality of Service, IWQoS 2008, pp. 229–238. IEEE (2008)
10. Lee, S., Kim, J.: Warningbird: a near real-time detection system for suspicious urls in twitter stream. IEEE Trans. Dependable Secure Comput. **10**(3), 183–195 (2013)
11. Klien, F., Strohmaier, M.: Short links under attack:geographical analysis of spam in a URL shortener network. In: Proceedings of the 23rd ACM Conference on Hypertext and Social Media, pp. 83–88. ACM (2012)

Recommender Systems

Nonparametric Bayesian Probabilistic Latent Factor Model for Group Recommender Systems

Nipa Chowdhury[1] and Xiongcai Cai[1,2(✉)]

[1] School of Computer Science and Engineering,
The University of New South Wales, Sydney, NSW 2052, Australia
{nipac,xcai}@cse.unsw.edu.au
[2] Techcul Research, Sydney, Australia
xiongcai.cai@techcul.com

Abstract. The explosion of the online web encourages online users to participate in group activities. Group recommender systems are essential for recommending items to a group of users based on their common preferences. However, existing group recommender systems do not exploit user interaction within a group and merely work on groups with fixed sizes of users and same levels of similarity among group members, which significantly limits its usage in real world scenarios. In this paper, we propose a novel nonparametric Bayesian probabilistic latent factor model to learn the collective users' tastes and preferences for group recommendation by exploiting user interaction within a group, which is able to well handle a variety of group sizes and similarity levels. We evaluate the developed model on three publicly available benchmark datasets. The experimental results demonstrate that our method outperforms all baseline methods for group recommendation.

Keywords: Group recommender systems · Collaborative filtering · Bayesian probabilistic matrix factorisation · Dirichlet prior

1 Introduction

Recommender systems (RS) [3,4,9] have been recognised as a personal assistant to online users to find relevant information or products to satisfy their personal preferences. Matrix factorisation (MF) is a popular method to build personalised recommender systems, which shows great success in the Netflix prize competition [9]. The major purpose of MF in RS is to obtain some forms of lower-rank approximation to the original feedback matrix for understanding the interaction between user preferences and item attractiveness in forms of latent factors.

Although the RS are primarily designed to satisfy individual preferences, there are scenarios where a group of users are interested in participating in a single activity, for example, in entertainment purpose, watching a movie, playing an online video game, listening music and visiting a tourist place. The group recommender systems (GRS) generate a single recommendation list for a group of users, aiming to maximise all users satisfaction [13]. The rapid explore of online

© Springer International Publishing AG 2016
W. Cellary et al. (Eds.): WISE 2016, Part I, LNCS 10041, pp. 61–76, 2016.
DOI: 10.1007/978-3-319-48740-3_5

data allows users to find people who have a common interest and participate in group activities, which draws the attention of researchers on GRS. However, the group recommendation scenario is more challenging than the individual one as users have different preferences and different levels of interaction within a group [1]. MF algorithms that are designed for personalised RS [9,15], failed to learn latent factors to model collective user tastes and preferences, thus are not suitable for group recommendation [6].

Existing research on GRS can be divided into two categories [1,2,4,7,8,13, 18]. The first one is based on profile aggregation where a virtual user is created by averaging all user profiles in a group [12,18]. The recommendation list for the group is generated by considering only the virtual user. The second category is based on rank aggregation where a recommendation list for the group is generated by merging all user individual recommendation lists [1,2,13]. Personalised RS methods are used to generate the individual recommendation. However, the contribution of both profile aggregation and rank aggregation-based group recommendation systems is limited on the aggregation technique. The performances of these methods also depend on the size of the group and the level of similarity among group members. Furthermore, the main drawback in the existing methods is that they always ignore the interaction among the group members in generating recommendations, whereas the interaction among members may make a great effect in group decisions.

In this paper, a novel method named nonparametric Bayesian probabilistic latent factor model (NBPLFM) is proposed for group recommender systems where the task is to generate recommendations for a group of users. The proposed method extends Bayesian probabilistic matrix factorisation (BPMF) [15] to consider tastes and preferences of groups of users, by applying a Dirichlet process mixture model to the prior of the user and item latent factors. Rather than relying on the aggregation strategy like existing GRS, the innovative NBPLFM models group user preferences by exploiting their interaction within the group. As a result, this proposed model explicitly takes into account collective users preferences in learning and generating recommendations. Further, the integration of nonparametric prior allows NBPLFM to model group with variable group size and similarity, which reflects the real world group recommendation scenario. To verify the performance, we apply the proposed method on three public datasets. The experimental results confirm the efficiency of the proposed method compared to other baseline methods for group recommendation.

The rest of the paper is arranged as follows: in Sect. 2, we summarise related work. Section 3 presents the proposed matrix factorisation method. Experimental results are presented in Sect. 4. Finally, we draw conclusions in Sect. 5.

2   Related Work

The approaches that are adopted to build GRS can be divided into two categories, i.e., profile aggregation and rank aggregation [1,2,4,7,8,13,18]. Rank aggregation methods come with better flexibility than profile aggregation methods and the results are explainable. Different merging strategies [4,11] from social theory such as additive utilitarian, multiplicative utilitarian, average, least misery, average without misery and most happiness are employed to aggregate individual recommendation lists.

The authors in [2] study different types of aggregation strategies and report that the performance of the GRS does not depend on the aggregation strategies, but depends on the similarity among users within a group and also on the group sizes. Polylens [13] is developed as a group recommender to recommend movies to Movielens[1] users. It uses the nearest neighbourhood algorithm to derive an individual recommendation list. The group recommendation list is generated by merging the individual recommendation lists according to least misery approach. Further, the research in [1] introduces the notion of consensus function, that aims at maximising item relevance and minimising disagreements between group members. It uses the average or least misery approach to compute group relevance. To compute group disagreement between members, it proposes two alternative ways: average of pairwise disagreements or score variance. Recently, some model-based approaches are proposed for GRS [16,17,19]. For example, a consensus model named COM is proposed for generating group recommendations by assuming that item selection in a group depends not only on a user's personal choice of the content factors but also on group topics [19]. The research in [17] believes that social influence of friends also contributes to one's item selection decision. Based on this assumption, it proposes a probabilistic generative model by exploiting the preferences of group members and their pairwise influence on each other to reach the final decision. The authors in [16] use the k-means algorithm on the user latent factors to detect user groups. They create group latent factors by averaging the user latent factor to cooperate with item latent factors to generate group recommendations. Because contextual and social information is sensitive and expensive to collect, we avoid to use such information and thus do not compare our method with other methods [17,19] that use contextual or social information in addition to user feedback.

Although, the above methods improve the group recommendation performances but these methods have limitations for widely commercial applications. For instance the profile aggregation methods [12,18] and rank aggregation methods [1,2,13] in group recommender systems rely on the aggregation strategies. They do not account user interaction within a group and thus failed to model collective user preferences. User preference within a group is different from individual user preference and the RS should capture the users combined preference to generate group recommendations. Moreover, most of the existing research perform well when the groups are small in size (consists of 2, 3, 5, 8 or 10 users) and have high similarity [1,2,13]. However, social influence [17] and topic influence [19] of group members do not hold when groups are large. Therefore it is important to learn the collective preferences of users, especially for a large group which is still a challenging open question in group recommendation.

To overcome these limitations, we propose a novel method named nonparametric Bayesian probabilistic latent factor model for group recommendation by

---

[1] https://movielens.org/.

learning user and item latent factors to reflect tastes and preferences of the group. We extend BPMF by imposing Dirichlet process mixture model as *a prior* of the latent factors. The user latent factors are learned to maximise the overall group satisfaction by exploiting user interaction within a group. Modelling user interaction in learning enables our model to generate recommendations for groups with any size and similarity levels. Thus, NBPLFM improves performance significantly compared to other approaches to group recommendation.

3   Nonparametric Bayesian Probabilistic Latent Factor Model

In this section, we firstly present a key component of our algorithm named Bayesian probabilistic matrix factorisation. Then we extend it with the nonparametric prior distribution of user and item latent factors to learn the factors that are best suitable for group recommendations.

3.1   Bayesian Probabilistic Matrix Factorisation (BPMF)

Assuming there are N users and M items in the data, let matrix $R \in \Re^{N \times M}$ be a user preference matrix. In MF, $R \in \Re^{N \times M}$ can be approximated by two low rank matrices $U \in \Re^{D \times N}$ and $V \in \Re^{D \times M}$ as $R \approx U'V$ by minimising the sum of squared errors, where $D \ll min(N, M)$ is the dimensionality of user latent factors and item latent factors [9].

BPMF [15] employs a probabilistic linear model with Gaussian noise to learn the user latent factor and item latent factor. It also places prior and hyperprior over the model parameters and hyperparameters, respectively. It assumes that the user parameters (U) and item parameters (V) follow the multivariate Gaussian distribution and the user hyperparameters (θ_u) and the item hyperparameters (θ_v) follow the Gaussian-Wishart distribution. Figure 1(a) shows the graphical model of BPMF. The conditional distribution over the observed ratings and the prior distributions over the user and item parameters and hyperparameters are given by:

$$\mathcal{P}(R \mid U, V, \sigma) = \prod_i \prod_j [\mathcal{N}(R_{ij} \mid U_i'V_j, \sigma)],$$

$$\mathcal{P}(U \mid \theta_u) = \prod_i \mathcal{N}(U_i \mid \mu_u, \Lambda_u),$$

$$\mathcal{P}(V \mid \theta_v) = \prod_j \mathcal{N}(V_j \mid \mu_v, \Lambda_v),$$

$$\mathcal{P}(\theta_u \mid \theta_0) = \mathcal{N}(\mu_u \mid \mu_0, (\beta_0 \Lambda_u)^{-1}) \mathcal{W}(\Lambda_u \mid \mathcal{W}_0, v_0),$$
$$\mathcal{P}(\theta_v \mid \theta_0) = \mathcal{N}(\mu_v \mid \mu_0, (\beta_0 \Lambda_v)^{-1}) \mathcal{W}(\Lambda_v \mid \mathcal{W}_0, v_0),$$

where $\theta_0 = \{\mu_0, \mathcal{W}_0, v_0, \beta_0\}$ and $\mathcal{W}(\Lambda | \mathcal{W}_0, v_0)$ is the Wishart distribution with v_0 degrees of freedom and scale matrix W_0. The posterior predictive distribution of a rating R_{ij}^* is obtained by marginalizing over model parameters and hyperparameters.

$$\mathcal{P}(R_{ij}^* \mid R, \theta_0) = \int \int \mathcal{P}(R_{ij}^* \mid U_i, V_j) \mathcal{P}(U, V \mid R, \theta_u, \theta_v) \mathcal{P}(\theta_u, \theta_v \mid \theta_0).$$

The computation of this predictive distribution is analytically intractable; thus, Markov chain Monte Carlo (MCMC) is used to approximate inference. Specifically, Gibbs sampler is used to calculate the approximation. BPMF shows promising results in personalised recommendation without heavily relying on parameter tuning. It assumes that the user parameter and item parameter are drawn from a single common multivariate normal distribution. However, in GRS, we assume that the users within a group are more similar than the population in general. Items that shares same genre are more similar than the rest of the items. Thus, using a common distribution over all user parameters and hyperparameters limits the performance of BPMF on group recommendation. In this regard, we propose NBPLFM for group recommendation.

3.2   Nonparametric Bayesian Probabilistic Latent Factor Model

NBPLFM is based on the observation that the users and items within a group should be more similar than other group or the rest of the population. It draws the user and item latent factors from their group specific distributions which enables the resultant model to capture group choice more effectively and coherently. In reality, the number of the groups and the size of each group should not be fixed in group recommendation because a recommender system does not know before hand the number of users who will form a group to watch a movie, music or to visit a tourist spot. Similarly, the system does not know in advance how many groups will be built to consume a product together. So the group recommendation model should allow the variability of group size as well as the flexibility for creating the number of groups. Existing research such as [1, 2, 13] assumes a fixed number of groups and fixed number of users within each group in group formulation, which does not comply with the real world scenarios of group recommendation. To integrate the variability of group size and the adaptability of group number into our model, we exploit a nonparametric distribution and apply it over the user and movie latent factors. Specifically, we use a Dirichlet process mixture model that does not require the prior knowledge of the size of the groups. The proposed model allows an infinite number of mixture components; thus, the model is able to introduce new mixture components when new users are added [5]. The research in [14] also uses a nonparametric prior for the user latent class and item latent class, which uses side information about users and items to achieve an improved personalised recommendation performance. However, the method in [14] does not model the interaction among users within each group, thus is unable to learn collective users tastes and preferences in group recommendation scenario.

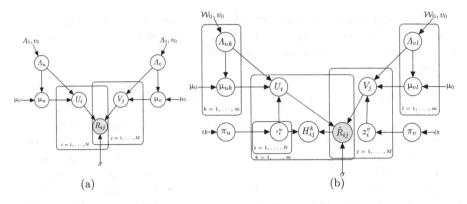

Fig. 1. Graphical model for (a) Bayesian probabilistic matrix factorisation and (b) Nonparametric Bayesian probabilistic latent factor model.

The probabilistic graphical model of our method is shown in Fig. 1(b). We assume that the users belong to k distinct cluster with means μ_u and items are distributed to l different cluster with means μ_v. The value of k could be infinite and automatically decided by our method. The prior probability of user cluster π_u and item cluster π_v are distributed according to a Dirichlet distribution $Dir(.)$. Cluster means μ_u and μ_v are generated from Gaussian distribution and cluster variances Λ_u and Λ_v are generated from Wishart distribution $\mathcal{W}(.)$. The user cluster label z^u and item cluster label z^v are sampled from their corresponding multinomial distribution $Multi(.)$. Please note that we transform the observed ratings R_{ij} into Gaussian distributions to reduce the rating variance across different user and item groups. The conditional distribution over the transformed observed ratings $\hat{R}_{ij}$ and the prior distribution over the user and item parameters and their group specific hyperparameters are given below:

$$P(\hat{R} \mid U, V, \sigma) = \prod_i \prod_j \mathcal{N}(\hat{R}_{ij} \mid U_i' V_j, \sigma), \qquad (1)$$

$$P(H \mid \hat{R}_{ij}, z_i^u) = \Delta(\hat{R}_{ij}; z_i^u), \qquad (2)$$

$$P(U \mid \theta_{uk}, z^u) = \prod_i \mathcal{N}(U_i \mid \mu_{z_i^u}, \Lambda_{z_i^u}), \qquad (3)$$

$$P(V \mid \theta_{vl}, z^v) = \prod_j \mathcal{N}(V_j \mid \mu_{z_j^v}, \Lambda_{z_j^v}), \qquad (4)$$

$$P(\theta_{uk} \mid \theta_0) = \mathcal{N}(\mu_{uk} \mid \mu_0, \Lambda_{uk}/\beta_0)\mathcal{W}(\Lambda_{uk} \mid \mathcal{W}_0, v_0), \qquad (5)$$

$$P(\theta_{vl} \mid \theta_0) = \mathcal{N}(\mu_{vl} \mid \mu_0, \Lambda_{vl}/\beta_0)\mathcal{W}(\Lambda_{vl} \mid \mathcal{W}_0, v_0), \qquad (6)$$

$$P(z^u \mid \pi_u) = Dir(\alpha/k), \qquad (7)$$

$$P(z^v \mid \pi_v) = Dir(\alpha/l), \qquad (8)$$

where $\theta_{uk} = \{\mu_{uk}, \Lambda_{uk}\}$ and $\{\theta_{vl} = \mu_{vl}, \Lambda_{vl}\}$. In a group, an item may be rated by many users. Average and least misery are commonly used aggregation techniques in group recommendation literature to learn the group's preferences for an item [7]. However, the least misery approach ignores the majority of users' information by concentrating on the user that has the smallest preference [4]. To account all users preferences for an item we use an average of user rating to learn the group's preference on that item. Thus, Δ refers as average function. We use Chinese restaurant process (CRP) to sample user membership and item membership from the DP mixture model with K components where the limit of the component K goes to infinity. According to CRP, the Nth user will be assigned to an existing group with probability $\frac{N_k}{N-1+\alpha}$ and it will be assigned to a new group with probability $\frac{\alpha}{N-1+\alpha}$, where α is a positive scaling parameter and N_k is the number of users already exists in group k [5]. The generative process for clusters is shown as follows:

(1) generates cluster distributions for each user and for each item:
$$\pi_u \sim Dir(\alpha/k),$$
$$\pi_v \sim Dir(\alpha/l);$$
(2) generates mean and variance for each user cluster:
$$\mu_{uk} \sim \mathcal{N}(\mu_0, \Lambda_{uk}/\beta_0),$$
$$\Lambda_{uk} \sim \mathcal{W}(v_0, \mathcal{W}_0);$$
(3) generates mean and variance for each item cluster:
$$\mu_{vl} \sim \mathcal{N}(\mu_0, \Lambda_{vl}/\beta_0),$$
$$\Lambda_{vl} \sim \mathcal{W}(v_0, \mathcal{W}_0);$$

where $k = 1, \ldots, K$ and $l = 1, \ldots, L$. After determining the cluster parameter, NBPLFM follows the following generative process to sample the cluster label parameter, the user and item latent factor parameter and the rating:

(1) for each user $i = 1, \ldots, N$ and for each item $j = 1, \ldots, M$, sample the user cluster label and item cluster label:
$$z_i^u \sim Multi(\pi_u),$$
$$z_j^v \sim Multi(\pi_v);$$
(2) for groups of users $k = 1, \ldots, K$, samples their vectors of parameters:
$$U_i^k \sim \mathcal{N}(U_i \mid \mu_{z_i^u}, \Lambda_{z_i^u});$$
(3) for each item $j = 1, \ldots, M$, sample a vector of parameters:
$$V_j \sim \mathcal{N}(V_j \mid \mu_{z_j^v}, \Lambda_{z_j^v});$$
(4) for users in group k, calculate
$$H_{ij}^k \sim \Delta(\hat{R}_{ij}^k; z^u);$$
(5) for each user i and movie j, sample a rating:
$$\hat{R}_{ij} \sim \mathcal{N}(\hat{R}_{ij} \mid U_i' V_j, \sigma) .$$

We formulate the user cycle group wise and item cycle item wise since our goal is to recommend items to various user groups. The posterior predictive distribution of a rating R_{ij}^* is obtained by marginalising over model parameters and group specific hyperparameters:

$$P(R_{ij}^* \mid R, \theta_0) = \int \int P(R_{ij}^* \mid U_i, V_j) P(U \mid R, H, V, \theta_{uk}) P(V \mid R, U, \theta_{vl}) P(\theta_{uk} \mid \theta_0)$$

$$P(\theta_{vl} \mid \theta_0) P(z^u \mid \pi_u) P(z^v \mid \pi_v).$$

Since the computation of above equation is expensive, following [15] we use Gibbs sampler to calculate an approximation. The Gibbs sampler cycles through the latent variables U_i, V_j, θ_{uk}, θ_{vl}, z_i^u and z_j^v are conditioned on current values of other variables. The conditional distribution of U_i is

$$P(U_i \mid \hat{R}, H, V, \theta_{uk}, z^u) \propto P(U_i \mid \theta_{uk}, z^u) P(H_{ij}^k \mid \hat{R}_{ij}; z^u) \prod_{ij} \mathcal{N}(H_{ij}^k \mid U_i' V_j, \sigma)$$

$$P(U_i \mid \hat{R}, H, V, \theta_{uk}, z^u) = \mathcal{N}(U_i \mid \mu_{uk}^*, \Lambda_{uk}^*), \tag{9}$$

where $\Lambda_{uk}^{*\,-1} = \Lambda_{uk}^{-1} + \frac{1}{\sigma^2} \sum_j V_j V_j'$, and $\mu_{uk}^* = \Lambda_{uk}^* (\Lambda_{uk}^{-1} \mu_{uk} + \frac{1}{\sigma^2} \sum_j [V_j H_{ij}])$. The conditional distribution over the group specific user hyperparameters conditioned on user factor matrix are given by the Gaussian-Wishart distribution:

$$P(\mu_{uk}, \Lambda_{uk} \mid U, \theta_0) = \mathcal{N}(\mu_{uk} \mid \mu_a, \Lambda_{uk}/\beta_a) \mathcal{W}^{-1}(\Lambda_{uk} \mid \mathcal{W}_a, v_a), \tag{10}$$

where $\mu_a = \frac{\beta_0}{\beta_0 + N_{uk}} \mu_0 + \frac{N_{uk}}{\beta_0 + N_{uk}} \overline{U}$, $\mathcal{W}_a = \Lambda_0 + S + \frac{\beta_0 N_{uk}}{\beta_0 + N_{uk}} (\overline{U} - \mu_0)(\overline{U} - \mu_0)'$, $\beta_a = \beta_0 + N_{uk}$, $v_a = v_0 + N_{uk}$ and $S = \sum_i^{N_{uk}} (U_i - \overline{U})(U_i - \overline{U})'$. The group membership of user and items are sampled according to the following equations:

$$P(z_i^u = k \mid z_{-i}^u, U_i, \theta_u, H) = b. \frac{N_{uk}^{-i}}{N_u - 1 + \alpha} \mathcal{N}(U_i \mid \theta_{uk}),$$

$$P(z_i^u \neq z_j^u \text{ for all } j \neq i \mid z_{-i}^u, U_i, \theta_u) = b. \frac{\alpha}{N_u - 1 + \alpha}$$

$$\int \mathcal{N}(U_i \mid \theta_u^*) \mathcal{N}(\mu_{nu}^* \mid \mu_0, \Lambda_v/\beta_0) \mathcal{W}^{-1}(\Lambda_{nv}^* \mid \mathcal{W}_0, v_0) d(\theta_u^*)), \tag{11}$$

where N_u is the number of users, N_{uk} the number of users assigned to group k i.e. $N_{uk} = \sum_i [z_i^u = k]$ and $\theta_u^* = \{\mu_{nu}^*, \Lambda_{nv}^*\}$. The conditional distribution of V_j is

$$P(V_j \mid \hat{R}, U, \theta_{vl}, z^v) \propto P(V_j \mid \theta_{vl}, z^v) \prod_{ij} \mathcal{N}(\hat{R}_{ij} \mid U_i' V_j, \sigma),$$

$$P(V_j \mid \hat{R}, U, \theta_{vl}, z^v) = \mathcal{N}(V_j \mid \mu_{vl}^*, \Lambda_{vl}^*), \tag{12}$$

where $\Lambda_{vl}^{*\,-1} = \Lambda_{vl}^{-1} + \frac{1}{\sigma^2} \sum_i U_i U_i'$ and $\mu_{vl}^* = \Lambda_{vl}^* (\Lambda_{vl}^{-1} \mu_{vl} + \frac{1}{\sigma^2} \sum_i [U_i \hat{R}_{ij}])$. The computation of $P(\theta_{vl} \mid V, \theta_0)$ follows the same equation as (10). The computation of z^v also follows (11), except it does not include H in its computation. The sampling algorithm of NBPLFM is given in Algorithm 1.

Algorithm 1. Nonparametric Bayesian probabilistic latent factor model

---

Initialise model parameter U^1, V^1
for t=1:T **do**

1   Sample group membership of users and items according to:
$$z_i^u \sim Multi(\pi_u),$$
$$z_j^v \sim Multi(\pi_v).$$

2   Sample the hyperparameters of user and item group according to (10):
$$\theta_{uk}^t \sim \mathcal{P}(\theta_{uk} \mid U, \theta_0),$$
$$\theta_{vl}^t \sim \mathcal{P}(\theta_{vl} \mid V, \theta_0).$$

3a   For users in a group k, sample U_i^k according to (9):
$$U_i^{t+1} \sim \mathcal{P}(U_i \mid \theta_{uk}, z^u, V, H, \hat{R}).$$

3b   For each item in a group, sample V_j according to (12):
$$V_j^{t+1} \sim \mathcal{P}(V_j \mid \theta_{vl}, z^v, \hat{R}, U).$$

end for

---

4 Experiments

4.1 Dataset

NBPLFM is evaluated on three publicly available benchmark datasets for group recommendation including (1) Movielens 100K,[2] denoted as ML, to recommend movies to a group of user; (2) Amazon video game dataset,[3] denoted as VG, to recommend video games to a group of players and (3) Amazon digital music dataset, (see Footnote 3) denoted as DM, to recommend music to a group of listeners. ML dataset consists of 100,000 ratings from 943 users on 1682 movies. VG dataset consists of 463,668 ratings from 228,570 users/reviewers on 21,025 video games. For 99,128 ratings in this dataset, we observe 'unknown' value in the 'userid' field. As 'unknown' does not mean any particular user, we remove those entries from the dataset. Again, for some video games we find multiple reviews from the same reviewer. In this case, we omit the preceding reviews and only count the latest review. Following the widely accepted procedure [10], the users who rated fewer than 5 movies and the items that have been rated by less than 3 users are removed. After the preprocessing, the dataset consists of total 60,913 ratings from 5644 users on 4568 video games. DM dataset consists of total 836,015 ratings from 478,243 users on 266,416 music. Following [10], we further remove users that review less than 10 songs and songs that are rated by fewer than 5 users resulting in 45,045 ratings from 4209 users on 2926 music. Users preferences on items are expressed as ratings on all datasets and scaled on integers 1–5.

In group recommendation literature, groups are formed considering different sizes and similarity levels. The majority of the research [1,2,8,11] consider groups containing a small number of users such as 2, 3, 5, 8 or 10 users and also account

---

[2] http://www.grouplens.org/node/73.

[3] McAuley, J., Pandey, R., Leskovec, J.: Inferring networks of substitutable and complementary products. In: KDD '15. pp. 785–794 (2015).

pre-defined similarity among the users. In reality, we cannot pre-define how many users will form a group. It may limit the method performance on groups with specific size and similarity by creating a restriction on group member size and similarity level. In our research, groups are formed without the restriction of the group size and pre-define similarity and only various numbers of groups are considered. More specifically, we use k-means algorithm to create G different groups. For example, when $G = 20$, there will be 20 different groups of users. In our experiment, G contains various ranges of values. The smaller value of G corresponds to less groups, where each group contains a large but indefinite number of users. Whereas the larger value of G refers to more groups, where each group includes smaller but indefinite number of users. For ML dataset we set $G \in \{4, 5, 10, 15, 20, 25, 30, 40, 50, 60, 70, 80, 90, 100\}$, and for VG and DM datasets we set $G \in \{10, 20, 30, 40, 50, 60, 70, 80, 90, 100, 110, 120, 130, 140, 150, 160\}$. As ML dataset contains fewer user than VG and DM dataset respectively, smaller ranges of values for G is used to carry experiment on this dataset. Each group has a variable number of users with varying levels of similarity.

Each dataset is divided into a training and a test set for each group. Each training set consists of 80 % ratings of members of a group and the remaining 20 % are used for test. Following [2], we only include items in the test set not presenting in the training set of any members for a group. We generate 10 versions of the dataset by randomly sampling items and the average result is reported. We also use 20 % data from training as the validation set to tune the model parameters.

4.2 Evaluation Metric

Following the standard evaluation metric used in [2,19], we use normalised discounted cumulative gain (NDCG) as IR performance measure for the testing and evaluation of our method. We compare the recommendation list generated for a group that generated from ground truth ratings for the group. As the datasets do not contain explicit group preference information, we adopt an aggregation method to compute the group's preference on items. We are not aware of any other research in group recommendation literature that generates recommendations for group that contains more than 10 users. However, the author in [8] reports that additive utilitarian is the winning strategy for the larger group when comparing various aggregation method vs. group size. Following [4,7,8,11], we also use the additive utilitarian strategy to compute group's preference on items. According to this strategy, each member's preference to test items are added and the recommendation list for the group is generated by sorting the summed score. Thus, NDCG score for a group g can be computed as

$$NDCG_g@n = \frac{DCG_g@n}{IDCG_g@n},$$

$$DCG_g@n = \sum_{j=1}^{n} \frac{2^{R_j^g}-1}{log_2(j+1)},$$

where j is the item that appears at the j th position in the top-n recommendation list for the group g and group score can be computed by $R_j^g = \sum_{u \in g} R_{uj}$.

4.3 Compared Methods

We compare the performance of NBPLFM with a number of group recommendation approaches as follows:

(1) UCF-AVG. User-based CF is the widely used approach which analyses the similarities among users to predict the recommendation preference on unseen user-item pairs. In group recommendation settings, it first calculates user's preferences on unseen items by following user-based CF method and averages of these predictions to generate group recommendation score for that item.

(2) UCF-MIS. The user-based CF method is used to generate individual predictions and least-misery approach to calculate group recommendation score.

(3) UCF-RD. This model calculates the recommendation score on an unseen item based on both relevance and disagreement between group members [1]. After computing individual prediction by user-based CF method, the average aggregation strategy is applied in the model to measure the relevance and score variance to measure the disagreement.

(4) LFM. This model uses a latent factor model for group recommendation [16].

(5) BPMF. It is the state-of-the-art method of recommendation [15].

(6) BPMFDP. BPMF with Dirichlet process mixture prior [14]. In our implementation of this method, we do not consider side information as such information is sensitive and expensive to collect as mentioned in Sect. 2.

4.4 Parameter Setting

For UCF-AVG, UCF-MIS and UCF-RD, we compute similarity only to the users who have at least 5 items in common. Following the best experimental results in [1], we also weighted relevance score by 0.2 and disagreement by 0.8 to generate group recommendation score. For each algorithm, we tune parameters separately on a validation set. Following the best performance of BPMF, BPMFDP and NBPLFM on validation set, we set latent factor dimension of user and item matrix $D = 10$, observation noise $\sigma = 0.9$, $\mu_0 = 0$, $v = 0$ and $\mathcal{W}_0 = 0$ for each methods on all datasets. On ML and DM datasets, we use $\alpha = 0.0013$ for both BPMFDP and NBPLFM. The value $\alpha = 0.013$ is used for both NBPLFM and BPMFDP on VG dataset. Since our method is based on Gibbs sampler, the first few samples from the Markov chain are discarded (burn-in), as they may not represent the desired distribution. From the validation burn-in samples are 20, 30 and 50 for ML, DM and VG, respectively. The average over the next 50, 70 and 150 samples from the posterior predictive distribution are used to generate results on ML, DM and VG datasets respectively.

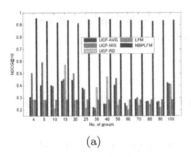
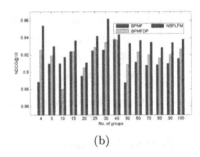

(a)   (b)

Fig. 2. NDCG@10 comparison on NBPLFM with (a) UCF-AVG,UCF-MIS,UCF-RD
and LFM and (b) with BPMF and BPMFDP on ML dataset.

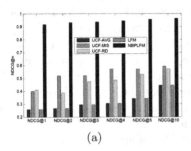
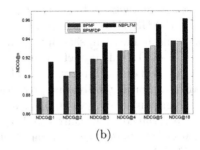

(a)   (b)

Fig. 3. Performance comparison on NBPLFM with (a) UCF-AVG,UCF-MIS,UCF-RD
and LFM and (b) with BPMF and BPMFDP for various top-n positions on ML dataset.

4.5   Results

To verify the proposed method, we apply it with compared methods on ML
dataset and the NDCG@10 performances results on this dataset are presented
in Fig. 2. We also examine the above methods performances by varying top-n
recommendation size. The best performance of these methods for NDCG@n =
1, 2, 3, 4, 5, 10 on ML dataset is reported in Fig. 3.

From this bar diagram as shown in Fig. 2, it is clear indicated that NBPLFM
achieves significant improvement overall performances than that of other methods for all group numbers. From Fig. 2(a), it reaches 33–74 % performance
improvement over both UCF-AVG and UCF-MIS. It also scores 36–72 % performance improvement when the comparison is made with both UCF-RD and
LFM. As shown in Fig. 2(b), our method also improves nearly 0.74–7 % performance improvement over BPMF. In comparison with BPMFDP, our method
achieves maximum improvement amount to 3.74 % for G=10. It also achieves
0.59–2.45 % improvement for all other group numbers. It is worthy noted that
both BPMF and BPMFDP achieves the best performance for $G = 40$, while the
proposed NBPLFM exhibits the best performance for $G = 30$. UCF-AVG, UCF-
MIS, LFM and UCF-RD achieves the best performance for group number 20, 5,
4 and 15, respectively. As shown in Fig. 3(a) NBPLFM achieves more than 50–
65 % improvement over UCF-AVG and LFM for different top-n positions. Over

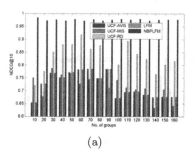 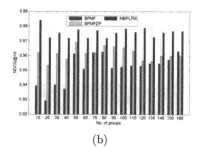

(a) (b)

Fig. 4. NDCG@10 comparison on NBPLFM with (a) UCF-AVG,UCF-MIS,UCF-RD and LFM and (b) with BPMF and BPMFDP on VG dataset.

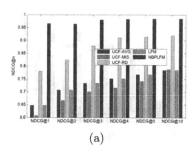 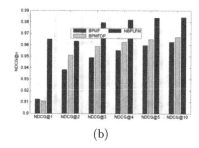

(a) (b)

Fig. 5. Performance comparison on NBPLFM with (a) UCF-AVG,UCF-MIS,UCF-RD and LFM and (b) with BPMF and BPMFDP for various top-n positions on VG dataset.

UCF-MIS and UCF-RD, it gains 36–51 % and 39–54 % improvement respectively. From Fig. 3(b), NBPLFM shows better performance among all other algorithms. In numbers, our method gains 1.64–3.85 % improvement over BPMF and 1.64–3.76 % improvement over BPMFDP for various top-n recommendation position. The reasons for NBPLFM outperforming all compared methods will be discussed at the end of this section.

The NDCG@10 performance of the methods on the VG dataset is shown in Fig. 4 and the best performance of the methods w.r.t varying top-n positions are shown in Fig. 5. From Fig. 4(a), it is found that NBPLFM achieves 18–33 % performance improvement over both UCF-AVG and UCF-MIS. Over UCF-RD, it achieves highest 26 % improvement for G=90 and the lowest improvement is 8 % for G=110. It also gains more than 65 % improvement for all group number over LFM. From Fig. 4(b), NBPLFM gains 0.92–4.5 % improvement over BPMF. Our method also improves the performance of BPMFDP by 0.5–2.2 %. It is obvious from Fig. 5 that our method has better performance than other baseline methods for all top-n positions. UCF-AVG, LFM, UCF-MIS, UCF-RD and NBPLFM achieves the best when the value of G is 90, 90, 30, 60 and 10, respectively. Over UCF-AVG and UCF-MIS, NBPLFM achieves 19–35 % improvement for all NDCG@n computations in Fig. 5(a). In comparison with LFM, our method achieves more than 65 % improvement for all top-n positions. It also gains 7–10 %

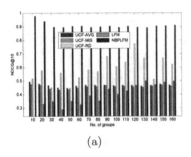 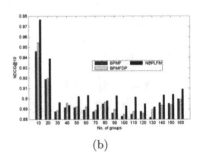

(a)   (b)

Fig. 6. NDCG@10 comparison on NBPLFM with (a) UCF-AVG,UCF-MIS,UCF-RD and LFM and (b) with BPMF and BPMFDP on DM dataset.

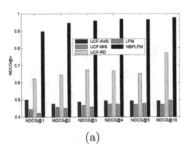 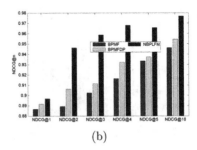

(a)   (b)

Fig. 7. Performance comparison on NBPLFM with (a) UCF-AVG,UCF-MIS,UCF-RD and LFM and (b) with BPMF and BPMFDP for various top-n positions on DM dataset.

improvement over UCF-RD. BPMF and BPMFDP achieve the best performance for $G = 80$ and $G = 50$, respectively. Comparing to BPMF and BPMFDP in Fig. 5(b), our method achieves more than 5 % performance improvement on NDCG@1 computations. For all other NDCG@n computations, it gains 2.15–3 % and 1.22–2.18 % improvement over BPMF and BPMFDP, respectively.

The results on DM dataset are plotted in Figs. 6 and 7. Over UCF-AVG, UCF-MIS and LFM, our method achieves more than 40 % improvement in Fig. 6(a) for all group number. In comparison with UCF-RD, NBPLFM achieves 22–46 % improvement. Comparing to BPMF, it achieves more than 0.3–3 % improvement in Fig. 6(b). Over BPMFDP, our method achieves nearly 2 % improvement for $G = 10$ and $G = 20$. Although for $G = 40$, our method performs slightly worse than BPMFDP and for all other cases it gains almost 0.25–1.4 % performance improvement. UCF-AVG, UCF-MIS, BPMF, BPMFDP and NBPLFM model achieves the best performance when the value of G is 10 on this dataset. LFM and UCF-RD show the best performance for $G = 120$ and $G = 160$, respectively. NBPLFM achieves more than 39 % improvement in top-n recommendation over UCF-AVG, UCF-MIS and LFM in Fig. 7(a). Over UCF-RD, it increases the performance more than 20 % for all top-n positions. From Fig. 7(b), it gains 0.5–4.72 % improvement over BPMFDP and 1–5.68 % improvement over BPMF for all top-n positions.

Overall, these results indicate the effectiveness of NBPLFM is the best as compared to the other state-of-the-art approaches. The reasons for the experimental results can be explained as follows. UCF-AVG and UCF-MIS use average and least misery strategy to generate group recommendations. They do not model group information in learning and the results are obvious. UCF-RD exploits both relevance and disagreement between group members to generate recommendations. As this information is not modelled into learning, the method does not score well. The results indicate their sensitivity with groups where user to user similarity varies from high to low and also with groups where the members size are not fixed and can vary from one group to another group. The LFM model uses a model-based approach to generate individual recommendation and average user's latent factors of a group to generate the group latent factor. However, this model does not consider the user interaction within a group and the average latent factor of a group may represent a virtual user who may not represent any members within the group. Thus, LFM is more sensitive for a group that consists of a large number of users. BPMF also does not exploit group information in learning, thus does not achieve satisfactory results in a group recommendation scenario. Although BPMFDP uses BPMF with Dirichlet prior, they do not exploit user interaction within a group in learning. To overcome these limitations the proposed method focuses on the user interaction within a group. It exploits such vital interactions by learning group-based latent factors that represent user preferences within each group and consider the collaborations within the group and from other groups. Thus, our model incorporates group preference in learning and also in generating recommendations. The use of infinite mixture model allows our model to learn group preferences with variable group sizes and similarities. While the performance of baseline methods are limited by the group size and similarity, it is shown that NBPLFM can generate recommendation regardless of any specific group size and similarity. Therefore the proposed model significantly improves the performance compared to the other state-of-the-art baseline methods for the task of top-n group recommendation.

5 Conclusion

In this paper, we propose a nonparametric Bayesian probabilistic latent factor model, namely NBPLFM, for group recommendations. Rather than relying on the aggregation method in generating recommendations, the proposed method learns user latent factors from their past interactions within the group. To fulfil the real-world group recommendation scenario, the proposed method has a freedom to formation of a group, i.e. with variable group sizes and similarities. As a result, the performance of NBPLFM is not limited by the group size and similarity. Whereas, existing group recommender systems focus on generating recommendations for groups that contain a smaller number of users which are fixed across all groups and also have the same level of similarity. NBPLFM follows a nonparametric approach, thus comes with the flexibility of an infinite number of groups. Comparative experimental results on three public benchmark datasets verify the efficiency of NBPLFM to generate recommendations for groups with variable sizes and similarities. In this paper, we assume that a user belongs to only one group. In future work, we will extend our model to handle the situation where a user could belong to multiple groups.

References

1. Amer-Yahia, S., Roy, S.B., Chawla, A., Das, G.: Group recommendation: semantics and efficiency. In: VLDB 2009 (2009)
2. Baltrunas, L., Makcinskas, T., Ricci, F.: Group recommendations with rank aggregation and collaborative filtering. In: RecSys 2010, pp. 119–126 (2010)
3. Cai, X., Bain, M., Krzywicki, A., Wobcke, W., Kim, Y.S., Compton, P., Mahidadia, A.: Collaborative filtering for people to people recommendation in social networks. In: Li, J. (ed.) AI 2010. LNCS (LNAI), vol. 6464, pp. 476–485. Springer, Heidelberg (2010). doi:10.1007/978-3-642-17432-2_48
4. Cantador, I., Castells, P.: Recommender systems for the social web (2012)
5. Gelman, A., Robert, C., Chopin, N., Rousseau, J.: Bayesian data analysis (1995)
6. Hu, L., Cao, J., Xu, G., Cao, L., Gu, Z., Cao, W.: Deep modeling of group preferences for group-based recommendation. In: AAAI 2014, pp. 1861–1867 (2014)
7. Jameson, Anthony, Smyth, Barry: Recommendation to Groups. In: Brusilovsky, Peter, Kobsa, Alfred, Nejdl, Wolfgang (eds.) The Adaptive Web. LNCS, vol. 4321, pp. 596–627. Springer, Heidelberg (2007). doi:10.1007/978-3-540-72079-9_20
8. Kompan, M., Bieliková, M.: Group recommendations: survey and perspectives. Comput. Inform. **33**(2), 446–476 (2014)
9. Koren, Y., Bell, R., Volinsky, C.: Matrix factorization techniques for recommender systems. Computer **42**, 30–37 (2009)
10. Marlin, B.: Modeling user rating profiles for collaborative filtering. In: NIPS (2003)
11. Masthoff, J.: Group modeling: selecting a sequence of television items to suit a group of viewers. User Model. User-Adap. Inter. **14**(1), 37–85 (2004)
12. McCarthy, J., Anagnost, T.: Musicfx: an arbiter of group preferences for computer supported collaborative workouts. In: CSCW 1998, pp. 363–372 (1998)
13. O'Connor, M., Cosley, D., Konstan, J.A., Riedl, J.: Polylens: A recommender system for groups of users. In: ECSCW 2001, pp. 199–218 (2001)
14. Porteous, I., Asuncion, A.U., Welling, M.: Bayesian matrix factorization with side information and dirichlet process mixtures. In: AAAI (2010)
15. Salakhutdinov, R., Mnih, A.: Bayesian probabilistic matrix factorization using markov chain monte carlo. In: ICML 2008, pp. 880–887. ACM (2008)
16. Shi, J., Wu, B., Lin, X.: A latent group model for group recommendation. In: IEEE MS 2015, pp. 233–238 (2015)
17. Ye, M., Liu, X., Lee, W.C.: Exploring social influence for recommendation: a generative model approach. In: SIGIR 2012, pp. 671–680 (2012)
18. Yu, Z., Zhou, X., Hao, Y., Gu, J.: Tv program recommendation for multiple viewers based on user profilemergin. In: UMAP 2006 (2006)
19. Yuan, Q., Cong, G., Lin, C.: Com: a generative model for group recommendation. In: KDD 2014, pp. 163–172 (2014)

Joint User Knowledge and Matrix Factorization for Recommender Systems

Yonghong Yu[1]([⊠]), Yang Gao[1], Hao Wang[1], and Ruili Wang[2]

[1] State Key Lab for Novel Software Technology, Nanjing University, Nanjing, People's Republic of China
yuyh.nju@gmail.com, {gaoy,wanghao}@nju.edu.cn
[2] School of Engineering and Advanced Technology, Massey University, Palmerston North, New Zealand
R.Wang@massey.ac.nz

Abstract. Currently, most of the existing recommendation methods treat social network users equally, which assume that the effect of recommendation on a user is decided by the user's own preferences and social influence. However, a user's own knowledge in a field has not been considered. In other words, to what extent does a user accept recommendations in social networks need to consider the user's own knowledge or expertise in the field. In this paper, we propose a novel matrix factorization recommendation algorithm based on integrating social network information such as trust relationships, rating information of users and users' own knowledge. Specifically, we first use a user's status (in this paper, status refers to the number of followers and the number of ratings one has done) in a social network to indicate a user's knowledge in a field since we cannot directly measure a user's knowledge in the field. Then, we model the final rating of decision-making as a linear combination of the user's own preferences, social influence and user's own knowledge. Experimental results on real world data sets show that our proposed approach generally outperforms the state-of-the-art recommendation algorithms that do not consider the knowledge level difference between the users.

Keywords: Recommender systems · Social networks · User status · Matrix factorization

1 Introduction

With the rapid growth of information available on the World Wide Web, users are confronted with a serious information overload problem. In order to alleviate this issue, recommender systems [1] are proposed to provide users with *personalized* information, products or services to satisfy their tastes and preferences. Because of such attractive features, recommender systems have become more and more popular and are widely deployed in modern e-commerce applications, such as Amazon, YouTube, Netflix, LinkedIn, etc.

© Springer International Publishing AG 2016
W. Cellary et al. (Eds.): WISE 2016, Part I, LNCS 10041, pp. 77–91, 2016.
DOI: 10.1007/978-3-319-48740-3_6

Collaborative filtering (CF) [2] is one of the most widely used techniques for building recommender systems and has achieved great success in e-commerce. CF methods discover hidden preferences of users from past activities of users, i.e., the user-item rating matrix, to make recommendations. However, CF approaches, including matrix factorization methods [3,4], suffer from *data sparsity* and *cold start* issues [1]. For example, matrix factorization techniques cannot effectively learn the latent feature vectors for users with only a few ratings or newly added items.

In order to improve the performance of traditional recommendation methods, several social-network-based recommendation approaches [5–9] have been proposed to extend basic matrix factorization methods by exploiting rich social information. The underlying assumption of social-based recommendation approaches is that friends share common interests. A user is more likely to adopt the item recommendations from her friends than those from non-friends. Yang et al. proposed a circle-based social network recommendation algorithm [10] to better characterize the domain-specific trust relationships among users. However, this circle-based method ignores the difference in users' knowledge levels in different categories. Intuitively, the factors that affect a user decision include (1) how much she trust her friends, and (2) the user's own knowledge in a field. For example, suppose user u is an expert in movies but has limited knowledge about cars, then user u may be less affected by other people's opinions/recommendations when she receives movie-related recommendations from her trusted friends. By contrast, the user u may be willing to accept car-related recommendations from her trusted friends who are familiar with cars. That is to say, the degree of social influence is strong for user u in the "Car" category since user u is layman in this field. Hence, besides the domain-specific trust circles, the user's knowledge also needs to be considered in recommender systems.

Based on the above intuitions, we assume that the processes of rating decision-making are affected by three factors: users' own preferences, social influence and users' own knowledge, and thus we propose a novel matrix factorization recommendation algorithm based on integrating social network information such as trust relationships, rating information of users as well as users' own knowledge. Specifically, we first use a user's status (e.g., the number of followers or the number of ratings one has done in a given field) in a social network to indicate a user's knowledge level in a given field since we cannot directly measure a user's knowledge in the field. Then, we model the final rating of decision-making as a linear combination of users' own preferences, social influence and users' own knowledge. Experimental results on real world data sets show that our proposed approach can model the decision-making processes of users better, and outperforms the state-of-the-art recommendation algorithms that do not consider the knowledge level difference between the users.

The key contributions of our work are summarized as follows:

– We propose a novel matrix factorization based recommendation algorithm by considering user's own preferences, social influence and user's own knowledge.

– In a social network, we utilized a user's status (i.e., the number of followers and the number of ratings one has done) to represent the user's knowledge level in a field since we cannot directly measure a user's knowledge in the field.
– We perform extensive experiments to evaluate our proposed method on real-life data sets. The results show that our proposed method outperforms the state-of-the-art recommendation algorithms.

The rest of this paper is organized as follows. Section 2 briefly reviews related work in recommender systems. Section 3 introduces some preliminary knowledge. Section 4 describes the details of our proposed item recommendation algorithm. Experiments are evaluated in Sect. 5. Finally, we conclude this paper and present some directions for future work in Sect. 6.

2 Related Work

CF approaches can be divided into three main categories [1]: memory-based algorithms, model-based algorithms and hybrid algorithms.

Memory-based filtering algorithms, also known as neighbor-based methods, use the entire user-item rating matrix to generate recommendations. Typical memory-based algorithms include user-based methods [2] and item-based methods [11,12]. In contrast to memory-based filtering approaches, model-based filtering methods first make use of statistical and machine learning techniques to learn a predictive model from training data. The predictive model characterizes the rating behaviors of target users. Then the model-based filtering approaches use the trained model rather than the original user-item matrix to compute predictions. Typical model-based filtering approaches include Bayesian networks [2], clustering model [13–15], latent semantic analysis [16,17], restricted Boltzmann machines [18] and association rules [19].

With the great success of the Netflix Prize competition, matrix factorization methods [3,4] have attracted a lot of attention since matrix factorization techniques can effectively and efficiently deal with a very large scale user-item rating matrix. Matrix factorization approaches make an assumption that only a few latent factors contribute to preferences of users and characteristics of items. Hence, matrix factorization approaches simultaneously embed both user and item feature vectors into a low-dimensional latent factor space, where the correlation between user preferences and item characteristics can be computed directly [4,20–23].

Recently, several social-based recommendation algorithms [5–9] have been proposed to extend basic matrix factorization methods by leveraging rich social relations. Ma et al. [5] proposed *SoRec*, which integrates a user-item rating matrix and users' social relationships. Specifically, *SoRec* bridges these two different data resources by sharing the user latent feature matrix between them. However, as reported in [6], *SoRec* does not intuitively reflect the real world recommendation process (i.e., is hardly interpretable). In order to improve *SoRec*, Ma et al. [6] proposed *RSTE*, which fuses users' own tastes and their trusted friends' favors together by an ensemble parameter. In [7], Jamali et al. proposed

SocialMF. Specifically, *SocialMF* incorporates trust propagation into probabilistic matrix factorization model [4] to improve the quality of recommendation. In order to model the multi-faceted friend relationships, Yang et al. [10] proposed a circle-based recommendation algorithm in social networks. In [8], Yang et al. proposed a novel social-based recommendation approach, named *TrustMF*, which fuses sparse ratings and sparse trust relationships. *TrustMF* is built on the observation that a user is likely to be affected by existing ratings or reviews from trusted others, and at the same time, this user's ratings or reviews will also influence others who trusts her. All of the above approaches report that social network information is beneficial to social-unaware matrix factorization based recommendation methods.

Unlike the aforementioned methods which ignore the difference between users' knowledge of different fields, our proposed method assumes that the processes for rating decision-making are affected by three factors: users' own preferences, social influence and users' own knowledge. In other words, a user's knowledge in a particular field decides how much the user will accept recommendations. Hence, the strength of influence from different friends is different. Incorporating users' knowledge into matrix factorization is capable of improving the performance of traditional social-network-based recommendation approaches.

3   Preliminary Knowledge

3.1   Problem Description

There are two different types of information used in social network based recommender systems: user-item rating matrix and social network information. User-item rating matrix R comprises two entity sets: the set of N users $U = \{u_1, u_2, ..., u_N\}$ and the set of M items $I = \{v_1, v_2, ..., v_M\}$. Each entry R_{ui} represents the rating given by user u on item i. In principle, R_{ui} can be any real number, but usually ratings are integers and fall into $[0, 5]$, in which 0 indicates that the user has not yet rated that item. A higher rating corresponds to better satisfaction.

A social network is generally represented as a directed social trust graph $G = (U, E)$, where U indicates the set of all users and E represents social trust relations between users. For each edge e_{ij} connecting user u_i to u_j, there is a corresponding trust weight $T_{ij} \in [0, 1]$ which is the extent to which user u_i trusts user u_j, and $T_{ij} = 0$ means that user u_i does not trust u_j at all. All the trust values between users are constructed as a trust matrix $T = [T_{ij}]_{N \times N}$. Note that the trust matrix T is often asymmetric since the trust relationships are not mutual in the real world.

The task of recommender systems is to predict the missing rating on the specified item i for an active user u, denoted by $\widehat{R}_{ui}$.

3.2   Matrix Factorization

The goal of matrix factorization is to learn the latent feature vectors of users and items from all known ratings, of which the inner products can be good estimators of unknown ratings. Formally, matrix factorization decomposes the user-item rating matrix R into two low-rank latent feature matrices $P \in \mathbb{R}^{K \times N}$ and $Q \in \mathbb{R}^{K \times M}$, where $K \ll \min\{N, M\}$, and then uses the product of P and Q to approximate the rating matrix R.

The latent feature matrices P and Q can be learned by minimizing the sum of squared errors with quadratic regularization terms. Formally,

$$\ell = \min_{P,Q} \frac{1}{2} \sum_{(u,i) \in \Omega} (R_{ui} - p_u^T q_i)^2 + \frac{\lambda_1}{2} ||P||_F^2 + \frac{\lambda_2}{2} ||Q||_F^2, \tag{1}$$

where $||\cdot||_F^2$ is the Frobenius norm [24], and Ω indicates the set of the (u, i) pairs for known ratings. λ_1 and λ_2 are regularization parameters. Usually, *stochastic gradient descent* (SGD) [25] is applied to seek a local minimum of the objective function given by Eq. (1).

4   Our Approach

In this section, we present our proposed recommendation approach that fuses three factors: users' own tastes, social influence of trusted friends, and users' own knowledge. We first introduce our proposed social network based recommendation model in Sect. 4.1. Then in Sect. 4.2, we describe how to compute user's knowledge level based on the user's status in a social network.

4.1   Social Based Recommendation with User Knowledge

In social networks, whether a user likes an item is determined by the user's own tastes and the influence of her trusted friends with respect to a specific field.

In general, user u's own tastes on item i in field c is represented as the inner product of user u latent feature vector $p_u^{(c)}$ and item i latent feature vector $q_i^{(c)}$:

$$\widehat{R}_{ui}^{(c)} = p_u^{(c)^T} q_i^{(c)}. \tag{2}$$

Since social network approaches assume that users are mutually influenced, the predicted rating $\widehat{R}_{ui}$ is computed as follows:

$$\widehat{R}_{ui}^{(c)} = p_u^{(c)^T} q_i^{(c)} + \sum_{k \in \mathcal{F}_u^{+(c)}} (T_{uk} \cdot \widehat{R}_{ki}) = p_u^{(c)^T} q_i^{(c)} + \sum_{k \in \mathcal{F}_u^{+(c)}} (T_{uk} \cdot p_k^{(c)^T} q_i^{(c)}), \tag{3}$$

where $\mathcal{F}_u^{+(c)}$ is the set of users whom user u trusts in the field c.

As described in Sect. 1, each user has different knowledge about different fields, which, to some extent, determines how much the user will accept recommendations. For a specific field, the social influence for users with limited knowledge will be larger than those with high knowledge level. Suppose the knowledge level of user u in field c is a_u^c, then the predicted rating $\widehat{R}_{ui}^{(c)}$ is a linear combination of user own tastes, the social influence of trusted friends and user own knowledge level a_u^c,

$$\widehat{R}_{ui}^{(c)} = (\tau + a_u^c){p_u^{(c)}}^T {q_i}^{(c)} + (1 - \tau - a_u^c) \sum_{k \in \mathcal{F}_u^{+(c)}} (T_{uk} \cdot {p_k^{(c)}}^T q_i^{(c)}). \qquad (4)$$

In Eq. (4), the predicted rating $R_{ui}^{(c)}$ is a trade-off among user u's own tastes, the social influence of her trusted friends in the field c, and user knowledge level. Parameter τ denotes the weight of user own tastes. User knowledge level a_u^c controls how much user u insists on her own opinions in the field c. In order to facilitate the description of our proposed approach, we set $\psi_u^c = \tau + a_u^c$.

For each field c, by integrating user own tastes, social influence of trusted friends and user own knowledge into a unified model, we have the following objective function given user-item matrix R and social trust matrix T.

$$\min_{P^{(c)}, Q^{(c)}} \frac{1}{2} \sum_{u=1}^{N} \sum_{i=1}^{M} I_{ui}^{(c)} \left(R_{ui}^{(c)} - \left(\psi_u^c {p_u^{(c)}}^T q_i^{(c)} + (1 - \psi_u^c) \sum_{k \in \mathcal{F}_u^{+(c)}} T_{uk} \cdot {p_k^{(c)}}^T q_i^{(c)} \right) \right)^2$$
$$+ \frac{\lambda_1}{2} \left\| P^{(c)} \right\|_F^2 + \frac{\lambda_2}{2} \left\| Q^{(c)} \right\|_F^2, \qquad (5)$$

where $I_{ui}^{(c)}$ is an indicator function that equals 1 if user u has rated item i and 0 if user u has not rated item i.

Note that in Eq. (5), feature vectors of users and items are learned only based on the observed ratings. However, most users in the social networks express very few or even no ratings. In other words, there are lots of cold start users in a social network. The objective function defined by Eq. (5) fails to work for these cold start users. In order to alleviate the cold start problem, we introduce a social regularization term, which is similar to the regularization term described in [26], to constrain the process of matrix factorization. The social regularization term is defined as:

$$\frac{\beta}{2} \sum_{u=1}^{N} \sum_{v \in \mathcal{F}_u^{+(c)}} T_{u,v} \left\| p_u^{(c)} - p_v^{(c)} \right\|_F^2, \qquad (6)$$

where β is a regularization parameter for controlling the effect of social trust. Specifically, a small value of $T_{u,v}$ means that the distance between two user latent feature vectors should be relatively large, and vice versa. Hence, this social regularization term makes two user latent feature vectors closer if they have more common interests.

In this paper, without loss of generality, we map the ratings $R_{ui}^{(c)}$ to the interval [0,1] using the function $f(x) = (x - minRating)/(maxRating - minRating)$, where $maxRating$ and $minRating$ are the maximum and minimum ratings in recommender systems, respectively. Meanwhile, we use logistic function $g(x) = 1/(1 + e^{-x})$ to limit the predicted ratings $\widehat{R}_{ui}^{(c)}$ within the range [0,1]. Finally, the objective function of our method is defined as follows.

$$
\mathcal{L}^* = \min_{P^{(c)}, Q^{(c)}} \frac{1}{2} \sum_{u=1}^{N} \sum_{i=1}^{M} I_{ui}^{(c)} \left(R_{ui}^{(c)} - g \left(\psi_u^c {p_u^{(c)}}^T q_i^{(c)} + (1 - \psi_u^c) \sum_{k \in \mathcal{F}_u^{+(c)}} T_{uk} \cdot {p_k^{(c)}}^T q_i^{(c)} \right) \right)^2
$$

$$
+ \frac{\beta}{2} \sum_{u=1}^{N} \sum_{v \in \mathcal{F}_u^{+(c)}} T_{u,v} \left\| p_u^{(c)} - p_v^{(c)} \right\|_F^2 + \frac{\lambda_1}{2} \left\| P^{(c)} \right\|_F^2 + \frac{\lambda_2}{2} \left\| Q^{(c)} \right\|_F^2 .
$$

$$(7)$$

We seek a local minimum of $\mathcal{L}^*$ by applying the SGD method. The derivatives of $\mathcal{L}^*$ with respect to $p_u^{(c)}$ and $q_i^{(c)}$ are computed as:

$$
\frac{\partial \mathcal{L}^*}{\partial p_u^{(c)}} = \psi_u^c \sum_{i=1}^{M} I_{ui}^{(c)} g'(y_{ui}) \left(g(y_{ui}) - R_{ui}^{(c)} \right) q_i^{(c)} + \lambda_1 p_u^{(c)}
$$

$$
+ (1 - \psi_w^c) \sum_{w \in \mathcal{F}_u^{-(c)}} \sum_{i=1}^{M} I_{wi}^{(c)} g'(y_{wi}) \left(g(y_{wi}) - R_{wi}^{(c)} \right) T_{wu} q_i^{(c)}
$$

$$
+ \beta \sum_{v \in \mathcal{F}_u^{+(c)}} T_{uv}(p_u^{(c)} - p_v^{(c)}) + \beta \sum_{g \in \mathcal{F}_u^{-(c)}} T_{gu}(p_u^{(c)} - p_g^{(c)}),
$$

$$(8)$$

$$
\frac{\partial \mathcal{L}^*}{\partial q_i^{(c)}} = \sum_{u=1}^{N} I_{ui}^{(c)} g'(y_{ui}) \left(g(y_{ui}) - R_{ui}^{(c)} \right) \times \left(\psi_u^c p_u^{(c)} + (1 - \psi_u^c) \sum_{k \in \mathcal{F}_u^{+(c)}} T_{uk} p_k^{(c)} \right) + \lambda_2 q_i^{(c)} .
$$

We set $y_{ui} = \psi_u^c {p_u^{(c)}}^T q_i^{(c)} + (1 - \psi_u^c) \sum_{k \in \mathcal{F}_u^{+(c)}} T_{uk} {p_k^{(c)}}^T q_i^{(c)}$ and $y_{wi} = \psi_w^c {p_w^{(c)}}^T q_i^{(c)} + (1 - \psi_w^c) \sum_{k \in \mathcal{F}_w^{+(c)}} T_{wk} {p_k^{(c)}}^T q_i^{(c)}$ for simplify presentation in Eq. (8), $g'(x) = e^{-x}/(1 + e^{-x})^2$ is the derivative of logistic function $g(x)$ and $\mathcal{F}_u^{-(c)}$ is the set of users who trust user u in field c.

4.2   User Knowledge Level Learning

User knowledge is a key component of our proposed approach since it determines the degree to which a user is influenced by her social relationships. In this section, we describe how to compute users' knowledge level based on users' status in a social network.

In social networks, a user u may be interested in several item fields, such as digital cameras, computer, books, music, etc. However, she may only be professional in a subset of fields of interest. In such fields, she will be relatively confident; as a consequence, she will tend to express more ratings in these fields than in the other ones. Moreover, a user may often have more followers in familiar fields than those in unfamiliar fields because she could provide valuable information for others. From this perspective, a user's status (e.g., the number of followers or the number of ratings one has done) reflects the user's knowledge level. Hence, we demonstrate a user's knowledge by the distributions of ratings and followers across fields.

The knowledge level of user u in the field c, denoted as a_u^c, consists of two components: the number of ratings that user u has expressed in field c and the number of followers who trust user u in field c. The rating distribution of user u over all fields is denoted as $\mathcal{DR}_u$,

$$\mathcal{DR}_u = \left(\frac{N_u^{(1)}}{N_u}, \frac{N_u^{(2)}}{N_u}, ..., \frac{N_u^{(n)}}{N_u} \right), \tag{9}$$

where N_u is the total number of ratings given by user u across all fields and $N_u^{(c)}$ indicates the number of ratings expressed by user u in the field c; n indicates the number of fields involved in recommender systems. From the definition of $\mathcal{DR}_u$, we can see that $\mathcal{DR}_u$ represents the knowledge level distribution of user u. Hence, the first component of user knowledge level a_u^c is defined as $\frac{N_u^{(c)}}{N_u}$.

From the perspective of social network structure, both the number of followers and the number of followees contribute to the status of users. The status of user u in field c should grow as the number of followers grows, and decline if user u has lots of followees. Formally, the knowledge level distribution of user u over all fields concerning social network structure is defined as,

$$\mathcal{DF}_u = \left(\frac{|\mathcal{F}_u^{-(1)}|}{|\mathcal{F}_u^{-(1)}| + |\mathcal{F}_u^{+(1)}|}, ..., \frac{|\mathcal{F}_u^{-(n)}|}{|\mathcal{F}_u^{-(n)}| + |\mathcal{F}_u^{+(n)}|} \right), \tag{10}$$

where $|\mathcal{F}_u^{-(c)}|$ and $|\mathcal{F}_u^{+(c)}|$ indicate the numbers of followers and followees of user u in field c, respectively. Thus, the second component of user knowledge level a_u^c is computed as $\frac{|\mathcal{F}_u^{-(c)}|}{|\mathcal{F}_u^{-(c)}| + |\mathcal{F}_u^{+(c)}|}$.

Fusing both components, the user knowledge level of user u in field c is the linear addition of these two components:

$$a_u^c = \frac{1}{2} \left(\frac{N_u^{(c)}}{N_u} + \frac{|\mathcal{F}_u^{-(c)}|}{|\mathcal{F}_u^{-(c)}| + |\mathcal{F}_u^{+(c)}|} \right). \tag{11}$$

Note that we only apply the linear addition to unify two user knowledge levels inferred from ratings and social network structure. Since this work focuses on incorporating users' own knowledge into recommender systems, we leave more complicated fusion methods to future work.

5   Experiments

In this section, we conduct several experiments on real data sets to compare the performance of our proposed recommendation algorithm with other state-of-the-art methods.

5.1   Data Sets and Evaluation Metrics

We choose the Epinions data set to evaluate the performance of our proposed method since Epinions contains rating information, social network relations and item field information related to items. The Epinions data set used in our experiments was published by the authors of [27]. It contains 922,267 ratings from 22,166 users and 296,277 items. We plot the distribution of the number of ratings in Fig. 1(a) and then observe that it generally exhibits a power-law distribution. Items are divided into 27 fields. In Fig. 1(b), the x-axis indicates the number of fields involved by users, and the y-axis is the accumulative percentage of users who involved with different number of fields. From Fig. 1(b), we can observe that only a small portion of users are interested in a large portion of fields. In Epinions, only 21.7 % of the users expressed ratings in at least 10 fields. In this paper, we select the top 10 fields in terms of field rating distribution to evaluate our proposed method.

Table 1. Statistics of Epinions and top 10 fields

DataSet	N	M	$|R|$	Density	$\bar{r}$	$|T|$	$\bar{t}$
Epinions	22,166	296,277	922,267	1.4×10^{-4}	41.6	355,813	16.05
Movies	14,180	28,616	167,261	4.1×10^{-4}	11.80	153,773	10.84
Books	10,731	59,129	102,975	1.6×10^{-4}	9.59	104,706	9.57
Music	9,010	34,541	85,419	2.7×10^{-4}	9.48	61,383	6.81
Kids family	8,606	24,124	85,113	4.0×10^{-4}	9.88	68,697	7.98
Hotels travel	10,660	15,421	61503	3.7×10^{-4}	5.77	99,997	9.38
Wellness beauty	7,209	22,575	55,087	3.3×10^{-4}	7.64	51,435	7.13
Restaurants gourmet	8,376	16642	48112	3.4×10^{-4}	5.74	62,638	7.47
Games	8,169	8,306	45,730	6.7×10^{-4}	5.59	50,840	6.22
Electronics	11,385	15,639	45,459	2.5×10^{-4}	3.99	102,844	9.03
Home garden	8,074	20,159	45,113	2.7×10^{-4}	5.58	62,805	7.77

The total number of trust statements issued is 355,813. The average number of trust links is 16.05 per user. Note that the field-specific social network is sparser than the original social network since the existence of a trust relation between two users does not mean that they have common interests in a specific field. General statistics about the Epinions and its top-10 subfields data sets are summarized in Table 1. In Table 1, $\bar{r}$ and $\bar{t}$ denote the average number of ratings per user and the average number of trust links per user, respectively.

We choose two popular metrics: *Mean Absolute Error (MAE)* and *Root Mean Squared Error (RMSE)*, to measure the recommendation quality of our proposed method compared with other recommendation algorithms.

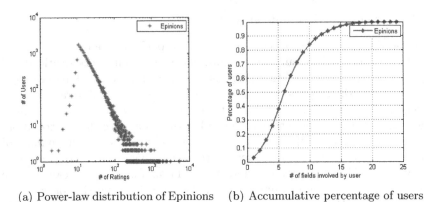

(a) Power-law distribution of Epinions   (b) Accumulative percentage of users

Fig. 1. Power-law distribution and accumulative percentage distribution of users

5.2 Baseline Approaches

In order to evaluate the effectiveness of our proposed method, we compare our method with the following state-of-the-art approaches:

1. PMF: This method is proposed by Mnih and Salakhutdinov [4] and can be viewed as a probabilistic extension of the SVD [20] model.
2. SoRec: This method is proposed by Ma et al. [5]. SoRec simultaneously factorizes user-item rating matrix and social network trust matrix by sharing the user latent feature matrix.
3. RSTE: This method is proposed in [6]. RSTE takes users' own tastes and their trusted friends' favors into account.
4. SocialMF: This method is presented by Jamali et al. [7]. It incorporates the mechanism of trust propagation into basic probabilistic matrix factorization to improve the quality of recommendation.
5. TrustMF: This method is presented in [8]. By factorizing the social trust matrix, TrustMF maps users into two latent low-dimensional spaces: truster space and trustee space.

5.3 Experiment Settings

The main parameters settings of all comparison methods are listed in Table 2. Note that, in order to make a fair comparison, we set parameters of each method according to respective references or based on our experiments. Under these parameters settings, each method achieves its best performance. In addition, since all comparison methods use a gradient descent algorithm to optimize respective objective functions, we set the learning rate η involved in the gradient descent algorithm to be 0.03 for PMF and 0.001 for other social-based recommendation approaches. The number of dimensions K of latent feature vectors is set to 10 in all our experiments.

Table 2. Parameter settings of comparison methods

Methods	Parameter settings
PMF	$\lambda_U = \lambda_V = 0.001$
SoRec	$\lambda_U = \lambda_V = \lambda_Z = 0.001, \lambda_C = 1$
RSTE	$\lambda_U = \lambda_V = 0.001, \alpha = 0.4$
SocialMF	$\lambda_U = \lambda_V = 0.001, \lambda_T = 1$
TrustMF	$\lambda = 0.001, \lambda_T = 1$
Our method	$\lambda_1 = \lambda_2 = 0.001, \tau = 0.6, \beta = 0.01$

For each field in Epinions, we conduct a five-fold cross validation by randomly extracting different training and test sets at each time, which accounts for 80 % and 20 %, respectively. Finally, we report the average results on test sets for each field.

5.4   Recommendation Quality

In this section, we compare the performance of our proposed method with all selected methods on the top 10 fields in Epinions data set.

Table 3 reports the results of recommendation quality for all comparison algorithms. We bold the best performance and use superscripts to denote the rank that our approach achieves among all comparison methods. From Table 3, we can observe that social network based approaches outperform PMF, which only utilizes user-item ratings to learn latent feature vectors. This observation is consistent with the results reported in [5–8]. Moreover, our approach is generally superior to other comparison methods. Specially, our approach achieves the best recommendation performance on 7 fields such as "Movies", "Kids and Family", "Hotel and Travel" in terms of MAE. Furthermore, taking RSTE as our main comparison method since the first part of our objective function is partly similar to that of RSTE, our approach is consistently better than RSTE on ten fields and the improvement over RSTE is significant. This observation confirms the assumption that ignoring users' knowledge can not accurately reflect the item adoption process, while incorporating users' own knowledge level into the process of recommendation decision-making can largely improve the recommendation quality.

5.5   Impact of Parameter β

In this section, we perform a group of experiments to investigate the impact of β on the accuracy of recommendation by changing the values of β from 0 to 1.

Figure 2 shows the impact of parameter β on MAE and RMSE for three fields, i.e., "Movies", "Wellness & Beauty", and "Home & Garden", which are representatives of large, middle and small scale data sets, respectively. From Fig. 2, we can see that the value of β affects the recommendation quality significantly, which indicates that integrating the social regularization term greatly

Table 3. Recommendation quality comparisons

Dataset	Metric	PMF	SoRec	RSTE	SocialMF	TrustMF	Our method
Movies	MAE	1.2584	0.8806	0.9468	0.8781	0.90703	**0.8770**[1]
	RMSE	1.6708	1.1380	1.2240	1.1310	1.1804	**1.1239**[1]
Books	MAE	2.3331	0.7504	0.8291	0.7268	**0.7261**	0.7299[3]
	RMSE	2.7019	0.9918	1.0520	0.9748	0.9863	**0.9598**[1]
Music	MAE	1.8545	0.7593	0.8353	**0.7300**	0.7359	0.7380[3]
	RMSE	2.3021	0.9995	1.0516	**0.9737**	0.9881	0.9890[3]
Kids family	MAE	1.5680	0.8389	0.8589	0.8103	0.8157	**0.8052**[1]
	RMSE	2.0408	1.1063	1.1132	1.0808	1.1058	**1.0738**[1]
Hotels travel	MAE	1.6338	0.8314	0.8372	0.8031	0.8165	**0.8020**[1]
	RMSE	2.0774	1.0987	1.0959	1.0698	1.1069	**1.0605**[1]
Wellness beauty	MAE	1.9871	0.9339	0.9512	0.9123	0.9577	**0.9079**[1]
	RMSE	2.4021	1.2250	1.2375	1.2148	1.3404	**1.1948**[1]
Restaurants gourmet	MAE	1.8798	0.8870	0.9150	0.8616	0.8727	**0.8583**[1]
	RMSE	2.2982	1.1562	1.1899	1.1344	1.1591	**1.1209**[1]
Games	MAE	1.3849	0.8408	0.8388	**0.8212**	0.8542	0.8252[2]
	RMSE	1.8189	1.1091	1.0901	**1.0882**	1.1370	1.0896[2]
Electronics	MAE	2.4476	0.8732	0.8764	0.8501	0.8785	**0.8385**[1]
	RMSE	2.7600	1.1831	1.1540	1.1599	1.2207	**1.1341**[1]
Home garden	MAE	2.3177	0.8619	0.8657	0.8259	0.8305	**0.8238**[1]
	RMSE	2.6986	1.1769	1.1633	1.1441	1.1857	**1.1439**[1]

improves the accuracy of recommendation. In detail, MAE and RMSE on three data sets show similar changing trends. As β increases, the values of MAE and RMSE firstly decrease, so the recommendation accuracy is accordingly improved. After β reaches a certain threshold, MAE and RMSE scores begin to increase as β increases, which means that the performance degrades when β is too large. These findings indicate that neither abandoning the social regularization term nor relying heavily on the social regularization term can produce high-quality recommendations. In addition, comparing Table 3 and Fig. 2, we also observe that our proposed method outperforms RSTE even when $\beta = 0$ on these three fields, which demonstrates the promising future of our proposed recommendation approach. On these three fields, our approach achieves the best recommendation performance when β is around 0.01.

5.6 Impact of Parameter τ

In order to explore the impact of τ on our proposed method, we conduct another group of experiments and observe the changes in recommendation quality by varying the value of τ from 0 to 1. In this group of experiments, we set $\beta = 0.01$ and $K = 10$. Since the experimental results on "Wellness & Beauty", "Home & Garden" and "Movies" show similar trends, we only plot the experimental results on "Movies" in Fig. 3.

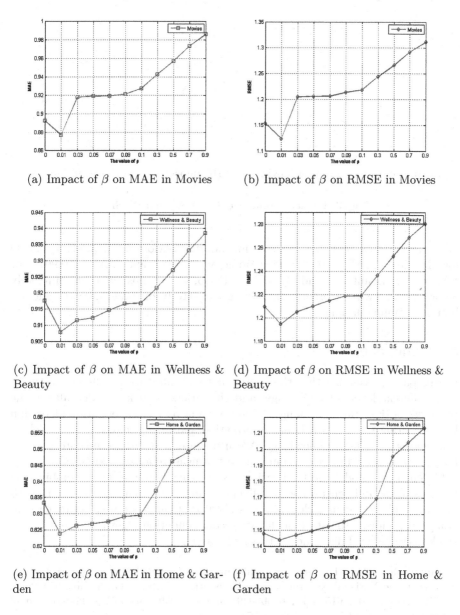

(a) Impact of β on MAE in Movies

(b) Impact of β on RMSE in Movies

(c) Impact of β on MAE in Wellness & Beauty

(d) Impact of β on RMSE in Wellness & Beauty

(e) Impact of β on MAE in Home & Garden

(f) Impact of β on RMSE in Home & Garden

Fig. 2. Impact of Different β on MAE and RMSE

As shown in Fig. 3, parameter τ does have a significant impact on our proposed method, which indicates that it is needed to trade off between user own preferences and social influence when we decide the rating for a target item. The recommendation quality of our method first moves toward optimal values, and then degrades as the values of τ continually increase. When τ is around 0.6,

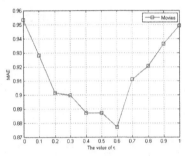
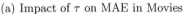

(a) Impact of τ on MAE in Movies

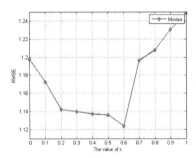

(b) Impact of τ on RMSE in Movies

Fig. 3. Impact of different β on MAE and RMSE

our approach achieves the best recommendation performance: MAE = 0.8769 for "Movies", MAE = 0.8238 for "Home & Garden" and MAE = 0.9079 for "Wellness & Beauty". In addition, from Fig. 3, we can observe that completely ignoring user own preferences or relying excessively on them cannot generate better recommendations.

6   Conclusion

In this paper, assuming that the degree of social influence is different for users with different levels of knowledge, and that users' own knowledge affects the processes of their rating-making, we propose a novel recommendation algorithm by integrating the social network information and rating information as well as considering each user's knowledge. We use a user's status to indicate a user's knowledge in a field since we cannot directly measure a user's knowledge in the field. Then, we model the final rating decision-making as a linear combination of users' own preferences, social influence and users' own knowledge. Experimental results on real-world data sets show that our proposed approach generally outperforms the state-of-the-art recommendation algorithms that ignore the knowledge level difference between the users.

Acknowledgments. The authors would like to acknowledge the support for this work from the National Natural Science Foundation of China (Grant Nos. 61432008, 61175042, 61403208, 61503178, 61303049) and the Natural Science Foundation of Jiangsu Province of China (BK20150587).

References

1. Adomavicius, G., Tuzhilin, A.: Toward the next generation of recommender systems: a survey of the state-of-the-art and possible extensions. TKDE **17**(6), 734–749 (2005)
2. Breese, J.S., Heckerman, D., Kadie, C.: Empirical analysis of predictive algorithms for collaborative filtering. In: UAI (1998)

3. Koren, Y., Bell, R., Volinsky, C.: Matrix factorization techniques for recommender systems. Computer **42**(8), 30–37 (2009)
4. Mnih, A., Salakhutdinov, R.: Probabilistic matrix factorization. In: NIPS (2007)
5. Ma, H., Yang, H., Lyu, M.R., King, I.: SoRec: social recommendation using probabilistic matrix factorization. In: CIKM (2008)
6. Ma, H., King, I., Lyu, M.R.: Learning to recommend with social trust ensemble. In: SIGIR, pp. 203–210 (2009)
7. Jamali, M., Ester, M.: A matrix factorization technique with trust propagation for recommendation in social networks. In: RecSys (2010)
8. Yang, B., Lei, Y., Liu, D., Liu, J.: Social collaborative filtering by trust. In: IJCAI, pp. 2747–2753 (2013)
9. Guo, G., Zhang, J., Yorke-Smith, N.: Trustsvd: collaborative filtering with both the explicit and implicit influence of user trust and of item ratings. In: AAAI (2015)
10. Yang, X., Steck, H., Liu, Y.: Circle-based recommendation in online social networks. In: KDD, pp. 1267–1275 (2012)
11. Sarwar, B., Karypis, G., Konstan, J., Riedl, J.: Item-based collaborative filtering recommendation algorithms. In: WWW (2001)
12. Linden, G., Smith, B., York, J.: Amazon.com recommendations: item-to-item collaborative filtering. IEEE Internet Comput. **7**(1), 76–80 (2003)
13. Ungar, L.H., Foster, D.P.: Clustering methods for collaborative filtering. In: AAAI Workshop on Recommendation Systems (1998)
14. Xue, G.R., Lin, C., Yang, Q., Xi, W., Zeng, H.J., Yu, Y., Chen, Z.: Scalable collaborative filtering using cluster-based smoothing. In: SIGIR (2005)
15. Yu, Y., Wang, C., Gao, Y., Cao, L., Chen, X.: A coupled clustering approach for items recommendation. In: Pei, J., Tseng, V.S., Cao, L., Motoda, H., Xu, G. (eds.) PAKDD 2013. LNCS (LNAI), vol. 7819, pp. 365–376. Springer, Heidelberg (2013). doi:10.1007/978-3-642-37456-2_31
16. Hofmann, T.: Latent semantic models for collaborative filtering. TOIS **22**(1), 89–115 (2004)
17. Hofmann, T.: Collaborative filtering via Gaussian probabilistic latent semantic analysis. In: SIGIR (2003)
18. Salakhutdinov, R., Mnih, A., Hinton, G.: Restricted Boltzmann machines for collaborative filtering. In: ICML (2007)
19. Sarwar, B., Karypis, G., Konstan, J., Riedl, J.: Analysis of recommendation algorithms for e-commerce. In: Proceedings of the 2nd ACM Conference on Electronic Commerce (2000)
20. Sarwar, B., Karypis, G., Konstan, J., Riedl, J.: Application of dimensionality reduction in recommender system - a case study. In: WebKDD Workshop (2000)
21. Seung, D., Lee, L.: Algorithms for non-negative matrix factorization. In: NIPS (2001)
22. Srebro, N., Rennie, J., Jaakkola, T.S.: Maximum-margin matrix factorization. In: NIPS (2004)
23. Rennie, J.D., Srebro, N.: Fast maximum margin matrix factorization for collaborative prediction. In: ICML (2005)
24. Golub, G.H., Van Loan, C.F.: Matrix computations, vol. 3. JHU Press (2012)
25. Nemirovski, A., Juditsky, A., Lan, G., Shapiro, A.: Robust stochastic approximation approach to stochastic programming. SIOPT **19**(4), 1574–1609 (2009)
26. Ma, H., Zhou, D., Liu, C., Lyu, M.R., King, I.: Recommender systems with social regularization. In: WSDM, pp. 287–296 (2011)
27. Tang, J., Gao, H., Liu, H., Das Sarma, A.: eTrust: understanding trust evolution in an online world. In: KDD, pp. 253–261 (2012)

GEMRec: A Graph-Based Emotion-Aware Music Recommendation Approach

Dongjing Wang[1,2], Shuiguang Deng[1(✉)], and Guandong Xu[2]

[1] College of Computer Science and Technology,
Zhejiang University, Hangzhou, China
{tokyol,dengsg}@zju.edu.cn
[2] Advanced Analytics Institute,
University of Technology Sydney, Sydney, Australia
Guandong.Xu@uts.edu.au

Abstract. Music recommendation has gained substantial attention in recent times. As one of the most important context features, user emotion has great potential to improve recommendations, but this has not yet been sufficiently explored due to the difficulty of emotion acquisition and incorporation. This paper proposes a graph-based emotion-aware music recommendation approach (GEMRec) by simultaneously taking a user's music listening history and emotion into consideration. The proposed approach models the relations between user, music, and emotion as a three-element tuple (user, music, emotion), upon which an Emotion Aware Graph (EAG) is built, and then a relevance propagation algorithm based on random walk is devised to rank the relevance of music items for recommendation. Evaluation experiments are conducted based on a real dataset collected from a Chinese microblog service in comparison to baselines. The results show that the emotional context from a user's microblogs contributes to improving the performance of music recommendation in terms of hitrate, precision, recall, and F1 score.

Keywords: Music recommendation · Emotion analysis · Random walk · Emotion aware

1 Introduction

Nowadays, the lower cost of hardware and advances in technology have led to the growth in digital music, and there is an enormous amount of music available on the Internet. For example, currently Apple Music offers over 30 million songs (http:// support.apple.com/en-us/HT204951). However, the vast amount of music has made it more difficult for users to find the music they really enjoy, which is known as the Paradox of Choice. Therefore, music recommendation has become an interesting topic both in research and application. Similar to recommender systems applied in various domains [1, 2], music recommendation has greatly benefited from the algorithmic advances of the recommender systems community, e.g., collaborative filtering, content-based and hybrid approaches, which have been predominantly adopted by simply treating music recommendation as a classic item, e.g., a book or movie, to solve

© Springer International Publishing AG 2016
W. Cellary et al. (Eds.): WISE 2016, Part I, LNCS 10041, pp. 92–106, 2016.
DOI: 10.1007/978-3-319-48740-3_7

the recommendation problem via user's long-term music preferences. Music is not a neutral item but a carrier of thoughts and emotions, so people will have different music tastes and preferences in different contexts [3]. For example, a user who likes both light music and rock & roll music may prefer the former when at rest. Therefore, incorporating contextual information like geo-location, emotional state, time, presence of other people, past and future events can help the recommender system to better understand and satisfy the users' real-time requirements [4].

Music is a specific type of content which can carry emotion features, hence our work focuses on utilizing emotion as a supplementary feature to make music recommendations. In particular, our work is inspired by the following three observations: (1) awareness: listening to music is a kind of emotion-sensitive behavior and direct associations exist between the users' mood and their favorite music [5]. For example, people who are experiencing a sad mood generally tend to listen to a different style of music compared to when they are experiencing a happy mood; (2) dynamics: people's emotions change constantly and dynamically so satisfactory music recommendation must reflect the real-time emotion status of the user; (3) indirect derivation: the emotional status of the user is a secondary context [6], since it cannot be measured directly, but needs to be derived from other types of contextual information. Thus, these observations make emotion-aware music recommendation a real challenge as to how to extract the users' emotion context and integrate this into recommendations, which constitute our major contributions. To the best of our knowledge, there are only a few existing studies which address music recommendation from this angle.

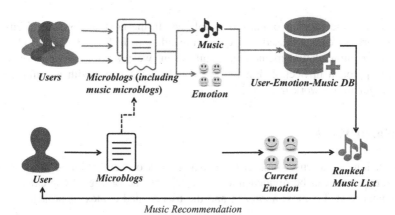

Fig. 1. The concepts of the proposed approach

In this paper, we focus on extracting and incorporating users' emotions into music recommendation. The key objective is to satisfy the users' music needs under certain emotional contexts with minimum user effort in relation to providing feedback. There are three key challenges to be tackled: (1) how to acquire users' current emotional contexts; (2) how to embed emotion in user-music relations; (3) how to infer personalized music preferences given a certain emotional context. To address these challenges, we propose a graph-based emotion-aware music recommendation method (GEMRec), which can (1) extract users' music listening history along with the corresponding emotion from their microblog texts, (2) model users' music records along with the corresponding emotional contexts by an Emotion Aware Graph (EAG), and (3) adopt the relevance propagation algorithm based on random walk in order to reveal the explicit and hidden associations between users and music items under certain emotion circumstances, then rank and recommend music items. The framework of the proposed approach is shown in Fig. 1. Firstly, we extract the users' music listening history with the accompanying emotion information when they are listening to music from their microblogs and store this in a database in the form of three-element tuple associations (user, music, emotion). Then, when a certain user posts new microblogs, the proposed system will determine the user's current emotional status and recommend appropriate music items by referring to the three-tuple association database. To validate the performance of the proposed system, we conduct comprehensive experiments with a real-world dataset collected from Sina Weibo (http://www.weibo.com, the most popular Chinese microblog service). In addition, we also investigate how emotion can influence recommendations in detail.

To summarize, the main contributions of this work are three-fold: (1) we propose an Emotion Aware Graph which integrates users, their music listening history with their various emotion statuses in a unified manner; (2) we devise a graph-based relevance propagation algorithm to capture the overall user preferences of music items given a specific emotion status; (3) we conduct extensive experiments to evaluate the proposed method on a real-world dataset in comparison to the state-of-the-art techniques.

The rest of the paper is organized as follows. In the next section, we discuss the existing work on context-based music recommendation. In Sect. 3, we introduce the proposed recommendation method in detail. Then, we provide an evaluation of the proposed approach in Sect. 4. Finally, we conclude the paper with a summary and directions for future work in Sect. 5.

2   Related Work

In this section, we discuss some of the existing work on context-based music recommendation, which can be divided into two categories according to context type: the environment-related context approach and the user-related context approach.

2.1   Environment-Related Context Approach

This type of research is based on the fact that the environment has an influence on the state-of-mind or emotional state of users, and therefore indirectly influences users' music preferences [5]. For instance, people usually enjoy reflective and complex music during fall/winter, while they may prefer more energetic music during spring/summer [7]. Consequently, music recommendation approaches with environment-related parameters will perform better. The environment-related context includes time [4, 7], location [8], and hybrid contexts [9]. Kaminskas and Ricci [8] explored the possibilities of adapting music to the place of interest (POIs) that the user is visiting. In detail, the author used emotional tags attached by users to both music and POIs and considered a set of similarity metrics for the tagged resources to establish a match between music tracks and POIs. Knees and Schedl [9] gave an overview of approaches to music recommendation which do not rely on the audio signal, but rather take into consideration various aspects of the context in which a music entity occurs. However, most of the prior approaches require researchers to label music with appropriate tags or map context with music, which restrict their application in the real world. Another limitation is that the prior methods have only been evaluated based upon synthetic or small-scale datasets.

2.2   User-Related Context Approach

Compared with the environment-related context, the user-related context has a closer relationship with the users, and can therefore influence the users' music preferences directly. The user-related contexts include activity [10], demographical information [11], and emotion state [12–14]. Wang et al. [10] described a novel probabilistic model for music recommendation that combines automated activity classification with automated music content analysis, with support for a rich set of activities and music content features. Chang et al. [12] proposed a correlation-coefficient-based approach to find emotional music sequences which may evoke a specific emotion in subjects. Then, they built a personal emotion-cognitive music database for an individual subject, and finally music from a personal emotion-cognitive music database is recommended to the user. Han et al. [13] proposed a context-aware music recommendation system in which music is recommended according to the user's current emotion state and the music's influence on changes in the user's emotion. In [14], the author proposed a novel framework for emotion-based music recommendation. Specifically, the core of this recommendation framework is the construction of the music emotion model by affinity discovery from film music, which plays an important role in conveying emotions in film. Deng et al. [15] presented another contextual music recommendation approach, which can infer users' emotion from her/his microblogs, and then recommend music pieces appropriate for users' emotion. However, there is not much work which attempts to utilize emotional information, and the performance of existing emotion-aware approaches still need to be improved greatly. In addition, most of the existing work requires the users to input contextual information, which restrains their application in practice.

3   Proposed Approach

Listening to music is a type of emotion-sensitive behavior and direct associations exist between a users' mood and their favorite music [5]. However, the emotion status of the user is a secondary context [6] which cannot be measured directly. In addition, people's emotions change constantly and dynamically. Therefore, the main challenges are how to acquire the users' real-time contexts of emotion and incorporate them into music recommendation.

As a social network service, microblogs have become increasingly popular in people's daily life. Users can share many kinds of information, such as news, knowledge, resources, opinions, status, and feelings, anywhere and anytime. Therefore, users' interests and contextual information can be inferred from their microblogs. In the meantime, users may share the music items they are listening to, which implies the underlying correlation between the posted microblogs and the shared music items. The idea behind our approach is that the association in users' microblogs, more specifically the music items, and the corresponding emotions in the microblogs, can reflect the users' emotion-aware music preferences.

As shown in Fig. 2, the proposed approach, the graph-based emotion-aware music recommendation method (GEMRec) consists of three main components: (1) *association extraction*, (2) *association modeling*, and (3) *music recommendation*. The *association extraction* component aims to extract the users' music listening history and the corresponding emotional contexts from their microblogs, and then obtains the associations between the user, the music items and the emotions, which can be represented as a three-element tuple (user, music, emotion). Then, in the *association modeling* component, all associations are modeled by an EAG, where the user, music, and emotion are represented as nodes and their relations are expressed as edges. Finally, in the *music recommendation* component, when a target user posts new microblogs or activates the recommendation manually, the music recommendation component performs recommendations on the Emotion Aware Graph using a random-walk-based relevance propagation algorithm and recommends appropriate music items according to the user's real-time emotion extracted from their recent microblogs.

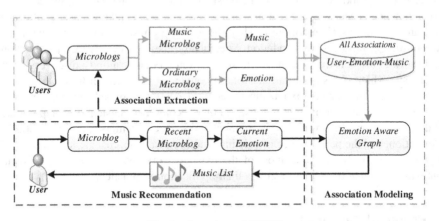

Fig. 2. Overview of GEMRec

3.1   Association Extraction

All the users' microblogs can be roughly categorized into two types, i.e., ordinary microblogs and music-sharing microblogs. As shown in Table 1, ordinary microblogs (the first two rows of Table 1) mainly give generic information on the users' opinions, feelings, status, and activities at a specific timestamp, while music-sharing microblogs contain information on music items, including the singer, music title, URL, comments and so on. We assume an inherent correlation exists between the music that the user listens to and the microblogs that they have recently posted. In other words, music listening behaviors are context-aware (emotion dependent). As for the example shown in Table 1, given that the polarity of emotion in microblogs posted before the music-sharing microblogs is positive in nature, we can speculate that this user (521156765) enjoys the music "Home" in a positive emotional context.

Table 1. Examples of Microblogs

User ID	Content	Time	Music
5211	[smile][smile]so excited about the good news!	2014-06-22 13: 25: 39	
5211	What a good day ~	2014-06-22 13: 29: 10	
5211	Share music Edward Sharpe & The..."Home": music appears in modern family ~ http://t.cn/zWNSPMS	2014-06-22 13: 30:01	Home

In this *association extraction* component, the pieces of music are given explicitly, so the most important task is extracting the users' emotion from their microblogs, which consists of three steps as follows.

Firstly, we construct a fine-grained Chinese emotion lexicon from three resources: (1) the original emotion lexicon from DUTIR (http://ir.dlut.edu.cn), and the emotion classification systems based on Ekman's classical psychological research [16], which consists of seven kinds of emotion, namely happy, like, surprise, angry, sad, fear, and hate; (2) the synonym lexicon from HITCIR (http://ir.hit.edu.cn/demo/ltp/Sharing_Plan.htm) to enhance the original emotion lexicon; (3) a list of emoticons collected from the microblog web site to extend the original emotion lexicon. In total, there are 48224 words in our emotion lexicon.

Secondly, we adopt a Chinese segmentation tool named ICTCLAS (http://www.ictclas.org) to segment Chinese microblog texts into words. Based on the constructed lexicon of emotion, we count the frequency of emotion words occurring in a text for each type of emotion, and the number of emotion words appearing in the text determines the emotion vector of the text. Then, each microblog can be represented as an emotion vector. In general, when we try to discover the user's emotion association with the music, we take into consideration only the microblogs before the music listening timestamp for emotion mining. In doing so, we obtain a list of the user's time-dependent microblogs by setting a certain time window around the timestamp when the music microblog was posted and compute the sum of all the emotion vectors of the microblogs within the time window as the emotion corresponding to the piece of music. Specifically, an appropriate time window is important because a narrow time window will result in too few microblogs from which to extract emotions, and a wide time window will cause emotional noise. In this paper, the time window is set as five hours.

Finally, we obtain all the emotion vectors at various music sharing timestamps, which forms the association data of emotion and music items for all users, represented by a three-element tuple (user, music, emotion).

3.2   Association Modeling

In this section, we illustrate how to model the association between the user, music, and the corresponding emotion on the graph by using different node types and treating edges with different weights.

As shown in Sect. 3.1, our association data are in the form of (user, music, emotion) triples which are usually modeled by a tri-partite graph or a tensor. However, both the tri-partite graph and tensor treat emotional information as a universal dimension shared by all users, while we argue that, in a recommender system, the dimension of emotions is a private effect and should not be shared and compared across all users arbitrarily. Specifically, we build an EAG to model our association data based on the following intuitions:

1. **Intuition 1:** Users have specific preferences to music, and tend to listen to their favorite music pieces [9].
2. **Intuition 2:** Music listening is a kind of emotion-sensitive behavior and direct associations exist between the users' emotional mood and their favorite music. Users' music preferences vary with differing emotions, and a user may listen to different music pieces when they experience different emotion [12].
3. **Intuition 3:** If two users have similar temporal preferences (in two emotions), they tend to listen to the same music pieces (when experiencing these two emotions) [13, 14].

As a bipartite graph, EAG is defined as:

$$\text{EAG} = G(U, EM, I, E, w) \tag{1}$$

where U denotes the set of user nodes, EM is the set of emotion nodes, I is the set of music item nodes, and $w: E \rightarrow R$ denotes a non-negative weight function for edges in E.

Nodes. In EAG, user node v_u connects to all items interacted by user u, denoted as $N(u)$, which represent u's long-term preferences under all emotional contexts; emotion node $v_{u,em}$ only connects to items user u viewed at emotion em, denoted as $N(u, em)$, representing u's specific preference at emotion em. Therefore, if we start walking from user node v_u, we will go through $N(u)$ and then reach unknown items similar to items in $N(u)$ (**Intuition 1**). Similarly, we will reach unknown items similar to items in $N(u, em)$ if we start from the emotion nodes (**Intuition 2**). In short, the user node represents the user's long-term preference and the emotion node represents the preference under a specific emotional context.

Edges. EAG is a weighted directed graph, and the edge weights are defined as:

$$w(v, v') = \begin{cases} 1 & v \in U, v' \in I \text{ or } v \in I, v' \in U \\ \varphi_{i,em} & v \in I, v' \in EM \text{ or } v \in EM, v' \in I \end{cases} \quad (2)$$

This definition means, given an edge $e(v, v')$ between user nodes and item nodes, its weight will be 1. As for the edges between item nodes and emotion nodes, its weight is the weight of the emotion when the user is accessing specific item i, which is $\varphi_{i,em}$. The edges from the user nodes to the music item nodes bridge users who share similar music preferences. Furthermore, the edges from the emotion nodes to the music item nodes bridge users who share similar music preferences in certain emotions. All these edges enable us to exploit other users' (emotional) music preferences for recommendation (**Intuition 3**).

Table 2 shows an example of association data which only contains two types of emotion for brevity. This example contains 3 user nodes, 6 item nodes, and 6 emotion nodes. It shows user A interacted with items a, b, c, user B interacted with items c, d, and user C interacted with items d, e, f. Furthermore, item a has an emotion vector (1.0, 0.0), which indicates that A interacted with this item under the context E_1. Figure 3 shows the corresponding EAG. In EAG, each user has their own emotion nodes, because the dimension of emotion is a private effect, which should not be shared and compared among all users, arbitrarily.

Table 2. Association example

User	Music	Emotion vector (E1, E2)
A	a	(1.0, 0.0)
A	b	(0.5, 0.5)
A	c	(0.0, 1.0)
B	c	(0.0, 1.0)
B	d	(1.0, 0.0)
C	d	(0.0, 1.0)
C	e	(0.0, 1.0)
C	f	(1.0, 0.0)

Then, we defined a weighted adjacent matrix M to represent the bipartite graph G (U, EM, I, E, w), and each element $m_{vv'}$ of matrix M is the weight of edge $e(v, v')$, whose value is $w(v,v')$. After this, we can compute the transition possibility matrix P by normalizing matrix M as follows:

$$p_{vv'} = \begin{cases} \dfrac{m_{vv'}}{\sum_{v' \in outnodes[v]} m_{vv'}} & \text{if } outnodes[v] \neq \emptyset \\ 0 & \text{otherwise} \end{cases} \quad (3)$$

Next, we explain how to perform music recommendation on EAG represented by the transition possibility matrix P.

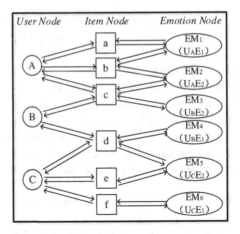

Fig. 3. An example of EAG

3.3 Music Recommendation on EAG

As shown in Sect. 3.1, our association data are modeled with the proposed EAG. We then devise a relevance propagation algorithm based on random walk to capture the overall user preference on music items given a specific emotion status and to recommend appropriate music pieces. Specifically, the relevance propagation algorithm for ranking entities on EAG is a variation of the Personalized PageRank algorithm [17], where a node will get a higher rank score if it is connected to more important nodes with less outgoing links. By applying the propagation algorithm to EAG, we can find the target user's potential preferences in certain emotions based on all emotion-aware music listening data. In general, the PageRank score can be calculated as follows:

$$\vec{r} = \alpha P^T \vec{r} + (1 - \alpha)\frac{1}{n}\vec{e} \tag{4}$$

where r_i is the rank score of node v_i, n is the number of nodes, $\vec{e} = (1, 1, \ldots, 1)^T$ represents the initial value for all nodes, and α is a damping factor constant that is normally set as 0.85. The algorithm is based on a random-walk with the restart process, which uniformly chooses a random node for restarting with possibility $(1 - \alpha)$ during the random walk process. However, in order to calculate the PageRank score of a node, we replace $(1/n)\vec{e}$ with a personalized initial vector $\vec{r}_0$. This personalized initial vector normally represents a user's interest. Thus, the corresponding PageRank score vector can be calculated as:

$$\vec{r} = cP^T \vec{r} + (1 - c)\vec{q} \tag{5}$$

Random walk with the restart process gives each node a higher ranking score when the node is more closely related to the nodes in the query vector, considering the importance of nodes at the same time. As a result, we can use the value of r_i as the rank value for each entity, including item nodes. Finally, the item nodes are sorted according to their ranking score $\vec{r}$ and the top-N items are recommended to the target user under certain emotional contexts.

There are several advantages of adapting the propagation algorithm to rank items in our approach. Firstly, we can utilize its propagation and attenuation properties [18]. The propagation property means the relatedness of the nodes propagates following the links (edges), and the attenuation property is that the propagation strength decreases as the propagation goes further from the initial node. Secondly, through the graph and with the random-walk process, we can find a rank score of target entities by considering the influence of various hybrid factors, which enables our approach to incorporate many kinds of contextual information in different forms, including emotional context. Thirdly, the propagation algorithm can be pre-computed and it scales for large size datasets. In our work, the formula of the rank score can be transformed to $\vec{r} = (1 - c)(1 - cP^T)^{-1}\vec{q}$, which shows that if we can pre-compute the matrix $(1 - c)(1 - cP^T)^{-1}$ independently, then we can simply multiply the initial query vector and the matrix $(1 - c)(1 - cP^T)^{-1}$ to compute the rank score for each node according to the query (the target user and their current emotion status) with acceptable time complexity $O(m*n)$. In our method, the query is given as a sparse vector with only several nonzero elements, which represent the target user and their emotional context. Thus, the matrix multiplication can be computed quickly and the recommendation can be performed online. In addition, although pre-computation requires matrix inversion which has very high time complexity, which is $O(n^3)$, [19] shows that we can estimate the inverse matrix with acceptable cost by using the dimension reduction approach.

4 Experimental Evaluation

In this section, we first investigate how the users' emotion affects the performance of music recommendation by adopting different emotion modeling methods. Then, we experimentally evaluate the performance of the proposed graph-based emotion-aware recommendation method and compare it against the performance of the baseline methods. All experiments were performed on a PC with Intel Core i3-2120 running at 3.30 GHz, along with 8 GByte of memory and 64-bit Windows 8 operating system.

4.1 Dataset

To evaluate the proposed approach, we collect microblog data from Sina Weibo (http://open.weibo.com/wiki/API). The detailed statistics of the ultimate dataset are shown in Table 3.

Table 3. Complete statistics of dataset

Users No	Music No	Listening No	Listening per user	Listening per music
1,008	2,552	92,283	91.6	36.2

4.2   Experimental Design and Metrics

Firstly, in order to evaluate the quality and performance of the top-N recommendation provided by our approaches, we split the dataset into a training and testing set according to the methodology of the 5-fold cross-validation. In all the experiments, we adopt four metrics [20] as follows:

HitRate. HitRate is the fraction of hits in all recommendations and hit means the recommendation list contains the piece of music in which the user is interested under their current emotional context. For example, as for an association (u, i, em) in the test data, if the recommended list of u under the emotional context em contains i, then it is a hit. The definition is given as follows:

$$HitRate = \#(hits)/\#(R)$$

where:

- $\#(hits)$ is the number of hits.
- $\#(R)$ is the number of recommendations.

Precision, Recall, and F1 Score. Precision is the fraction of recommended music items in which the target user is actually interested. Recall is the fraction of the target user's interested music that is recommended. F1 score is the harmonic mean of precision and recall. The definitions are given as follows:

$$Precision = \sum_{u \in U} |R(u) \cap T(u)| \Big/ \sum_{u \in U} |R(u)|$$

$$Recall = \sum_{u \in U} |R(u) \cap T(u)| \Big/ \sum_{u \in U} |T(u)|$$

$$F1 = 2 \times Precision \times Recall/(Precision + Recall)$$

where:

- U is the user set in the test data.
- $R(u)$ is the recommended music list.
- $T(u)$ is the set of all music items u actually listened to in the test data.

4.3   Effect of Emotion Modeling Methods

In order to investigate how users' emotion can affect the performance of music rec-ommendation, we evaluate three recommendation methods, namely EAGRec-Vector, EAGRec-Single-Label and EAGRec-Multi-Label, which adopt vector, single-label and multi-label, respectively, as their emotion modeling methods. As shown in Table 4, the vector can model all the emotions extracted from the users' microblogs along with the emotion weight, the multi-label can model all emotions without a corresponding weight, and the single-label only incorporates the emotion with the highest weight value.

In other words, the vector can represent all details of the original extracted emotion, while the multi-label can express all kinds of emotion appearing with the same weight and the single-label only considers the main emotion with the highest value.

As shown in Fig. 4, we can conclude that (1) EAGRec-Vector has the best performance in all four metrics, namely hitrate, precision, recall, and F1 Score. This is because the vector can represent all of the emotional information extracted from the users' microblogs and incorporates all kinds of emotion along with their weights into recommendation. In other words, all emotions play important roles in emotion-aware music recommendation; (2) EAGRec-Single-Label outperforms EAGRec-Multi-Label and the relative improvement in terms of precision with n set as 15 is 4.6 %. In our view, the reason for this is that primary emotions and secondary emotions make a different contribution to the recommendation and should not be treated equally. In addition, primary emotions are more important than secondary emotions in users'

Table 4. Emotion modeling methods example

Modeling methods	Emotion Value (E1, E2, E3, E4, E5, E6, E7)
Extracted emotion	(E1 = 0.7, E2 = 0.3, E3 = 0.0, E4 = 0.0, E5 = 0.0, E6 = 0.0, E7 = 0.0)
Vector	(0.7, 0.3, 0.0, 0.0, 0.0, 0.0, 0.0)
Multi-label	(0.5, 0.5, 0.0, 0.0, 0.0, 0.0, 0.0)
Single-label	(1.0, 0.0, 0.0, 0.0, 0.0, 0.0, 0.0)

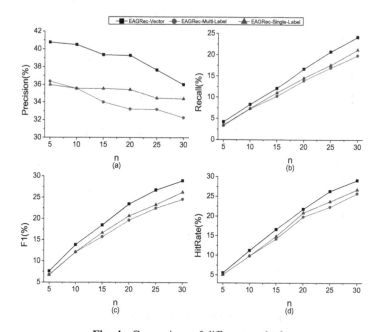

Fig. 4. Comparison of different methods

emotion-aware music prediction and recommendation, and secondary emotions may have a negative impact if not utilized properly.

4.4 Comparison with Baselines

Our second experiment is designed to compare the proposed approach with different baseline recommendation methods, namely Bayesian Personalized Ranking (BPR) [21] and Alternating Least Square for Personalized Ranking (RankALS) [22] which do not incorporate contextual information, together with User Collaborative Filtering with Emotion (UCFE) [15] which considers users' emotional context. We choose the best performing EAGRec-Vector as the representative for clear comparison.

As shown in Fig. 5, EAGRec-Vector outperforms the other baseline methods, including traditional and emotion-aware ones in all four metrics. Taking precision as an example, when compared with BPR, RankALS, and UCFE with recommendation number n set as 15, the relative performance improvement by EAGRec-Vector is around 62.6 %, 70.3 %, and 41.8 %, respectively. The improvement over the baseline methods shows that: (1) emotional context plays an important role in the users' music preference, and the users' music listening behaviors and preferences are relevant to their emotional context; (2) extracting users' emotion from their microblogs is feasible and useful; (3) our proposed EAG is more effective for incorporating emotional data into the graph than UCFE and the random-walk-based recommendation approach is also effective for emotion aware music recommendation.

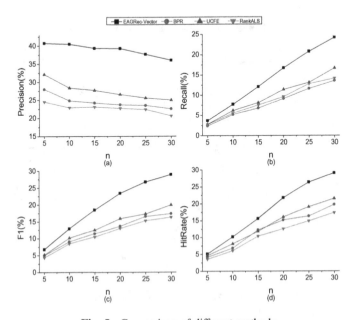

Fig. 5. Comparison of different methods

In conclusion, the emotional context information from the user's microblogs influences the users' music preferences. In addition, our proposed graph-based emotion-aware music recommendation methods can effectively incorporate users' emotional context and improve the performance of music recommendation.

5 Conclusion

This paper presents a graph-based approach for emotion-aware music recommendation, which can recommend music items appropriate for the users' current emotional context to satisfy their real-time requirements. The proposed approach models the relations between user, music, and emotion as a three-element tuple (user, music, emotion), upon which an EAG was built, and then a relevance propagation algorithm based on random walk was devised to rank the relevance of music items for recommendation. The experimental evaluation shows that the proposed method outperforms the traditional approach which does not consider the user's emotion. In addition, we study in detail how users' emotional context extracted from their microblogs can affect the performance of music recommendation. The results show that all parts of the extracted emotions are useful for music recommendation, and incorporating emotions along with their weight into the approach can improve the recommendation results to the maximum extent possible.

Based on our current work, there are two possible future directions. First, we will attempt to connect a microblog service (such as Twitter, Sina Weibo) with music service websites (such as Xiami Music, Last.fm) to extract more tuples (user, music, emotion), and provide better recommendation results, especially when the user does not have many music microblogs. Secondly, we will explore if the users' satisfaction can be increased when users listen to recommended music using online experiments.

Acknowledgements. This research work is supported in part by the National Key Technology Research and Development Program of China under Grant 2014BAD10B02, the Fundamental Research Funds for the Central Universities 2016FZA5012, and Australian Research Council (ARC) Linkage Project under No. LP140100937.

References

1. Yuan, B., Xu, B., Chung, T., Shuai, K., Liu, Y.: Mobile phone recommendation based on phone interest. In: Benatallah, B., Bestavros, A., Manolopoulos, Y., Vakali, A., Zhang, Y. (eds.) WISE 2014. LNCS, vol. 8786, pp. 308–323. Springer, Heidelberg (2014). doi:10. 1007/978-3-319-11749-2_24
2. Wu, H., Shao, J., Yin, H., Shen, H.T., Zhou, X.: Geographical constraint and temporal similarity modeling for point-of-interest recommendation. In: Wang, J., Cellary, W., Wang, D., Wang, H., Chen, S.-C., Li, T., Zhang, Y. (eds.) WISE 2015. LNCS, vol. 9419, pp. 426–441. Springer, Heidelberg (2015). doi:10.1007/978-3-319-26187-4_40
3. Kaminskas, M., Ricci, F.: Contextual music information retrieval and recommendation: state of the art and challenges. Comput. Sci. Rev. **6**, 89–119 (2012)

4. Wang, K., Zhang, R., Liu, X., Guo, X., Sun, H., Huai, J.: Time-aware travel attraction recommendation. In: Lin, X., Manolopoulos, Y., Srivastava, D., Huang, G. (eds.) WISE 2013. LNCS, vol. 8180, pp. 175–188. Springer, Heidelberg (2013). doi:10.1007/978-3-642-41230-1_15

5. North, A.C., Hargreaves, D.J.: Situational influences on reported musical preference. Psychomusicology Music Mind Brain 15, 30–45 (1996)

6. Abowd, G.D., Dey, A.K., Brown, P.J., Davies, N., Smith, M., Steggles, P.: Towards a better understanding of context and context-awareness. In: Handheld and Ubiquitous Computing, pp. 304–307 (1999)

7. Pettijohn II, T., Williams, G., Carter, T.: Music for the seasons: seasonal music preferences in college students. Curr. Psychol. 29, 328–345 (2010)

8. Kaminskas, M., Ricci, F.: Location-adapted music recommendation using tags. In: Konstan, J.A., Conejo, R., Marzo, J.L., Oliver, N. (eds.) UMAP 2011. LNCS, vol. 6787, pp. 183–194. Springer, Heidelberg (2011). doi:10.1007/978-3-642-22362-4_16

9. Knees, P., Schedl, M.: A survey of music similarity and recommendation from music context data. ACM Trans. Multimedia Comput. Commun. Appl. (TOMM) 10, 2 (2013)

10. Wang, X., Rosenblum, D., Wang, Y.: Context-aware mobile music recommendation for daily activities. In: Proceedings of the 20th ACM International Conference on Multimedia, pp. 99–108. ACM (2012)

11. North, A., Hargreaves, D.: The social and applied psychology of music. Oxford University Press (2008)

12. Chang, C., Lo, C., Wang, C., Chung, P.: A music recommendation system with consideration of personal emotion. In: International Computer Symposium, pp. 18–23 (2010)

13. Han, B., Rho, S., Jun, S., Hwang, E.: Music emotion classification and context-based music recommendation. Multimedia Tools Appl. 47, 433–460 (2010)

14. Shan, M., Kuo, F., Chiang, M., Lee, S.: Emotion-based music recommendation by affinity discovery from film music. Expert Syst. Appl. 36, 7666–7674 (2009)

15. Deng, S., Wang, D., Li, X., Xu, G.: Exploring user emotion in microblogs for music recommendation. Expert Syst. Appl. 42, 9284–9293 (2015)

16. Ekman, P.E., Davidson, R.J.: The nature of emotion: Fundamental questions. Oxford University Press (1994)

17. Haveliwala, T.H.: Topic-sensitive pagerank: A context-sensitive ranking algorithm for web search. IEEE Trans. Knowl. Data Eng. 15, 784–796 (2003)

18. Gori, M., Pucci, A.: ItemRank: a random-walk based scoring algorithm for recommender engines. In: Proceedings of the 20th International Joint Conference on Artifical Intelligence, pp. 2766–2771. Morgan Kaufmann Publishers Inc., Hyderabad (2007)

19. Tong, H., Faloutsos, C., Pan, J.-Y.: Fast random walk with restart and its applications. In: Proceedings of the Sixth International Conference on Data Mining, pp. 613–622. IEEE Computer Society (2006)

20. Cremonesi, P., Koren, Y., Turrin, R.: Performance of recommender algorithms on top-n recommendation tasks. In: Proceedings of the Fourth ACM Conference on Recommender Systems, pp. 39–46. ACM, Barcelona (2010)

21. Rendle, S., Freudenthaler, C., Gantner, Z., Schmidt-Thieme, L.: BPR: Bayesian personalized ranking from implicit feedback. In: Proceedings of the Twenty-Fifth Conference on Uncertainty in Artificial Intelligence, pp. 452–461. AUAI Press (2009)

22. Takács, G., Tikk, D.: Alternating least squares for personalized ranking. In: Proceedings of the Sixth ACM Conference on Recommender Systems, pp. 83–90. ACM (2012)

Topic Modeling

Domain Dictionary-Based Topic Modeling for Social Text

Bo Jiang, Jiguang Liang, Ying Sha[(✉)], Rui Li, and Lihong Wang

National Engineering Laboratory for Information Security Technologies,
Institute of Information Engineering, Chinese Academy of Sciences,
Beijing 100093, China
{jiangbo,liangjiguang,shaying,lirui,wanglihong}@iie.ac.cn

Abstract. Online social networks are becoming increasingly popular and posting large volumes of unstructured social text documents every day. Inferring topics from large-scale social texts is a significant but challenging task for many text mining applications. Conventional topic models has been shown unsatisfactory results due to the sparsity and noise of content in short texts. Besides, the learned topics are very difficult to understand the semantic information only by the top weighted terms. In this paper, we propose a novel social text topic modeling method to deal with the problems. The proposed model utilizes topic domain dictionary to construct a weakly supervised matrix, which can play a role of making reference matrix and the learned topic matrix become similar. Experimental results on the constructed social text dataset from Twitter demonstrate that our proposed method can outperform the state-of-the art baselines significantly and also improve the semantic relevancy of the learned topic.

Keywords: Social text · Topic modeling · Domain dictionary · Matrix factorization

1 Introduction

Social networks have overtaken traditional content distribute systems as one of the most popular online information diffusion platforms. People can freely share and browse a variety of information which involves various topics such as daily chatting, business promotions, and news stories. According to the latest official toll, about 500 million tweets are produced every day in Twitter. Thus, with the emerging large scale social texts, discovering the latent topics discussed by users is an important and valuable task for many content analysis applications, such as personalized recommend, user interest inferring, and public opinion analysis.

Topic modeling is an effective method which can automatically discover the hidden semantic structure in large collection of documents. Conventional topic models, such as PLSA [11] and LDA [3], reveal topics within a text corpus by implicitly capturing the document-level word co-occurrence patterns [4,30]. The main advantage of these models is that they do not require any prior information

© Springer International Publishing AG 2016
W. Cellary et al. (Eds.): WISE 2016, Part I, LNCS 10041, pp. 109–123, 2016.
DOI: 10.1007/978-3-319-48740-3_8

and can be easily be extended. Many research work has been demonstrated that these topic modeling methods have extensively been used for text mining, document classification, and information retrieval.

Against the backdrop of social texts emerging, conventional topic models show absence of good performance due to the sparsity and noise of content in social text topic identification [34]. Thus, some novel topic modeling methods have been proposed for social texts, such as unsupervised methods [6,35] and supervised methods [22–24,26]. In general, completely unsupervised method for identifying topic ability is always limited. Supervised method usually requires high-quality labeled data, and use of human annotation is too expensive. Recently, some work has been explored automatic labelling techniques [5,13,16,20]. These approaches assign a label to a topic either by using top weighted terms based on different ranking mechanisms or using external data sources (e.g. Wikipedia, WordNet). However, these mentioned methods are not precise enough, and sometimes find closely related auxiliary data with the original data may be expensive or even impossible.

In this paper, we propose a framework of social text topic modeling by using topic domain dictionary to solve these problems. Specifically, we propose a matrix factorization model to identify hidden topics from social texts. Different from previous works, in the proposed model, we introduce external topic domain dictionary as the topic prior information, and construct a reference matrix based on topic domain dictionary. The reference matrix can play a role of making learned latent topic matrix become similar each other. In addition, we can observe that social texts usually contain a larger proportion of useless terms which could mislead document topic identification that rely on the frequency of terms, and a single social text is more likely to only refer to a single topic. Therefore, the proposed model applies a sparsity-inducing prior to limit the number of terms in word distributions. This increases recognizable accuracy and robustness to limited training data. Our experimental results on a real-world social text dataset show that our proposed method can discover better coherence of topics compared to other baseline methods.

Our main contributions in this work are summarized as follows:

- We incorporate seed words from domain dictionary as priors information for topics based on matrix factorization model. The unsupervised learning process become supervised and improve topic readability for social text.
- We introduce L_1 norm to learn more sparse representation for topics and achieve automatic feature selection, while eliminating useless words information, thereby increasing topic readability.
- The experimental results show that the proposed model can achieve better performance compared to the state-of-the-art topic modeling methods.

The rest of this paper is organized as follows: In Sect. 2, we review some related works on topic modeling methods. In Sect. 3, we explain the detail of the proposed method. In Sect. 4, we illustrate our experimental results conducted on short texts dataset. The conclusion and future work are presented in Sect. 5.

2   Related Work

Considerable literature has grown up relating to topic modeling of texts in the past couple of decades. Generally speaking, these methods can be classified into two categories: probabilistic approaches and non-probabilistic approaches. The former defines topics as a probability distribution over a vocabulary and documents as data generated from mixtures of topics. The representative models are PLSI [11] and LDA [3]. Specifically, LDA has become well-known topic model due to nice generalization ability and extensibility in text mining. The latter represents each document as a vector of topics, and the documents are projected the latent topic space. The widely-used approaches have LSA [7], NMF [18,19] and RLSI [29].

Our work is more closely related with short text topic modeling. For example, Phan et al. [22] inferred the topics of short texts based on a conventional topic model estimated on another large scale dataset for short text classification. Jin et al. [15] proposed a model based on LDA that jointly learns topics over short texts and related long texts. It is expected to leverage the topical knowledge learned from long texts to help the topic learning task over short texts. Recently, Ramage et al. [24] used labeled-LDA to Twitter, but the model relies on hashtags in Twitter, which may not include all topics. Zhao et al. [35] developed a new Twitter-LDA model for modeling each tweet in the way of mixture of unigrams and showed its effectiveness compared with existing models. Yan et al. [33] developed methods based on non-negative matrix factorization for short text topic learning by exploiting global word co-occurrence information. Cheng et al. proposed [6] learns topics by directly modeling the generation of word co-occurrence patterns in the short text corpus, making the inference effective with the rich corpus-level information. Yang et al. [34] presented a deployed large-scale topic modeling system that infers topics of tweets over an ontology of hundreds of topics in real-time and at stringently high precision. Cano et al. [5] proposed to generate topic labels by applying summarisation algorithms.

In addition, in [8,28,32,36], they also extended the conventional MF methods and incorporated the external information into their models which perform topic mining task in document collection. There is also a lot of related work on seeding topics with dictionaries and with adding sparsity to topics. For example, [1,12,14,21] incorporated dictionaries and seed words as priors for inferring topics. [9,27] applied a sparsity-inducing prior to limit the number of terms in word distributions. This increases predictive accuracy and robustness to limited training data. [2,31] presented sparsity pattern in topic distributions of documents in a manner.

In all methods presented above each document in the collection of documents is modelled as a multinomial distribution over topics. Instead, a single short text is more likely to talk about one topic inferred by only a small number of words. Inspired by this, our model is devised by using topic domain dictionary as prior information to supervise the learning process of topics and employing sparse regularization on topics to remove useless information.

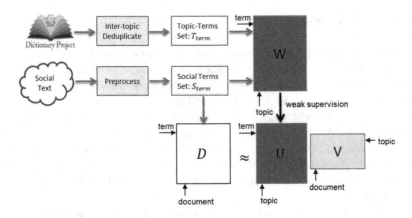

Fig. 1. Overview of social text topic model with topic domain dictionary.

3 Domain Dictionary Based Topic Model

In this section, we first give a formal definition of the problem and then present the Domain Dictionary based Topic Model (DDTM) and its optimization algorithm. As illustrated in Fig. 1, DDTM framework provides a weakly-supervised non-probabilistic topic model using topic domain dictionary.

3.1 Problem Formulation

Suppose that we have N documents with the jth document denoted as d_j, and M terms with the ith term denoted as t_i. Let $D = (d_{ij})$ be represented in an $M \times N$ term-document matrix $D = [d_1, \cdots, d_N]$, in which each row corresponds to a term and each column corresponds to a document. d_{ij} is the tf-idf weight of term t_i in document d_j.

We first factorize D into matrix $U \in \mathbb{R}^{M \times K}$ and $V \in \mathbb{R}^{K \times N}$ where $U = [u_1, \cdots, u_K]$ is a term-topic matrix, in which each column corresponds to an M-dimensional topic vector u_k, where the ith entry denotes the weight of the ith term in the topic, and $V = [v_1, \cdots, v_N]$ is a topic-document matrix, in which where column v_j stands for the representation of document d_j in the latent topic space. $K \ll min(M, N)$ denotes the number of latent topics.

Topic modeling can discover the latent topics in the document collection as well as represent the documents by distribution of the topics. Specifically, given a text collection $D = [d_1, \cdots, d_N]$, our goal is to obtain a low-rank approximation $D \approx UV$, where the quality of the approximation is described by the squared error as the reconstruction loss function. Hence, we formulate as follows:

$$\min_{U,V} \frac{1}{2} \|D - UV\|_F^2 \tag{1}$$

where $\|\cdot\|_F$ is the Frobenius norm fitting constraint.

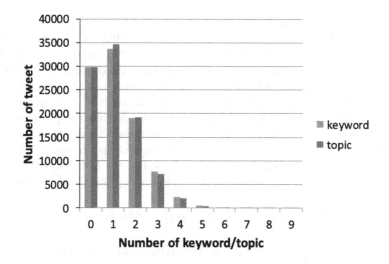

Fig. 2. Percentage of tweet when number of keywords/topics increases.

Social text, such as tweet from Twitter and microblog from Weibo, is a length-limited document. Recent research has shown that the average document length for short text is less than 6 [6]. Thus, most short texts have no any meaningful topic (i.e., daily talk) or only refer to a single topic. Besides, we also perform an investigation to estimate the percentage that short texts hit the number of domain dictionary's keywords/topics using a large collection of documents from Twitter. The result of statistical analysis is shown in Fig. 2. From the figure, we can see conclude that nearly 85 % short texts involve equal or less than 2 keywords/topics. Meanwhile, the short texts usually contain a larger proportion of useless terms which could mislead document topic identification that rely on the frequency of terms. Therefore, we propose using ℓ_1 regularization norm on term-topic matrix U and ℓ_2 regularization norm on topic-document matrix V to avoid overfitting. The benefits of this method that ℓ_1 regularization norm on topic can extract high quality of terms and ℓ_2 regularization norm on document representations can better describe document topic distribution. Thus, on the basis of (1), we add regularization norms on U and V, respectively:

$$\min_{U,V} \frac{1}{2} \|D - UV\|_F^2 + \beta \|U\|_1 + \frac{\gamma}{2} \|V\|_F^2 \tag{2}$$

where $\beta \geq 0$ and $\gamma \geq 0$ are the parameters controlling the sparsity representation on topic and the shrinkage on document, respectively. Scalar parameters β and γ are used to control the strength of regularization.

Topic modeling methods usually select top weighted terms to represent hidden latent semantics involved in each topic. The learned processes have been commonly approached with unsupervised clustering algorithms. However, the learned topics are usually no good topic readability and are very difficult to align to a predefined topic taxonomy. To solve the problem, we introduce a weakly supervised matrix so that the learning process of latent topic matrix U can support controllability and workability. The intuition behind is that the stronger the correlation terms have, the more closely the high-level semantic information of latent topics become. The introduced weakly supervised matrix is represented as $W \in \mathbb{R}^{M \times K}$, which has the same dimension with learned latent term-topic matrix U. Let $W = [w_1, \cdots, w_K]$, like u_k, each column of W corresponds to an M-dimensional topic vector w_k, where the ith entry denotes the presence of the ith term in the topic. We assume that $w_{ik} = 1$ if the ith term at the same time appear in W and D and $w_{ik} = 0$ otherwise. The details on how to construct referenced term-topic matrix W will be explained in the last subsection. Given the above, DDTM amounts to solving the following optimization problem:

$$\min_{U,V} \frac{1}{2} \|D - UV\|_F^2 + \frac{\alpha}{2} \|W - U\|_F^2 + \beta \|U\|_1 + \frac{\gamma}{2} \|V\|_F^2 \qquad (3)$$

where $\alpha \geq 0$ is a parameter controlling the strength of similarity between reference matrix and learned latent topic matrix.

3.2   Optimization Algorithm

The optimization problem in Eq. (3) is not jointly convex with respect to the two variables U and V. However, it is convex with respect to one of them, when the other one is fixed. Therefore, we optimize the function in Eq. (3) by alternately minimizing it with respect to term-topic matrix U and topic-document matrix V, respectively.

Update of Term-topic Matrix. Keeping topic-document matrix $V = [v_1, \cdots, v_N]$ fixed, the updating rule of U equivalent to the following optimization problem:

$$\min_{U} \frac{1}{2} \|D - UV\|_F^2 + \frac{\alpha}{2} \|W - U\|_F^2 + \beta \|U\|_1 \qquad (4)$$

Let $d_i = (d_{i1}, \cdots, d_{iN})^T$, $u_i = (u_{i1}, \cdots, u_{iK})^T$ and $w_i = (w_{i1}, \cdots, w_{iK})^T$ be the column vectors whose entries are those of the ith row of D, U and W respectively. Thus, the above optimization problem further can be rewritten as:

$$\min_{\{u_i\}} \frac{1}{2} \sum_{i=1}^{M} \left\|d_i - V^T u_i\right\|_F^2 + \frac{\alpha}{2} \sum_{i=1}^{M} \|w_i - u_i\|_F^2 + \beta \sum_{i=1}^{M} \|u_i\|_1 \qquad (5)$$

where the rows of U are independent, the optimization problem in Eq. (5) can be decomposed into a set of optimization problems of the form:

$$\min_{u_i} \frac{1}{2} \left\|d_i - V^T u_i\right\|_2^2 + \frac{\alpha}{2} \|w_i - u_i\|_2^2 + \beta \|u_i\|_1 \qquad (6)$$

for $i = 1, \cdots, M$.

Equation (6) is an ℓ_1-regularized least squares problem, we can't directly use gradient descent since the gradient of the Lasso objective function doesn't exist everywhere. Here, the optimization problem can be solved using coordinate descent and soft thresholding [10]. They optimize (exactly or approximately) the objective with respect to one variable at a time while all others are kept fixed. More specifically, we view u_{ik} as the variable, and minimize the objective function in Eq. (6) with respect to u_{ik} while holding all the u_{ij} fixed for which $j \neq k, k = 1, \cdots, K$ for each iteration.

To reformulate the Lasso objective function as a function of the u_{ik}, let $r_i = d_i - \sum_{j=1}^{K} u_{ij}v_j - u_{i0}\mathbf{1}$ for $j \neq k$, Then we can rewrite the objective in Eq. (6) as a function with respect to u_{ik}:

$$f(u_{ik}) = \frac{1}{2} \|r_i - v_k u_{ik}\|_2^2 + \frac{\alpha}{2}(w_{ik} - u_{ik})^2 + \beta \|u_i\|_1$$

$$= \frac{1}{2}(\|v_k\|^2 + \alpha)u_{ik}^2 - (\langle r_i, v_k \rangle + \alpha w_{ik})u_{ik} \qquad (7)$$

$$+ \beta |u_{ik}| + C$$

where $\langle , \rangle$ is the inner product of vectors and C is a constant with respect to u_{ik}. This allows us to rewrite the problem as follows:

$$f(u_{ik}) = \frac{1}{2}L\left[(u_{ik} - \frac{\langle r_i, v_k \rangle + \alpha w_{ik}}{L})^2 + \frac{\beta}{L}|u_{ik}|\right] + \Omega \qquad (8)$$

where $L = \|v_k\|^2 + \alpha$, and Ω is a constant with respect to u_{ik}. Now that we can directly apply the soft-thresholding operator S from the univariate Lasso problem in Eq. (8) as follows:

$$arg \min_{u_{ik}} f(u_{ik}) = S(\frac{\langle r_i, v_k \rangle + \alpha w_{ik}}{L}, \frac{\beta}{L})$$

$$= \frac{1}{L}S(\langle r_i, v_k \rangle + \alpha w_{ik}, \beta) \qquad (9)$$

Update of Topic-document Matrix. The update of V with U fixed is minimizer of the residual sum of squared errors with a penalty term proportional to the squared values of v_j. This is referred to as ℓ_2 regularization. The problem can be decomposed into N optimization problems. For each v_j, we can be solved as follow:

$$\min_{v_j} f(v_j) = \frac{1}{2}\|d_j - Uv_j\|_2^2 + \frac{\gamma}{2}\|v_j\|_2^2 \qquad (10)$$

for $j = 1, \cdots, N$. We use stochastic gradient descent method to solve the problem. Specifically, the gradients of Eq. (10) with respect to v_j is:

$$\frac{\partial f}{\partial v_j} = (Uv_j - d_i)U + \gamma v_j \qquad (11)$$

3.3   Domain Dictionary Matrix Construction

As the above section mentioned, we introduce a domain dictionary based term-topic matrix W, which has the property of weakly-supervised for automatic labelling topic and the same dimension with learned latent topic matrix U. To construct W, we opt to choose Longman Dictionary[1] as our topic domain dictionary due to a wide range of topic field and obviously distinguished terms of topics. Moreover, selected topics that are frequently discussed on social networks and that almost all the short texts can be grouped into one of chosen topic categories. More precisely, we first collect all categories of topics corresponding with all entries for each topic. We obtain the total topics to 207 (e.g., Education, Music, Sport, etc.), in which topics have at least 1 term and at most 472 terms. We then notice that the same term maybe fall into different topic categories. To make sure a term only belongs to a single topic, we perform a process of deduplicate terms for inter-topics which removes approximately 158 terms, and then also discard topics whose number of terms are less than 10. After preprocessing, we get a new set of topic categories represented as T, which has 169 topics. Further, in order to learn better quality of topic semantic representation, we select more important terms in the collection of short texts to learn high quality of topic representation. Specifically, we take cf-idf to measure the importance of a term in the given document collection. The actual cf-idf formula we used is

$$cf - idf(t, d, D) = cf(t, D) \times log \frac{|D|}{|\{d \in D : t \in d\}|} \qquad (12)$$

where t represents a term, d represents a document, D represents a document collection, $cf(t, D)$ is the number of times that term t appears in document collection D, $|D|$ is the total number of documents in the collection, and $|\{d \in D : t \in d\}|$ is the number of documents where term t appears.

We use $\mathcal{D}$ to represent the set of weighted terms measured by Eq. (12). We run an inverse operator by the weight of term on $\mathcal{D}$. We then perform the intersection of $\mathcal{D}$ and T to obtain top n important terms and topic categories existing the document collection. Finally, we construct domain dictionary based term-topic matrix W with the same dimension with learned latent topic matrix U by known topic terms and known topic categories.

4   Experiments and Discussions

In the section, we conduct plentiful experimental work to compare our proposed method with existing topic modeling methods, and to test effect of parameter for the performance of the proposed model.

---

[1] http://www.ldoceonline.com/.

Table 1. Summary of the short texts collection.

Dataset	#Tweets	#Terms	AvgDocLen	Language
Twitter	98,251	39,511	5.74	English

4.1 Data Preparation

In order to show the effectiveness of our approach over short texts dataset, we collect a standard short texts collection from Twitter service using Twitter API, which provides 16572 users and 16 million tweets sampled from 1th March to 30th June 2014.

Due to noisy of the short texts collection, we perform preprocessing to get high quality text content. Specifically, we first remove all mentioned username represented via @ symbol and URL, and stop words in a standard list was removed. We then employ linguistic algorithms to both stemming and lemmatization for short texts. We further discard any term that appears less than two times and any document that contains less than three terms, and remove all tweets that are determined not to be English. Finally, we randomly select a subset of preprocessed short texts, which approximately 100000 documents, as topic modeling experimental data. Table 1 shows a more detailed information for evaluation.

4.2 Comparison Methods

For comparison with our proposed model, we select conventional topic modeling methods and fresh short text topic modeling methods as baseline methods:

- **LSA:** Latent semantic analysis (LSA) is a technique of describing the occurrences of terms using a term-document matrix in documents [7].
- **PLSA:** Probabilistic latent semantic analysis (PLSA) is a probabilistic approach for the analysis of a low-dimensional representation of the observed variables in terms of their affinity to latent variables [11].
- **LDA:** Latent Dirichlet allocation (LDA) is a generative probabilistic model added Dirichlet process based on PLSA, and represents documents as mixtures of topics that spit out words with certain probabilities [3].
- **NMF:** Non-negative matrix factorization (NMF) is a recently developed dimension reduction method for finding parts-based, linear representations of non-negative data [18].
- **RLSI:** Biterm Topic Model (BTM) is a word co-occurrence based topic model that learns topics by modeling word-word co-occurrences patterns which uses the aggregated patterns in the whole corpus to solve the problem of sparse at document-level [29].
- **T-LDA:** Twitter-LDA (T-LDA) is a topic modeling method for modeling each tweet in the way of mixture of unigrams [35].
- **BTM:** Biterm Topic Model (BTM) is a word co-occurrence based topic model that learns topics by modeling word-word co-occurrences patterns [6].

The number of topics K is set to equal the number of W's column for all the methods (here $K = 100$). For convenience, we employ the publicly available topic models when running the baselines. Specifically, in RSLI, we use the optimal parameters ($\lambda_1 = 0.5, \lambda_2 = 1.0$) used in [29]. The rest of methods are set to default. For our proposed model, parameters α, β, and γ are set in ranges of $[0.1, 1]$, 0.5 and 1.0 in our experiments, respectively. We run all the methods in 100 iterations to confirm convergence.

4.3   Evaluation Metrics

We aim to evaluate the effectiveness and efficiency of the proposed method on short texts. Note that the evaluation of a topic model is not a trivial problem. In this paper, we quantitatively measure learned latent topic and the representation of document in the latent topic space, respectively. More specifically, we employee both topic coherence C_V and word intrusion methods to measure the quality of latent topic. The matric C_V is based on a sliding window, a one-set segmentation of the top words and an indirect confirmation measure [25]. Word intrusion measures topic interpretability indirectly, by computing the fraction of annotators who successfully identify the intruder word [17].

We also use classification method to evaluate how accurate and discriminative of the learned document representations from different models are. More precisely, we apply the same strategy [23] to build a ground truth topics of short texts corresponding to the taxonomy in Longman Dictionary by manual curations. We then divide the constructed data set into training and testing data, and perform 10-fold cross validation. We leverage four common used metrics to evaluate the performance of short text topic identification, namely Precision, Recall, F_1-score, and Accuracy.

4.4   Analysis and Discussions

Topic Coherence. The overall performance of each topic modeling method with different top words N and different topics K is illustrated in Fig. 3. From the figure, we can see conclude that our proposed DDTM method significantly outperforms all the other methods consistently, which shows our proposed method can identify more representative words of latent topics. Besides, we have the following observations. First, we can observe that the conventional topic modeling methods can't learn the latent topics very well from the short texts, and new short text topic modeling methods slightly improve the coherence of learned topics. Second, along with the increase of top terms for each topic, coherence of topics gradually decrease.

Word Intrusion. The word intrusion approach attempt to compute the semantic interpretability of topic models. In this paper, we first present each topic with a set of six terms, which consists of top five weighted topic terms and a randomly selected intruder word with low probability in the current topic but high probability in some other topics. We then use automatic evaluation

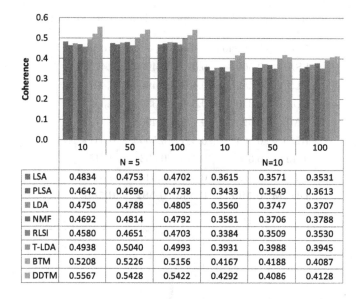

	10	50	100	10	50	100
		N = 5			N=10	
■ LSA	0.4834	0.4753	0.4702	0.3615	0.3571	0.3531
■ PLSA	0.4642	0.4696	0.4738	0.3433	0.3549	0.3613
■ LDA	0.4750	0.4788	0.4805	0.3560	0.3747	0.3707
■ NMF	0.4692	0.4814	0.4792	0.3581	0.3706	0.3788
■ RLSI	0.4580	0.4651	0.4703	0.3384	0.3509	0.3530
■ T-LDA	0.4938	0.5040	0.4993	0.3931	0.3988	0.3945
■ BTM	0.5208	0.5226	0.5156	0.4167	0.4188	0.4087
■ DDTM	0.5567	0.5428	0.5422	0.4292	0.4086	0.4128

Fig. 3. Coherence measures with different topic modeling methods.

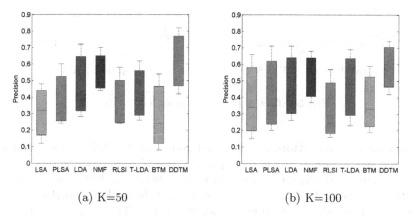

(a) K=50   (b) K=100

Fig. 4. Compare the precision of the topic models with word intrusion.

method proposed in [17] to measure the human-interpretability of topic models. Note that the whole process of successfully identifying the intruder word is fully automatic, without the need for hand-labelled training data. The result is summarized in Fig. 4. The experimental results clearly show that the proposed method has better topic interpretability than the other baselines with a significant margin. Meanwhile, we observe phenomena that the state-of-the-art short text topic modeling methods can not consistently outperform traditional topic modeling methods, with some room for improvement.

Table 2. The performances of different topic modeling methods in the test dataset.

Model	Precision	Recall	F_1-score	Accuracy
LSA	0.509	0.472	0.490	0.472
PLSA	0.489	0.491	0.490	0.491
LDA	0.330	0.349	0.339	0.349
NMF	0.523	0.472	0.496	0.472
RLSI	0.564	0.547	0.555	0.547
T-LDA	0.356	0.415	0.383	0.415
BTM	0.571	0.528	0.549	0.528
DDTM	**0.599**	**0.566**	**0.582**	**0.566**

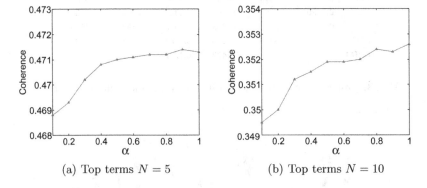

(a) Top terms $N = 5$ (b) Top terms $N = 10$

Fig. 5. Effect of parameter α with $K = 100$.

Document Classification. As discussed above, we carefully choose produced consistent results across the three topic spaces evaluated with human annotators. We then use the learned document representations as feature for document classification with Liblinear[2] as classifier. Table 2 shows the performance of the comparison methods using 10-fold cross validation. The experimental results demonstrate that the proposed method can achieve better performance than all other topic modeling methods. Meanwhile, the demonstration that BTM slightly below the performance of RLSI in terms of F_1-score. A reasonable explanation is that RLSI using sparsity topic regularization norm can better capture a useful bit of information, and thus is more suitable for inferring topics of short texts than BTM. Besides, LDA and T-LDA showed bad performance on document classification task even lower about 15 points compared to LSA. The phenomenon can be explained as that word co-occurrence patterns based matrix factorization topic modeling approaches generally outperform probabilistic approaches.

Effect of Parameter α. We set the parameter α in our proposed DDTM model as $\alpha = 0.5$. We also experimentally verify how the parameter affects the model

---

[2] http://www.csie.ntu.edu.tw/cjlin/liblinear/.

performance. The parameter α is set in range of $[0.1, 1]$ for our proposed DDTM model. The experimental results presented in Fig. 5 show that the model is not very sensitive with α since the performances of DDTM with different values for α are subtle changes. Therefore, considering the calculation effect and time efficiency, we choose $\alpha = 0.5$ as the best value in our experiments.

5   Conclusion and Future Work

In this study, we put forward a domain dictionary based social text topic modeling method, which is a weakly-supervised topic learning method. The proposed model is constructed on the basis of the assumption of terms semantic similarity with both domain dictionary reference matrix and learned latent topic matrix. We construct extensive experiments to validate our proposed models. The experimental results reveal that the proposed method can outperform conventional topic modeling methods and state-of-the-art short text topic modeling methods.

There are several interesting directions to further extend this work. For example, we will explore incorporating the information of user feedback into our topic model for further research. Here, we only define a broad level categories without hierarchical semantics of topics, so we will rebuild a hierarchy topic taxonomy on the short texts which is another challenging but interesting problem.

Acknowledgments. This work was supported by National Key Technology R&D Program(No. 2012BAH46B03), and the Strategic Leading Science and Technology Projects of Chinese Academy of Sciences(No. XDA06030200).

References

1. Andrzejewski, D., Zhu, X., Craven, M.: Incorporating domain knowledge into topic modeling via dirichlet forest priors. In: Proceedings of the 26th Annual International Conference on Machine Learning, pp. 25–32. ACM (2009)
2. Balasubramanyan, R., Cohen, W.W.: Regularization of latent variable models to obtain sparsity. In: SDM, pp. 414–422. SIAM (2013)
3. Blei, D.M., Ng, A.Y., Jordan, M.I.: Latent dirichlet allocation. J. Mach. Learn. Res. **3**, 993–1022 (2003)
4. Boyd-Graber, J.L., Blei, D.M.: Syntactic topic models. In: Advances in Neural Information Processing Systems, pp. 185–192 (2009)
5. Basave, A.E.C. He, Y., Xu, R.: Automatic labelling of topic models learned from twitter by summarisation. Association for Computational Linguistics (ACL) (2014)
6. Cheng, X., Yan, X., Lan, Y., Guo, J.: Btm: topic modeling over short texts. IEEE Trans. Knowl. Data Eng. **26**(12), 2928–2941 (2014)
7. Deerwester, S.C., Dumais, S.T., Landauer, T.K., Furnas, G.W., Harshman, R.A.: Indexing by latent semantic analysis. JAsIs **41**(6), 391–407 (1990)
8. Dredze, M., Wallach, H.M., Puller, D., Pereira, F.: Generating summary keywords for emails using topics. In: Proceedings of the 13th International Conference on Intelligent User Interfaces, pp. 199–206. ACM (2008)

9. Eisenstein, J., Ahmed, A., Xing, E.P.: Sparse additive generative models of text (2011)
10. Friedman, J., Hastie, T., Tibshirani, R.: Regularization paths for generalized linear models via coordinate descent. J. Stat. Softw. **33**(1), 1 (2010)
11. Hofmann, T.: Probabilistic latent semantic indexing. In: Proceedings of the 22nd Annual International ACM SIGIR Conference on Research and Development in Information Retrieval, pp. 50–57. ACM (1999)
12. Yuening, H., Boyd-Graber, J., Satinoff, B., Smith, A.: Interactive topic modeling. Mach. Learn. **95**(3), 423–469 (2014)
13. Hulpus, I., Hayes, C., Karnstedt, M., Greene, D.: Unsupervised graph-based topic labelling using dbpedia. In: Proceedings of the Sixth ACM International Conference on Web Search and Data Mining, pp. 465–474. ACM (2013)
14. Jagarlamudi, J., Daumé III, H., Udupa, R.: Incorporating lexical priors into topic models. In: Proceedings of the 13th Conference of the European Chapter of the Association for Computational Linguistics, pp. 204–213. Association for Computational Linguistics (2012)
15. Jin, O., Liu, N.N., Zhao, K., Yu, Y., Yang, Q.: Transferring topical knowledge from auxiliary long texts for short text clustering. In: Proceedings of the 20th ACM International Conference on Information and Knowledge Management, pp. 775–784. ACM (2011)
16. Lau, J.H., Grieser, K., Newman, D., Baldwin, T.: Automatic labelling of topic models. In: Proceedings of the 49th Annual Meeting of the Association for Computational Linguistics: Human Language Technologies, vol. 1, pp. 1536–1545. Association for Computational Linguistics (2011)
17. Lau, J.H., Newman, D., Baldwin, T.: Machine reading tea leaves: automatically evaluating topic coherence and topic model quality. In: Proceedings of the Association for Computational Linguistics, pp. 530–539 (2014)
18. Lee, D.D., Seung, H.S.: Learning the parts of objects by non-negative matrix factorization. Nature **401**(6755), 788–791 (1999)
19. Lee, D.D., Seung, H.S.: Algorithms for non-negative matrix factorization. In: Advances in Neural Information Processing Systems, pp. 556–562 (2001)
20. Mei, Q., Shen, X., Zhai, C.: Automatic labeling of multinomial topic models. In: Proceedings of the 13th ACM SIGKDD International Conference on Knowledge Discovery and Data Mining, pp. 490–499. ACM (2007)
21. Paul, M.J., Dredze, M.: You are what you tweet: analyzing twitter for public health. In: ICWSM, pp. 265–272 (2011)
22. Phan, X.-H., Nguyen, L.-M., Horiguchi, S.: Learning to classify short and sparse text & web with hidden topics from large-scale data collections. In: Proceedings of the 17th International Conference on World Wide Web, pp. 91–100. ACM (2008)
23. Quercia, D., Askham, H., Crowcroft, J.: Tweetlda: supervised topic classification and link prediction in twitter. In: Proceedings of the 4th Annual ACM Web Science Conference, pp. 247–250. ACM (2012)
24. Ramage, D., Dumais, S.T., Liebling, D.J.: Characterizing microblogs with topic models. In: ICWSM, vol. 10, p. 1 (2010)
25. Röder, M., Both, A., Hinneburg, A.: Exploring the space of topic coherence measures. In: Proceedings of the Eighth ACM International Conference on Web Search and Data Mining, pp. 399–408. ACM (2015)
26. Sriram, B., Fuhry, D., Demir, E., Ferhatosmanoglu, H., Demirbas, M.: Short text classification in twitter to improve information filtering. In: Proceedings of the 33rd International ACM SIGIR Conference on Research and Development in Information Retrieval, pp. 841–842. ACM (2010)

27. Wang, C., Blei, D.M.: Decoupling sparsity and smoothness in the discrete hier-archicaldirichlet process. In: Advances in Neural Information Processing Systems, pp. 1982–1989 (2009)
28. Wang, D., Li, T., Zhu, S., Ding, C.: Multi-document summarization via sentence-level semantic analysis and symmetric matrix factorization. In: Proceedings of the 31st Annual International ACM SIGIR Conference on Research and Development in Information Retrieval, pp. 307–314. ACM (2008)
29. Wang, Q., Jun, X., Li, H., Craswell, N.: Regularized latent semantic indexing: a new approach to large-scale topic modeling. ACM Trans. Inf. Syst. (TOIS) 31(1), 5 (2013)
30. Wang, X., McCallum, A.: Topics over time: a non-markov continuous-time model of topicaltrends. In: Proceedings of the 12th ACM SIGKDD International Confer-enceon Knowledge Discovery and Data Mining, pp. 424–433. ACM (2006)
31. Williamson, S., Wang, C., Heller, K.A., Blei, D.M.: The ibp compound dirichlet process and its application to focused topic modeling. In: Proceedings of the 27th International Conference on Machine Learning (ICML 2010), pp. 1151–1158 (2010)
32. Wu, Y., Wu, W., Li, Z., Zhou, M.: Mining query subtopics from questions in community question answering. In: Twenty-Ninth AAAI Conference on Artificial Intelligence (2015)
33. Yan, X., Guo, J., Liu, S., Cheng, X., Wang, Y.: Learning topics in short texts by non-negative matrix factorization on term correlation matrix. In: Proceedings of the SIAM International Conference on Data Mining (2013)
34. Yang, S.-H., Kolcz, A., Schlaikjer, A., Gupta, P.: Large-scale high-precision topic modeling on twitter. In: Proceedings of the 20th ACM SIGKDD International Conference on Knowledge Discovery and Data Mining, pp. 1907–1916. ACM (2014)
35. Zhao, W.X., Jiang, J., Weng, J., He, J., Lim, E.-P., Yan, H., Li, X.: Comparing twitter and traditional media using topic models. In: Clough, P., Foley, C., Gurrin, C., Jones, G.J.F., Kraaij, W., Lee, H., Mudoch, V. (eds.) ECIR 2011. LNCS, vol. 6611, pp. 338–349. Springer, Heidelberg (2011). doi:10.1007/978-3-642-20161-5_34
36. Zhu, S., Yu, K., Chi, Y., Gong, Y.: Combining content and link for classification using matrix factorization. In: Proceedings of the 30th Annual International ACM SIGIR Conference on Research and Development in Information Retrieval, pp. 487–494. ACM (2007)

Towards an Impact-Driven Quality Control Model for Imbalanced Crowdsourcing Tasks

Kinda El Maarry[✉] and Wolf-Tilo Balke

IFIS, TU Braunschweig, Brunswick, Germany
{elmaarry,balke}@ifis.cs.tu-bs.de

Abstract. Crowdsourcing have been gaining increasing popularity as a highly distributed digital solution that surpasses both borders and time-zones. Moreover, it extends economic opportunities to developing countries, thus answering the call of impact sourcing in alleviating the welfare of poor labor in need. Nevertheless, it is constantly criticized for the associated quality problems and risks. Attempting to mitigate these risks, a rich body of research has been dedicated to design countermeasures against free riders and spammers, who compromise the overall quality of the results, and whose undetected presence ruins the financial prospects for other honest workers. Such quality risks materialize even more severely with imbalanced crowdsourcing tasks. In fact, while surveying this literature, a common rule of thumb can be indeed derived: the easier it is to cheat the system and go undetected, the more restrictive and across-the-board discriminating countermeasures are taken. Hence, also honest yet low-skilled workers will be placed on par with spammers, and consequently exposed and deprived of much-needed earnings. Therefore in this paper, we argue for an impact-driven quality control model, which fulfills the impact-sourcing vision, thus materializing the social responsibility aspect of crowdsourcing, while ensuring high quality results.

Keywords: Crowdsourcing · Impact sourcing · Quality control · Fraudulent workers · Fraud detection

1 Introduction

Crowdsourcing has evolved as a method to tap into an unprecedented international agile work force that is only a key stroke away. Via crowd sourcing platforms like: Amazon's Mechanical Turk, CrowdFlower, or Samasource, companies can hire a readily available workforce at any time. Generally, such hiring seeks intelligent information processing skills for numerous tasks, ranging from content annotation [1], information extraction [2], to more complex tasks like sentiment analysis [3] and crowd-enabled database retrieval [4]. Unfortunately, there are no free lunches: The very flexible and agile nature of crowdsourcing's virtual workspace presents free-riders with an open invitation for quick monetary gains. If undetected, a random-answering mechanism can simply cheat the system and may be indeed very profitable. Of course such cheating cannot be tolerated. It does not only mean lost money for the task provider, but on top of that, it exposes:

© Springer International Publishing AG 2016
W. Cellary et al. (Eds.): WISE 2016, Part I, LNCS 10041, pp. 124–139, 2016.
DOI: 10.1007/978-3-319-48740-3_9

(1) The entire task (and thus the provider's services built on this task) to *severe quality problems*. Accordingly, a rich body of research has focused on devising automatic, yet light-weight countermeasures to ensure a satisfying result quality. Among the more established countermeasures are *gold questions*, where questions with known answers are blended into the workload and used as a quality threshold test. That is, if a worker fails to correctly answer for example 30 % of the gold questions, he/she is excluded from the crowdsourcing task. *Majority voting*, where quality is attained through the aggregation of different workers' submissions. And finally, *reputation-based systems*, where a worker's quality is incrementally computed as a function of past submissions' feedback. Yet, all of these countermeasures have their fair share of drawbacks (see Related work).

(2) The *honest workers* in the workforce, some of whom are in dire need of every possible financial prospect, to *being excluded from the labor-pool*. Actually, this is more severe than it sounds, since the social aspect of the crowd sourcing solution can be immense, with 1.8 billion people unable to access a formal job and half of the world's population living on less than $2.50 a day[1]. In acknowledgment of this significant impact, many crowdsourcing platforms have been founded with the sole mission of supporting the underprivileged, as well as connecting them to the global economy e.g. Samasource, RuralShores, or ImpactHub. Moreover, some socially-responsible quality control measures were developed to identify biased or low-skilled honest workers, such as: an algorithm separating unrecoverable error rates from recoverable bias [5], adaptive gold questions, which we presented in [6], that adapts the difficulty of the tasks to the detected skills of the worker, and our Rasch-model-based framework, which mines the underlying workers' answer-patterns for irregularities implying fraud [7].

Building a robust and widely applicable quality control measure becomes even more complicated, when answer sets for crowdsourcing tasks are inherently *skewed,* where the attribute value distributions in the underlying data sources are skewed [8]. This is often the case with web data, where Zipf distributions are quite common (e.g., Web data annotation or taxonomic metadata generation). Given such an imbalanced task, strategic spammers can easily exploit the inherent skewness to avoid detection. The simple, yet effective idea of such an attack model is to get highly accurate results by always submitting the frequent class label (see Example 1).

Example 1 – Nepal Aerial Clicker (see Footnote 1) (Aerial Image Analysis). Images taken by UAVs (unmanned aerial vehicles) are increasingly being sent to the crowd for disaster and damage assessment. One particular project: Nepal Aerial Clicker assessed the population number after the Gorkha earthquake in April 2015. Crowd workers were assigned to count the number of persons they see in the picture (see Fig. 1). As could be expected, pictures taken in remote regions tended to be empty, and rarely featured anyone. Assuming that only 15 % of the pictures comprised people, a strategic spammer always submitting a population count of 0, would have an average accuracy of 85 %.

---

[1] http://clickers.micromappers.org/.

Fig. 1. Aerial image analysis

In imbalanced crowdsourcing scenarios, all of the aforementioned countermeasures that were developed for quality control on crowdsourcing platforms – whether tailored to fend of spammers, or socially-tailored to identify biased or low-skilled honest workers and support them – fail to identify strategic spammers since they are outwardly doing a very good job. In fact, the question of finding strategic spammers have been raised in Ipeirotis's work [5, 9] (see related work).

In previous work [1], we conducted a theoretical analysis of medical test theory and its classical measures (see related work): sensitivity, *specificity* and *positive predictive values* for handling imbalanced tasks. Based on the surprising insights of this analysis, it became evident that a partially redundant-based quality control model, need only hire a second worker for a given task, when a first worker's submission for that task is a frequent class label. Moreover, a less frequent class label is always taken to be conclusive. In other words, in Example 1, if the first worker annotates the picture with zero count, only then, is a second worker given the same task. If both workers agree on the frequent class label (i.e. zero count), it becomes the picture's final label. Otherwise, the other less frequent class label submission will be taken as the picture's final label. While this technique proved to be robust with imbalanced tasks, like many countermeasures, it makes no distinction between strategic spammers and honest worker. In fact, strategic spammers' submissions are still accepted and recognized as valid contributions. Moreover, it's discriminative against honest workers, since the core assumption is that all errors committed by the workers are random, irrespective of the underlying difficulty of the task. Such a statistical independence assumption does not hold in reality however, since an honest worker's error rate will always be higher for more difficult tasks.

Accordingly, in this paper, we present an impact-driven model, which ensures high quality results within imbalanced tasks, while promoting honest workers, who might often still show lower accuracy rates than strategic spammers, due to the inherent skewness of the task. To that end, in combination with the partially redundant technique, we introduce a skewly-biased reputation score that gives a higher weight to less frequent class label submissions and balances the inherent skewness of the task, such that the strategic spammers' accuracy rate is actively diminished and that of the honest workers' amplified. These score are additionally used to adapt the assignment of easier tasks to low-skilled honest workers, thus supporting them in developing their skills while at the same time ensuring higher quality results. Our evaluation on both synthetic and real world datasets attest to both the cost and quality efficiency of our model.

2   Related Work

In this section, we provide an overview over the current automatic quality control measures used in crowdsourcing, as well as the classical measures of medical test theory, upon which our partially redundant-based technique is based on.

Crowdsourcing and its Quality Control Measures. A lot of studies and effort has been put into devising countermeasures against spammers, in order to control the resulting quality. These different measures can be roughly categorized based on how they approach the problem of acquiring and/or maintaining high quality levels as follows:

Pessimistic approaches focus on identifying as many spammers in the workforce as possible and subsequently exclude them. Thus, only honest workers are retained. A typical method in this approach are *gold questions,* which are added to the crowd-sourcing tasks. When used as a control measure, they are covertly added to the task to catch the spammers off guard. Upon failing to correctly answer a certain percentage of these questions, the worker is declared to be fraudulent (or at least unfit) and is accordingly discarded from the workforce. Thus, while result qualities are usually good, a severe drawback is the immediate exclusion of less-skilled workers, who could still perform easier tasks [10]. Moreover, gold questions are only valid for factual tasks and can't be employed e.g., in opinion-based tasks or individual perceptions and sentiments. In fact, their applicability has been doubted in typical crowdsourcing tasks, see e.g. [2].

Optimistic approaches shift the focus from individual workers to the submitted responses and aim at aggregating the results, such that the final output exhibits a high overall quality. The family of aggregation methods is rather extensive, yet the most commonly used method is *majority voting.* Basically, majority voting exploits the low-cost nature of crowdsourcing and assigns the same task to multiple workers in parallel, thus following the notion of the *wisdom of the crowd* to defeat the individual fraudulent misuse. Other weighted aggregation methods in the literature include the expectation maximization (EM) algorithm [10], which considers a submission's quality based on the corresponding worker and ultimately utilizes this computed quality for a weighted aggregation. However, EM would fail in imbalanced tasks, due to its inability to identify strategic spammers submitting only frequent class labels [5]. Other approaches that consider workers' error rate include a Bayesian version of the EM algorithm [11] and the probabilistic approach in [12], which considers both the worker's skill and the task's difficulty. Unlike Gold questions, such aggregation methods can be applied to a wider class of tasks, including opinion-based tasks, where correctness depends on consensual agreement. Still, this redundancy approach incurs more costs and has its limitations, especially when the percentage of spammers in the workforce is high, see e.g. [13].

Feedback-based approaches combine optimistic and pessimistic traits. One good example is *reputation-based systems,* whether based on a reputation model [10], or on deterministic approaches [15]. These systems often use a worker's reputation or skill score as a measure of confidence about the individual result correctness and only exclude fraudulent workers based on reputation scores, which are computed over longer time scales. Usually the workers' scores are determined in a complex process as a function of the customers' satisfaction & feedback [16], task completion rates, etc. In [17], they proposed a model, CDAS, which computes an estimated accuracy for each generated result based on the worker's historical performances. Reputation systems can be however exploited [18], and suffer from both the cold start problem [19] and the challenge of computing robust, yet reliable aggregated reputation scores for workers.

Finally the *Incentive-based* approaches covers the various motivational incentives that are used to encourage workers, or rather discourage them from cheating the system. There are basically two types of incentives: intrinsic and extrinsic [20]. An *intrinsic* motivation is inspired from the underlying task itself, where the task itself appears important and personally drives the workers to carry it out without expecting any return, e.g. a crowd-driven project by Tomnod[2] searching for the missing Malaysian airlines flight 370 that had disappeared en route. An *extrinsic* motivation materializes as a rewarding mechanism like: monetary compensations, status, recognition, etc. In fact, monetary incentives do not always help. According to a study that was conducted to see how monetary compensations can be manipulated to motivate workers, low paid jobs yielded sloppy work, while high paid jobs attracted spammers, which proved such a mechanism tricky to put into effect [21].

Our model falls under both the optimistic and the feedback-based approaches, as it combines a partially redundant-based technique with a skewly-biased reputation score. All the traditional feedback-based approach would always compute a higher reputation score for strategic spammers, since they are in practice doing a good job. Accordingly, we evaluate our model against the optimistic approaches family, more specifically, the front-runner approach: Majority voting. Moreover, our model promotes a dynamic assignment of tasks with the goal of supporting the social aspect of the crowd sourcing solution and leveraging the low-skilled honest workers. Such an online assignment strategy isn't novel and has been investigated by many others, though their ultimate goal differs: given a limited budget and questions with varying difficulty levels, how many times should a question be assigned to the crowd to ensure high quality results? This issue is rather interesting since easy questions require fewer assignments than more difficult questions. The number of assignment would also differ based on the skills of the given worker. Prominent strategies include CDAS [17] and AskIt! [22].

Identifying Strategic spammers. Ipeirotis et al. [5, 9] raised the question of identifying strategic spammers, whose attack model is based on identifying the prevalent class with the highest class prior and always submitting that frequent label. The proposed algorithm focused on separating the unrecoverable error rates from those recoverable biases of workers by means of generating scalar scores for workers. Such a score would be then used as a worker's quality metric. In the paper, they investigated the following case:

---

[2] http://www.tomnod.com/.

Example 2: (Adult website classifier). Assume two workers who are assigned to label websites into two classes: porn and not porn, upon which an adult classifier can be later trained on. For a skewed class distribution with 95 % of the websites being not porn and only 5 % being porn, a strategic spammer classifying all the website as not-porn would have a low error-rate of 5 %.

Although, the proposed algorithm considers the uncertainty associated with the strategic spammers' prevalent answers, it only performs well with approximately 5 labels per questions and for workers with histories of 20 or 30 labels. In fact, [5] reports that obtaining fewer labels per question would dramatically decrease the resulting quality, and that having just a few strategic spammers would also be completely detrimental to the resulting quality. Our model requires 2 labels at most and can handle the existence of many strategic spammers without significant quality degradation.

Medical Test Theory's Classical Measures. Since many of the crowdsourcing tasks rely on binary decisions (e.g., recognizing whether a certain entity is contained in some image or classifying documents or articles), the classical measures from medical test theory, in particular, sensitivity, specificity and prevalence [23] apply. Sensitivity and Specificity are used to measure the performance of a binary classification test (e.g., is a patient diagnosed as HIV positive?), reflecting the true positive rate and the true negative rate respectively (see Definition 1). The prevalence on the other hand, defines the percentage of population shown to have the tested classification condition (i.e. percentage of people found to actually have a disease). Moreover, the usefulness of a test [24] can be measured by the positive predictive value $ppv(T)$: the probability that a positive answer (i.e. is HIV positive?) is indeed correct (see Definition 5). When mapping these measure from medical test theory to crowdsourcing platforms, the workers become the tests, whose usefulness and performance needs to be checked and boosted.

Definition 1 – Sensitivity, Specificity and Positive predictive value: the positive predictive value $ppv(T)$, sensitivity $\sigma(T)$, and specificity $\tau(T)$ for some test T is defined as follows:

$$ppv(T) := \frac{1}{1 + \frac{1 - \tau}{p}}, \ for \text{close to 1 and p close to 0}$$

$$\sigma(T) := \frac{|\text{true positives}|}{|\text{positively tested}|}$$

Where *positively tested*, refers to those being predicted as positive

$$\tau(T) := \frac{|\text{true negatives}|}{|\text{negatively tested}|}$$

Where *negatively tested*, refers to those being predicated as negative.

3   Higher Quality in Imbalanced Tasks

As of now, all current countermeasures employed within crowdsourcing platforms would fail when dealing with imbalanced tasks, where a spammer with the strategy of always submitting the frequent class label, will always seem to be doing a very good job. In previous work [8], we demonstrated how a partially redundant-based technique can be built for imbalanced tasks, where only a maximum of two workers per task is required. Next, we explain why this technique would be a good basis for an impact-driven model and investigate the weaknesses, which needs to be addressed.

A Partially Redundant-based Technique. Typically, redundancy-based approaches utilize aggregation as a quality control measure e.g. Majority voting. Here, a minimum of three submissions from three different workers is required. Yet of course, a majority of wrongs, doesn't make it right. Particularly, for imbalanced tasks, the underlying skewness should be taken into account and a straight forward majority count should and can be indeed avoided at increasing quality and decreasing incurred costs as will be shown next.

A closer examination at medical test theory's measures: sensitivity σ, specificity τ and prevalence p (see Related work) revealed that solely focusing on the specificity measure would improve a test's performance by enhancing the positive predictive value $ppv(T)$ i.e. decreasing the number of false positives. In other words, for a binary classification task, where the negative class label is the frequent class, both false positives and false negatives constitute the quality risk (see Fig. 2). But due to the skewness in the class labels, the resulting prevalence's imbalance adds in most of the weight and the mass of quality risk within the false positive branch (frequent class label), leaving the false negative branch (i.e. less frequent class label) unworthy to focus on.

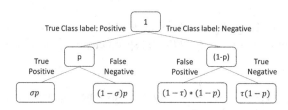

Fig. 2. Testing error diagram with prevalence p, sensitivity σ, and specificity τ

Accordingly, we focus on the frequent class label submissions. That is, a second worker is hired for the same task, only when the first worker's submitted label is the frequent class label. Eventually, the final class label is assigned as per the following rules of aggregation given in Definition 2:

Definition 2 – Rules of Aggregation: given a crowdsourcing task t, where $t \in T$, such that T is an imbalanced label annotation crowdsourcing task, also known as a human intelligence task (HIT), with the negative class N being the frequent class label and the positive class P being the less frequent label. Let $w_1(t) = N$, denote that the first worker has assigned a negative label N for crowdsourcing task t. Similarly, $w_2(t) = P$ denotes that the second worker assigned a positive label P for the same crowdsourcing task t. Then the final label $L(t)$ to be assigned to task t is aggregated as follows:

(a) $w_1(t) = P$, *then* $L(t) = P$
(b) $w_1(t) = N \wedge w_2(t) = P$, *then* $L(t) = P$
(c) $w_1(t) = N \wedge w_2(t) = N$, *then* $L(t) = N$

Note that, for $w_1(t) = P$, no second worker was sought out, because the first worker submitted the less frequent class label. Extensive evaluation has proven this partially redundant-base technique to be more cost and time efficient than its traditional redundancy-based quality control peer, where three submissions are needed, rather than just one or two submissions. Moreover, the partially redundant-based technique outperforms Majority voting in terms of resulting quality. Our experiments [8] showed consistent quality gains, even when the percentage of strategic spammers increases.

Towards an impartial quality-control model. Indeed, the partially redundant-based technique enhances the final resulting quality by following the defined rules of aggregation (see Definition 2), which respects our analytical findings from medical test theory. In practice however, it doesn't distinguish between strategic spammers and honest workers. Strategic spammers' submissions are in fact still accepted. Additionally, the assumption that all errors committed by the workers are random shows prejudice against low-skilled honest workers, where an honest worker's error rate will always depend on the underlying difficulty of a task and shouldn't be assessed in isolation. That is, if worker w_1 fails in answering task t, also the average error probability of an honest worker w_2 for task t increases. An impact-driven model would ideally distinguish difficult tasks, and assign it to more skilled workers, which shields less-skilled workers from incurred error-rates, which would negatively reflect in their reputation score on the crowdsourcing platforms. For the partially redundant technique, the consequences of not adopting this assumption can be illustrated as follows:

Example 3: Given a crowd of honest workers W, the error probability of a second worker $w_2 \in W$ to be assigned to the same task t, to which a previous worker $w_1 \in W$ failed to answer, increases.

In order to see how $ppv(T)$ is affected when seeking a second opinion, we use the tree diagram in Fig. 3. Assume $p = 10^{-3}$ and start with a population $N = 10^7$. The combined positive predictive value $ppv_{1,2}$ can be computed accordingly,

$$ppv_{1,2} = \frac{9810}{9810 + 9990} = 0.495$$

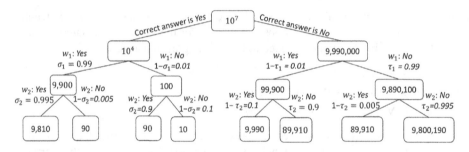

Fig. 3. Tree Diagram illustrating the error probability of a second worker

As illustrated in Fig. 3, the combined positive predictive value $ppv_{1,2}$ (i.e. after seeking a second worker submission for the same task) improves when compared to $ppv_1 = 0.09$. However, the rate of improvement will be lesser without the assumption of statistical independence of questions, since the dependency would lead to more false positives. That is, as soon as worker w_1 produces a false positive, which might be due to the question's difficulty, worker w_2 will with a higher probability reproduce the same mistake. And the higher the number of false positives, the lower the combined positive predictive value $ppv_{1,2}$ will be. At which point, not only does the partially redundant technique fail to distinguish between honest workers and spammers but also has its own limitation in terms of the final resulting quality.

4   An Impact-Driven Model

There are two central design aspects for building an impact-driven model. First, it has to ensure high quality. Second, it has to fulfill the social responsibility of crowdsourcing, and show biasness to honest workers, especially the low-skilled ones. This can be tricky, since low-skilled workers' contributions would arguably lower the final quality, unless the assignment of difficult tasks for these workers is avoided. Next we show how our impact-driven model satisfies both aspects.

Ensuring High Quality.   To ensure high quality for imbalanced tasks, while taking into consideration the consequences of varying task difficulties, the combined positive predictive value has to be improved i.e. decrease the false positives. Accordingly, more difficult tasks has to be assigned to honest workers who are also more skilled. Therefore, solely relying on an aggregation technique doesn't cut the mustard, since it cannot distinguish between spammers and honest workers. Moreover, it cannot distinguish between highly-skilled honest workers, who should be assigned to more difficult questions, and low-skilled honest workers. To that end, in combination with the partially redundant technique, we introduce a skewly-biased reputation score that gives a higher weight to less frequent class label submissions. The higher importance given to the less frequent class labels follows from our medical test theory analysis's insight: the prevalence's imbalance due to the skewness in the data adds the mass of the quality risks within the false positives branches. That is, the false negative branch becomes unworthy to focus on, and our assumption becomes that the less frequent class label will be more often than not a true positive. The skewly-biased reputation scores for an imbalanced crowdsourcing tasks, where the negative class N is the frequent class label and the positive class P is the less frequent label, can be ranked accordingly (see Definition 3).

Definition 3 – Skewly-biased Reputation scores ranking: For worker $w(t)$, let his/her reputation score $R(w)$ be incremented accordingly:

(a) if $\left(w_1(t) = P\right)$, then $R\left(w_1\right) = 3$

(b) if $\left(w_1(t) = N \wedge w_2(t) = N\right)$, then $R\left(w_1\right) = R\left(w_2\right) = 2$

(c) if $\left(w_1(t) = N \wedge w_2(t) = P\right)$, then $R\left(w_1\right) = 1$ and $R\left(w_2\right) = 3$

The highest reputation score is assigned to a less frequent class label submission. The next highest reputation score is assigned to a unanimous agreement on a frequent class label. And lastly, the lowest reputation score is given to a worker submitting the frequent class label, yet is contested by a second worker submitting the less frequent class label. As will be illustrated later, this last score is negative and thus acts as a penalty.

In order to balance the inherent skewness of the task, such that the strategic spammers' accuracy rate is actively diminished and that of the honest workers' amplified, we compute the skewly–biased reputation scores in a reverse fashion with respect to the underlying skewness (see Definition 4). That is, the upper bound of a strategic spammer's reputation score would be equal to the percentage of the less frequent class. On the other hand, an honest worker, annotating all of the less frequent class labels correctly, would get a reputation score equivalent to the frequent class percentage. Though we always assume the less frequent class label to be often a true positive, nevertheless, the final label is determined by the more reputable worker's submission i.e. the worker having the higher reputation score. It's important to note that over time, these reputation scores will indeed stabilize and strategic spammers will have lower reputation scores than that of the honest workers.

Definition 4 – Computing Skewly-biased Reputation scores: given a crowdsourcing task t, where $t \in T$, such that T is a skewed-label binary annotation data and $\|T\|$ is the total number of classification problems in your data. Let the percentage of the less frequent class P labels be a and that of the frequent class N be $(1 - a)$. Then, the reputation score R is computed accordingly

(a) Each unanimously frequent class label submission,

$$R = a * 100/(1 - a) * \|T\|$$

(b) For each less frequent class label submission,

$$R = (1 - a) * 100/a * \|T\|$$

(c) For each split decision on a less frequent class submission, the worker with the frequent class submission,

$$R = -((1 - a) * 100/a * \|T\|)$$

Favoring Honest Workers. The second central design aspect that our model should fulfill is supporting the vision of impact sourcing. This is already materialized to some extent with the skewly-biased reputation score, since it distinguishes between spammers and honest workers. In order to support low-skilled honest workers, an adaptive assignment of easier tasks would not only ensure higher quality results, but would shield the low-skilled honest workers from definite error rates, which would negatively reflect on their reputation scores, thus hindering them from higher paid tasks and opportunities to further develop their skills. To that end, we use the already exiting reputation scores for this adaptive assignment as shown in Definition 5. The final label is then determined by the more reputable worker's submission i.e. the worker having the higher reputation score.

Definition 5 – Adaptive Assignment of Tasks: given a crowdsourcing task t, where $t \in T$, such that T is a skewed-label binary annotation data. Again, let the percentage of the less frequent class label P be the less frequent class label and N the frequent class label. Given a crowd of honest workers W, let $w_1 \in W$, with reputation score $R(w_1)$. Assume $w_1(t) = N$ submitted the frequent class label for task t. Then, task t is assigned to a second worker, $w_2 \in W$, such that

$$R(w_2) > R(w_1)$$

5 Evaluation

In this section, we conduct a comprehensive set of experiments to evaluate the efficiency of our impact-driven model in terms of cost and quality, where quality refers to the correctness accuracy rate. Our model falls under both the feedback and optimistic approaches. For the feedback approach, none of the reputation systems would work well, since the strategic spammers' attack model would amount to them getting higher reputation scores. Accordingly, we focus on comparing our model to the optimistic approach's front-runner: Majority voting. Worthy to consider as well is Ipeirotis's algorithm (see related work) to detect strategic spammers. However, as they report in their own work [5], a minimum of 5 labels per question is required, as opposed to a maximum of 2 labels with our model. Moreover, they report that having just a few strategic spammers would immensely degrade the resulting quality, unlike our model, which can handle high percentages of strategic spammers in the workforce, as will be shown next.

Since we need to induce and control spammers, we simulate both the crowdsourcing tasks and the crowd. Only through simulation can we: (1) correctly identify spammers (in the sense of ground truth), (2) control the error rates of the honest workers, (3) adapt the percentage of spammers in the crowd, (4) control the skewness percentage Š of the crowdsourcing tasks, and (5) control the attack model of the spammers. Such a flexible and deterministic parameter control allows us to extensively experiment with varying parameter combinations. For our experiments, we used two data sets:

1. A real world dataset: Zentralblatt MATH[3] Corpus (ZB), which is one of the biggest mathematical digital libraries. Figure 4 illustrates the Zipfian distribution of the class labels in the corpus. The dataset is made up of 2,679,550 publications (journals, books, conference proceedings, etc.), each annotated according to the Mathematics Subject Classification (MSC) taxonomy. Our crowdsourcing tasks are based on a subset of the dataset – 247,063 publications, which comprises the following classes to be later classified through the crowdsourcing tasks: 208,754 Computer science publications, 30,388 History and Biography publications, 4,312 Geophysics publications, and 3609 Astronomy and Astrophysics publications. Accordingly, we end up at a skewness percentage $\check{S} = 15\%$, i.e. 85 % of the total dataset belongs to the Computer science class label. The crowdsourcing tasks are designed to reflect the same imbalance.

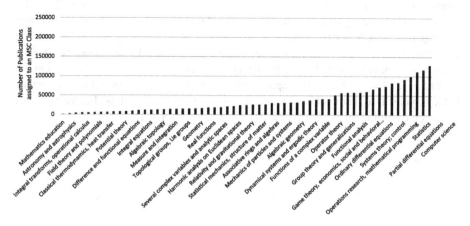

Fig. 4. Number of publications per MSC level 1 category label

2. A synthetic dataset for a binary classification task, where 1 is the frequent class label and 0 the less frequent label. The dataset's skewness percentage $\check{S}$ can be altered as needed. The experimental results are an aggregation over 1000 binary classification problems, which is given to a workforce comprising 1000 workers.

For both datasets, a strategic spammer's strategy is always to submit the frequent class label, which would strategically only add up to 15 % error rate for the ZB Corpus. On the other hand, an honest work's error rate spans over the less frequent class.i.e. the error rate of an honest worker has an upper bound corresponding to the skewness of the less frequent class(es), that is, $\check{S}$. The error rate of the honest workers are randomly assigned, such that the error rates have a lower bound based on an input error rate parameter that we assign per experiment. The HITs are designed to comprise 20 binary classification problems. To avoid the cold start problem, which is always associated with reputation-based systems, each worker is preassigned a total of 10 HITs, upon which their reputation scores are computed.

---

[3] http://www.zentralblatt-math.org/zbmath/.

Accuracy under varying Parameters. Next, we illustrate how different external parameters affect the resulting correctness quality of the impact-driven model and compare it to that of the majority voting.

1. *Impact of percentage of Spammers*

In Figs. 5 and 6, we illustrate the impact of varying percentage of spammers in the workforce on the quality for each of: the impact-driven model and majority voting. Regardless of the percentage of skewness Š set, the impact-driven model beats majority voting with a margin; As the percentage of spammers increases, the quality decreases. The lower bound of the quality corresponds to the complement of Š. That is, when Š = 15 %, the worst quality attained when the entire workforce is made up of spammers equals to 85 %, which adheres to the spammer's strategy of always giving the high frequent answer, in order to exhibit only 15 % error rate. Similarly, for Š = 35 %, the lowest quality resides at 65 %. The bigger the skewness the better the impact-driven model performs at lower percentage of spammers, fitting perfectly for highly skewed answer sets. At Š = 35 % and 20 % spammers, the impact-driven model achieves a correctness quality of 94.22 %, in contrast 88.15 % achieved by majority voting. On the other hand, at Š = 15 % and 20 % spammers, it achieves a lower correctness quality of 92.68 %, yet still higher than the 88.21 % achieved by majority voting. Figure 5, shows an expected similar behavior with the ZB dataset, where Š = 15 %.

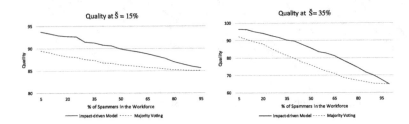

Fig. 5. Impact of the % of Spammers – Synthetic dataset

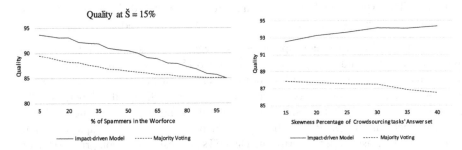

Fig. 6. Impact of the % of Spammers – ZB dataset

Fig. 7. Impact of Skewness percentage– Synthetic dataset

2. Impact of Answer sets' Skewness percentage

Figure 7 confirms the earlier findings, that is, the higher the skewness level, the higher the quality. Whereas, the impact-driven model can handle higher skewness levels, majority voting's performance degrades. For our model, at $Š = 20$ %, quality rises from 93.27 % to 94.38 % at $Š = 40$. As for Majority Voting, at $Š = 20$ %, quality drops from 87.73 % to 86.54 % at $Š = 40$ %. Furthermore, at majority voting's quality peaks, its correctness quality at 87.89 % is lower than 92.56 %, which corresponds to that achieved by the impact-driven model. This experiment was only ran on the synthetic dataset, since the ZB dataset has a fixed $Š = 15$ %.

3. Impact of Honest Worker's Error Rate

Next, we examine the impact of varying honest worker's error rate on the quality for both impact-driven model and majority voting. Figures 8 and 9, illustrates the effects on both the synthetic ($Š = 35$ %) and ZB ($Š = 15$ %) datasets respectively, with 20 % strategic spammers in the force. As the error rate increases, the overall quality decreases. Since we assume that the honest workers' error rate spans over the less frequent class, the acceptable lower bounds of their error rate corresponds to the size of the less frequent class in each dataset i.e. the value of $Š$ for both datasets. The quality attained through the impact-driven model is significantly better than that resulting from majority voting.

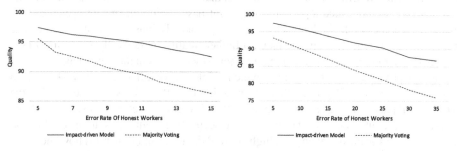

Fig. 8. Impact of Honest Worker's Error Rate– ZB dataset

Fig. 9. Impact of the Honest Workers' Error Rate– Synthetic dataset

Crowdsourcing Cost. Lastly, we investigate the incurred cost as the percentage of spammers increases. We design our HIT to comprise 20 tasks, each of which costs 0.05$ i.e. 1$ a HIT. For 5 crowdsourced tasks (100 tasks), majority voting will always incur 15$ costs, regardless of the percentage of spammers in the workforce. Our model seeks a second opinion, when the first worker annotates the task with the frequent label. Accordingly, as the percentage of spammers increase, more frequent class labels will be submitted, which consequently triggers the underlying partially-redundancy mechanism of seeking a second opinion, and increases the costs (see Figs. 10 and 11). The upper bound is reached, when the entire workforce is made up of spammers and all tasks are labeled twice, at which point, the cost lies at 10$. On the other hand, the lower bound is higher than 5$ (HIT being submitted only once), since even in the absence of spammers, frequent class annotations will be correctly submitted. For less skewed datasets (e.g. ZB dataset, $Š = 15$ %), more frequent class labels are bound to be correctly

Fig. 10. Impact of the % of Spammers on the   **Fig. 11.** Impact of the % of Spammers on the
costs – ZB dataset   costs – Synthetic dataset

submitted, accordingly a relatively higher cost of 9.7\$ at 1 % spammers is incurred as opposed to higher skewed datasets (e.g. Synthetic dataset, $\check{S} = 35$ %), which incurs a cost of 8.7\$ at 1 % spammers.

6   Conclusion

In this paper we focused on the challenges of simultaneously (1) overcoming the quality risk associated with imbalanced crowdsourcing tasks, which arises with the presence of strategic spammers, and (2) fulfilling the social responsibility of crowdsourcing by designing a non-discriminating quality model that supports low-skilled honest workers. To that end, we presented a partially redundant-based solution: Impact-driven model, which overcomes both challenges by employing a skewly-biased reputation score. The reputation score ensures, on one hand, a higher quality by correctly assigning lower scores to strategic spammers and higher to honest workers. On the other hand, it allows for an adaptive assignment of tasks, which shields low-skilled honest workers from certain error-rates, while giving them room to improve their skills, thus supporting the vision of impact sourcing. Both synthetic and real world datasets were used to evaluate our impact-driven model's cost efficiency and accuracy. Results attest to the quality and monetary gains achieved even with high skewed datasets, and illustrates how the partially redundant impact-driven model outperforms majority voting, which remains the contending favorite of the redundancy-based family measures.

References

1. Sorokin, A., Forsyth, D.: Utility data annotation with Amazon Mechanical Turk. In: Computer Society Conference on Computer Vision and Pattern Recognition Workshops (CVPRW), Anchorage, AK. IEEE (2008)
2. Lofi, C., Selke, J., Balke, W.T.: Information extraction meets crowdsourcing: a promising couple. Datenbank-Spektrum **12**(2), 109–120 (2012)
3. Kouloumpis, E., Wilson, T., Moore, J.: Twitter sentiment analysis: the good the bad and the OMG! In: International Conference on Weblogs & Social Media, Barcelona, Spain (2011)
4. Selke, J., Lofi, C., Balke, W.-T.: Pushing the boundaries of crowd-enabled databases with query-driven schema expansion. In: International Conference on Very Large Data Bases (VLDB), Istanbul, Turkey (2012)

5. Ipeirotis, P.G., Provost, F., Wang, J.: Quality management on Amazon mechanical turk. In: ACM SIGKDD on Human Computation Workshop (HCOMP), New York, USA (2010)
6. El Maarry, K., Güntzer, U., Balke, W.-T.: Realizing impact sourcing by adaptive gold questions: a socially responsible measure for workers' trustworthiness. In: International Conference on Web-Age Information Management (WAIM), Qingdao, Shandong, China (2015)
7. El Maarry, K., Balke, W.-T.: Retaining rough diamonds: towards a fairer elimination of low-skilled workers. In: International Conference on Database Systems for Advanced Applications (DASFAA), Hanoi, Vietnam (2015)
8. El Maarry, K., Güntzer, U., Balke, W.-T.: A majority of wrongs doesn't make it right. In: Conference on Web Information Systems Engineering (WISE), Miami, USA (2015)
9. Wang, J., Ipeirotis, P.G., Provost, F.: Managing crowdsourced workers. In: winter Conference on Business Intelligence, Salt Lake City, Utah, USA (2011)
10. Dawid, P., Skene, A.M.: Maximum likelihood estimation of observer error-rates using the EM algorithm. J. Royal Stat. Soc. **28**(1) (1979)
11. Raykar, V.C., Yu, S., Zhao, L.H., Valadez, G.H., Florin, C., Bogoni, L., Moy, L.P.: Learning from crowds. J. Mach. Learn. Res. **11** (2010)
12. Whitehill, J., Ruvolo, P., Wu, T., Bergsma, J., Movellan, J.: Whose vote should count more: optimal integration of labels from labelers of unknown expertise. In: Advanced Neural Information Processing Systems (NIPS), Vancouver, Canada (2009)
13. Kuncheva, L.I., Whitaker, C.J., Shipp, C.A., Duin, R.P.W.: Limits on the majority vote accuracy in classifier fusion. J. Pattern Anal. Appl. **6**(1), 22–31 (2003)
14. El Maarry, K., Balke, W.-T., Cho, H., Hwang, S., Baba, Y.: Skill ontology-based model for Quality Assurance in Crowdsourcing. In: International Conference on Database Systems for Advanced Applications (DASFAA), Uncrowd Workshop, Bali, Indonesia (2014)
15. Noorian, Z., Ulieru, M.: The state of the art in trust and reputation systems: a framework for comparison. Journal of theoretical and applied electronic commerce research **5**(2), 97–117 (2010)
16. Ignjatovic, A., Foo, N., Lee, C.T.: An analytic approach to reputation ranking of participants in online transactions. In: IEEE/WIC/ACM International Conference on Web Intelligence and Intelligent Agent Technology, Sydney, Australia (2008)
17. Liu, X., Lu, M., Ooi, B.C., Shen, Y., Wu, S., Zhang, M.: CDAS: a crowdsourcing data analytics system. VLDB Endowment **5**(10), 1040–1051 (2012)
18. Yu, B., Singh, M.P.: Detecting deception in reputation management. In: International Joint Conference on Autonomous Agents and Multiagent Systems, Melbourne, VIC, Australia (2003)
19. Daltayanni, M., de Alfaro, L., Papadimitriou, P.: WorkerRank: Using employer implicit judgements to infer worker reputation. In: ACM International Conference on Web Search and Data Mining (WSDM), Shanghai, China (2015)
20. Hossain, M.: Users' motivation to participate in online crowdsourcing platforms. In: Conference on Innovation, Management and Technology Research, Malacca, Malaysia (2012)
21. Kazai, G.: In search of quality in crowdsourcing for search engine evaluation. In: European Conference on Advances in Information Retrieval, Dublin, Ireland (2011)
22. Boim, R., Greenshpan, O., Milo, T., Novgorodov, S., Polyzotis, N., Tan, W.C.: Asking the right questions in crowd data sourcing. In: International Conference on Data Engineering, Washington, DC, USA (2012)
23. Altman, G., Bland, J.M.: Diagnostic tests. 1: sensitivity and specificity. British Med. J. **308**(6943), 1552 (1994). (Clinical research edition)
24. Altman, G., Bland, J.M.: Diagnostic tests 2: predictive values. British Med. J. **309**(6947), 102 (1994). (Clinical research edition)

Modeling and Analyzing Engagement
in Social Network Challenges

Marco Brambilla, Stefano Ceri, Chiara Leonardi, Andrea Mauri,
and Riccardo Volonterio[✉]

Dipartimento di Elettronica, Informazione e Bioingegneria (DEIB),
Politecnico di Milano, Piazza Leonardo da Vinci, 32, 20133 Milano, Italy
{marco.brambilla,stefano.ceri,chiara.leonardi,
andrea.mauri,riccardo.volonterio}@polimi.it

Abstract. Participation to challenges within social networks is a very
effective instrument for promoting a brand or event. In this paper, we
take the challenge organizer's perspective, and we study how to raise
the engagement of players in challenges where the players are stimulated
to create and evaluate content, thereby indirectly raising the awareness
about the brand or event itself. We illustrate a comprehensive model of
the actions and strategies that can be exploited for progressively boosting the social engagement during the challenge evolution. The model
studies the organizer-driven management of interactions among players,
and evaluates the effectiveness of each action in light of several other
factors (time, repetition, third party actions, interplay between different social networks, and so on). We evaluate the model through a set
of experiment upon a real case, the YourExpo2015 challenge. Overall,
our experiments lasted 9 weeks and mobilized hundreds of thousands of
users on two different social platforms; our quantitative analysis assesses
the validity of the model.

1 Introduction

Social networks are an essential aspect of communication strategies for promoting events and brands; effectiveness of communication is typically measured in
terms of the intensity and quality of engagement of its members. This paper is
concerned with the modeling and analysis of social challenges whose main purpose is to collect content produced by users in connection to a specific brand or
event.

In this context, the objective of this paper is to *define a model for social
challenges, which allows challenge managers to define the appropriate strategies
for increasing both participation and content creation, using multiple social networks, over a planned lifetime, in the presence of players, fans, and owners.*
More precisely, the research questions we want to investigate are the following:

1. Can we identify different roles in an online challenge, based on their activity?
2. Can we determine which are the most effective actions that a manager can
   perform to maximize the engagement of participants?

© Springer International Publishing AG 2016
W. Cellary et al. (Eds.): WISE 2016, Part I, LNCS 10041, pp. 140–154, 2016.
DOI: 10.1007/978-3-319-48740-3_10

3. Can we build different profiles of users based on their behavior during the challenge?

We applied our model by performing a large set of experiments in the context of a real-life scenario, called YourExpo2015, which featured about 750,000 social activities over a period of 9 weeks, with 9 corresponding challenges launched. We developed the challenge progression in a mixed top-down and bottom-up approach: on one hand, in every challenge we had the freedom of deciding engagement policies; on the other hand, direct experience in one challenge suggested how to define or modify key elements in the next one. Observation of the challenges allow us to draw interesting conclusions on the effectiveness of the various actions that we performed (as managers) or induced (on the players, fans, and owners).

The paper is structured as follows. We start with the related work analysis; then we define a model of the actors, the actions, and their interplay. Then we briefly present our technological framework and we extensively discuss our experiments within the YourExpo2015 challenge, and we show the effect of the various engagement policies. Finally, we discuss the results and conclude.

2 Related Work

Several works focused on analyzing and mining users behavior on social networks [4,8,21]. We are not aware of work specifically dedicated to the engagement is social network challenges; we therefore describe social engagement in other contexts. We use the gaming paradigm of a photo challenge for engaging users and increase awareness; this is in line with the gamification approach [5]. Reward and reputation systems are at the core of gamified applications, which builds upon incentive-centered design. This is studied in persuasive technology [7], where games are seen as means to shape user behavior [12,14], or to instill desired values [1]. Factors that influence human behavior are fundamentally of two types, namely, incentive and cost [13,16]. The former increases and the latter decreases the motivation to complete an action. In this paper we experimented with purely external and immaterial incentives, i.e., visibility gain [15]. User motivations are very heterogeneous, and in particular they vary a lot depending on the scenario [20] and on incentives can be dynamically tuned [2] or combined [17].

Engagement has been often measured in the context of online applications measuring the actions performed by the users, such as page views, click-through rates and return rates [11]. These metrics have been used as basis for our measurement and mapped to the concepts provided by the different platforms.

Various works studied the incentives and conditions that favor participation in In the context of social networks, Irena et al. [3] study which are the most important factors for customer engagement in the domain of a Facebook brand page. Yogo et al. study incentives that stimulate activities in social networking services [22], while [19] examine the conditions that encourage users to participate more intensively in social networking. [6] studied the trade-off between the *cost* represented by the concern about personal data privacy and the *incentive* of sharing personal facts, records or content.

Similarly to what we do, Kumar et al. classify social network users into roles [10,18] study specific roles for the content tagging activity; however, our roles are specific for challenge scenarios. Finally, some studies on social media try to understand at which time the most social activity occurs [9].

3   Modeling Social Network Challenges

In the paper we focus on the specific case of content production challenges, i.e., any kind of online games where users are requested to submit original content (photos, videos, text or any other media). We start by describing a model of social network challenge, as shown in Fig. 1: the model includes the actors, the actions, their interplay in content production games, and external factors that influence engagement.

3.1   Actors

A social challenge requires the interplay of four kinds of actors.

- **Manager.** Sets the rules of the game, typically encoded in the *game regulations*, and then performs the activities which are prescribed by such rules. In addition, the organizer performs activities targeted to enhancing social participation, such as boosting visibility of the top players.
- **Player.** Autonomously decides to play the game, typically for obtaining visibility for themselves or for their content. Visibility is granted either by other members of the social network or by the organizers of the challenge (or both).
- **Fan.** Follows the evolution of the game and decides the outcome by voting on the content produced by the players.
- **Owner.** Initiates the challenge based on some business or marketing need. According to that, he assigns the practical execution of the challenge to the manager. Owners include popular social accounts (actors, celebrities, institutions, well known brands), or real world events that have a strong impact on the public (such as TV programs, commercial advertising, endorsement by government or large companies). Their high visibility supports the managers in boosting the social participation; their actions may or may not be under the control of the organizers, and thus the consequences may be hard to predict.

3.2   Activities and Actions

We identify four main kinds of activities which contribute to a social challenge, each embodied in several actions.

- **Invite.** The purpose of inviting is to make the challenge known on the social network and to convince potential players to participate. These actions may imply publishing of direct descriptions of the challenge, or evocative contents for the topic or focus of the challenge.

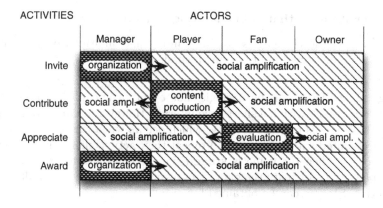

Fig. 1. Model of actors, activities and amplification effect in a social challenge.

- **Contribute.** The purpose is to participate to the game on the social platform of choice. For content production games, this implies to submit original content (photos, videos or text) produced by the player, specifically devised for the challenge and prepared during the challenge period. Some challenges may allow the player to post content that was pre-existing to the challenge. This entails both content produced by other people (e.g., photos or videos found on the web) or content produced by the player himself for other reasons (typically in the past) and reused for the challenge. If the challenge is asking to enrich existing content, the contribution may consist in expanding the content descriptions, through adding tags, location, text, people or other entities to the content; or by identifying similar items or listing the contents in some specific order or ranking.
- **Appreciate.** The purpose is to express and share appreciation for the activities or content of a player, or for the participant himself, for instance by liking the content, i.e., annotating the content with a "like" or "preference" tag (typical in social networks like Facebook, Instagram or Twitter), or by *following* the player, i.e., declaring interest in the person and following his activities.

Invite and *award* actions are typically performed by the manager; *contribute* is performed by players; while *appreciate* actions are performed by the fans, but can be tactically performed by organizers and players too, in order to enhance the social participation. Typically, owners do not engage directly into appreciation actions, because they must appear as neutral, and their size and visibility is incomparable to those of players.

The peculiar case of social-network based games enables an additional behaviour, which is to generate **Social Amplification** of the actions, through social sharing. Social amplification can be performed by any actor, upon the activity of any other. The effect is that any activity can be made more visible and appreciated by a wider audience, based on the visibility of the actor that performs the sharing. A crucial role in social amplification is played by the owner, whose sharing activity can dramatically boost visibility.

3.3 Other Aspects that Influence Engagement

Aspects such as the challenge's staging, timing and multiplatform execution should be also considered in organizing a challenge.

Staging. A challenge can be a single-shot event, where players make their actions and fans vote within a short time interval. However, an important aspect of social challenges is to build **loyalty**, which occurs when the players and fans become acquainted with the game and repeat their actions many times. Loyalty can only occur when the game is **staged**, i.e., it is structured in a way that allows a player to anticipate the game progression and engage in a multi-action participation. Staging can be obtained by:

- **Repeating the challenge periodically**, with a winner for each period.
- **Sub–structuring the challenge into phases**, and giving to the player different task to perform at each stage.

Influence of Daytime. Social amplification of actions is extremely different depending on the time at which the action is performed and also of the particular day (e.g., workday/weekend/vacation). Time-dependent effects must be studied for each challenge, e.g. certain challenges may attract higher participation during night time or weekends.

Cross-Social Network Fertilization. Cross-network fertilization is made possible because most actors participate in multiple social networks at the same time (e.g., they share content on Instagram and Twitter, and have friends on Facebook and collaborators on LinkedIn). Thus, it is possible to influence their behaviour on one social network through actions that occur on another one.

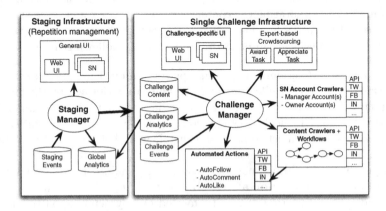

Fig. 2. Architecture of the system that manages and monitors the challenge.

4 Implementation

We designed and implemented an architecture for analyzing and managing staged social challenges, shown in Fig. 2. The architecture is layered, with an external layer, called **staging infrastructure**, which generates several versions of the internal layer, called **single challenge infrastructure**. They both are associated with several resources dedicated to user interaction. Normally, these include a **Web Interface** which publishes the rules of the challenge, and several **social network pages**, describing the specific portion of the challenge which takes place on each social network. These resources may be present only at the stage level, if each stage is managed as an independent challenge.

The **staging manager** holds information about the series of events which are associated with each stage, and in addition it manages global data analytics obtained as the summary of the various stages. The organization of the challenge core is performed by the **challenge manager**. It instantiates and manages several crawlers belonging to predefined classes, that are addressed to specific social networks and perform specific tasks, which are either **content-specifications** or **account monitoring** on each social networks; some of them are **automatic actions** generated in response to the task's output (e.g., following accounts or liking contents). Some of the content-specific tasks need to be organized through mini-workflows (e.g., the monitoring of *likes* given to a specific content must follow the *post* of that content.) Crawlers activities are timed so as to respect the constraints on API usage which are imposed by each social network; content is accessed through the given challenge **hashtag**, generated by the staging infrastructure and used as parameter in the calls to the social network APIs.

Data collected by crawlers are stored into stage-specific data analytics. Thanks to several sensing techniques over the challenge evolution, it is possible to understand and classify the behavior of players in terms of amount of activity, continuity of the actions in time, reactivity to solicitation from organizers, willingness to share challenge content and messages of the challenge, and extent of social amplification. Finalists and winners can be automatically determined by the system based upon such analytics.

5 Experiments on YourExpo2015

The Universal Exhibition **Expo 2015** was hosted in Milano; it received over 20 million visitors in its 1.1 million square meters of exhibition area. Over a six-month period, more than 140 participating countries run their pavillons around the theme of guaranteeing healthy, safe and sufficient food for everyone. Expo 2015 created a number of marketing campaigns on traditional and social channels. In such framework, we performed a set of experiments for increasing brand awareness of Expo 2015 *before* the event, through the development of a photo challenge, called YourExpo2015. The challenge has been independently managed by us, but has benefited of some interaction with the official social accounts of Expo 2015; their actions can be considered as *owner activities* according to our model. The objective of the experiment was two-fold: first, we wanted to verify the effectiveness of our model in a real world scenario; second, we wanted to respond to our research questions, and thus get insights on how the various actors behave, in order to get information useful for organizing future challenges.

5.1 Purpose and Actions of the Challenge

The game `YourExpo2015` is based on the social production of photos on *Instagram*, in response to specific hashtags which are published every week by Expo 2015. The purpose of the challenge is twofold: to engage users for increasing the visibility of the Expo 2015 brand, and to collect relevant content associated to the Expo idea and purpose. The challenge proposes at every stage two hashtags, paired in a way that hints to a contrast (e.g., `Fast/Slow`, `Art/Fun`, `Land/Sea`, and so on). Although most posted photos show food, they can be on arbitrary subjects. The best photo for each of the two hashtags, separately selected, compose a **postcard** where the two photos and hashtags are shown together; as the main challenge reward, Expo 2015 used the postcards as header of its Facebook page. The challenge run for 9 weeks between Dec. 7, 2014 and February 21, 2015, with one pair of hashtag published every week.

Within the challenge we applied the model described in Sect. 3, in particular, we devised two types of *Invite* actions:

- **Announce:** explicitly declares the start of the challenge, through posting of rules of engagement, deadlines, or aim of the game.
- **Recall:** sends our reminders and repetitive messages about the topics, the game rules, and the duration of the challenge.

We also covered *Awards* by mentioning players in three ways:

- **Composite:** a selection of four good photos posted until a given moment, with explicit mention of the social accounts of the respective authors.
- **Finalists:** the selection of the four best photos of the current week, with explicit mention of the authors. The selection is posted on two different social networks. These photos enter the voting for the winner of that week.
- **Winners:** the post advertising the winners of the week.

Figure 3 shows a typical history of votes (likes) expressed upon few photos, from the initial posting of photos to the definition of the winner.

A few days after the start of the stage (**Announce** action), we repost the most voted photos in a **Composite**, that consists of four photos. Then, we select the **Finalists** and we publish them on Facebook; with a small delay we republish a **Finalist** post on Instagram too, so that the finalists (whose identity on Facebook is not known to us) come to know about the selection, and they start a second round of promotion through their Facebook friends, thereby performing cross-platform engagement. Eventually, the winners are selected (by counting its Facebook votes) and advertised through specific **Winner** posts.

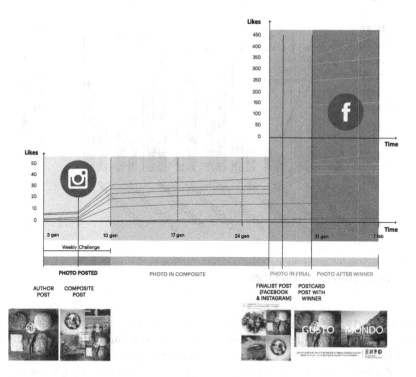

Fig. 3. Storyboard of a winning photo and monitoring of appreciation across social network.

5.2   Structure and Rationale of the Experiments

We run 9 different experiments, corresponding to the 9 weeks of the challenge, instantiated according to 4 different configurations, so that each experiment was repeated at least 2 times within each configuration. When setting up each experiment, we relied on the insight gathered during the previous runs in order to define significant variations of the configuration. The four configurations devised for our experiments were:

- **Amplification by Owner** (*first configuration*): each challenge is run on Instagram and starts with an **announcement** post. During the week additional **recall** posts are issued, to remind about the ongoing challenge theme. The **social amplification** action of the owner is scheduled on the recall post on the third day of the challenge.
- **Amplification by Manager** (*second configuration*): with respect to the first configuration, we added the **composite** posts. They were issued two times a day, so as to create competition and engagement around the selection.
- **Cross–platform Amplification** (*third configuration*): we expanded the action to multiple social network, namely including Facebook: every week we selected the four most liked photos as **finalists** on Facebook (and subsequently on Instagram), asking to vote for the best. Eventually, the **winners**

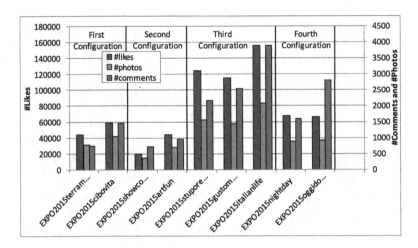

Fig. 4. Total number of likes, photos and comments for each experiment, grouped based on the different configurations.

are selected. In this configuration we also added an automatic like (by the account of the challenge manager) to every photo submitted to the challenge. Additionally, with respect to the second configuration, we increased the number of award actions, by adding the selection of finalists and winners. The actions of the owner were scheduled on the fourth, sixth and seventh days of challenge (Thursday, Saturday and Sunday) and consisted in a share of our **announcement**, **recall** and **award** posts.

- **Amplification without Owner** (*fourth configuration*): in the final configuration, we run the challenge without the help of the owner, i.e., with no social amplification by the owner. We also added an automatic action by the manager, i.e., the automatic follow of all the players in the current challenge.

The expected result of this structure of experiment is to obtain a coherent pattern of engagement, based on our actions. We consider the role of the owne asr particularly relevant. Therefore, our expectation is to have a pattern with significant peaks of engagement for every major social amplification performed by the owner.

6   Experiment Results and Discussion

6.1   Players Participation

The complete cycle of nine weeks of challenges generated more than 600,000 actions (post, like, comment) on Instagram and about 150,000 contacts on Facebook, with more than 3,000 followers on Instagram and more than 2,000 followers on Facebook.

Figure 4 shows the distribution of likes, comments, and photo posts on Instagram, for each run of the challenge. The figure shows that in the third configuration, thanks to the actions we put in place, we obtained a very good level of engagement. Conversely, in the second configuration the engagement was rather low, but this was due to external conditions (Christmas and New Year's holiday weeks). The relative drop of engagement in the fourth configuration (where we removed the actions of the owner) confirms the prominence of the owner's role.

6.2   Reactions to Actions of Organizers

Besides monitoring the overall success of the challenge in terms of photos posted, liked and commented by players and fans, we also analyzed the response of the audience to specific individual solicitations.

We verified that the different types of posts by the challenge manager generated significantly different reactions of participants. For this test, we separated our actions in classes and performed the t-test on the classes, comparing the distribution of likes and comments of each group. The null hypothesis is that the different types of posts generate the same amount of interaction. We used the Welch's t-test, since the limited non-normality of our data and because the assumption of homogeneity of variances failed. We also applied the Holm-Bonferroni correction for multiple tests. We run our tests on two levels of granularity. Based on our model (see Sect. 3.2), we first contrasted *invite* actions (*announcement* and *recall*) with *award* actions (*composite*, *finalists*, and *winners*). Results of the test are acceptable, with p-value <0.001 both for likes and comments.

Subsequently, we tested more in detail how the reactions to the different actions differ from each other, by using the six classes defined above. Using the number of likes as observed feature, the p-value is acceptable for most pairs of classes (11 over a total of 15 pairs), confirming that the actions generated significantly different reactions. For what concerns comments instead, the t-test failed in most of the cases. Indeed, comments differently from likes, they are hardly triggered by external events, because people decide to comment on content when they are driven by very specific personal motivations.

Figure 5(a) shows the number of likes and comments to our posts on Instagram, based on the type of post; the diagram shows that composites and winners got the maximum number of reactions. Most of reactions take place in the 12 h immediately following the posts.

An interesting feature of our challenge architecture is the ability to automatically perform actions in response to the activities of players, that we monitor through crawlers. In particular, in the third configuration we generated automatic likes to every posts, and we inspected the player's response to such automatic *likes* in the form of a *follow* action to the official challenge account; in Fig. 5(b) the reaction is plotted as a function of the total number of likes received by each user. It turns out that the first and especially the second *like* got the strongest reaction (respectively with around 25 % and 60 % of follow backs), while subsequent *likes* were less effective.

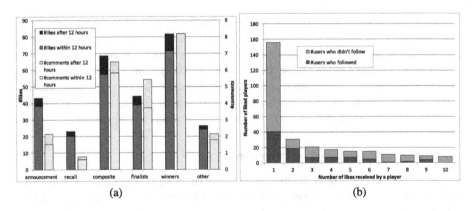

Fig. 5. Average number of likes and comments received on the different types of our posts on Instagram (a); and number of players who followed the official account of the challenge versus the number of *likes* received (b).

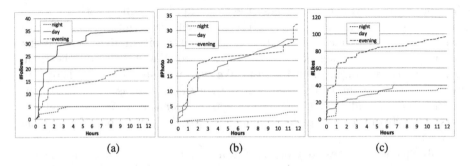

Fig. 6. Numbers new follows (a), new photos (b) and likes (c) as reactions to our automatic *follow* actions, for each time slot when our action was issued. The night time slot is between midnight and 9 am. Day is between 9 am and 5 pm, while evening corresponds to the time between 5 pm and midnight.

In the fourth configuration we also organized an automatic *follow* action to players, and we monitored the player's responses to our action; Fig. 6 shows such responses. In general, we can see that most reactions occur within the first two-three hours. In particular, as shown in Fig. 6(a), follow-back reactions are fast if the user are followed during the day, while they are very slow during the night. Figure 6(b) shows the number of photos posted after our follow. This type of reaction is very fast both during the day and the evening. Finally Fig. 6(a) shows the number of like reactions. In this case the reactions are fast in all the time slots of the day, as this action is less demanding.

In terms of volume, we noticed that the actions of players and participants are heavily influenced by the daytime, with the majority of activities taking place from 7 am to midnight, as expected, with peaks at lunch time and after dinner.

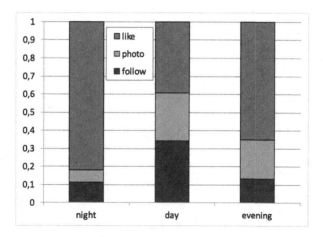

Fig. 7. Distribution of reactions to automatic follow actions upon posting of a photo, during three different daytime slots.

Figure 7 shows the distribution of the reactions in the different time slots (*night,* from midnight to 9 am; *day,* from 9 am to 5 pm; *evening,* from 5 pm to midnight). During night hours the majority of reactions are of type *like*, with very limited amount of posts and follow actions. During the day and evening the share of new photos (and follow actions) increases significantly.

6.3   Cluster Analysis of Players

We applied cluster analysis to determine how users are involved in the challenge, either as players or as voters; to evaluate their behavior, we selected the number of likes and number of posted photos as dimensions. We first removed the outliers that were qualitatively detected as the account of the organizer (YourExpo2015) and all the users that liked only one image and never posted a photo (100k over 160k users), then we run the k-mean cluster algorithms, which produced the clusters, shown in Fig. 8. The number of clusters was determined using the Elbow Method, that consists in stopping the k-mean algorithm when adding a new cluster does not decrease the squared distance between each member of the cluster and its centroid. Clusters 7 and 1 group users who specialized as players and voters respectively; then, clusters 6 and 4 group users who were less active but still specialized. The other clusters include users which were active in both roles, with decreasing activity going from cluster 3 to clusters 2 and 5. Clusters 6 and 7 identify the most active players (158 users), while cluster 5 represents the least active users (57504 users).

Clustering of users could be useful for increasing the success of a challenge, by addressing each type of message to a specific cluster of users. For instance, if the objective is to increase the collected content in a creation challenge, the organizer should leverage clusters 7 and 6; conversely, solicitation of votes or likes would should be addressed to users of cluster 1.

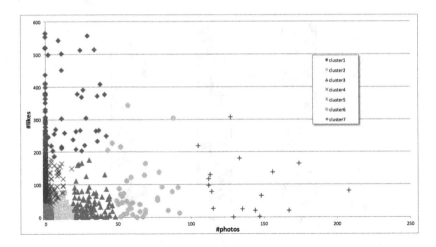

Fig. 8. Clusters of users.

6.4   Discussion

We close this section by summarizing the insights obtained with this experiment. It is clearly important to **use each social network at its best**: in our experience we used Instagram for posting and Facebook for final voting, while our attempts to propagate attention using Twitter were less successful. Indeed, our challenge had no significant/unexpected events, and thus it was not a suitable target for Twitter, that is notoriously very effective in spreading information about big events, such as terrorist attacks, but also comparatively minor events, such as TED conferences.

It is quite important to produce **regular challenges**, e.g., through periodic staging, so that many players may repeat their actions several times (e.g., several players of YourExpo2015 posted tens of photos); repetition and long duration also helps in growing a large and fidelized audience. It is also important to be fully aware of temporal factors, e.g., daytime or festivities, in order to properly plan automatic and owner actions that may engage new players. Temporal factors must be considered for any kind of action, including the automatic ones which are useless/negative if posted when people don't react.

Mere announcements of challenges and call to actions fail to engage people, but **active management by managers** pays. In particular, mentioning players is viral, especially when several of them are mentioned together, because this action stirs interactions among them, even if they do not know each other in advance. Although players want to win, the development of social relationship is perhaps an even stronger driver, therefore organizers should spend lots of efforts in creating mutual engagement. In our experiments, composite mentioning of several players created a sub-network of participants who positively interacted, both by contributing content and by mobilizing voters. This constitutes a notable difference with respect to traditional crowdsourcing platforms, where instead workers are not engaged by mutual actions or interactions.

Visibility provided by **owner actions** is fundamental to boost the challenge and to keep it alive; assuming that owner actions are scarce/expensive resources, they must be programmed in a way that provides maximum effect upon the players.

7   Conclusions

In this paper, we defined a model of the actors and activities involved in a social network challenge; we demonstrated that social platforms are effective in engaging players in games for content generation or enrichment, and we illustrated how the different choices of game organizers impact on the quality and quantity of player's engagement. Referring to our initial research questions, we can claim that:

1. We modeled different roles, which indeed perform different activities in the challenge, as reported also in our experiments;
2. We studied the different types of actions performed, and monitored which ones are most effective action types with respect to the engagement;
3. We cluste red the users in profile "groups" based on their behavior.

Of course, the experimental results reported in this paper are relative to our context, but many of them (e.g., those on individual engagement and on social reaction rules and their timings) have general validity. The main objective of social challenges is to boost social interest (on events, brands, etc.), but as a side effect valuable content is produced or improved; thus, our model applies to an interesting class of *games with a purpose*, where users produce results by being moved by non-monetary rewards. We are currently planning an application of the method to brand promotion with large companies, e.g. in fashion.

References

1. Barr, P., Noble, J., Biddle, R.: Video game values: human-computer interaction and games. Interact. Comput. **19**(2), 180–195 (2007)
2. Cheng, R., Vassileva, J.: Design and evaluation of an adaptive incentive mechanism for sustained educational online communities. User Model. User-Adapt. Interact. **16**(3–4), 321–348 (2006)
3. Cvijikj, I.P., Michahelles, F.: Online engagement factors on facebook brand pages. Soc. Netw. Anal. Min. **3**(4), 843–861 (2013)
4. Das, A., Gollapudi, S., Munagala, K.: Modeling opinion dynamics in social networks. In: 7th ACM Conference on Web Search and Data Mining, WSDM 2014, pp. 403–412 (2014)
5. Deterding, S., Sicart, M., Nacke, L., O'Hara, K., Dixon, D.: Gamification using game-design elements in non-gaming contexts. In: International Conference on Human Factors in Computing Systems, CHI 2011, Vancouver, BC, Canada, 7–12 May, pp. 2425–2428 (2011)

6. Dwyer, C.: Digital relationships in the "myspace" generation: results from a qualitative study. In: 40th Hawaii International International Conference on Systems Science (HICSS-40 2007), USA, p. 19 (2007)

7. Fogg, B.J.: Persuasive technology: using computers to change what we think and do. In: Ubiquity, December 2002

8. Gao, S., Ma, J., Chen, Z.: Modeling and predicting retweeting dynamics on microblogging platforms. In: 8th ACM Conference on Web Search and Data Mining, WSDM 2015, pp. 107–116 (2015)

9. Golder, S., Wilkinson, D., Huberman, B.: Rhythms of social interaction: messaging within a massive online network. In: Steinfield, C., Pentland, B., Ackerman, M., Contractor, N. (eds.) Communities and Technologies 2007, pp. 41–66. Springer, London (2007)

10. Kumar, R., Novak, J., Tomkins, A.: Structure and evolution of online social networks. In: 12th ACM SIGKDD International Conference on Knowledge Discovery and Data Mining, Philadelphia, PA, pp. 611–617 (2006)

11. Lalmas, M., O'Brien, H., Yom-Tov, E.: Measuring User Engagement. Synthesis Lectures on Information Concepts, Retrieval, and Services. Morgan & Claypool Publishers (2014)

12. Lockton, D., Harrison, D., Stanton, N.A.: The design with intent method: a design tool for influencing user behaviour. Appl. Ergon. **41**(3), 382–392 (2010)

13. Milgrom, P., Roberts, J.: Economics, organization and management (1992)

14. Niebuhr, S., Kerkow, D.: Captivating patterns - a first validation. In: Second International Conference on Persuasive Technology, PERSUASIVE 2007, Palo Alto, CA, USA, pp. 48–54 (2007)

15. Rashid, A.M., Ling, K.S., Tassone, R.D., Resnick, P., Kraut, R.E., Riedl, J.: Motivating participation by displaying the value of contribution. In: Conference on Human Factors in Computing Systems, CHI 2006, Montréal, Canada, 22–27 April, pp. 955–958 (2006)

16. Sato, K., Hashimoto, R., Yoshino, M., Shinkuma, R., Takahashi, T.: Incentive mechanism considering variety of user cost in p2p content sharing. In: IEEE Global Telecommunications Conference (GLOBECOM) 2008, pp. 1–5 (2008)

17. Scekic, O., Truong, H.L., Dustdar, S.: Incentives and rewarding in social computing. Commun. ACM **56**(6), 72–82 (2013)

18. Thom-Santelli, J., Muller, M.J., Millen, D.R.: Social tagging roles: publishers, evangelists, leaders. In: Conference on Human Factors in Computing Systems, CHI 2008, Florence, Italy, pp. 1041–1044 (2008)

19. Toriumi, F., Ishida, K., Ishii, K.: Encouragement methods for small social network services. In: 2008 IEEE/WIC/ACM International Conference on Web Intelligence, WI 2008, Sydney, pp. 84–90 (2008)

20. Wasko, M.M., Faraj, S.: Why should I share? examining social capital and knowledge contribution in electronic networks of practice. MIS Q. **29**(1), 35–57 (2005)

21. Wu, S., Sarma, A.D., Fabrikant, A., Lattanzi, S., Tomkins, A.: Arrival and departure dynamics in social networks. In: 6th ACM Conference on Web Search and Data Mining, WSDM 2013, pp. 233–242 (2013)

22. Yogo, K., Shinkuma, R., Konishi, T., Itaya, S., Doi, S., Yamada, K., Takahashi, T.: Incentive-rewarding mechanism to stimulate activities in social networking services. Int. J. Netw. Manage. **22**(1), 1–11 (2012)

Data Diversity

Select, Link and Rank: Diversified Query Expansion and Entity Ranking Using Wikipedia

Adit Krishnan[1(✉)], Deepak Padmanabhan[2], Sayan Ranu[1], and Sameep Mehta[3]

[1] IIT Madras, Chennai, India
{adit,sayan}@cse.iitm.ac.in
[2] Queen's University Belfast, Belfast, Northern Ireland, UK
D.Padmanabhan@qub.ac.uk
[3] IBM Research, New Delhi, India
sameepmehta@in.ibm.com

Abstract. A search query, being a very concise grounding of user intent, could potentially have many possible interpretations. Search engines hedge their bets by diversifying top results to cover multiple such possibilities so that the user is likely to be satisfied, whatever be her intended interpretation. Diversified Query Expansion is the problem of diversifying query expansion suggestions, so that the user can specialize the query to better suit her intent, even before perusing search results. We propose a method, Select-Link-Rank, that exploits semantic information from Wikipedia to generate diversified query expansions. SLR does collective processing of terms and Wikipedia entities in an integrated framework, simultaneously diversifying query expansions and entity recommendations. SLR starts with selecting informative terms from search results of the initial query, links them to Wikipedia entities, performs a diversity-conscious entity scoring and transfers such scoring to the term space to arrive at query expansion suggestions. Through an extensive empirical analysis and user study, we show that our method outperforms the state-of-the-art diversified query expansion and diversified entity recommendation techniques.

1 Introduction

Users of a search system may choose the same initial search query for varying information needs. This is most evident in the case of *ambiguous queries* that are estimated to make up one-sixth of all queries [24]. Consider the example of a user searching with the query *python*. It may be observed that this is a perfectly reasonable starting query for a zoologist interested in learning about the species of large non-venomous reptiles[1], or for a comedy-enthusiast interested in learning about the British comedy group *Monty Python*[2]. However, search results would most likely be dominated by pages relating the programming language[3],

---

[1] https://en.wikipedia.org/wiki/Pythonidae.
[2] https://en.wikipedia.org/wiki/Monty_Python.
[3] https://en.wikipedia.org/wiki/Python_(programming_language).

© Springer International Publishing AG 2016
W. Cellary et al. (Eds.): WISE 2016, Part I, LNCS 10041, pp. 157–173, 2016.
DOI: 10.1007/978-3-319-48740-3_11

that being the dominant interpretation (aka *aspect*) in the web. *Search Result Diversification (SRD)* [5,29] refers to the task of selecting and/or re-ranking search results so that many *aspects* of the query are covered in the top results; this would ensure that the zoologist and comedy-fan in our example are not disappointed with the results. If the British group is to be covered among the top results in a re-ranking based SRD approach for our example, the approach should consider documents that are as deep in the un-diversified ranked list as the rank of the first result that relates to the group. In our exploration, we could not find a result relating to *Monty Python* among the first five pages of search results for *python* on Bing. Such difficulties in covering long tail aspects, as noted in [2], led to research interest in a slightly different task attacking the same larger goal, that of Diversified Query Expansion (DQE). Note that techniques to ensure coverage of diverse aspects among the top results are relevant for apparently unambiguous queries too, though the need is more pronounced in inherently ambiguous ones. For an unambiguous query: *python programming*, there are many aspects based on whether the user is interested in *books, software* or *courses*.

DQE is the task of identifying a (small) set of terms (i.e., words) to extend the search query with, wherein the extended search query could be used in the search system to retrieve results covering a diverse set of aspects. For our *python* example, desirable top DQE expansion terms would include those relating to the programming language aspect such as *language* and *programming* as well as those relating to the reptile-aspect such as *pythonidae* and *reptile*. In existing work, the extension terms have been identified from sources such as corpus documents [26], query logs [17], external ontologies [2,3] or the results of the initial query [26]. The aspect-affinity of each term is modeled either explicitly [17,26] or implicitly [2] followed by selection of a subset of candidate words using the *Maximum Marginal Relevance (MMR)* principle [5]. This ensures that terms related to many aspects find a place in the extended set. Diversified Entity Recommendations (DER) is the analogous problem where the output of interest is a ranked list of entities from a knowledge base such that diverse query aspects are covered among the top entities.

In this paper, we address the DQE and DER problems and develop a novel method, *Select-Link-Rank* (**SLR**). Our main contributions are:

- A novel technique, *SLR*, for diversified query expansion and entity recommendation that harvests terms from initial query results and prioritizes terms and entities using the Wikipedia graph in a diversity conscious fashion. Our method does not require query logs or supervision and thus is immune to cold start issues.
- We present an empirical evaluation including a user study that illustrates that SLR's DQE results as well as the entity ranking results are much superior than those of the respective baselines. This establishes SLR as the method of choice for DQE and DER.

2   Related Work

We will start by scanning the space of SRD methods, followed by a detailed analysis of techniques for DQE/DER.

SRD: Search Result Diversification is the task of producing a result set such that most aspects of the query are covered. The pioneering SRD work [5] proposed the usage of the MMR principle in a technique that targets to reduce the redundancy among the top-results as a method to implicitly improve aspect representation:

$$\arg\max_{d} \quad \lambda \times S_1(d, Q) - (1 - \lambda) \times \max_{d' \in S} S_2(d, d')$$

In MMR, the next document to be added to the result set, S, is determined as that maximizing a score modeled as the relevance to the query (S_1) penalized by the similarity (S_2) to already chosen results in S. A more recent SRD method uses Markov Chains to reduce redundancy [29]. Since then, there have been methods to explicitly model query aspects and diversify search results using query reformulations [20], query logs [11] and click logs [15], many of which use MMR-style diversification.

DQE/DER: Diversified Query Expansion, a more recent task as well as the problem addressed in this paper, starts from a query and identifies a set of terms that could be used to extend the query that would then yield a more aspect-diverse result set; thus, DQE is the diversity-conscious variant of the well-studied Query Expansion problem [8]. Table 1 summarizes the various DQE methods in literature. Drawing inspiration from recent interest in linking text with knowledge-base entities (notably, since ESA [13]), BHN [2] proposes to choose expansion terms from the names of entities in the ConceptNet ontology, thus generating expansion terms that are focused on entities. BLN [3] extends BHN to use Wikipedia and query logs in addition to ConceptNet; the Wikipedia part relies on being able to associate the query with one or more Wikipedia pages, and uses entity names and representative terms as candidate expansion terms from Wikipedia. While such choices of expansion terms make BHN and BLN methods suitable for entity recommendations (i.e., DER), the limited vocabulary of expansion terms makes it a rather weak query expansion method. For example, though *courses* might be a reasonable expansion term for *python* under the computing aspect, BHN/BLN will be unable to choose such words since *python courses* is not an encyclopaedic concept to be an entity in the ConceptNet or Wikipedia. The authors in [3] note that the BLN-Wiki is competitive with BHN in cases where the query corresponds to a known Wikipedia concept, and that BHN performs better in general cases. We will use BHN as an entity ranking (DER) baseline in our experiments.

LBSN [17] gets candidate expansion terms from query logs. Such direct reuse of search history is not feasible in cold start scenarios and cases where the search engine is specialized enough to not have a large enough user base (e.g., single-user desktop search) to accumulate enough redundancy in query logs; our method, *SLR*, targets more general scenarios where query logs may not be available.

Table 1. Techniques for Diversified Query Expansion

Method[a]	User Data Reqd	External Resource Reqd	Source of Exp. Terms	Remarks
BHN [2] (DER Baseline)	–	ConceptNet	Entity Names	Expansion terms from the small vocabulary of entity names
ts_{xQuAD} [26] (DQE Baseline)	Sub-topics (i.e., aspects) and sub-topic level relevance judgements	–	**Documents**	Relevance judgements are often impractical to get, in real systems
LBSN [17]	Query Logs	–	Query Logs	Cold start issue, also inapplicable for small-scale systems
BLN [3]	Query Logs	ConceptNet Wikipedia	Entity Names, Categories, Query Logs etc	Expansion terms from small vocabulary as BHN and query log usage as LBSN
SLR (Ours)	–	Wikipedia	Documents	

[a] When the authors have not used a name for a method, we will refer to it using the combination of first characters of author names

ts_{xQuAD} [26], another DQE method, is designed to use terms from corpus documents to expand the query, making it immune to the small vocabulary problem and useful in a wide range of scenarios, much like the focus of SLR. However, ts_{xQuAD} works only for queries where the set of relevant documents are available at the aspect level. Given that, if each result document retrieved for the initial query may be deemed relevant to at least one aspect, a topic learner such as LDA [1] may be used to partition the results into topical groups by assigning each document to the topic with which it has the highest affinity. Since such topical groups are likely to be aspect-pure, such result partitions can be fed to ts_{xQuAD} to generate expansion terms without usage of relevance judgments. We will use the LDA-based ts_{xQuAD} as the baseline DQE technique for our experiments. Another related work is that of enhancing queries using entity features and links to entities [9], which may then be processed using search engines that have capabilities to leverage such information; we, however, target the DQE/DER problem where the result is a simple ordered list of expansion terms or entities.

Wikipedia for Query Expansion: Apart from BLN, there has been previous work on using Wikipedia for Query expansion, such as [28]. This work uses Wikipedia documents, differently weighted by the structure of Wikipedia documents, in a pseudo-relevance feedback framework; it may be particularly noted that, unlike the approaches discussed so far, this work does not address the diversity factor.

DQE Uptake Model: The suggested uptake model for DQE as used in most methods (e.g., [2]) is that the original search query (e.g., *python*) be appended with all the terms[4] in the result (e.g., *language, monty*) to form a single large query that is expected to produce a result set encompassing multiple aspects. While this may be a good model for search engines that work on a small corpus, we observe that such extended queries are not likely to be of high utility for large-scale search engines. This is so since there is a likelihood of a very rare aspect in the intersection of multiple terms in the extended query that would most likely end up being the focus of the search since search engines do not consider terms as being independent. Figure 1 illustrates a couple of such examples, where very rare and non-noteworthy aspects form part of the top results. Thus, we focus on the model where terms in the DQE result set be separately appended to the initial query to create multiple *aspect-pure* queries.

Fig. 1. Sample results from extended queries

3   Problem Formulation

Given a document corpus $\mathbb{D}$ and a query phrase $\mathcal{Q}$, the *diversified query expansion* (DQE) problem requires that we generate an ordered (i.e., ranked) list of *expansion terms* $\mathbb{E}$. Each of the terms in $\mathbb{E}$ may be appended to $\mathcal{Q}$ to create an extended query phrase that could be processed by a search engine operating over $\mathbb{D}$ using a relevance function such as BM25 [27] or PageRank [18]. The ideal $\mathbb{E}$ is

---

[4] Terms may have associated weights.

that ordering of terms such that the separate extended queries formed using the top *few* terms in $\mathbb{E}$ are capable of eliciting documents relevant to *most* aspects of Q from the search engine. Typically, users are interested in perusing only a few expansion possibilities; thus, a quality measure for DQE is the aspect coverage achieved over the top-k terms for an appropriate value of k such as 5. *Diversified entity recommendation* (DER) is the analogous problem of generating an ordered list of entities, $\mathcal{E}$, from an ontology (Wikipedia, ConceptNet etc.) such that most diverse aspects of the query are covered among the top few entities.

4   Select-Link-Rank: Our Method

Figure 2 outlines the flowchart of SLR. Given a search query, SLR starts by selecting informative terms (i.e., words or tokens) from the results returned by the search engine using a statistical measure. Since we use a large number of search results in the select phase to derive informative terms from, we expect to cover terms related to most aspects of the query. A semantic footprint of these terms is achieved by mapping them to Wikipedia entities in the Link Phase. The sub-graph of Wikipedia encompassing linked entities and their neighbors is then formed. The Rank phase starts by performing a diversity-conscious scoring of entities in the entity sub-graph. Specifically, since distinct query aspects are expected to be semantically diverse, the Wikipedia entity sub-graph would likely comprise clusters of entities that roughly map to distinct query aspects. The *vertex-reinforced random walk (VRRW)* ensures that only a few representatives of each cluster, and hence aspect, would get high scores; this produces an aspect-diversified scoring of entities. Such a diversified entity scoring is then transferred to the term space in the last step, achieving a diversified term ranking. In the following sections, we will describe the various phases in SLR. We will use the ambigious query *jaguar* as an example to illustrate the steps in SLR; jaguar has multiple aspects corresponding to many entities bearing the same name. These include an animal species[5], a luxury car manufacturer[6], a formula one competitor[7], a video game console[8] and an American professional football franchise[9] as well as many others.

4.1   Select: Selecting Candidate Expansion Terms

We first start by retrieving the top-K relevant documents to the initial query Q, denoted by $Res_K(Q, \mathbb{D})$ from a search engine operating on $\mathbb{D}$. From those documents, we then choose T terms whose distribution among the top-K documents contrasts well from their distribution across documents in the corpus. This divergence is estimated using the Bo1 model [14], a popular informativeness measure

---

[5] https://en.wikipedia.org/wiki/Jaguar.

[6] http://www.jaguar.co.uk/.

[7] https://en.wikipedia.org/wiki/Jaguar_Racing.

[8] http://www.retrogamer.net/profiles/hardware/atari-jaguar-2/.

[9] http://www.jaguars.com/.

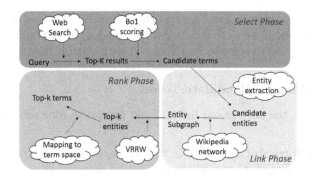

Fig. 2. Pipeline of the SLR algorithm.

that uses Bose-Einstein statistics to quantify divergence from randomness as below:

$$Bo1(t) = f(t, Res_K(Q, \mathbb{D})) \times log_2 \frac{1 + (f(t, \mathbb{D})/|\mathbb{D}|)}{f(t, \mathbb{D})/|\mathbb{D}|} + log_2(1 + (f(t, \mathbb{D})/|\mathbb{D}|))$$

where $f(a, B)$ denotes the frequency of the term a in the document collection represented by B. Thus, $f(t, \mathbb{D})/|\mathbb{D}|$ denotes the normalized frequency of t in $\mathbb{D}$. To ensure all aspects of Q have a representation in $Res_K(Q, \mathbb{D})$, K needs to be set to a large value; we set both K and T to 1000 in our method. The selected candidate terms are denoted as $Cand(Q, \mathbb{D})$. The top Bo1 words for our example query *jaguar* included words such as *panthera* (relating to animal), *cars*, *racing*, *atari* (video game) and *jacksonville* (American football).

Remarks: Starting with the top documents from a standard search engine allows our approach to operate as a layer on top of standard search engines. This is important from a practical perspective since disturbing the standard document scoring mechanism within search engines would require addressal of indexing challenges entailed, in order to achieve acceptable response times. Such considerations have made re-ranking of results from a baseline relevance-only scoring mechanism a popular paradigm towards improving retrieval [5,23].

4.2   Link: Linking to Wikipedia Entities

In this phase, we link each term in $Cand(Q, \mathbb{D})$ to one or more related Wikipedia entities. Since our candidate terms are targeted towards extending the original query, we form an extended query for each candidate term by appending the term to Q. We leverage entity linking methods, such as TagMe [12] and [10], which match small text fragments with entity descriptions in Wikipedia to identify top-related entities. At the end of this phase each term t in $Cand(Q, \mathbb{D})$ is associated with a set of entities, $t.E$. We use $r(t, e)$ to denote the relatedness score between term t and entity e (in $t.E$) as estimated by the entity linking technique.

For our example, *panthera* got linked to the *Jaguar* and *Panthera* entities whereas *cars* brought in entities such as *Jaguar Cars* and *Jaguar E-type*. The *racing* related entities were *Jaguar Racing* and *Tom Walkinshaw Racing*. Jaguar E-type was observed to be a type of Jaguar car, whereas Tom Walkinshaw Racing is an auto-racing team very closely associated with Jaguar Racing.

4.3   Rank: Ranking Candidate Terms

This phase forms the crux of our method and comprises four sub-phases.

Wikipedia Subgraph Creation: In this phase, we first construct a subgraph $G(\mathcal{Q}) = \{V(\mathcal{Q}), E(\mathcal{Q})\}$ of the Wikipedia entity network $W = \{V_W, E_W\}$. In W, each Wikipedia page (entity) is a node in V_W and there is a directed edge $(e, e') \in E_W$ if an outward hyperlink from $e \in V_W$ to $e' \in V_W$ exists. $G(\mathcal{Q})$ is a subgraph of W spanning entities that are linked to terms in $Cand(Q, \mathbb{D})$ and their directly related neighbors. More specifically, $V(\mathcal{Q}) = N_1 \cup N_2$ where

$$N_1 = \{\cup_{t \in Cand(Q,\mathbb{D})} t.E\} \tag{1}$$
$$N_2 = \{e \mid \exists e' \in N_1,\ e \notin N_1,\ (e', e) \in E_W\} \tag{2}$$

The edge set $E(\mathcal{Q})$ is the set of all edges (i.e., Wikipedia links) between nodes in $V(\mathcal{Q})$. Here, N_1 captures entities linked to candidate terms. N_2 brings in their one-hop outward neighbors. In other words, N_2 contains entities that are directly related to the linked entities and could therefore enrich our understanding of the aspects related to the query. The inclusion of one-hop neighbors, while being a natural first step towards expanding the concept graph, subsumes inclusion of all nodes along two-hop paths between nodes in N_1; the latter heuristic has been used in knowledge graph expansion in [22]. For the *jaguar* example, N_2 was seen to comprise entities such as *Formula One* that was found to connect to both *Jaguar Racing* and *Jaguar Cars* entities, thus uncovering the connection between their respective aspects.

Entity Importance Weights: In this sub-phase, we set a weight to each node (i.e., entity) in $G(\mathcal{Q})$ based on its estimated importance. We start with assigning weights to entities that are directly linked to terms in $Cand(Q, \mathbb{D})$:

$$wt'(e \in N_1) = \frac{\sum_{t \in Cand(Q,\mathbb{D})} I(e \in t.E) \times r(t, e)}{\sum_{e' \in N_1} \sum_{t \in Cand(Q,\mathbb{D})} I(e' \in t.E) \times r(t, e')}$$

where $I(.)$ is the identity function. Thus, the weight of each entity in N_1 is set to be the sum of the relatedness scores from each term that links to it. This is normalized by the sum of weights across entities in N_1 to yield a distribution that sums to 1.0. The weights for those in N_2 uses the weights of N_1 and is defined as follows:

$$wt'(e \in N_2) = \frac{max\{wt(e')|e' \in N_1,\ (e', e) \in E(\mathcal{Q})\}}{\sum_{e'' \in N_2} max\{wt(e')|e' \in N_1,\ (e', e'') \in E(\mathcal{Q})\}}$$

Thus, the weight of nodes in N_2 is set to that of their highest scored[10] inward neighbor in N_1, followed by normalization. In the interest of arriving at an importance probability distribution over all nodes in $G(\mathcal{Q})$, we do the following transformation to estimate the final weights:

$$wt(e) = \begin{cases} \alpha \times wt'(e) & e \in N_1 \\ (1 - \alpha) \times wt'(e) & e \in N_2 \end{cases} \tag{3}$$

where $\alpha \in [0, 1]$ is a parameter that determines the relative importance between directly linked entities and their one-hop neighbors. Intuitively, this would be set to a high value to ensure directly linked entities have higher weights.

Vertex Reinforced Random Walk: Our goal in this step is to rank the linked entities based on their diversity and relevance. For that purpose, the nodes in $G(\mathcal{Q})$ are scored using a diversity-conscious adaptation of PageRank that does a *vertex reinforced random walk (VRRW)* [19]. While in PageRank the transition probability $p(e, e')$ between any two nodes $e,\ e'$ is static, in VRRW, the transition probability to a node (entity) e' is reinforced by the number of previous visits to e'. The impact of this reinforcement can be seen in Fig. 3, wherein the weights are redistributed to a more mutually diverse set of nodes.

To formalize VRRW, let $p_0(e, e')$ be the transition probability from e to e' at timestamp 0, which is the start of the random walk. In our problem, $p_0(e, e') \propto wt(e')$. Now, let $N_T(v)$ be the number of times the walk has visited e' up to time T. Then, VRRW is defined sequentially as follows. Initially, $\forall e \in V(\mathcal{Q})$, $N_0(e) = 1$. Suppose the random walker is at node e at the current time T. Then, at time $T + 1$, the random walk moves to some node e' with probability $p_T(e, e) \propto p_0(e, e')N_T(e')$. Furthermore, for each node in $V(\mathcal{Q})$, we also add a self edge. VRRW is therefore generalized as follows.

$$p_T(e, e') = \lambda\, wt(e') + (1 - \lambda)\frac{wt(e')N_T(e')}{D_T(e)} \tag{4}$$

Entity network   PageRank   VRRW

Fig. 3. The three nodes (shaded) with the highest scores in PageRank vis-a-vis VRRW.

---

[10] The other option, using *sum* instead of *max*, could cause some highly connected nodes in N_2 to have much higher weights than those in N_1.

where $D_T(e) = \sum_{(e,e') \in E(\mathcal{Q})} wt(e')N_T(v)$ is the normalizing term. Here, λ is the teleportation probability, which is also present in PageRank. $(1 - \lambda)$ represents the probability of choosing one of the neighboring nodes based on the reinforced transition probability. However, with probability λ the random walk chooses to restart from a random node based on the initial scores of the nodes. If the network is ergodic, VRRW converges to some stationary distribution $S(\cdot)$ after a large T, i.e., $S(e') = \sum_{e \in V(\mathcal{Q})} p_T(e, e')S(e)$ [19]. Furthermore, $\sum_{\forall e \in V(\mathcal{Q})} S(e) = 1$. The higher the value of $S(e)$ of an entity e, the more important e is. *The top scored entities (nodes) at the end of this phase, $\mathcal{E}$, form the entity recommendation (DER) output of SLR.* The top-5 entities for our example query were found to be: *Jaguar Cars, Jaguar* (the entity corresponding to the animal species), *Atari Jaguar* (video game), *Jaguar Racing* and *Jacksonville Jaguars*.

Why does VRRW favor representativeness? As in PageRank, nodes with higher centralities get higher weights due to the flow arriving at these nodes. This, in turn results in larger visit counts $(N_T(v))$. When the random walk proceeds, the nodes that already have high visit counts tend to get an even higher weight. In other words, a high-weighted node starts dominating all other nodes in its neighborhood; such vertex reinforcement induces a competition between nodes in a highly connected cluster leading to an emergence of a few clear leaders per cluster as illustrated in Fig. 3.

Diversified Term Ranking: The DQE output, $\mathbb{E}$, is now constructed using the entity scores in $S(.)$. In the process of constructing $\mathbb{E}$, we maintain a set of entities that have already been *covered* by terms already chosen in $\mathbb{E}$ as $\mathbb{E}.E$. At each step, the next term to be added to $\mathbb{E}$ is chosen as follows:

$$t^* = \underset{t \in Cand(\mathcal{Q},\mathbb{D})}{\arg\max} \sum_{e \in t.E} I(e \notin \mathbb{E}.E) \times r(t,e) \times S(e)$$

Informally, we choose terms based on the sum of the scores of linked entities weighted by relatedness (i.e., $r(t,e)$), while excluding entities that have been *covered* by terms already in $\mathbb{E}$ to ensure diversification. The generation of $\mathbb{E}$, the DQE output, completes the SLR pipeline. The top-5 expansion terms for the *jaguar* query were found to be: *car, onca*[11], *atari, jacksonville, racing*. It is notable that despite *cars* and *racing* aspects being most popular on the web, other aspects are prioritized higher than *racing* when it comes to expansion terms. This is so due to the presence of entities such as *Formula One* in the entity neighborhood (i.e., N_2) that uncover the latent connection between the *racing* and *cars* aspects; VRRW accordingly uses the diversity criterion to attend to other aspects after choosing *cars*, before coming back to the related *racing* aspect.

---

[11] P. Onca is the scientific name of the wild cat called Jaguar.

Algorithm 1. *Select-Link-Rank*

Input: Query $\mathcal{Q}$, corpus $\mathbb{D}$
Output: List of diversified expansion terms, $\mathbb{E}$, and diversified entities, $\mathcal{E}$
Select Phase
1. Retrieve K result documents for search query $\mathcal{Q}$
2. Select T informative terms from them as $Cand(\mathcal{Q}, \mathbb{D})$
Link Phase
3. Link each term t in $Cand(\mathcal{Q}, \mathbb{D})$ to Wikipedia
4. Let linked entities be $t.E$ and relatedness score be $r(t, e)$
Rank Phase
5. Construct $G(\mathcal{Q})$, graph of linked entities and neighbors
6. Score each entity using relatedness to linked terms
7. Perform VRRW on $G(\mathcal{Q})$, entity scores initialized using (6)
8. Collect the top-scored entities based on VRRW scores as $\mathcal{E}$
9. Construct $\mathbb{E}$, a diversified term ranking using entity scores and term-entity relatedness.

4.4 Summary and Remarks

The various steps in SLR and their sequence of operation are outlined in the pseudocode in Algorithm 1. It may be noted that we do not make use of wikipedia disambiguation pages in SLR.

5 Experiments

Experimental Setup. We use the ClueWeb09 [7] Category B dataset comprising 50 million web pages in our experiments. In SLR, we use the publicly accessible Indri interactive search interface for procuring initial results. This was followed by usage of a simple custom entity linker based on Apache Lucene [16]; specifically, all entities were indexed by text fields. For parameters, we set $K = K' = 1000$, $\alpha = 0.65$ and $\lambda = 0.25$ unless mentioned otherwise. We consistently use a query set of 15 queries gathered across motivating examples in papers on SRD and DQE.

We compare our DQE results against LDA-based ts_{xQuAD} [26] where we set the #topics to 5. SLR's DER results are compared against that of BHN [2]. For both ts_{xQuAD} and BHN, all parameters are set to values recommended in the respective papers.

Our primary evaluation is based on a user study where users are requested to choose from between our method and the baseline when shown the top-5 results from both. The user study was rolled out to an audience of up to 100 technical people (grad students and researchers) of whom around 50 % responded. All questions were optional; thus, some users only entered responses to a few of the queries. Since the user study was intended to collect responses at the result-set level to reduce the number of entries in the feedback form, we are unable to use evaluation measures such as α-NDCG that require relevance judgements at the level of each result-aspect combination. Apart from the user study, we also perform an automated diversity evaluation focused on the DQE task.

Table 2. #Votes from User Study: Expansions (SLR vs.**ts$_{xQuAD}$**) &Entities (SLR vs. BHN)

Query Information		DQE Expansions Eval.		DER Entities Eval.	
Sl#	Query	SLR	ts_{xQuAD}	SLR	BHN
1	coke	**37**	6	**40**	11
2	fifa 2006	**40**	3	**33**	18
3	batman	**32**	11	**49**	2
4	jennifer actress	**40**	3	**48**	3
5	phoenix	**39**	4	**42**	10
6	valve	**38**	5	**40**	12
7	rock and roll	**40**	3	**46**	4
8	amazon	**39**	4	**39**	13
9	washington	**37**	6	**38**	12
10	jaguar	**37**	6	**46**	5
11	apple	**30**	14	**41**	9
12	world cup	**36**	8	**50**	1
13	michael jordan	**39**	4	**36**	13
14	java	**41**	2	**41**	9
15	python	**39**	4	25	**26**
Average		**37.6**	5.53	**40.9**	9.87
Percentage		**87 %**	13 %	**81 %**	19 %

5.1 User Study

Expansion Quality Evaluation (DQE). First, we compare the quality of SLR results against those of ts_{xQuAD} over the dataset of 15 queries. For each query, we generate the top-5 recommended expansions by both methods and request users to choose the method providing better recommendations. The number of votes gathered by each technique is shown in Table 2. The exact recommended expansions, along with all details of the user study, can be found at a web page[12]. SLR is seen to be preferred over ts_{xQuAD} across all queries.

Entity Quality Evaluation (DER). We compare the DER output from SLR against the entity ranking from BHN. We follow a similar approach as in the expansion evaluation to elicit user preferences. Table 2 suggests that users strongly prefer SLR over BHN on 14 queries while being ambivalent about the query "python". Our analysis revealed that BHN had entities focused on the reptile and the programming language, while our method also had results pertaining to a British comedy group, *Monty Python*; we suspect most users were

---

[12] https://sites.google.com/site/slrcompanion2016/.

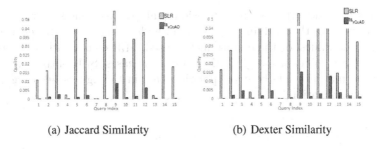

(a) Jaccard Similarity   (b) Dexter Similarity

Fig. 4. Diversity Analysis, SLR vs ts_{xQuAD}

unaware of that aspect for python, and thus did not credit SLR for considering that.

5.2 Automated Diversity Evaluation

We further evaluate the performance of SLR with respect to the diversity of the aspects represented by the expansion terms and their relevance. Since all previous efforts on DQE use evaluation measures that are based on expensive human-inputs in the form of releveance judgements (e.g., [4,21]), we now devise an intuitive and automated metric to evaluate the diversity of DQE results by mapping them to the entity space where external entity relatedness measures can be exploited. Consider the top-k query expansions as $\mathbb{E}$; we start by finding the set of entity nodes associated with those expansions, $\mathbb{N}$. We then define an entity-node relevance score $r_{\mathbb{E}}(n)$ as the sum of its relevance scores across its associated expansions; i.e., $r_{\mathbb{E}}(n) = \sum_{e \in \mathbb{E}} r(e, n)$. Let $S(n_i, n_j)$ denote an entity-pair semantic relatedness estimate from an external oracle; our quality measure is:

$$Q(\mathbb{E}, \mathbb{N}) = \frac{1}{\binom{|\mathbb{N}|}{2}} \sum_{(n_i, n_j) \in \mathbb{N}} r_{\mathbb{E}}(n_i) \times r_{\mathbb{E}}(n_j) \times exp(-S(n_i, n_j))$$

where $exp(-S(n_i, n_j))$ is a positive value inversely related to similarity between the corresponding entities. Intuitively, it is good to have highly relevant entities to be less related to ensure that entity-nodes in $\mathbb{N}$ are diverse. Thus, *higher values* of the $Q(., .)$ metric are desirable. We use two versions of Q by separately plugging in two different estimates of semantic similarity to stand for the oracle:

$$S_J(n_i, n_j) = \frac{n_i.neighbors \cap n_j.neighbors}{n_i.neighbors \cup n_j.neighbors}$$

$$S_D(n_i, n_j) = Dexter(n_i, n_j)$$

where $n.neighbors$ indicate the neighbors of the node n according to the Wikipedia graph, and $Dexter(., .)$ denotes the semantic similarity from Dexter [6].

Figures 4(a) and (b) show the expansion qualities based on Jaccard and Dexter respectively for the SLR and ts_{xQuAD} methods. As can be seen, regardless of the parameter values, or the quality metric used, SLR consistently outperforms ts_{xQuAD} by a significant margin. Infact, for some cases, SLR outperforms it by such a large margin that the corresponding bars in the figures been segmented for better visualization.

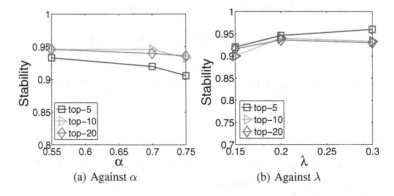

(a) Against α (b) Against λ

Fig. 5. Stability of the SLR algorithm.

5.3 SLR Parameter Sensitivity Analysis

Finally, we analyze the *stability* of SLR DQE against the two parameters that it requires: the teleportation probability λ, and the weighting factor α. Stability is defined as the fraction of common recommendations in the top-20 expansions produced at two different parameter values. We consider the default setting as reference, and measure stability against of results at altered parameter values against the reference. The results in Figs. 5(a) and (b) indicate that SLR is stable across wide variations of both parameters, achieving a stability of up to 0.95. Similar trends were recorded for SLR DER.

5.4 Computational Cost Analysis

Although computational efficiency is not the focus of this work, we attempt to provide a brief analysis of the computational costs of our algorithm.

- The **Select** phase uses the Indri Search Engine to run the queries, which combines language modeling and inference network approaches to perform the search. Interested readers may refer [25] for performance numbers. Selection of K' terms from K retrieved documents can be performed using a heap, at a cost of $K.L_{avg} + W_u.log(K')$, where L_{avg} is the mean count of non stop-words per document and W_u is the total number of unique words.

- In the **Link** phase, each of the K' chosen terms from the previous phase are used to expand queries and link to entities. This is performed using a reverse index from words to Wiki pages and a scoring mechanism such as TF-IDF. Computational costs depend on the number of candidate pages, which is roughly proportional to the total number of pages (with a very small constant), and inversely to the vocabulary of the corpus (number of unique words).
- Under the **Rank** phase, let us consider a subgraph of size $|S|$ nodes, on which DivRank is executed. With the matrix implementation of DivRank, the total computational cost is $\propto |S|^2$ per iteration. In practice, we found all our subgraphs to reasonably converge in less than 15 iterations, leading to very fast computations in the order of a few seconds.

5.5   Discussion

Our user study on both expansions and entities indicate that SLR results outperform other methods. SLR is also seen to perform better on automated diversity evaluation measures. These results establish two key properties of the proposed technique. First, the Wikipedia entity network is a meaningful resource to understand the various aspects of a query. Second, VRRW is effective in mining accurate representatives of the various aspects related to the query. Overall, the empirical analysis establishes that entities may be leveraged towards providing good term-level abstractions of diverse user intents.

6   Conclusions and Future Work

In this paper, we considered the problem of Diversified Query Expansions and developed a method that leverages semantic information networks such as Wikipedia towards providing diverse and relevant query expansions. Our method, SLR, exploits recent technical advancements across fields such as entity analysis, NLP and graph traversals using a simple 3-phase select-link-rank framework. The SLR query expansion and entity recommendations were seen to outperform respective baselines by large margins, on a user study as well as on an automated diversity evaluation. These establish SLR as the method of choice for DQE and diversified entity recommendations. As future work, we intend to look at extending SLR to exploit structured domain-specific knowledge sources to enhance usability for specialized scenarios such as intranet search. We are currently exploring integrated graph-based visualization of DQE results and entity recommendations.

References

1. Blei, D.M., Ng, A.Y., Jordan, M.I.: Latent dirichlet allocation. J. Mach. Learn. Res. **3**, 993–1022 (2003)

2. Bouchoucha, A., He, J., Nie, J.Y.: Diversified query expansion using conceptnet. In: Proceedings of the 22nd ACM International Conference on Conference on Information and Knowledge Management, pp. 1861–1864. ACM (2013)

3. Bouchoucha, A., Liu, X., Nie, J.-Y.: Integrating multiple resources for diversified query expansion. In: Rijke, M., Kenter, T., Vries, A.P., Zhai, C.X., Jong, F., Radinsky, K., Hofmann, K. (eds.) ECIR 2014. LNCS, vol. 8416, pp. 437–442. Springer, Heidelberg (2014). doi:10.1007/978-3-319-06028-6_38

4. Bouchoucha, A., Liu, X., Nie, J.-Y.: Towards query level resource weighting for diversified query expansion. In: Hanbury, A., Kazai, G., Rauber, A., Fuhr, N. (eds.) ECIR 2015. LNCS, vol. 9022, pp. 1–12. Springer, Heidelberg (2015). doi:10.1007/978-3-319-16354-3_1

5. Carbonell, J., Goldstein, J.: The use of mmr, diversity-based reranking for reordering documents and producing summaries. In: Proceedings of the 21st Annual International ACM SIGIR Conference on Research and Development in Information Retrieval, pp. 335–336. ACM (1998)

6. Ceccarelli, D., Lucchese, C., Orlando, S., Perego, R., Trani, S.: Dexter 2.0 - an open source tool for semantically enriching data. In: Proceedings of the ISWC 2014 Posters and Demonstrations Track a Track within the 13th International Semantic Web Conference, ISWC 2014, Riva del Garda, Italy, October 21, 2014, pp. 417–420 (2014)

7. Clueweb: (2009). http://lemurproject.org/clueweb09/

8. Collins-Thompson, K.: Estimating robust query models with convex optimization. In: Advances in Neural Information Processing Systems, pp. 329–336 (2009)

9. Dalton, J., Dietz, L., Allan, J.: Entity query feature expansion using knowledge base links. In: Proceedings of the 37th International ACM SIGIR Conference on Research and Development in Information Retrieval, pp. 365–374. ACM (2014)

10. Deepak, P., Ranu, S., Banerjee, P., Mehta, S.: Entity linking for web search queries. In: Hanbury, A., Kazai, G., Rauber, A., Fuhr, N. (eds.) ECIR 2015. LNCS, vol. 9022, pp. 394–399. Springer, Heidelberg (2015). doi:10.1007/978-3-319-16354-3_43

11. Dou, Z., Hu, S., Chen, K., Song, R., Wen, J.R.: Multi-dimensional search result diversification. In: Proceedings of the Fourth ACM International Conference on Web Search and Data Mining, pp. 475–484. ACM (2011)

12. Ferragina, P., Scaiella, U.: Tagme: on-the-fly annotation of short text fragments (by wikipedia entities). In: Proceedings of the 19th ACM International Conference on Information and Knowledge Management, pp. 1625–1628. ACM (2010)

13. Gabrilovich, E., Markovitch, S.: Computing semantic relatedness using wikipedia-based explicit semantic analysis. IJCAI **7**, 1606–1611 (2007)

14. He, B., Ounis, I.: Combining fields for query expansion and adaptive query expansion. Inf. Process. Manage. **43**(5), 1294–1307 (2007)

15. He, J., Hollink, V., de Vries, A.: Combining implicit and explicit topic representations for result diversification. In: Proceedings of the 35th International ACM SIGIR Conference on Research and Development in Information Retrieval, pp. 851–860. ACM (2012)

16. Jakarta, A.: Apache lucene-a high-performance, full-featured text search engine library (2004)

17. Liu, X., Bouchoucha, A., Sordoni, A., Nie, J.Y.: Compact aspect embedding for diversified query expansions. Proc. AAAI **14**, 115–121 (2014)

18. Page, L., Brin, S., Motwani, R., Winograd, T.: The pagerank citation ranking: Bringing order to the web. In: Proceedings of the 7th International World Wide Web Conference, pp. 161–172 (1998)

19. Pemantle, R.: Vertex-reinforced random walk. Probab. Theor. Relat. Fields **92**(1), 117–136 (1992)
20. Santos, R.L., Macdonald, C., Ounis, I.: Exploiting query reformulations for web search result diversification. In: Proceedings of the 19th International Conference on World Wide Web, pp. 881–890. ACM (2010)
21. Santos, R.L.T., Peng, J., Macdonald, C., Ounis, I.: Explicit search result diversification through sub-queries. In: Gurrin, C., He, Y., Kazai, G., Kruschwitz, U., Little, S., Roelleke, T., Rüger, S., Rijsbergen, K. (eds.) ECIR 2010. LNCS, vol. 5993, pp. 87–99. Springer, Heidelberg (2010). doi:10.1007/978-3-642-12275-0_11
22. Schuhmacher, M., Ponzetto, S.P.: Knowledge-based graph document modeling. In: Proceedings of the 7th ACM International Conference on Web Search and Data Mining, pp. 543–552. ACM (2014)
23. Singh, A., Raghu, D., et al.: Retrieving similar discussion forum threads: a structure based approach. In: Proceedings of the 35th International ACM SIGIR Conference on Research and Development in Information Retrieval, pp. 135–144. ACM (2012)
24. Song, R., Luo, Z., Wen, J.R., Yu, Y., Hon, H.W.: Identifying ambiguous queries in web search. In: Proceedings of the 16th International Conference on World Wide Web, pp. 1169–1170. ACM (2007)
25. Strohman, T., Metzler, D., Turtle, H., Croft, W.B.: Indri: A language model-based search engine for complex queries. In: Proceedings of the International Conference on Intelligent Analysis. vol. 2, pp. 2–6. Citeseer (2005)
26. Vargas, S., Santos, R.L., Macdonald, C., Ounis, I.: Selecting effective expansion terms for diversity. In: Proceedings of the 10th Conference on Open Research Areas in Information Retrieval, pp. 69–76 (2013)
27. Whissell, J.S., Clarke, C.L.: Improving document clustering using okapi bm25 feature weighting. Inf. Retr. **14**(5), 466–487 (2011)
28. Xu, Y., Jones, G.J., Wang, B.: Query dependent pseudo-relevance feedback based on wikipedia. In: Proceedings of the 32nd International ACM SIGIR Conference on Research and Development in Information Retrieval, pp. 59–66. ACM (2009)
29. Zhu, X., Goldberg, A.B., Van Gael, J., Andrzejewski, D.: Improving diversity in ranking using absorbing random walks. In: HLT-NAACL, pp. 97–104. Citeseer (2007)

Multi-dimension Diversification
in Legal Information Retrieval

Marios Koniaris[1]([⊠]), Ioannis Anagnostopoulos[2], and Yannis Vassiliou[1]

[1] KDBS Lab, School of ECE,
National Technical University of Athens, Athens, Greece
mkoniari@dblab.ece.ntua.gr
[2] Department of Computer Science and Biomedical Informatics,
University of Thessaly, Lamia, Greece

Abstract. The number of freely available legal data sets is increasing at high speed. Citizens can easily access a lot of information about regulations, court orders, statutes, opinions and analytical documents. Such openness brings undeniable benefits in terms of transparency, participation and availability of new services. However, legal information overload poses new challenges, especially in the field of Legal Information Retrieval. Search result diversification has gained attention as a way to increase user satisfaction in web search. We hypothesize that such a strategy will also be beneficial for search on legal data sets. We address diversification of results in legal search by introducing legal domain specific diversification criteria and adopting several state of the art methods from the web search, network analysis and text summarization domains. We evaluate our diversification framework using a real data set from the Common Law domain that we subjectively annotated with relevance judgments for this purpose. Our findings reveal that web search diversification techniques outperform other approaches (e.g. summarization-based, graph-based methods) in the context of legal diversification, as well as that the diversity criteria we introduce provide distinctively diverse subsets of resulting documents, thus differentiating our proposal in respect to traditional diversification techniques.

1 Introduction

Over the last years, as a result of the momentum of Open data initiatives, there has been a vast increase on the number of freely available legal data sets. Portals that allow users to search for legislation, using keywords, titles, etc. are now a common place. In such portals, legal documents are not stored as plain text, but in a more structured format with a rich set of meta data. Thus, it is possible for the end users to navigate to a specific section of a document or to inquiry information about the documents, such as date of enactment, date of repeal, jurisdiction, etc. Furthermore, with the advent of methods for the semantic indexing of Legal documents [31], several orthogonal categorization schemes can help users to find the information they need via navigation. To alleviate the

© Springer International Publishing AG 2016
W. Cellary et al. (Eds.): WISE 2016, Part I, LNCS 10041, pp. 174–189, 2016.
DOI: 10.1007/978-3-319-48740-3_12

data overload problem, in this paper we propose a novel way to efficiently and effectively diversify legal documents.

Legal text retrieval, in contrary to web retrieval, is primarily based upon concepts and not the explicit wording in documents texts. Earlier works essentially focus on classifing sources of law according to legal concepts. A complementary issue, over-looked in the legal text retrieval literature, is the diversification of the search results, i.e., covering different intents of the query in the top-ranked results. Consider, for example, a lawyer preparing his/her arguments for a given case who submits a user query to retrieve information. He/she has to iteratively browse an enormous number of judgments selecting, through knowledge and experience, relative documents in order to acquire a broad and in-depth context understanding. A diverse result, i.e. a result covering a wide range of possible legal interpretations is intuitively more informative and helpful than a set of homogeneous results that contain only relevant cases with similar features.

In order to satisfy a wide range of users, query results diversification has attracted a lot of attention in the field of text mining. IR systems attempt to diversify search results, so that they cover a wide range of possible interpretations (aspects, intents or subtopics) of a query. In consequence, the number of redundant items in a search result list should decrease, while the likelihood that a user will be satisfied with any of the displayed results should increase. There has been extensive work on query results diversification, see related work Sect. 2, where the key idea is to select a small set of results that are sufficiently dissimilar, according to an appropriate similarity metric.

In this work we address result diversification in the legal IR. To this end, we adopt various methods from the literature that are introduced for text summarization (LexRank [6] and Biased LexRank [27]), graph-based ranking (GrassHopper [37] and DivRank [22]) and web search result diversification (MMR [3], Max-Sum [13], Max-Min [13] and MonoObjective [13]). While investigating the performance of these approaches, we analyze the impact of various features in computing the query-document relevance and document-document similarity scores. We evaluate the performance of the above methods on a legal corpus subjectively annotated with relevance judgments using metrics employed in TREC Diversity Tasks. To the best of our knowledge none of these methods were employed in the context of diversification in legal IR and evaluated using diversity-aware evaluation metrics.

Our findings reveal that (i) web search diversification techniques outperform other evaluated approaches (e.g. summarization-based, graph-based methods) in the context of providing diversified results in the legal domain, and (ii) the diversification criteria we introduce provide distinctively diverse subsets of resulting documents, as opposed to other approaches that are based only on textual similarity.

The remainder of this paper is organized as follows: Sect. 2 reviews previous work in query result diversification, diversified ranking on graphs and in the field of legal text retrieval. Section 3 introduces the concepts of search diversification and presents diversification algorithms, while Sect. 4 describes our experimental framework and evaluation results. Finally, we draw our conclusions and future work aspects in Sect. 5.

2   Related Work

We first present related work on query result diversification, afterwards on diversified ranking on graphs and then on legal text retrieval techniques.

2.1   Query Result Diversification

Users of (Web) search engines typically employ keyword-based queries to express their information needs. These queries are often underspecified or ambiguous to some extent [5]. Different users who pose exactly the same query may have very different query intents. Simultaneously the documents retrieved by an IR system may reflect superfluous information. Search result diversification aims to solve this problem, by returning diverse results that can fulfill as many different information needs as possible. The published literature on search result diversification is reviewed in [28]. One of the earliest works on diversification is the maximal marginal relevance [3]. It envolves re-ranking search results as the combination of two metrics, one measuring the similarity among documents and the other the similarity between documents and the query. [13] introduced a general framework for result diversification with a set of diversification axioms and three diversification objectives, which we utilize in our work. Other researchers [33] utilized the correlation between documents as a measure of their similarity in the pursuit of diversification and risk minimization in document ranking. Diverfication heuristics that explicitly leverage external information, computed through probabilistic methods also have been proposed in [1,16,29]. In contrary to the above methods, given the fact that these methods utilize proprietary information, we do rely only on implicit knowledge of the legal corpus.

2.2   Diversified Ranking on Graphs

Many network-based ranking approaches have been proposed to rank objects according to different criteria [19] and recently diversification of the results has attracted attention. Research is currently focused on two directions: a greedy vertex selection procedure and a vertex reinforced random walk. The greedy vertex selection procedure, at each iteration, selects and removes from the graph the vertex with maximum random walk based ranking score. One of the earlier algorithms that address diversified ranking on graphs by vertex selection with absorbing random walks is Grasshopper [37]. A diversity-focused ranking methodology, based on reinforced random walks, was introduced in [22]. Their proposed model, DivRank, incorporates the rich-gets-richer mechanism to PageRank with reinforcements on transition probabilities between vertices. We utilize these approaches in our diversification framework considering the connectivity matrix of the citation network between documents that are relevant for a given user query.

2.3   Legal Text Retrieval

Legal text retrieval traditionally relies on external knowledge sources, such as thesauri and classification schemes. [25] presents various techniques used in legal text retrieval. Several supervised learning methods have been proposed to classify sources of law according to legal concepts [2,14,23]. Ontologies and thesaurus have been employed to facilitate information retrieval [12,17,30,32] or to enable the interchange of knowledge between existing legal knowledge systems [15]. Legal document summarization [7,8,24] has been used as a way to make the content of the legal documents, notably cases, more easily accessible. We also utilize state of the art summarizations algorithms but under a different objective: we aim to maximize diversity of the result set for a given query.

In another line of work citation analysis has been used in the field of law to construct case law citation networks [21][1]. Case law citation networks contain valuable information, capable of measuring legal authority [26], identifying authoritative precedent[2] [10], evaluating the relevance of court decisions [9] or even assisting summarizing legal cases [11], thus showing the effectiveness of citation analysis in the Case law domain. While the American legal system has been the one that has undergone the widest series of studies in this direction, recently various researchers applied network analysis in the Civil law domain as well. The authors of [18] propose a network-based approach to model the law. Network analysis techniques where also employed in [34] to identify context networks in dutch legislation and in [35] to recommend relevant sources of law given a focus document. In this work we also utilize citation analysis techniques and construct the Legislation Network, as to cover a wide range of possible aspects of a query.

3   Legal Document Diversification

At first, we define the problem addressed in this paper and provide an overview of the diversification process. Afterwards, legal document's features relevant for our work are introduced and distance functions are defined. Finally, we describe the diversification algorithms used in this work.

3.1   Problem Formulation

Result diversification is a trade-off between finding relevant to the user query documents and diverse documents in the result set. Given a set of legal documents and a query, our aim is to find a set of relevant and representative documents and to select these documents in such a way that the diversity of the set is maximized. More specifically, the problem is formalized as follows:

---

[1] Case documents usually cite previous cases, which in turn may have cited other cases and thus a network is formed over time with these citations between cases.

[2] Legal norm inherited from English common law that encourages judges to follow precedent by letting the past decision stand.

Definition 1 (Legal document diversification). *Let q be a user query and N a set of documents relevant to the user query. Find a subset $S \subseteq N$ of documents that maximize an objective function f that quantifies the diversity of documents in S.*

$$S = \underset{\substack{|S| = k \\ S \subseteq N}}{\operatorname{argmax}} f(N) \qquad (1)$$

3.2 Diversfication Overview

Figure 1, illustrates the overall workflow of the diversification process. At the highest level, the user express his/her information need, the user query. Relevant, with the information need, documents are retrieved. Diversification aims to find a subset of those documents that maximize an objective function that quantifies the diversity of documents. Significant components of the process include:

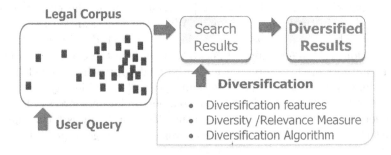

Fig. 1. Diversification overview

- *Ranking Features*, features of legal documents that will be used in the ranking process.
- *Distance Measures*, functions to measure the similarity between two legal documents and the relevance of a query to a given document.
- *Diversification Heuristics*, heuristics to produce a subset of diverse results.

3.3 Ranking Features

Under the Vector Space model, which we employ in this work, each document u can be represented as a term vector $U = (is_{w1u}, is_{w2u}, ..., is_{wmu})^T$, where $w_1, w_2, ..., w_m$ are all the available terms, and is can be any popular indexing schema e.g. $tf, tf - idf, logtf - idf$. User queries are represented in the same manner as documents.

Typically diversification techniques measure diversity in terms of content, where only textual similarity between items is used in order to quantify information similarity. In this work, we extend the notion of diversity on supplementary features/dimensions, besides textual similarity. In order to identify these features we examine the unique characteristics of the legal documents. Documents in the legal domain possess some noteworthy characteristics, such as being intrinsically multi-topical, relying on well crafted, domain-specific language, and possessing a broad and unevenly distributed coverage of legal issues. [20].

- **Content**. Various well-known functions from the literature (e.g. Jaccard, cosine similarity etc.) can be employed at computing the textual similarity of legal documents. In this work, we choose cosine similarity as a similarity measure, thus the textual similarity between documents u and v, with term vectors U and V is:

$$S_c(u,v) = \cos(u,v) = \frac{U \cdot V}{\| U \| \| V \|} \tag{2}$$

- **Topical Taxonomies**. We consider the selection of categories that cover many different interpretations in respect to legal users' information needs. Topical similarity of two documents having topical sets X_u and X_v is calculated using the Jacard similarity

$$S_x(u,v) = \frac{|X_u \cap X_v|}{|X_u \cup X_v|} \tag{3}$$

- **Time**. Time is a valuable diversification dimension, since in many cases, subtopics associated to queries in the legal domain are temporally ambiguous due to dynamic evolution and dependencies across the legislation system. Time similarity, between documents u and v, having timestamps t_u and t_v is calculated on the difference of their normalized timestamps with Min-Max Normalization.

$$S_t(u,v) = 1 - |t_{norm}(u) - t_{norm}(v)| \tag{4}$$

- **Readability**. A document's writing quality is a diversification factor, since it expresses comprehensibility of the document itself. The most influential quantitative measure of text quality is the Flesch Reading Ease Score[3], which produces a numerical score, with higher numbers indicating easier texts. Readability similarity, between documents u and v, having readability scores r_u and r_v, is calculated on the difference of their normalized scores with Min-Max Normalization.

$$S_r(u,v) = 1 - |r_{norm}(u) - r_{norm}(v)| \tag{5}$$

Following diversification features formalization we define:

- **Document Similarity**. The final similarity score of two documents u, v is calculated as a linear weighted function of the Content, Topical Taxonomies, Time and Readability score

---

[3] http://en.wikipedia.org/wiki/Readability.

$$sim(u,v) = \sum_{i=1}^{|4|} w_i \; feat_i(u,v) = w_1 \; S_c(u,v) + w_2 \; S_x(u,v) + w_3 \; S_t(u,v) + w_4 \; S_r(u,v) \quad (6)$$

with weights $\sum_{i=1}^{|4|} w_i = 1$.
- **Document Distance.** The distance of two documents is

$$d(u,v) = 1 - sim(u,v) \quad (7)$$

- **Query Document Similarity.** The relevance of a query q to a given document u can be assigned as the initial ranking score obtained from the IR system, or calculated using the similarity measure e.g. cosine similarity on the corresponding term vectors

$$r(q,u) = S_c(q,u) \quad (8)$$

3.4 Diversification Heuristics

Most of existing diversification methods first retrieve a set of documents based on their relevance scores, and then re-rank the documents so that the top-ranked documents are diversified to cover more query subtopics. Since the problem of finding an optimum set of diversified documents is NP-hard, a greedy algorithm is often used to iteratively select the diversified document. Let N the document set, $u,v \in N$, $r(q,u)$ the relevance of u to the query q, $d(u,v)$ the distance of u and v, $S \subseteq N$ with $|S| = k$ the number of documents to be collected and $\lambda \in [0..1]$ a parameter used for setting trade-off between relevance and similarity. In this paper, we focus on the following representative diversification methods discussed in the previous section.

- **MMR:** Maximal Marginal Relevance [3], a greedy method to combine query relevance and information novelty, iteratively constructs the result set S by selecting documents that maximizes the following objective function

$$f_{MMR}(u,q) = (1 - \lambda) \; r(u,q) + \lambda \sum_{v \in S} d(u,v) \quad (9)$$

MMR incrementally computes the standard relevance-ranked list when the parameter $\lambda = 0$, and computes a maximal diversity ranking among the documents in N when $\lambda = 1$. For intermediate values of $\lambda \in [0..1]$, a linear combination of both criteria is optimized. The set S is usually initialized with the document that has the highest relevance to the query. Since the selection of the first element has a high impact on the quality of the result, MMR often fails to achieve optimum results.
- **MaxSum:** The Max-sum diversification objective function [13] aims at maximizing the sum of the relevance and diversity in the final result set. This is achieved by a greedy approximation algorithm that selects a pair of documents that maximizes Eq. 10 in each iteration.

$$f_{MAXSUM}(u,v,q) = (1 - \lambda) \; (r(u,q) + r(v,q)) + 2\lambda \; d(u,v) \quad (10)$$

where (u, v) is a pair of documents, since this objective considers document pairs for insertion. When $|S|$ is odd, in the final phase of the algorithm an arbitrary element in N is chosen to be inserted in the result set S.

- **MaxMin:** The Max-Min diversification objective function [13] aims at maximizing the minimum relevance and dissimilarity of the selected set. This is achieved by a greedy approximation algorithm that select a document that maximizes Eq. 11 in each iteration.

$$f_{MAXMIN}(u, q) = (1 - \lambda) \ r(u, q) + \lambda \min_{v \in S} d(u, v) \tag{11}$$

where $\min_{v \in S} d(u, v)$ is the minimum distance of u to the already selected documents in S.

- **MonoObjective:** MonoObjective [13] combines the relevance and the similarity values into a single value for each document. It is defined as:

$$f_{MONO}(u, q) = r(u, q) + \frac{\lambda}{|N| - 1} \sum_{v \in N} d(u, v) \tag{12}$$

- **LexRank:** LexRank [6], is a stochastic graph-based method for computing relative importance of textual units. A document is represented as a network of inter-related sentences, and a connectivity matrix based on intra-sentence similarity is used as the adjacency matrix of the graph representation of sentences. In LexRank scoring formula 13, Matrix B captures pairwise similarities of the sentences and square matrix A, which represents the probability of jumping to a random node in the graph, has all elements set to $1 = M$, where M is the number of sentences.

$$p = [\lambda \ A + (1 - \lambda) \ B]^T p \tag{13}$$

In our setting, instead of sentences, we use documents that are in the initial retrieval set N for a given query and thus set Matrix B as the connectivity matrix based on document similarity.

- **Biased LexRank:** Biased LexRank [27] provides for a LexRank extension that takes into account a prior document probability distribution e.g. the relevance of documents to a given query.

$$p = [\lambda \ A + (1 - \lambda) \ B]^T p \tag{14}$$

In Biased LexRank scoring formula 14, we set Matrix B as the connectivity matrix based on document similarity for all documents that are in the initial retrieval set N for a given query and Matrix A elements proportional to the query document relevance.

- **DivRank:** DivRank balances popularity and diversity in ranking, based on a time-variant random walk. In contrast to PageRank which is based on stationary probabilities, DivRank assumes that transition probabilities change over time, they are reinforced by the number of previous visits to the target vertex. If $p_T(u, v)$ is the transition probability from any vertex u to vertex v

at time T, $p^*(d_j)$ is the prior distribution that determines the preference of visiting vertex d_j, and $p_0(u,v)$ is the transition probability from u to v prior to any reinforcement then,

$$p_T(d_i,d_j) = (1-\lambda).p^*(d_j) + \lambda.\frac{p_0(d_i,d_j).N_T(d_j)}{D_T(d_i)} \qquad (15)$$

where $N_T(d_j)$ is the number of times the walk has visited d_j up to time T and,

$$D_T(d_i) = \sum_{d_j \in V} p_0(d_i,d_j)N_T(d_j) \qquad (16)$$

Since DivRank is a query independent ranking model, we introduce a query dependent prior and thus utilize DivRank into a query dependent ranking schema. In our setting, we use documents that are in the initial retrieval set N for a given query q, create the citation network between those documents and apply DivRank algorithm to select top-k divers documents in S.

– **Grasshopper:** A similar with DivRank ranking algorithm, is described in [37]. This model starts with a regular time-homogeneous random walk and in each step the vertex with the highest weight is set as an absorbing state.

$$p_T(d_i,d_j) = (1-\lambda).p^*(d_j) + \lambda.\frac{p_0(d_i,d_j).N_T(d_j)}{D_T(d_i)} \qquad (17)$$

where $N_T(d_j)$ is the number of times the walk has visited d_j up to time T and,

Since Grasshopper and DivRank utilize a similar approach and will ultimately present similar results we utilized Grasshopper distinctively from DivRank. In particularly, instead of creating the citation network of documents belonging to the initial result set, we form the adjacency matrix based on document similarity.

4   Experimental Setup

In this section, we describe the legal corpus we use, the set of query topics, the respective methodology for subjectively annotating our corpus with relevance judgments for each query, as well as the metrics employed for the evaluation assessment. Finally, we provide our diversification results along with a short discussion.

4.1   Legal Corpus

Our corpus contains 63,742 precedential legal cases from the Supreme Court of the United States[4]. The cases were originally downloaded from CourtListener[5].

---

[4] http://www.supremecourt.gov/.

[5] http://www.courtlistener.com, a free legal research website containing legal opinions from federal and state courts.

The legal corpus contains all cases from the Supreme Court of the United States, covering more than two centuries of legal history, spanning from 1754 up to 2015. We extracted from the cases text all the necessary information for our feature selection framework e.g. relationships to other documents, date of Judgment. Since our corpus was initially unclassified, we acquired topical taxonomies from the Supreme Court Database[6] using commonly shared unique identification variable SCDB Case ID. Topical taxonomies within Supreme Court Database are the outcome of a manual analysis and interpretation of the legal provisions considered in each case. Our text pre-processing step involved standard stop word removal and porter stemming. Finally our index, build with log based $tf - idf$ indexing technique contains a total of 63,742 documents, 174,370 unique terms and 54,243,977 terms in total. Overall we believe that the corpus is of size to demonstrate the effectiveness of our proposed approach.

4.2   Evaluation Metrics

We evaluate diversification methods using metrics employed in TREC Diversity Tasks[7]. In particular we report

- **a-nDCG:** a-Normalized Discounted Cumulative Gain [4] metric quantifies the amount of unique aspects of the query q that are covered by the $top - k$ ranked documents. We use $a = 0.5$, as typical in TREC evaluation.
- **Precision-IA:.** Precision-Intent Aware [1] accounts for the ratio of relevant documents for different subtopics within the $top - k$ items.
- **Subtopic-Recall:** Subtopic-Recall [36] quantifies the amount of unique aspects of the query q that are covered by the $top - k$ ranked documents

4.3   Relevance Judgements

One of the difficulties in evaluating methods designed to introduce diversity in the legal document ranking process is the lack of standard testing data. Evaluating diversification requires a data corpus, a set of query topics and a set of relevance judgments, preferably made by human assessors for each query. While TREC added a diversity task to the Web track in 2009, this dataset was designed assuming a general web search, and so it not possible to adapt it to our setting. In the absence of a standard dataset specifically tailored for this purpose and since it was not feasible to involve legal experts in this sort of exploratory study, we looked for an subjective way to evaluate and assess the performances of various diversification methods on our corpus. We do acknowledge the fact that the process of automatic query generation is at best an imperfect approximation of what a real person would do. To this end we employed the following method:

User Profiles/Queries. We used West Law Digest Topics[8] as candidates user queries. Each topic was issued as candidate query to our retrieval system.

---

[6] http://scdb.wustl.edu.
[7] http://trec.nist.gov/data/web10.html.
[8] A taxonomy of identifying points of law from reported cases and organizing them by topic and key number. It is used to organize the entire body of American law.

Table 1. West Law Digest Topics as user queries

31:	Antitrust and Trade Regulation	61:	Breach of Marriage Promise
84:	Commodity Futures Trading Regulation	199:	Implied and Constructive Contracts
376:	Unemployment Compensation	398:	Merit Systems Protection

Outlier queries, whether too specific/rare or too general, where removed using the interquartile range, below or above values $Q1$ and $Q3$, sequentially in terms of number of hits in the result set and score distribution for the hits, demanding in parallel a minimum cover of $min|N|$ results. In total, we kept 330 queries The following Table 1 provides a sample of the topics we further consider as user queries.

Query assessments and ground-truth. For each topic/query we kept the $top - n$ results. An LDA topic model, using an open source implementation[9], was trained on the $top - n$ results for each query. From the resulting topic distributions for each document, with an acceptance threshold of 15 %, we consider relevance judgments for each query/ document and subtopic. In other words, we consider the topics created from LDA as aspects of each query, and based on the topic/ document distribution we can infer whether a document is relevant for an aspect. In total, we acquired 1,650 subtopics for all the 330 queries. We have made available[10] our complete dataset, ground-truth data, queries and relevance assessments in standard qrel format, as to encourage progress on the diversification in legal IR.

4.4   Results

As a baseline to compare diversification methods, we consider the simple ranking produced from an IR system using cosine similarity and log based $tf - idf$ indexing schema. For each query, our initial set N contains the $top - n$ query results. For all variations that apply diversity, we set a fixed weight for the diversity score to $\lambda = 0.5$ and, thus, the weight for query-to-document similarity is $1 - \lambda = 0.5$. We present the evaluation results for the methods employed, using the aforementioned evaluation metrics, at cut-off values of 5, 10 and 20, as typical in TREC evaluations. Note that each of the diversification variations, is applied in combination with each of the diversification algorithms and for each user query. Table 2 summarizes testing parameters and their corresponding ranges.

We firstly employed the diversification methods using only content similarity as used in most works handling diversification, e.g. in web search results diversification. That is, weights on features time, readability and topical categories were set to zero. Table 3 presents results of the diversification methods.

---

[9] http://mallet.cs.umass.edu/.
[10] https://github.com/mkoniari/MultiLegalDiv.

Table 2. Parameters tested in the experiments

Parameter	Range
Algorithms tested	MMR, MaxMin, MaxSum, Mono, LexRank, BiasedLexRank, DivRank, GrassHopper
Tradeoff λ values	0.5
Candidate set size n $= \|N\|$	100
Result set size k $= \|S\|$	5, 10, 20
# of sample queries	330
Exp. 1 Feature weights	Content 1.0, Time, 0 Readability 0, Topical Taxonomies, 0
Exp. 2 Feature weights	Content 0.6, Time, 0.13, Readability 0.13, Topical Taxonomies, 0.14

Statistically significant values, using the paired two-sided t-test with $p_{value} < 0.05$ are denoted with $°$ and with $p_{value} < 0.01$ with $*$.

MMR and DivRank are the best diversification strategies for different evaluation metrics for $N = 100$ and $k = 30$. In particular, MMR outperforms all other methods in terms of the nDCG and Subtopic-Recall metrics, whereas DivRank achieves the highest score for the Precision IA metric. Interestingly, text summarization methods (LexRank, Biased LexRank and GrassHopper, as it was utilized without a network citation graph) failed to improve the baseline ranking. They actually constantly perform lower than the baseline ranking at all levels across all metrics. From web search result diversification methods, MMR almost constantly achieves better results in respect to the rest methods for all metrics, with the exception of nDCG@5 where MaxMin performs better. Graph diversification method, DivRank, outperforms other methods in Precision IA metric at all levels, but generally fails to improve over the baseline ranking for nDCG and Subtopic-Recall metrics.

As a second experiment, we incorporate all ranking features into the diversification methods while computing the similarity scores for the documents pairs, except DivRank where the citation network between documents in the result set for each query is utilized. In particular we set the following weights on ranking features: Content 0.6, Time 0.13, Readability 0.13 and Topical Taxonomies 0.14. In Table 4 we present results of the second experiment, alongside with indicators for statistically significant values.

It is clear that with the incorporation of the suggested ranking features all of the approaches tend to perform better than using only content similarity. We also notice a similar trending behavior with the one discussed for Table 3. MMR and DivRank are the best diversification strategies for different evaluation metrics. Text summarization methods, although with better scores, once again fail to improve over the baseline ranking. MMR almost constantly achieves better results in respect to the rest methods for all metrics, with the exception of Precision IA where MaxMin and DivRank perform better.

Table 3. Retrieval Performance of the diversification algorithms using only content similarity for $N = 100$ and $k = 30$. Highest scores are shown in bold. Statistically significant values, using the paired two-sided t-test with $p_{value} < 0.05$ are denoted with $°$ and with $p_{value} < 0.01$ with $*$

Method	a-nDCG			Precision IA			ST recall		
	@5	@10	@20	@5	@10	@20	@5	@10	@20
IR	0,532	0,595	0,656	0,314	0,313	0,314	0,688	0,833	0,948
MMR	**0,571***	**0,643***	**0,695***	0,315°	0,321	0,322*	**0,783***	**0,923***	**0,977***
MaxSum	0,549	0,620*	0,675*	0,300*	0,305°	0,303*	0,744*	0,880*	0,969°
MaxMin	0,568*	0,633*	0,686*	0,319	0,319*	0,319*	0,777*	0,907*	0,976*
MonoObjective	0,541°	0,602°	0,664*	0,313	0,310	0,312	0,713*	0,844	0,960°
LexRank	0,487	0,532*	0,586*	0,308*	0,313	0,320	0,584*	0,705*	0,820*
BiasedLexRank	0,488*	0,533*	0,587*	0,309	0,314	0,320	0,585*	0,708*	0,821*
DivRank	0,533	0,589	0,635	**0,320**	**0,326°**	**0,326°**	0,667	0,803	0,888*
GrassHopper	0,492*	0,542*	0,598*	0,310	0,316	0,322°	0,592*	0,725*	0,846*

Table 4. Retrieval Performance of the diversification algorithms using all ranking features for $N = 100$ and $k = 30$. Highest scores are shown in bold. Statistically significant values, using the paired two-sided t-test with $p_{value} < 0.05$ are denoted with $°$ and with $p_{value} < 0.01$ with $*$

Method	a-nDCG			Precision IA			ST recall		
	@5	@10	@20	@5	@10	@20	@5	@10	@20
IR	0,532	0,595	0,656	0,314	0,313	0,314	0,688	0,833	0,948
MMR	**0,586***	**0,657***	**0,709***	0,321	0,321°	0,325*	**0,815***	**0,939***	**0,989***
MaxSum	0,564*	0,636*	0,689*	0,306	0,306°	0,308°	0,779*	0,913*	0,977*
MaxMin	0,581*	0,650*	0,702*	**0,322**	0,322	0,321*	0,793*	0,931*	0,983*
MonoObjective	0,550*	0,612*	0,673*	0,321°	0,313	0,314	0,716*	0,857°	0,968*
LexRank	0,484*	0,532	0,587*	0,304*	0,306	0,316	0,604*	0,724*	0,839*
BiasedLexRank	0,488*	0,537*	0,592*	0,304	0,308	0,316	0,607*	0,731*	0,845*
DivRank	0,533	0,589	0,635	0,320	**0,326***	**0,326°**	0,667	0,803	0,888*
GrassHopper	0,504°	0,555*	0,612*	0,306	0,308	0,317	0,649	0,760*	0,880*

Overall it is demonstrated that more refined criteria than plain content similarity can improve the effectiveness of the diversification process. Furthermore web search diversification techniques outperform other approaches (e.g. summarization-based, graph-based methods) in the context of legal search diversification. Graph based diversification, DivRank generally fails to improve over the baseline ranking but outperforms other methods in terms of Precision IA metric. We do plan to further examine the performance of graph based diversification heuristics, in terms of citation network criteria and ranking features, as to enrich search results with otherwise hidden aspects of the legal query space.

5   Conclusions

In this paper, we studied the novel problem of diversifying legal documents by incorporating diversity in four dimensions: content, time, topical taxonomies and readability. We adopted and compared the performance of several state of the art methods from the web search, network analysis and text summarization domains as to handle the problems' challenges. We evaluated all the methods/ dimensions using a real data set from the Common Law domain that we subjectively annotated with relevance judgments for this purpose. Our findings demonstrate the effectiveness of our proposed method, as opposed to applying plain content diversity on legal search results.

A challenge we faced in this work was the lack of ground-truth. We hope on an increase of the size of truth-labeled data set in the future, which would enable us to draw further conclusions about the diversification techniques. In the future we plan to perform an exhaustive evaluation of all the methods as to provide insights for legal IR systems between reinforcing relevant documents, result set similarity, or sampling the information space around the legal query, result set diversity.·

References

1. Agrawal, R., Gollapudi, S., Halverson, A., Ieong, S.: Diversifying search results. In: Proceedings of WSDM 2009, pp. 5–14 (2009)
2. Biagioli, C., Francesconi, E., Passerini, A., Montemagni, S., Soria, C.: Automatic semantics extraction in law documents. In: Proceedings of ICAIL 2005 (2005)
3. Carbonell, J., Goldstein, J.: The use of mmr, diversity-based reranking for reordering documents and producing summaries. In: Proceedings of SIGIR 1998, pp. 335–336 (1998)
4. Clarke, C.L.A., Kolla, M., Cormack, G.V., Vechtomova, O., Ashkan, A., Büttcher, S., MacKinnon, I.: Novelty and diversity in information retrieval evaluation. In: Proceedings of SIGIR 2008 (2008)
5. Cronen-Townsend, S., Croft, W.B.: Quantifying query ambiguity. In: Proceedings of Human Language Technology Research 2002 (2002)
6. Erkan, G., Radev, D.R.: LexRank: graph-based lexical centrality as salience in text summarization. J. Artif. Int. Res. **22**(1), 457–479 (2004)
7. Farzindar, A., Lapalme, G.: Legal text summarization by exploration of the thematic structures and argumentative roles. In: Text Summarization Branches Out Workshop Held in Conjunction with ACL, pp. 27–34 (2004)
8. Farzindar, A., Lapalme, G.: Letsum, an automatic legal text summarizing system. In: Proceedings of JURIX 2004, pp. 11–18 (2004)
9. Fowler, J.H., Johnson, T.R., Spriggs, J.F., Jeon, S., Wahlbeck, P.J.: Network analysis and the law: measuring the legal importance of precedents at the U.S. Supreme Court. Polit. Anal. **15**(3), 324–346 (2006)
10. Fowler, J.H., Jeon, S.: The authority of Supreme Court precedent. Soc. Netw. **30**(1), 16–30 (2008)
11. Galgani, F., Compton, P., Hoffmann, A.: Citation based summarisation of legal texts. In: Anthony, P., Ishizuka, M., Lukose, D. (eds.) PRICAI 2012. LNCS (LNAI), vol. 7458, pp. 40–52. Springer, Heidelberg (2012). doi:10.1007/978-3-642-32695-0_6

12. Gangemi, A., Sagri, M.T., Tiscornia, D.: Metadata for content description in legal information. In: Proceedings of LegOnt Workshop on Legal Ontologies (2003)
13. Gollapudi, S., Sharma, A.: An axiomatic approach for result diversification. In: Proceedings of WWW 2009, pp. 381–390 (2009)
14. Grabmair, M., Ashley, K.D., Chen, R., Sureshkumar, P., Wang, C., Nyberg, E., Walker, V.R.: Introducing LUIMA. In: Proceedings of ICAIL 2015 (2015)
15. Hoekstra, R., Breuker, J., di Bello, M., Boer, A.: The lkif core ontology of basic legal concepts. In: Proceedings of the Workshop on Legal Ontologies and Artificial Intelligence Techniques (LOAIT 2007) (2007)
16. Hu, S., Dou, Z., Wang, X., Sakai, T., Wen, J.R.: Search result diversification based on hierarchical intents. In: Proceedings of CIKM 2015, pp. 63–72 (2015)
17. Klein, M.C., Van Steenbergen, W., Uijttenbroek, E.M., Lodder, A.R., van Harmelen, F.: Thesaurus-based retrieval of case law. In: Proceedings of JURIX 2006, vol. 152, p. 61 (2006)
18. Koniaris, M., Anagnostopoulos, I., Vassiliou, Y.: Network analysis in the legal domain: a complex model for european union legal sources. In: Physics and Society, Cornell University Library, arXiv (2015). http://arxiv.org/abs/1501.05237
19. Langville, A.N., Meyer, C.D.: A survey of eigenvector methods for web information retrieval. SIAM Rev. 47(1), 135–161 (2005)
20. Lu, Q., Conrad, J.G., Al-Kofahi, K., Keenan, W.: Legal document clustering with built-in topic segmentation. In: Proceedings of CIKM 2011, p. 383 (2011)
21. Marx, S.M.: Citation networks in the law. Jurimetrics J. 10(4), 121–137 (1970)
22. Mei, Q., Guo, J., Radev, D.: Divrank: the interplay of prestige and diversity in information networks. In: Proceedings of KDD 2010, pp. 1009–1018 (2010)
23. Loza Mencía, E., Fürnkranz, J.: Efficient pairwise multilabel classification for large-scale problems in the legal domain. In: Daelemans, W., Goethals, B., Morik, K. (eds.) ECML PKDD 2008. LNCS, vol. 5212, pp. 50–65. Springer, Heidelberg (2008). doi:10.1007/978-3-540-87481-2_4
24. Moens, M.F.: Summarizing court decisions. Inf. Process. Manage. 43(6), 1748–1764 (2007)
25. Moens, M.: Innovative techniques for legal text retrieval. Artif. Intell. Law 9(1), 29–57 (2001)
26. van Opijnen, M.: Citation analysis and beyond: in search of indicators measuring case law importance. In: Proceedings of JURIX 2012, pp. 95–104 (2012)
27. Otterbacher, J., Erkan, G., Radev, D.R.: Biased LexRank: passage retrieval using random walks with question-based priors. Inf. Process. Manage. 45(1), 42–54 (2009)
28. Santos, R.L.T., Macdonald, C., Ounis, I.: Search result diversification. Found. Trends Inf. Retrieval 9(1), 1–90 (2015)
29. Santos, R.L., Macdonald, C., Ounis, I.: Exploiting query reformulations for web search result diversification. In: Proceedings of WWW 2010, pp. 881–890 (2010)
30. Saravanan, M., Ravindran, B., Raman, S.: Improving legal information retrieval using an ontological framework. Artif. Intell. Law 17(2), 101–124 (2009)
31. Schweighofer, E.: Semantic indexing of legal documents. In: Francesconi, E., Montemagni, S., Peters, W., Tiscornia, D. (eds.) Semantic Processing of Legal Texts. LNCS, vol. 6036, pp. 157–169. Springer, Heidelberg (2010). doi:10.1007/978-3-642-12837-0_9
32. Schweighofer, E., Liebwald, D.: Advanced lexical ontologies and hybrid knowledge based systems: first steps to a dynamic legal electronic commentary. Artif. Intell. Law 15(2), 103–115 (2007)

33. Wang, J., Zhu, J.: Portfolio theory of information retrieval. In: Proceedings of SIGIR 2009 (2009)
34. Winkels, R., Boer, A., Plantevin, I.: Creating context networks in dutch legislation. In: Proceedings of JURIX 2013, vol. 259, p. 155 (2013)
35. Winkels, R., Boer, A., Vredebregt, B., van Someren, A.: Towards a legal recommender system. In: Proceedings of JURIX 2014, pp. 169–178 (2014)
36. Zhai, C.X., Cohen, W.W., Lafferty, J.: Beyond independent relevance. In: Proceedings of SIGIR 2003 (2003)
37. Zhu, X., Goldberg, A.B., Van Gael, J., Andrzejewski, D.: Improving diversity in ranking using absorbing random walks. In: HLT-NAACL, pp. 97–104 (2007)

Generating Multiple Diverse Summaries

Natwar Modani, Balaji Vasan Srinivasan[⊠], and Harsh Jhamtani

BigData Experience Lab, Adobe Research, Bangalore, India
{nmodani,balsrini,jhamtani}@adobe.com

Abstract. Authors often re-purpose existing content to create shorter versions for other channels. Automatic summarization techniques can be used to generate a candidate content that can be further fine-tuned by the author. Existing work in automatic summarization primarily focus on providing a single succinct summary. However, this may not suit the needs of a content author or curator, who may want to repurpose/select the content from several alternative candidates. In this paper, we propose an approach to generate multiple diverse summaries, so that authors can choose an appropriate summary without compromising on the summary quality. Our approach can be utilized in conjunction with a large class of extractive summarization techniques, and we illustrate our approach with several summarization techniques. We experimentally show that our approach results in fairly diverse summaries, without compromising the quality of the summaries with respect to the single summary generated by the corresponding base methods.

1 Introduction

Summarization of text content is an important problem that has been studied extensively in the literature. The primary use for summarization techniques is to save time for the end users by providing them with a quick overview of the article. Another potential use is for re-purposing existing content by an author. Starting from a longer version, the author may use the summarization techniques to generate a candidate piece of content of a desired length, and then fine-tune it to suit the specific purpose. The content authors would typically have several criteria in consideration while creating the summary, including the representativeness of the summary with respect to the original content, the diversity of the content in the summary, coherence of the content in summary and suitability of the summary for certain audience segments.

Existing summarization techniques produce a single summary, which attempts to optimize the representativeness and diversity [6,8,10]. While this may suit the needs of content consumption, it is desirable to generate multiple candidate summaries for better serving the author needs so that the author can choose one to repurpose. Given that the documents typically have high degree of redundancy (which is the reason in the first place that a good summary can be generated), one may substitute each of the selected sentences with alternatives available, and the quality of the summary may not suffer by a large amount.

© Springer International Publishing AG 2016
W. Cellary et al. (Eds.): WISE 2016, Part I, LNCS 10041, pp. 190–198, 2016.
DOI: 10.1007/978-3-319-48740-3_13

One may use simple ways, such as running multiple summarization algorithms and taking the summaries produced by them, or running the probabilistic algorithms (e.g., LexRank [2], DivRank [6], Dragon [10], etc.) multiple times with different random seeds. However, in our experiments, we found that the diversity of summaries produced in this manner is not satisfactory.

In this paper, we propose a system which will generate multiple extractive summaries of a given text corpus that are diverse (i.e., do not have large degree of overlap in the set of selected sentences). The proposed approach can be used with any iterative extractive summarization technique that assigns a score to each text unit for selecting the summary, e.g., [1,2,6,8], etc.

There are two key steps in our proposed approach. First, we convert the scores into a probability distribution (e.g., by normalizing with respect to the sum of scores for all the text units). We select the next text unit to include in the summary by sampling from the resulting probabilistic distribution. After including a text unit in summary, we may update the scores for each of the remaining sentences. When we have selected sufficient sentences for the summary, we stop this process. This generates one summary.

The second step is to multiply the scores of the sentences selected in the previous summaries by a damping factor, which reduces the probability of the same sentences to be selected again, without completely precluding them from inclusion. Also, these sentences would still contribute to the scores of the sentences similar to them (in several approaches), thus ensuring that the representativeness of the next summary is not compromised. We now repeat the process of finding the summary desired number of times to produce multiple summaries.

2   Related Work

There has been considerable prior research on summarization, identifying and ranking relevant content and diversifying the ranking results. Text summarization could be broadly categorized into two types: abstractive and extractive. In abstractive summary [3], new content is generated from the available input text, while in extractive summary, segments from input text are use to construct the summary. In this paper, we will focus on extractive summarization techniques.

Automatic extractive text summarization has been widely studied by researchers. Graph based algorithms, such as TextRank [7] and LexRank [2] outperformed SVM regression based models used in social communities [4] as shown by Wu et al. [11]. These methods build a graph similar to link-based approaches for ranking web pages using the similarity relationships among sentences. Solving the summarization problem for product reviews, [8] proposed a graph based formulation which uses a fast and scalable greedy algorithm. They considered the informativeness and diversity of the sentences to select the summary of the reviews.

Another class of approaches optimizes a cost function based on several parameters subject to constraints around the summary. For example, Maximum Marginal Relevance (MMR) [1] based summarization generates a summary that is relevant to input content subject to constraints that the selected content ensures diversity. Later DivRank [6] used Vertex Reinforced Random Walks [9] to introduce diversity in the prestige ranking.

However, none of these approaches generate multiple summaries for a given article or corpus of text. In this paper, we have developed an approach that can be used in conjunction with several of these extractive summarization techniques where a score is assigned to each units and sentence with the top score is selected iteratively. The algorithm can either repeatedly select the units to meet a length requirement (e.g. LexRank [2], DivRank [6]) or alternatively modify the scores after the selection of the top text unit to select the subsequent units (e.g. MMR [1] and graph based approach [8]). The proposed approach works in conjunction with both these class of approaches.

3   Algorithms

An extractive summarization framework takes an input content along with the desired size of summary in appropriate units, e.g., 3 sentences, or 60 words, etc. We will refer to this as available budget. The summarization routine will divide the article into units (where the units could be sentences or sentence segments, etc.) and each unit is assigned a goodness score indicating the value of including this unit into the summary. This value would typically be based on the information content of the unit, its similarity to the other units in the article, and may also depend on the cost of the unit. The score is then used to select the desired summary. In our approach, once we have the score for each unit in the input content, we follow the steps below to achieve multiple summaries.

1. We associate a cost with each unit. If the available budget becomes less than the cost of the unit, then this unit is discarded.
2. We assign a probability score to each unit by normalizing its score (or any function thereof) by the sum of scores of all units (or any function thereof, correspondingly).
3. We select a unit by drawing a sample from this probability distribution.
4. We add the selected unit to the current summary, update the available budget by subtracting its cost, and mark this unit as 'not available'.
5. The scores of the other units are adjusted appropriately. Typically, the score for a "text unit" would be decreased if it is related to the selected "text unit" to ensure summary diversity.

The series of steps are repeated as long as there is a text unit within the available budget that is not yet included in the summary. This results in generating a single summary. To produce next summary, we multiply the scores by a damping level which is set to 1 initially for all sentences (i.e., before starting to generate the first summary). When a sentence is included in a summary, the damping level for that sentence is multiplied by a damping factor γ (where $0 \leq \gamma \leq 1$) to reduce the chance of selecting the same sentence in multiple summaries. The choice of value of γ controls the tradeoff between inter-summary diversity and the quality of the summaries chosen. A low value of γ (close to 0) would lead to significant reduction in probability of a sentence being selected in multiple summaries, leading to a high degree of inter-summary diversity. At the same time, if the level of redundancy level in the corpus is low, it may lead to deterioration in the quality of subsequent summaries. On the other hand, a high value of γ (close to 1) will not reduce the probability of the same sentence being selected in multiple summaries significantly, hence the quality may not deteriorate much, but the inter-summary diversity may be low. We will illustrate how to apply this framework with two specific approaches.

Graph based Approach: In this approach [8], every sentence is represented as a node $v \in V$ in a graph $G = (V, E, W)$. A reward r_i indicating the number of meaningful entities in a text unit is assigned to every node initially. We take nouns, verbs, adjectives, adverbs and numbers as the meaningful entities in all our methods. The edge $e = (u, v) \in E$ indicates that the sentence u captures some part of the information provided by the sentence v. The extent to which u captures the information present in v is indicated by the edge weight $w \in W$, defined as, $w_{uv} = |u \cap v|/|v|$, where the norm $|v|$ indicates the number of meaningful terms in sentence v. When a sentence, say u, is included in the summary in l^{th} step, the reward for a sentence v that has an incoming edge from it (i.e., u captures some information about v) is reduced. The updated reward for v is given by, $r_u^l = r_u^{l-1} * (1 - w_{uv})$ i.e., the reward of each neighbor of a selected node in step l, is its previous reward multiplied by the dissimilarity with the selected node. We call the current value of reward as the discounted reward for the sentence.

The gain G_v of including a node (i.e., text unit) v in the summary at step l is defined as the weighted sum of the current discounted individual reward values of all the neighbors of v, $G_v^l = r_v^{l-1} + \sum_{u \in N_v} r_u^{l-1} * w_{uv}$, where N_v is the set of neighboring nodes for v. The gain score is divided by the cost of the sentence to get the goodness score. While in [8], the sentence that yields the maximum gain to cost ratio (or goodness score as we are calling it here) is chosen, we convert these goodness scores into a probability distribution by normalizing them with respect to the sum of all such goodness scores given by, $p_v^l = (G_v^l/c_v)/\sum_{\forall u}(G_u^l/c_u)$. Once a text unit is selected, the rewards of the neighbor nodes is reduced as mentioned above. This way, inclusion of similar text units is avoided and (within summary) diversity is ensured. We continue selecting sentences till there is no other (unselected) sentence available with a cost no more than remaining budget. To select subsequent summaries we damp the scores generated by a damping factor (as described before) and continue the selection as described here. Subsequent summaries are generated by damping the scores of the selected text units.

MMR Based Approach: MMR [1] is an iterative algorithm, such that at each iteration the highest scored sentence is selected from the set of unselected sentences. The score is based on the relevance of the sentence to the query, with a penalty term for its similarity to the already chosen summary. In the absence of a query, we modify the relevance to number of meaningful entities in the text unit similar to the Graph based approach. The goodness score for MMR is thus given by, $G(v) = \lambda * R(v) - (1 - \lambda) * \max_{u \in S}(w_{uv}) \ \forall v \notin S$ where $R(v)$ is the reward of sentence v and is given by the number of meaningful entities in the sentence v, normalized by the length of the sentence. Also, w_{uv} represent the amount of information about v captured by u, computed in the same way as the edge weights in the Graph based approach. Parameter $0 \leq \lambda \leq 1$ controls the trade-off between importance of reward $R(v)$ and penalty for overlap with already selected sentences. Higher the value of λ, higher would be the importance given to reward.

At each iteration, every unselected sentence is assigned a score between -1 and 1. The score is converted into a probability distribution by scaling and normalizing the scores. The candidate sentence is selected based on this distribution. Now, we select a sentence according to this probability distribution. Again, subsequent summaries are generated by damping the scores of the selected text units.

DivRank: DivRank [6] balances between centrality and diversity and is based on a vertex reinforced random walk. Starting with a very similar setup to LexRank, DivRank modifies the definition of transition probability (edge weight) between 2 nodes as, $p_T(u, v) = (1 - \lambda)p^*(v) + \lambda \frac{p(u,v)N_T(u)}{D_T(u)}$, where, $D_T(u) = \sum_{v \in V} p(u, v)N_T(u)$. Here $p^*(v)$ is the prior preference of visiting the node v and is motivated by the reinforced random walk. $p(u, v)$ is the organic transition probability computed similar to LexRank. $N_T(v)$ is the number of times the random walk has visited v. The ranks obtained are converted into a probability distribution to sample the selection for summary.

4   Measuring Diversity Across Multiple Summaries

Here, we propose a metric to capture the level of diversity formally. As the basic unit of text is a sentence, we measure the diversity based on non-overlap amongst the summaries at the sentence level. Note that the overlap at word level is not penalized in this measure. Let T_i be the set of text units in the ith summary. The diversity of the information content across the multiple summaries is then given by,

$$D_s = \frac{\left| \bigcup_{i \in S} T_i \right|}{\sum_{i \in S} |T_i|} \tag{1}$$

where D is the measure of sentence level diversity and S is the set of all the summaries. The numerator is the cardinality of the distinct text units chosen across all the summaries. The denominator measures the sum of the text units selected in each summary. Equation 1 measures the amount of overlap in the text units selected across the different summaries. The diversity value is 1 when distinct text units are chosen by the algorithm and hence the summaries are most diverse. If all the summaries are the same, the diversity value is (approximately) $1/m$, where m is the number of summaries generated.

5   Experimental Results

To test the proposed approach, we generate summaries from a corpus of text from helpx.adobe.com. The corpus corresponds to help documentation present on Adobe.com website. The corpus consists of 334 articles. To ensure that a meaningful summary can be generated, we selected only the articles with minimum 300 words. This resulted in a corpus with 103 articles with article lengths varying between 301 words to 2745 words with average number of words as 653.1. The number of sentences varied between 8 to 155, with average number of sentences being 35.87.

First, we find the degree of diversity achieved in the DivRank method by using a different seed for different runs. Figure 1a shows our results. We can see that the diversity level is low. We conducted experiments with LexRank as well and found that for almost all articles, the multiple summaries generated were not different from each other. Without the modifications proposed in this paper, one would not expect any change to the summary across the runs for the graph based and MMR based summarization approaches as they are deterministic in nature, and hence we did not conduct experiments on these. Now, we examine the diversity achieved by taking the summaries generated by these four methods independently. Figure 1b shows our results. One can see that the diversity is comparable to multiple runs of DivRank algorithms. The reason is that there are enough sentences that are deemed to be highly important from summarization point-of-view, that all the summarization methods would include them. Hence, taking the summaries generated by multiple methods is also not a good solution.

Now, we will evaluate the quality and diversity of the multiple summaries generated by the Graph based approach and MMR based approach. The quality of the summary is measured by retention rate, JS divergence, popular metrics in the summarization literature [5]. Our objective is not to evaluate the absolute quality of the summaries, but to see how it changes when we modify the method to generate multiple summaries. Hence, we take the average of the quality metric

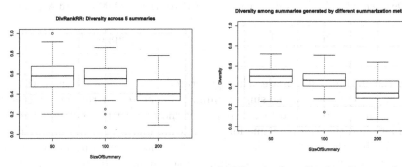

(a) Diversity for multiple runs of DivRank

(b) Diversity for collection of summaries from distinct methods

Fig. 1. Diversity for naive approaches

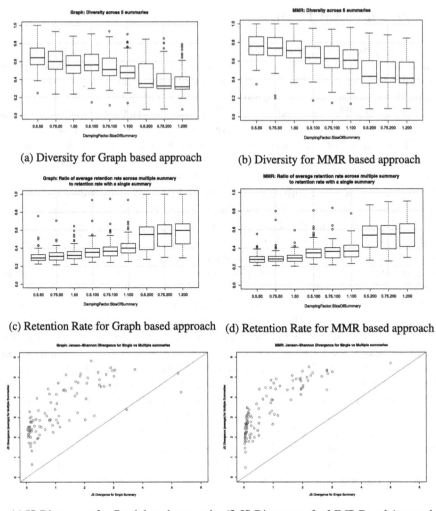

(a) Diversity for Graph based approach   (b) Diversity for MMR based approach

(c) Retention Rate for Graph based approach   (d) Retention Rate for MMR based approach

(e) JS-Divergence for Graph based approach   (f) JS-Divergence for MMR Based Approach

Fig. 2. Various quality metrics for MMR based approach

value for the multiple summaries thus generated, and take its ratio with the
value of the same metric for the single summary generated by the same method
without the changes proposed in this paper. If the ratio of the metric value for
one summary and the average value for multiple summaries is close to one, that
would imply that the quality of the summary does not deteriorate much, and a
value away from 1 would imply large change in the quality. Given that a smaller
value indicates better quality in case of JS divergence, we would expect the ratio
to be more than 1 for these two. On the other hand, for retention rate, a higher

value indicates a better quality, and hence, we would expect the ratio to be lower than 1 for this metric.

Figures 2a and b show the diversity, c and d show the retention rate and e and f show the JS divergences for the Graph and MMR based approaches, respectively.

First, we discuss the diversity for the various methods. The first part of the legends (0.5, 0.75 or 1) indicates the damping factor applied (1 is equivalent of not applying damping, and 0.5 is stronger damping than 0.75). The second part of the legend (50, 100 or 200) indicate the size of the summary in words. Hence, to see the effect of the damping, one can compare the adjacent box-and-whiskers plots. The same convention is followed in the retention rate figures also. As we increase the value of damping factor, the diversity level reduces as expected. Also, as the size of summary increases, the diversity value again goes down. This is probably due to fewer choices available in a longer summary in the later stages of sentence selection.

Now we discuss the results for retention rate. Here, as the damping factor value increases, the quality of multiple summaries become better (at the cost of lower diversity, as noted in the discussion relating to diversity. Also, as the summary size increases, again the quality ratio improves, probably due to the fewer choices available for selecting sentences for later part of summary.

Due to space restrictions, the JS divergence is shown only for summary size 200 with damping 0.75. The line represents the $X = Y$ boundary, which implies that the points close to this line are having the average quality of multiple summaries to be close to the quality of the single summary. As one can see, most of the points are in the close vicinity of the line, implying that the loss of quality is not significant.

To summarize, higher level of damping (lower value of damping factor) leads to higher level of diversity, but at the same time, the quality also goes down. For longer summaries, the diversity is lower than for shorter summaries.

The plots show that the overall quality is not compromised greatly in the task of generating multiple summaries, since the ratio is close to 1 for both the metrics. The damping reduces the performance a bit, but it is not very significant.

Our algorithm does not rule out a text unit after its first selection and hence the diversity score is usually lower than 1. However, the diversity is better for the damped case against the un-damped version (median, first and third quartile of damped version is better than the undamped version), indicating the generation of more diverse summaries with damping.

6   Conclusion

In this paper we have proposed an algorithm to generate multiple diverse summaries based on an input text. Such an algorithm will benefit an author creating a succinct version of a previously created content by giving him several options to choose from. The proposed algorithm works off a class of extractive summarization technique that provides a ranking score to the units of the input content. We show from our experiments that our algorithm results in summaries that are diverse from each other, but does not compromise on the overall quality of the generated summaries.

References

1. Carbonell, J.G., Goldstein, J.: The use of MMR, diversity-based reranking for reordering documents and producing summaries. In: Research and Development in Information Retrieval, pp. 335–336 (1998)
2. Erkan, G., Radev, D.R.: LexRank: graph-based lexical centrality as salience in text summarization. J. Artif. Intell. Res. **22**, 457–479 (2004)
3. Ganesan, K., Zhai, C., Han, J.: A graph-based approach to abstractive summarization of highly redundant opinions. In: Association for Computational Linguistics, pp. 340–348 (2010)
4. Khabiri, E., Hsu, C., Caverlee, J.: Analyzing and predicting community preference of socially generated metadata: a case study on comments in the digg community. In: ICWSM (2009)
5. Louis, A., Nenkova, A.: Automatically evaluating content selection in summarization without human models. In: EMNLP, pp. 306–314 (2009)
6. Mei, Q., Guo, J., Radev, D.R.: Divrank: the interplay of prestige and diversity in information networks. In: Rao, B., Krishnapuram, B., Tomkins, A., Yang, Q. (eds.) KDD, pp. 1009–1018. ACM (2010)
7. Mihalcea, R.: Language independent extractive summarization. ACLdemo, pp. 49–52 (2005)
8. Modani, N., Khabiri, E., Srinivasan, H., Caverlee, J.: Creating diverse product review summaries: a graph approach. In: Wang, J., Cellary, W., Wang, D., Wang, H., Chen, S.-C., Li, T., Zhang, Y. (eds.) WISE 2015. LNCS, vol. 9418, pp. 169–184. Springer, Heidelberg (2015). doi:10.1007/978-3-319-26190-4_12
9. Pemantle, R.: Vertex-reinforced random walk. Probab. Theory Relat. Fields **92**(1), 117–136 (1992)
10. Tong, H., He, J., Wen, Z., Konuru, R., Lin, C.Y.: Diversified ranking on large graphs: an optimization viewpoint. In: KDD, pp. 1028–1036. ACM (2011)
11. Wu, J., Xu, B., Li, S.: An unsupervised approach to rank product reviews. In: FSKD, pp. 1769–1772 (2011)

Diversifying the Results of Keyword Queries on Linked Data

Ananya Dass[1], Cem Aksoy[1], Aggeliki Dimitriou[2],
Dimitri Theodoratos[1(✉)], and Xiaoying Wu[3]

[1] New Jersey Institute of Technology, Newark, USA
dth@ujit.edu
[2] National Technical University of Athens, Athens, Greece
[3] Wuhan University, Wuhan, China

Abstract. Keyword search is a popular technique for retrieving information from the ever growing repositories of RDF graph data on the Web. However, keyword queries are inherently ambiguous, resulting in an overwhelming number of candidate results. These results correspond to different interpretations of the query. Most of the current keyword search approaches ignore the diversity of the result interpretations and might fail to provide a broad overview of the query aspects to the users who are interested in exploratory search. To address this issue, we introduce in this paper, a novel technique for diversifying keyword search results on RDF graph data. We generate pattern graphs which are structured queries corresponding to alternative interpretations of the given keyword query. We model the problem as an optimization problem aiming at selecting a set of k pattern graphs with maximum diversity. We devise a metric to estimate the diversity of a set of pattern graphs, and we design an algorithm that employs a greedy heuristic to generate a diverse list of k pattern graphs for a given keyword query.

1 Introduction

Keyword search is the most popular technique for querying data on the Web because it frees the user from knowing a complex structured query language (e.g., XQuery, SPARQL) and allows querying the data without having full or even partial knowledge of its structure/schema. The convenience and flexibility of keyword search comes with a cost. Keyword queries are ambiguous. As a consequence, there is usually a huge number of candidate results. These results correspond to different interpretations of the query and/or represent different aspects of a specific query interpretation. Most keyword search approaches try to capture the user intention by exploiting structural or semantic characteristics of the data and query results, or by ranking the results in descending order of their estimated popularity [8]. However, a user issuing a keyword query might not be always interested in the most popular interpretation of this query. For instance, a user issuing the query "apple" could be interested in searching about the fruit "apple", the American multinational technology company selling consumer

© Springer International Publishing AG 2016
W. Cellary et al. (Eds.): WISE 2016, Part I, LNCS 10041, pp. 199–207, 2016.
DOI: 10.1007/978-3-319-48740-3_14

goods and computer products, or the second largest chartered savings bank in New York state. If results are returned to the user based on the most plausible interpretation of the query (in this case "apple" as technology company), there is an inherent risk of leaving the user who is interested in "apple" bank or in "apple" fruit unsatisfied. This problem is known as *over-specialization*. Unfortunately, the aforementioned approaches do not consider diversifying the result set [5–7] and therefore, they might dissatisfy the users who look for less popular results or are interested in exploratory search [1]. In the latter case, if a diverse result set that covers different aspects of the entire result space is returned to the users, they will be able to explore and find the desired result.

In this paper, we study the problem of keyword search results diversification on RDF data graphs. In recent years, there is a proliferation of RDF repositories on the Web and keyword search is a popular technique for querying linked data. Therefore, designing effective and scalable approaches for diversifying the results of keyword queries on linked data is of great importance for current keyword search systems.

Our Approach. We propose a novel technique for diversifying keyword search results on RDF graph data. We formulate the diversification problem as an optimization problem over pattern graphs. Pattern graphs are structured queries which cluster together results with the same structural and semantic characteristics and represent alternate interpretations of a keyword query. By diversifying pattern graphs instead of query results which can be too numerous and redundantly represented we have two benefits: first, we can diversify more effectively the results since we diversify the alternative interpretations of the query; second, we address the data scalability problem of diversification since pattern graphs can be computed efficiently by exploiting a structural summary of the RDF data without exhaustively computing the query results. Our diversification approach aims at selecting a diverse set of pattern graphs.

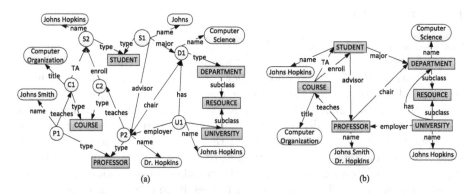

Fig. 1. (a) An RDF graph D, (b) The Structural summary of S.

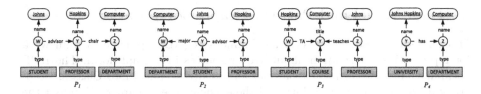

Fig. 2. Patterns graphs of the keyword query $Q = \{$Johns, Hopkins, Computer$\}$.

For example, consider the RDF data graph D of Fig. 1(a) and its structural summary S in Fig. 1(b). Figure 2 shows the pattern graphs P_1, P_2, P_3 and P_4 for the keyword query $Q = \{$Johns, Hopkins, Computer$\}$ computed over S which correspond to alternative interpretations of Q. For instance, the pattern graph P_1 states that "Johns" is a student advised by Professor "Hopkins" who is the chair of a department whose name involves "Computer". Let's assume that one has to select a diverse set of only three pattern graphs out of the four pattern graphs of Fig. 2. In order to characterize the diversity of a set of pattern graphs it is important to quantify the semantic dissimilarity of pairs of pattern graphs. In this case, comparing pattern graphs P_1 and P_2, one can see that the keywords in these two patterns graphs have the same semantics, that is, "Johns" is a student, "Hopkins" is a professor and "Computer" is a department. On the other hand, comparing the pattern graphs P_1 and P_3, we see that none of the keywords in the two pattern graphs has the same meaning. Therefore the pattern graph P_3 is more dissimilar to P_1 than P_2 is. In the same direction, the pattern graph P_4, although it interprets "Computer" in the same way as pattern graph P_1, it interprets "Johns" and "Hopkins" differently: "Johns Hopkins" is a name of a University in P_4. Hence, the pattern graphs P_1, P_3 and P_4 are a possible selection for a diverse set of pattern graphs. These remarks provide some intuition on how we proceed to diversify pattern graph sets. The details of the approach are presented in the next sections.

Contribution. The main contributions of the paper are the following:

- We address the problem of diversifying keyword query results on RDF graph data. Our approach applies a diversification scheme on pattern graphs (query interpretations) instead of results in order to avoid the expensive computation of all the results and to guarantee scalability.
- We devise a metric for assessing the diversity of sets of pattern graphs based on the semantic distance between two pattern graphs.
- We formulate the diversification problem as an optimization problem which maximizes the diversity of a set of pattern graphs.
- To cope with the high complexity of the diversification problem, we design a greedy heuristic algorithm for computing a diverse set of k pattern graphs.

2   Data Model and Pattern Graph Computation

Data Model. Resource Description Framework (RDF) provides a framework
for representing information about Web resources in a graph form. The RDF
vocabulary includes elements that can be broadly classified into Classes, Properties, Entities and Relationships. All the elements are resources. We adopt the
data model definition and query language semantics as described in [3, 4].

Our data model is an RDF graph represented as a quadruple $G = (V, E, L, l)$
where, V is a finite set of *entity*, *class* and *value* vertices, E is a finite set of
relationship edges (between entity vertices), *property* edges (between entity and
value vertices) and *type* edges (between entity and class vertices), L is a finite
set of labels for the elements of V and E and l is a function that assigns labels
to V and E. Figure 3(a) shows an example RDF graph. For simplicity, vertex
and edge identifiers are not shown in this example.

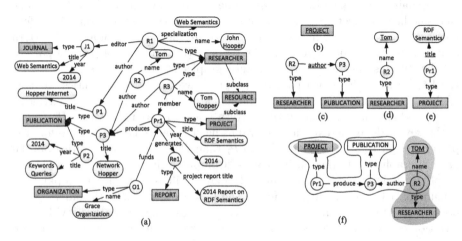

Fig. 3. (a) An RDF graph, (b), (c), (d) and (e) class, relationship, value and property
matching constructs, respectively, (f) inter-construct connection and result graph.

Query Language Semantics. A *query* Q on an RDF graph G is a set of
keywords. A *keyword instance* of a keyword k in Q is a vertex or edge label in
G containing k. The *answer* of Q on G is a set of result graphs of Q on G. Each
result graph is a minimal subgraph of G involving at least one instance of every
keyword in Q. In order to facilitate the interpretation of the semantics of the
keyword instances, every instance of a keyword in Q is matched against a small
subgraph of G which involves this keyword instance and the corresponding class
vertices. This subgraph is called *matching construct*. Figures 3(b), (c), (d) and (e)
show a class, relationship, value and property matching construct, respectively,
for different keyword instances in the RDF graph of Fig. 3(a). Underlined labels
in a matching construct denote the keyword instances.

A *signature* of Q is a function that matches every keyword k in Q to a
matching construct of k in G. Given a query signature S, an *inter-construct*

connection between two distinct matching constructs C_1 and C_2 in S is a simple path augmented with the class vertices of the intermediate entity vertices in the path (if not already in the path). Figure 3(f) shows an inter-construct connection between the matching constructs for keywords Project and Tom in the RDF graph of Fig. 3(a). The matching constructs are shaded and the inter-construct connection is circumscribed.

Given a signature S for Q on G, a *result graph* of S on G is a connected subgraph of G which contains only the matching constructs in S and possibly inter-construct connections between them. Figure 3(f) shows a result graph for the query {Project, Tom} on the RDF graph of Fig. 3(a).

The Structural Summary and Pattern Graphs. In order to construct pattern graphs we use the structural summary of the RDF graph. Intuitively, the structural summary of an RDF graph G is a special type of graph which summarizes the data graph showing vertices and edges corresponding to the class vertices and property, relationship and subclass edges in G. Roughly speaking, the structural summary can be constructed by merging all the entity vertices of a class with their class vertex and then by merging together all the property edges and all the relationship edges that are of the same kind. The formal definition of a structural summary can be found in [3]. Figure 4(a) shows the structural summary for the RDF graph G of Fig. 3(a).

Similarly to the concept of matching constructs on the data graph we define matching constructs on the structural summary. Since a structural summary does not contain entity vertices, a matching construct on a structural summary possesses one distinct entity variable vertex for every class vertex and a distinct value variable for every value vertex label which does not contain a keyword instance. Figure 4(b), (c), (d), and (e) show the class, relationship, value and property matching constructs for the keywords "Project", "author", "Tom", and "title", respectively, on the structural summary of Fig. 4(a).

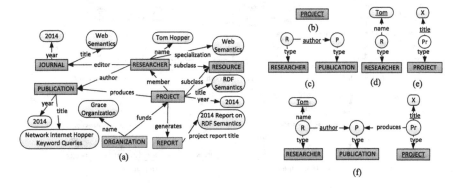

Fig. 4. (a) Structural summary G', (b), (c), (d) and (e) Matching constructs for keywords in Q1 = {Tom, author, Project, title} on G' (f) Pattern graph of Q on G'.

Pattern graphs are subgraphs of the structural summary, strictly consisting of one matching construct for every keyword in the query and the connections between them. Figure 4(f) shows a pattern graph, for $Q = \{$Tom, author, project, title$\}$ on the RDF graph of Fig. 3(a). Labels R, P, and Pr are entity variables.

Given a keyword query Q over an RDF data graph G, we first find all the matching constructs for all the keywords in Q on the structural summary G' and then generate all the pattern graphs on G' for all possible signatures of Q. We use the algorithm in [3] to generate pattern graphs as r-radius Steiner graphs.

3   Pattern Graph Set Diversification

Problem Statement. Our goal is to provide the user with a set of pattern graphs which are diverse. Let G denote an RDF data graph, Q be a keyword query on G, $\mathcal{P}$ be the set of pattern graphs of Q on G and k be a positive integer. Given a subset $\mathcal{S}$ of $\mathcal{P}$, let $diversity(\mathcal{S})$ denote the diversity of set $\mathcal{S}$. We aim at selecting a subset $\mathcal{S}$ of $\mathcal{P}$ which maximizes the diversity of $\mathcal{S}$. In other words,

$$\mathcal{S} \in \underset{\mathcal{S}' \subseteq \mathcal{P},\, |\mathcal{S}'|=k}{\arg\max} \; (diversity(\mathcal{S}'))$$

We formalize the diversity of a set $\mathcal{S}$ of pattern graphs of size k as the average semantic distance between two pattern graphs in $\mathcal{S}$:

$$diversity(\mathcal{S}) = \sum_{P_i, P_j \in \mathcal{S},\, i<j} dist(P_i, P_j)/(k(k-1)/2)$$

where $dist(P_i, P_j)$ denotes the semantic distance between the pattern graphs P_i and P_j.

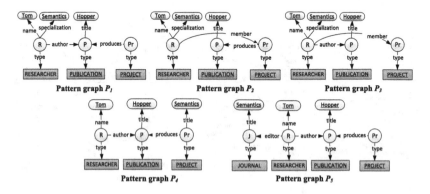

Fig. 5. Five pattern graphs for $Q = \{$Tom, semantics, publication, Hopper, project$\}$.

Assessing the Semantic Distance between two Pattern Graphs. In order to measure the diversity of a set of pattern graphs for a keyword query, we introduce a distance metric to measure the similarity of two pattern graphs.

The first factor we consider in assessing the distance of two pattern graphs is the similarity of their matching constructs. Given a pattern graph P for a keyword query $Q = \{k_1, \ldots k_n\}$, let $mc(P)$ denote the set of matching constructs of Q in P—one for every keyword in Q. The similarity of the matching constructs in the two pattern graphs is given by the formula $mc\_sim(P_1, P_2) = (|mc(P_1) \cap mc(P_2)|)/n$, where n is the number of matching constructs in $mc(P_1)$ or $mc(P_2)$.

For instance, Fig. 5 shows 5 pattern graphs of a query with 5 keywords. Intuitively, P_2 and P_3 are more similar to P_1 than P_4 and P_5 because P_4 and P_5 interpret the keyword `semantics` differently. The metric mc_sim catches this intuition since $mc\_sim(P_1, P_2) = mc\_sim(P_1, P_3) = 1$ while $mc\_sim(P_1, P_4) = mc\_sim(P_1, P_5) = 0.8$.

Although P_2 and P_3 have the same common matching constructs with P_1, P_2 looks more similar to P_1 than P_3 does. Therefore, the second factor we consider is to what extent matching constructs for the same keywords are connected in the same way in the two pattern graphs. Let z be the number of unordered pairs of query keywords which have the same connections in the two pattern graphs. The similarity of the keyword pair connections in P_1 and P_2 is given by: $conn\_sim(P_1, P_2) = z / (n(n-1)/2)$, where n is the number of keywords in Q. The denominator reflects the number of unordered keyword pairs for the keywords in Q.

In the example of Fig. 5, both pattern graphs P_2 and P_3 have five common matching constructs with P_1. However, $conn\_sim(P_1, P_2) = 0.6$ and $conn\_sim(P_1, P_3) = 0.4$. Intuitively, P_2 looks more similar to P_1 than P_3 to P_1 does.

Measuring the similarity of two pattern graphs P_1 and P_2 based solely on the similarity of matching constructs ($mc\_sim(P_1, P_2)$) and matching construct connections ($conn\_sim(P_1, P_2)$) cannot entirely capture their semantic closeness. Compare, for instance, the pattern graphs P_4 and P_5 with the pattern graph P_1 in Fig. 3. Both P_4 and P_5 have 4 keyword matching constructs and 6 pairs of matching construct connections in common with P_1. However, our intuition suggests that P_5 is less similar (more dissimilar) to P_1 than P_4 is as it has the class vertex (concept) "Journal" which does not appear in P_1. In contrast, P_4 and P_1 have the same class vertices. Therefore, we introduce the metric of concept dissimilarity to capture the dissimilarity of two pattern graphs. Let $c(P)$ denote the set of class vertices in a pattern graph P. Given two pattern graphs P_1 and P_2 of a keyword query, $conc\_dsim(P_1, P_2) = |(c(P_1) \cup c(P_2)) - (c(P_1) \cap c(P_2))| / |c(P_1) \cup c(P_2)|$.

Taking into account all the factors, we define the distance $dist(P_1, P_2)$ of two pattern graphs P_1 and P_2 as follows. Note that $concept\_dsim(P_1, P_2)$ is considered with a negative sign since it expresses dissimilarity.

$$dist(P_1, P_2) = \frac{1 - [(mc\_sim(P_1, P_2) + conn\_sim(P_1, P_2))/2 - conc\_dsim(P_1, P_2)]}{2}$$

Algorithm 1. PGDiversification (Pattern Graph Diversification)

Require: $Q = \{k_1, \ldots, k_n\}$: a keyword query with n keywords, S: Structural Summary
       of the data graph, k: size of the output list.
Ensure: $\mathcal{P}_{div}$: set of diversified pattern graphs of size k.
 1: **for all** $k_i \in Q$ **do**
 2:    $L_i \leftarrow \{$set of all matching constructs of k_i on $S\}$;
 3: $\mathcal{P} \leftarrow ComputePatternGraphs(L_1, \ldots, L_n, S)$;   ▷ Compute pattern graphs for Q.
 4: $\mathcal{P}_{div} \leftarrow InitializeWithPGSeed(\mathcal{P})$;   ▷ Initialize $\mathcal{P}_{div}$
 5: $\mathcal{P} \leftarrow \mathcal{P} - \mathcal{P}_{div}$;
 6: **while** $|\mathcal{P}_{div}| \neq k$ **do**
 7:    $PG_{next} \leftarrow null$;   ▷ next pattern graph to be inserted in $\mathcal{P}_{div}$.
 8:    $MaxDistance = 0$;
 9:    **for all** $pg \in \mathcal{P}$ **do**
10:      $distance = 0$;
11:      **for all** $p \in \mathcal{P}_{div}$ **do**
12:        $distance = distance + dist(pg, p)$
13:      **if** $distance > MaxDistance$ **then**
14:        $MaxDistance = distance$;
15:        $PG_{next} = pg$;
16:    $\mathcal{P}_{div}.add(PG_{next})$; $\mathcal{P}.remove(PG_{next})$;

---

4   Algorithm

Exhaustively generating all size-k subsets of a set of pattern graphs for a keyword query and computing their diversity in order to find an optimal one has exponential complexity in the number of pattern graphs and in fact is a NP-hard problem [2, 6].

Therefore, we design a heuristic algorithm, called *PGDiversification*, which greedily selects a new pattern graph at every iteration and incrementally computes the diversity of pattern graph sets. Algorithm *PGDiversification* takes as input a keyword query Q, the structural summary S of an RDF graph and a positive integer k. The output is a subset of the set of pattern graphs of Q on S of size k.

The algorithm starts by finding all the matching constructs of the keywords in query Q on S (lines 1–2) and then generates the set $\mathcal{P}$ of r-radius Steiner pattern graphs using the algorithm in [3] for all possible signatures of Q (line 3). Variable $\mathcal{P}_{div}$ represents the output set of size k which is a subset of the set of pattern graphs $\mathcal{P}$. The set $\mathcal{P}_{div}$ is initialized in line 4. We considered two versions of this algorithm by initializing $\mathcal{P}_{div}$ in two different ways. In the first version, the two pattern graphs having the largest distance are used as seed pattern graphs in $\mathcal{P}_{div}$. The other version uses the most popular pattern graph based on statistical information as in [3, 8]. Subsequently, at every iteration, a pattern graph is chosen for inclusion in $\mathcal{P}_{div}$ so that the new $\mathcal{P}_{div}$ set maximizes the diversity of $\mathcal{P}_{div}$ (line 6–17). The process terminates when $|\mathcal{P}_{div}| = k$.

5   Conclusion

We presented a novel technique for diversifying the results of keyword queries on RDF data graphs. Our diversification scheme has been applied to pattern graphs which are clusters of result graphs having the same structural and semantic features and represent alternative interpretations of a keyword query. In doing so, we ensure diverse query interpretations in the result set. We introduced a metric for assessing the diversity of pattern graph sets and formally defined the problem of diversification as an optimization problem. We designed a greedy algorithm for incrementally selecting a set of pattern graphs which maximizes the diversity. Our experiments which are omitted here because of lack of space, showed that our algorithm efficiently generates a diverse set of pattern graphs.

References

1. Achiezra, H., Golenberg, K., Kimelfeld, B., Sagiv, Y.: Exploratory keyword search on data graphs. In: SIGMOD, pp. 1163–1166 (2010)
2. Agrawal, R., Gollapudi, S., Halverson, A., Leong, S.: Diversifying search results. In: WSDM, pp. 5–14. ACM (2009)
3. Dass, A., Aksoy, C., Dimitriou, A., Theodoratos, D.: Exploiting semantic result clustering to support keyword search on linked data. In: Benatallah, B., Bestavros, A., Manolopoulos, Y., Vakali, A., Zhang, Y. (eds.) WISE 2014. LNCS, vol. 8786, pp. 448–463. Springer, Heidelberg (2014). doi:10.1007/978-3-319-11749-2_34
4. Dass, A., Aksoy, C., Dimitriou, A., Theodoratos, D.: Keyword pattern graph relaxation for selective result space expansion on linked data. In: Cimiano, P., Frasincar, F., Houben, G.-J., Schwabe, D. (eds.) ICWE 2015. LNCS, vol. 9114, pp. 287–306. Springer, Heidelberg (2015). doi:10.1007/978-3-319-19890-3_19
5. Demidova, E., Fankhauser, P., Zhou, X., Nejdl, W., Divq: diversification for keyword search over structured databases. In: SIGIR, pp. 331–338. ACM (2010)
6. Drosou, M., Pitoura, E.: Search result diversification. ACM SIGMOD Rec. **39**(1), 41–47 (2010)
7. Hasan, M., Mueen, A., Tsotras, V., Keogh, E.: Diversifying query results on semi-structured data. In: CIKM, pp. 2099–2103. ACM (2012)
8. Tran, T., Wang, H., Rudolph, S., Cimiano, P.: Top-k exploration of query candidates for efficient keyword search on graph-shaped (RDF) data. In:ICDE (2009)

Data Similarity

Semantic Similarity of Workflow Traces with Various Granularities

Qing Liu[1(⊠)], Quan Bai[2], and Yi Yang[2]

[1] Software and Computational Systems, Data61, CSIRO, Eveleigh NSW, Australia
Q.Liu@csiro.au
[2] Auckland University of Technology, Auckland, New Zealand
{Quan.Bai,Yi.Yang}@aut.edu.nz

Abstract. A workflow trace describes provenance information of a particular workflow execution. Understanding workflow traces and their similarity have many applications in both scientific research and business world. Given workflow traces generated by heterogeneous systems with difference granularities, it is a challenge for users to understand their similarities. In this work, we investigate workflow traces' granularity problem and their similarity method. Algorithms are developed to transform a trace into its multi-granularity forms assisting by a workflow trace ontology. A novel generic semantic similarity algorithm is proposed that not only considers the structural similarity but also the semantics coverage embedded in traces during transformation. Furthermore, theoretical analysis is presented to compute the maximum semantic similarity. Our approach enables that two workflow traces can be compared with any granularity. The experiment using real world workflow traces demonstrates the effectiveness of the proposed methods.

Keywords: Proveance · Workflow trace · Granularity · Workflow trace similarity

1 Introduction

Workflow provenance records the processes, process dependencies and data input and output of processes during a workflow execution. The provenance of a particular workflow execution is often referred as *workflow trace* [2]. Workflow trace can be seen as the meta-data of a particular workflow. It provides information on how a workflow is executed and how its results are derived from original data. Workflow provenance helps scientists/business to validate the processes involved in a workflow and examine its data quality.

Different users may have different needs on what to analyze and how detailed the provenance should be [7]. This refers to the research question of provenance granularity. Provenance granularity provides an abstraction of processes on different levels of details involved in a workflow. For example, a data scientist may want to view a detailed workflow provenance to analyze the performance of the workflow and improve the existing workflow. People at the managerial level may

© Springer International Publishing AG 2016
W. Cellary et al. (Eds.): WISE 2016, Part I, LNCS 10041, pp. 211–226, 2016.
DOI: 10.1007/978-3-319-48740-3_15

want to view a workflow provenance from a coarse level to check if certain components are included in the workflow. Furthermore, it is often required by users to compare workflow traces to understand why one trace generates "better" results than another for advanced scientific discovery. Workflow trace similarity also has many other applications, such as clone detection, trustworthiness measurement and storage size reduction etc. Since workflow traces may be generated by heterogeneous systems and may have different abstraction level, the challenge lies in knowing how to compute the similarity of two traces if they are with different granularities.

There are some existing solutions to workflow trace multi-granularity representation and comparison. In these works, the granularities are determined by user-specific heuristics. In other words, different users could define different granularities given the same process trace. This leads to the problem that a specific multi-granularity workflow trace may not be understood by others and cannot be generalized for use with other applications. Most of the existing trace comparison methods model traces using graphs and their similarities are computed only based on their graph structural similarity. [6] proposed a granularity transformation method using a workflow trace ontology and also developed similarity algorithms with semantics in mind. However, the semantic coverage during trace transformation is not considered.

In this paper, we investigate workflow traces' granularity problem and their similarity method. Our approach enables two workflow traces can be compared with any granularities. Specifically, the contributions of the paper are: (1) given a workflow trace, *Disperse* algorithm is developed to transform a trace with more abstract concepts into its detailed forms assisting by a workflow trace ontology; (2) a novel generic semantic similarity algorithm is proposed that not only considers the structural similarity but also the semantics coverage similarity embedded in traces; and (3) theoretical analysis is presented to compute the maximum semantic similarity. The rest of the paper is organized as follows. In Sect. 2, some existing works in the field of provenance modelling and provenance similarity analysis are reviewed. Section 3 introduces the definition of concept trace and its transformation methods. In Sect. 4, we propose the concept of semantic similarity of two workflow traces. A computation method and the analysis of semantic similarity are also presented. Section 5 evaluates the proposed techniques using some real workflow traces. Finally, the paper is concluded in Sect. 6.

2   Related Work

A provenance model is a representation of artefacts, processes and their relations involved in the information life cycle of data [6]. In recent years, with the maturity of Semantic Web technologies, several Semantic Web-based provenance models have been proposed. Examples include Provenance Data Model (PROV) [5] and Open Provenance Model (OPM) [8]. Benefited from Semantic Web, these models can support data linkage and multi-granularity provenance generated in heterogeneous environments. Provenance granularity has been studied in [4, 6, 8–10, 12]. Granularities are constructed from users' perspectives, and limited to specified application domains. Liu et al. modeled workflow trace granularities using *Workflow Trace Ontology* (WTO) [6]. WTO extends the *opmo:Process* class in OPMO [8] by defining sub-classes to describe different levels of semantics of executed processes [8]. In addition, an annotation property, *hasDepth*, is defined for all the classes in WTO to describe abstraction levels (depth) across granularities. Similar to the four-tier model in [4], the classes with small value of depth in WTO carry coarse semantic information. In this paper, we will use WTO to describe the multi-granularity workflow trace.

Provenance similarity analysis is another important problem in provenance research. Xie et al. presented a provenance compression algorithm to compress provenance graphs. It builds the provenance data into a name-identified reference list and the similarity of processes in the provenance is measured by the process's successors' similarity [11]. Chapman et al. developed a set of provenance factorization algorithms to reduce the provenance storage. The factorization algorithms are based on the pre-defined provenance node similarity functions, and two nodes in the provenance data are considered to be similar if they are specifically similar under the similarity function [3]. In [1], the authors used *graph edit distance* method to define the similarity between provenance graphs. One major limitation from edit distance is that it requires pre-defined cost functions for each elementary operations to calculate similarity, which is not generic and flexible. Liu et al. defined a similarity on the provenance data across different granularities [6]. They used the Maximum Common Subgraph (MCS) to compare two traces with the same depth. However, the semantic coverage among concept traces with different granularities are not captured using the traditional MCS algorithm.

3   Concept Trace Transformation

Workflow trace ontology is domain specific and normally defined by domain experts. In this paper, we construct a WTO for Montage[1] dataset as an example to explain the idea. In Fig. 1, all vertices in ellipses represent executed processes (ontology instances) in a workflow; and rectangles represent high level abstractions of executed processes (ontology classes). For example, *mjpeg45* is an executed process which generates JPEG images from FITS files. Each class in WTO model has a property *hasDepth*, and the value of *hasDepth* denotes the actual level of a class in WTO. We use both *depth* and *conceptual abstraction level* interchangeably to describe a class level in WTO. The larger a depth is, the closer a class approaching to an instance level. On the other hand, the coarser a class's conceptual abstraction level is, the closer a class to the *Process*. Without ambiguity, in the rest of the paper, a label's super-class/sub-class could describe its parent/children class or predecessor/descendant classes in WTO. The rest

---

[1] https://confluence.pegasus.isi.edu/display/pegasus/WorkflowGenerator.

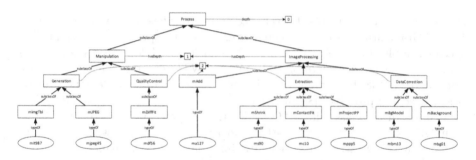

Fig. 1. Montage workflow trace ontology

of this section presents the definition of concept trace and the algorithms for transforming a workflow trace into a concept trace with a required granularity.

3.1   Concept Trace Definition

As discussed in Sect. 2, a workflow trace can be modeled as a directed graph. Each vertex in a graph represents an executed process and each edge in a graph represents a data flow between two executed processes. Given a WTO, processes in a workflow trace can be mapped into different classes to present different levels of conceptual abstractions.

Definition 1 (Strict Concept Trace at Depth n). *Given a workflow trace p, a strict concept trace at depth n is a directed labelled graph $C_p^n = (V, E, L)$ where V is a set of vertices, each vertex representing a conceptual process v_i; $E \subseteq V \times V$ is a set of directed edges, each edge (v_i, v_j) representing data flow from v_i to v_j; L is a labelling function, each v_i has a label $L(v_i) \in \{WTO_{class}^n\}$, where WTO_{class}^n are the classes in WTO whose property "hasDepth" is $\leq n$.*

In some cases, an executed process can only be mapped to a class with depth $n' < n$. By the above definition, a concept trace with depth n represents a conceptual abstraction of a workflow trace using a defined granularity n. Then two concept traces having the same level of conceptual abstraction are comparable.

It is possible that an execution process could be mapped to a class in WTO with any depth. When mapping a workflow trace into a concept trace with depth n, if the depth of a mapped class is larger than n, it means that the current class is too detailed and a coarse abstraction is required. Therefore, we need to transform the class into its super-class with depth n. We call this procedure as a converge procedure. On the other hand, if the depth of a mapped class is smaller than n, it means that the current class is too abstract. We need to transform the class into its subclass in WTO with depth n. We call this procedure as a disperse procedure. If the depth of a mapped class is equal to n, no further operation is required. After all the labels have class labels with depth n, the last step is to merge adjacent vertices with the same labels into one vertex. This is because a concept trace represents a workflow trace on a conceptual abstraction view.

Adjacent vertices having the same labels mean that the vertices carry the same conceptual function.

In summary, to transform a workflow trace into its concept trace with depth n, there are three steps involved: *Converge, Disperse* and *Merge*. Since the merge procedure is straight forward, next we present two algorithms that provide converge and disperse functions.

Converge. Given a workflow trace and a target depth n, for every vertex with its label's depth is larger than n, we need to change its label using its super-class in WTO with depth n. The converge algorithm used in this paper is similar to that in [6]. Due to space limitation, interested readers may refer that paper for details. An example will be given at the end of this section.

Disperse. *Disperse* procedure can be understood as the reverse of *converge* procedure. During *converge*, vertices with detailed conceptual abstraction are converted to the vertices with coarse conceptual abstraction. In other words, the vertices' labels are changed from sub-classes to their corresponding superclasses. Since every class has one and only one super-class with depth n, every p can be transformed to one and only one concept trace, C'^n_p, after converging process. On the contrast, *disperse* procedure is to disperse a class in WTO to its sub-classes. As one class may have more than one sub-class in WTO, a dispersed trace is not unique given a converged trace.

Algorithm 1. *Disperse*

    Input   : a converged concept trace C'^n_p; n is the depth required; WTO is the
               workflow trace ontology
    Output: p's concept trace C^n_p
1 **while** *traversing C'^n_p* **do**
2   **if** *the label of a visiting vertex v hasDepth $< n$ in WTO* **then**
3   Put v's origin and destination into S_{source} and S_{des} respectively;
4   Remove v from V;
5   Create set V' in which it has random x number of vertices v' that
              $L(v').hasDepth = n$ and v' is a sub-class of v;
6   Randomly partition V' into l disjoint sets $V'_1, V'_2...V'_l$ and connect
              neighbor sets to make a directed graph G';
7   link S_{source}, G' and S_{des} and update V and E;
8   **end if**
9 **end while**

Algorithm 1 describes how to disperse a converged concept trace to its concept trace at depth n. Given a converged trace C'^n_p, a target depth n and a workflow trace ontology WTO, the algorithm produces a random number of concept traces of p at depth n. While traversing C'^n_p, if the depth of a vertex label v is less than the target depth n, it creates empty sets S_{source}/S_{des} and puts all the in/out-neighbors of v into them respectively (Line 3). Then v is removed from C'^n_p (Line 4) because it needs to be dispersed. To do that, the algorithm creates a temporal graph G' to maintain the dispersed vertex's structure. It randomly creates x new

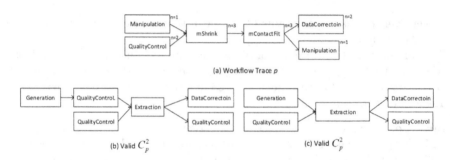

Fig. 2. An example of concept trace's converge and disperse procedure

vertices V' with labels assigned by L where L can assign a subclasses with depth n of v's label to a vertex (Line 5). The algorithm partitions V' into l disjoint sets and connects the vertices in between each disjoint set (Line 6). It ensures that G' is fully connected and acyclic. Next the elements in G' is treated as the dispersed structure of v and linked to the rest of V. If adjacent vertices have the same labels, they are merged as one vertex (Line 7). Combined with Algorithm *Converge* and Algorithm 1, we can generate a strict concept trace with depth n given a workflow trace with any granularity. Figure 2 shows an example of Converge and Disperse procedure.

Example 1. Given $n = 2$ and a workflow trace in Fig. 2(a) with depth listed on top of each vertex, *mShrink* and *mConcatFit* are converged and merged as *Extraction* in C_p^2 by WTO in Fig. 2(b). Semantically the two processes provide the same conceptual function *Extraction* at depth 2. In WTO, class *Manipulation* has two sub-classes *Generation* and *QualityControl* at depth 2. Therefore, *Manipulation* can be dispersed into different graphs, such as *Generation* → *QualitycControl* in Fig. 2(b) or *Generation* in Fig. 2(c) and *QualityControl* in Fig. 2(b–c). Both C_p^2s are valid.

4   Semantic Similarity of Workflow Traces

In this section, first, we define the similarity of two workflow traces. Then the concept of semantic coverage that captures the semantic movement during trace transformation as well as the concept of semantic similarity and its computation are introduced. This is followed by the analysis on how to compute the maximum semantic similarity.

4.1   Structure Similarity

As we mentioned before, workflow traces generated by heterogeneous systems may be represented using various granularities. By transforming two workflow traces into their strict concept traces with depth n, they have the same conceptual abstraction level. Therefore, their similarity at depth n can be computed by applying traditional graph similarity methods. In this paper, we use the *Maximum Common Sub-graph* (MCS) method to calculate the similarity (see Eq. (1)).

$$s(C_{p_1}^n, C_{p_2}^n) = \frac{|MCS(C_{p_1}^n, C_{p_2}^n)|}{|C_{p_1}^n| + |C_{p_2}^n| - |MCS(C_{p_1}^n, C_{p_2}^n)|} \tag{1}$$

In Eq. (1), $|C_{p_1}^n|$ and $|C_{p_2}^n|$ are the sizes of two strict concept traces with depth n respectively and $|MCS(C_{p_1}^n, C_{p_2}^n)|$ is the size of maximal common subgraph between $C_{p_1}^n$ and $C_{p_2}^n$.

The larger the s is, the more similar the two traces are. Since we are more interested in the similarity of two workflow traces as a whole but not with a particular depth that is proposed in [6], the similarity of two workflow traces can be calculated by averaging the similarity of concept traces with all depths:

$$Similarity(p_1, p_2) = \frac{\sum_{n=1}^{depth(WTO)} s(C_{p_1}^n, C_{p_2}^n)}{depth(WTO)} \tag{2}$$

where $depth(WTO)$ is the maximum depth of WTO.

Strict concept traces at depth 0 does not contribute to the similarity because all workflow traces can be converged to *Process* at depth 0 that does not express any differences. The benefit of computing an overall similarity between two workflow traces is that it enables users to understand the overall relationship between the two traces better. However, general graph similarity methods only evaluate graphs' structure similarity without considering the semantic abstraction that graphs represent. For concept traces, each vertex represents a high level conceptual abstraction of an executed process. If we apply general graph similarity methods directly, it would ignore the semantic information involved in its concept trace. Next, we will explain what is the problem and how we approach it.

4.2   Semantic Similarity

In this sub-section, first we introduce the concept of Semantic Coverage. Then the Semantic Similarity (SS) is defined and its computation method is presented.

Semantic Coverage. As discussed, disperse procedure substitutes a vertex representing a class to a graph which is composed by its sub-classes defined in WTO. Since a vertex can be dispersed into many different graphs, we have shown that a strict concept trace with depth n may not be unique if a workflow trace contains vertices with depth smaller than n ($1 \leq n \leq depth(WTO)$).

A vertex with depth $< n$ can be dispersed into a graph with a large number ($\rightarrow \infty$) of vertices at depth n. While this is true as a graph, it is not practical in real world data and the structure of a workflow trace will also be destructed. On the other hand, disperse is a procedure of concretization. The semantic coverage of a vertex with depth n is not as complete as that of its ancestors. Therefore, we need to capture the movement of semantic coverage during converge and disperse procedure while not concerning the number of new vertices generated by disperse procedure. Formally, we define the semantic coverage as follows:

$$SC(g(v^n)) = \begin{cases} 1 & \text{if } depth(v) \geq n \\ \frac{|L(g(v^n))|}{|WTO_v^n|} & otherwise, \end{cases} \tag{3}$$

where $g(v^n)$ is a graph generated by dispersing/converging v to depth n, $|L(g(v^n))|$ is the number of concepts involved in graph $g(v^n)$ and $|WTO_v^n|$ is v's number of descendent concepts at depth n defined in WTO.

Since we do not care the actual graph structure $g(v^n)$, dispersed by v, but the semantic coverage as discussed before, we use a virtual v to represent $g(v^n)$. At this point, a concept trace can be represented using a vertex-weighted graph in which a vertex weight describes $v's$ semantic coverage at depth n compared with that in its original workflow trace.

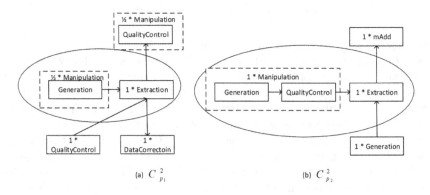

Fig. 3. An example of weighed concept trace and semantic similarity

Example 2. Fig. 3(a) is the concept trace with depth 2 by transforming Fig. 2(a). The first *Manipulation* in its original workflow trace is dispersed into a graph *Generation* which cannot fully represent the concept *Manipulation*. Since in WTO, *Manipulation* has two children, the weight of virtual vertex *Manipulation* in $C_{p_1}^2$ is 1/2. However, in Fig. 3(b), *Manipulation* is dispersed into a graph *Generation* → *QualityControl* and the number of concepts involved is 2. Therefore, the weight for *Manipulation* in $C_{p_2}^2$, representing subgraph *Generation* → *QualityControl*, is 2/2 = 1. Since *Extraction* is generated by converging processes $mShink$ → $mContactFit$ with depth 3 which is ≥ 2, the weight of *Extraction* after transformation is 1 as defined in Eq. (3). Furthermore, $C_{p_1}^2$ and $C_{p_2}^2$ can be described as vertex-weighted graphs that contain virtual vertex *Manipulation* respectively but not including any actual structure graph such as *Generation* → *QualityControl* in $C_{p_2}^2$.

Semantic Similarity Definition. Based on Eq. 3, the Semantic Similarity (SS) of two strict concept traces with depth n can be computed by computing the similarity of their corresponding vertex-weighted graphs. Formally, Eq. 4 re-defines $|MCS|$ in Eq. (1) by incorporating the concept of semantic coverage into it:

$$|MCS^s(C_{p_1}^n, C_{p_2}^n)| = \sum_{v \in MCS(C_{p_1}^n, C_{p_2}^n)} min(SC(g(v_{p_1}^n)), SC(g(v_{p_2}^n))), \quad (4)$$

where $min(SC(g(v_{p_1}^n)), SC(g(v_{p_2}^n)))$ returns $v's$ possible maximum common semantic coverage of two concept traces with depth n. Then the SS of two workflow traces at depth n can be defined as follows:

$$s^s(C_{p_1}^n, C_{p_2}^n) = \frac{|MCS^s(C_{p_1}^n, C_{p_2}^n)|}{|C_{p_1}^n| + |C_{p_2}^n| - |MCS^s(C_{p_1}^n, C_{p_2}^n)|} \quad (5)$$

From the above definition, it can be seen that the traditional MCS similarity, Eq. (1), is a special case of SS in which the weight of every vertex always equals to 1. This is because that the semantic coverage is always assumed as 1 which may not be true for workflow traces with various granularities involved. The SS definition is able to capture the semantic similarity between two concept traces to be compared.

Example 3. Fig. 3 shows two concept traces with depth 2. Their MCS (circled) is $1/2 * Manipulation \rightarrow 1 * Extraction$ by Eq. (4). The semantic meaning behind this is that *Manipulation* is half semantically covered in MCS but *Extraction* is fully presented. Therefore, its $|MCS^s(C_{p_1}^2, C_{p_2}^2)| = 1/2 + 1 = 1.5$.

There are two properties held by Eq. (5):

Property 1. If $s^s(C_{p_1}^n, C_{p_2}^n) \neq 0$, where $1 \leq n \leq depth(WTO)$, then $s^s(C_{p_1}^{n'}, C_{p_2}^{n'}) \neq 0$ where $1 \leq n' < n$.

Property 2. If $s^s(C_{p_1}^n, C_{p_2}^n) = 0$, where $1 \leq n \leq depth(WTO)$, then $s^s(C_{p_1}^{n''}, C_{p_2}^{n''}) = 0$ where $n < n'' \leq depth(WTO)$.

By converge procedure, we can see that if two vertices $v_1^{n''}$ and $v_2^{n''}$ are isomorphically mapped to each other in $MCS^s(C_{p_1}^{n''}, C_{p_2}^{n''})$, their corresponding semantic-parent vertices v_1^n and v_2^n ($n < n''$) must also be isomorphically mapped to each other in $MCS^s(C_{p_1}^n, C_{p_2}^n)$ because their converged vertices' labels are the same. Therefore, the following proposition is held (the proof is omitted due to space limitation).

Proposition 1. If $MCS^s(C_{p_1}^{n''}, C_{p_2}^{n''}) \neq \varnothing$, for any vertex $v_1^{n''} \in C_{p_1}^{n''}$, $v_2^{n''} \in C_{p_2}^{n''}$ and $v_1^{n''}/v_2^{n''} \in MCS^s(C_{p_1}^{n''}, C_{p_2}^{n''})$, if $f(v_1^{n''}) \rightarrow v_2^{n''}$, there must have $f(v_1^n) \rightarrow v_2^n$ where $v_1^n \in C_{p_1}^n$, $v_2^n \in C_{p_2}^n$, $v_1^n/v_2^n \in MCS^s(C_{p_1}^n, C_{p_2}^n)$, where v_1^n/v_2^n is the super-class of $v_1^{n''}/v_2^{n''}$ respectively, $1 \leq n < n''$.

Here $f(x) \rightarrow y$ represents the corresponding mapping between x and y.

Semantic Similarity Computation. To compute the semantic similarity of two workflow traces p_1 and p_2, first, p_1 and p_2 are transformed to $C_{p_1}^1$ and $C_{p_2}^1$, respectively. The reason we do depth 1 transformation is that every vertex has the minimum depth in a concept trace with depth 1. It implies that there is no disperse procedure required during transformation. Therefore, p_1 and p_2 can be transformed to one and only one $C_{p_1}^1$ and $C_{P_2}^1$ respectively. Then $MCS^s(C_{p_1}^1, C_{p_2}^1)$ is computed. If $MCS^s(C_{p_1}^1, C_{p_2}^1) = \varnothing$, $s^s(C_{p_1}^n, C_{p_2}^n) = 0$. By Property 2, we can conclude that $Similarity^s(p_1, p_2) = 0$. If $MCS^s(C_{p_1}^1, C_{p_2}^1) \neq \varnothing$, mappings $f(v_{p_1}^1) \rightarrow v_{p_2}^1$ ($v_{p_1}^1 \in C_{p_1}^1$, $v_{p_2}^1 \in C_{p_2}^1$, $v_{p_1}^1/v_{p_2}^1 \in MCS^s(C_{p_1}^1, C_{p_2}^1)$) can be obtained.

The above step is important because it provides mappings $f()$ between vertices of p_1 and p_2 that contribute to MCS with no ambiguity. The mapping will guide the MCS computation at depth 2. By Proposition 1, only vertices in p_1 and p_2 that contribute to $MCS^s(C_{p_1}^1, C_{p_2}^1)$ may contribute to $MCS^s(C_{p_1}^2, C_{p_2}^2)$. Therefore, we only need to check those vertices' transformation to compute $MCS^s(C_{p_1}^2, C_{p_2}^2)$. We will show an example later on to explain the idea. Similarly, the semantic similarity at depth n can be computed by using the MCS mapping at $n - 1$. The overall semantic similarity between two workflow traces can be computed by Eq. (6):

$$Similarity^s(p_1, p_2) = \frac{\sum_{n=1}^{depth(WTO)} s^s(C_{p_1}^n, C_{p_2}^n)}{depth(WTO)} \tag{6}$$

By identifying vertices' semantic coverages of two strict concept traces with the same depth, we are able to compare two workflow traces semantically with any granularities. Since there is a lot of possible strict concept traces with depth n that can be generated given a workflow trace, if we just compare two random generated concept traces with depth n, their similarity, $s^s(C_{p_1}^n, C_{p_2}^n)$, is arbitrary and therefore, the overall similarity, $Similarity^s(p_1, p_2)$, is also arbitrary that may not make sense for users. In the next sub-section, we will analyse SS and discuss how to compute the maximum SS of two workflow traces with any granularities that may provide users a better understanding of the relationship between two workflow traces.

4.3 Semantic Similarity Analysis

The maximum $Similarity^s(p_1, p_2)$ can be achieved if $s^s(C_{p_1}^n, C_{p_2}^n)$ is maximized for $\forall n \in [1..depth(WTO)]$. It is clear that to find the maximum $s^s(C_{p_1}^n, C_{p_2}^n)$, we need to maximize $|MCS^s(C_{p_1}^n, C_{p_2}^n)|$ and minimize $|C_{p_1}^n|$ and $|C_{p_2}^n|$ respectively.

Given two workflow traces p_1, p_2 and a depth n, our goal is to generate their corresponding concept traces $C_{p_1}^n$ and $C_{p_2}^n$ that $s(C_{p_1}^n, C_{p_2}^n)$ is maximized. By Proposition 1, in converge procedure, a vertex $v^{n''}$ ($n'' \geq n$) can only be transformed to one and only one vertex v^n. However, a vertex $v^{n'}$ $(n' \leq n)$ may have many representations after disperse procedure. Therefore, the key is to control the disperse procedure to achieve the maximum similarity. Next we analyze all the possible cases of how vertices v $(v \in p_k, k \in [1,2])$ can be transformed and demonstrate how to control the transformation procedure to reach the maximum similarity.

In the SS computation algorithm, if $MCS^s(C_{p_1}^1, C_{p_2}^1) \neq \varnothing$, the mapping $f(v_{c_{p_1}^1}) \rightarrow v_{c_{p_2}^1}$ $(v_{c_{p_1}^1} \in C_{p_1}^1, v_{c_{p_2}^1} \in C_{p_2}^1, v_{c_{p_1}^1}/v_{c_{p_2}^1} \in MCS^s(C_{p_1}^1, C_{p_2}^1))$ can be obtained. For each $v_{c_{p_k}^1}$ $(k \in [1,2])$, we use $ref(v_{c_{p_k}^1})$ to represent the corresponding vertices in original workflow trace p_k that contribute to $v_{c_{p_k}^1}$. At this point, we can generalize the problem as: Given p_1, p_2, $C_{p_1}^{n-1}$, $C_{p_2}^{n-1}$, $ref(v_{c_{p_k}^{n-1}})$ and the mapping $f(v_{c_{p_1}^{n-1}}) \rightarrow v_{c_{p_2}^{n-1}}$ $(n \geq 2)$, how to generate $C_{p_1}^n$ and $C_{p_2}^n$ so that $s^s(C_{p_1}^n, C_{p_2}^n)$ is maximized. To transform p_k to $C_{p_k}^n$, each $v_k \in p_k$ must sit in one of the following cases and for each case, a rule is defined with principles of maximizing the $MCS^s(C_{p_1}^n, C_{p_2}^n)$ and minimizing $|C_{p_k}^n|$.

- Case 1: for any vertex $v_k \in p_k$ not contributing to $MCS^s(C_{p_1}^{n-1}, C_{p_2}^{n-1})$.
  - Case 1.1: If $depth(v_k) = n$, v_k is not changed;
  - Case 1.2: If $depth(v_k) > n$, by Proposition 1, we can use v_k's super-class with depth n to replace v_k;
  - Case 1.3: If $depth(v_k) < n$, by Eq. (5), we need to disperse v_k by any one of its sub-classes with depth n in WTO to minimize the size of $C_{p_k}^n$ that leads to maximize $s^s(C_{p_1}^n, C_{p_2}^n)$;
- Case 2: vertices in p_k that contribute to $MCS(C_{p_1}^{n-1}, C_{p_2}^{n-1})$. This case is the most complex case since sometimes we need to consider both p_1 and p_2 during transformation to generate their concept traces with depth n for the maximum similarity. Specifically, given $f(v_{c_{p_1}^{n-1}}) \rightarrow v_{c_{p_2}^{n-1}}$ $(v_{c_{p_1}^{n-1}} \in C_{p_1}^{n-1}, v_{c_{p_2}^{n-1}} \in C_{p_2}^{n-1}$, $v_{c_{p_1}^{n-1}}/v_{c_{p_2}^{n-1}} \in MCS(C_{p_1}^{n-1}, C_{p_2}^{n-1}))$, we need to study the relationship between $ref(v_{c_{p_1}^{n-1}})$ and $ref(v_{c_{p_2}^{n-1}})$. There are 4 sub-cases involved.
  - Case 2.1: If $v_k \in ref(v_{c_{p_k}^{n-1}})$ and $depth(v_k) = n$, nothing needs to be changed since it is proper for $C_{p_k}^n$;
  - Case 2.2: If $v_k \in ref(v_{c_{p_k}^{n-1}})$ and $depth(v_k) > n$, converge procedure applied and use v_k's super-class with depth n to replace v_k;
  - Case 2.3: If $v_i \in ref(v_{c_{p_1}^{n-1}})$, $v_j \in ref(v_{c_{p_2}^{n-1}})$, $f(v_i) \rightarrow v_j$, $depth(v_i) < n$ and $depth(v_j) \geq n$, it means v_i needs to be dispersed. To reach the maximum similarity, $g(v_i^n)$ dispersed by v_i must be the same as $g(v_j^n)$ converged by v_j. By this means, they can contribute to $MCS^s(C_{p_1}^n, C_{p_2}^n)$ as they did for $MCS^s(C_{p_1}^{n-1}, C_{p_2}^{n-1})$;
  - Case 2.4: If $v_i \in ref(v_{c_{p_1}^{n-1}})$, $v_j \in ref(v_{c_{p_2}^{n-1}})$, $f(v_i) \rightarrow v_j$ and $depth(v_i/v_j) < n$, then both v_i and v_j have to be dispersed. In Eq. (4), if both the semantic coverage $SC(g(v_i^n))$ and $SC(g(v_j^n))$ are maximized as 1, their contribution to $|MCS^s(C_{p_1}^n, C_{p_2}^n)|$ is maximized.

By the above transformation rules, the maximum SS can be achieved. Figure 4 shows an example.

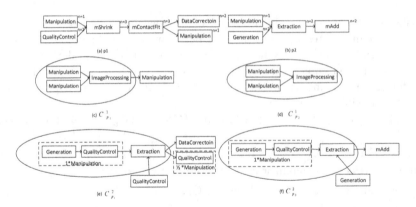

Fig. 4. An example of maximum similarity

Example 4. Figure 4(a) and (b) show two workflow traces in which vertices presented have various granularities. Figure 4(c) and (d) are their corresponding concept traces at depth 1. Since there is no disperse procedure involved, its maximum similarity is computed by applying Eq. (1) directly. The $MCS^s(C_{p_1}^1, C_{p_2}^1)$ is circled. To transform two traces into depth 2, the transformation case 1 is applied to all vertices in p_1 and p_2 that are not contributing to in the circle of Fig. 4(c) and (d). For *ImageProcessing* in the circle, since the $ref(ImageProcessing_{p_1}^1)$ is $mShrink \rightarrow mContactFit \rightarrow DataCorrection$ and $ref(ImageProcessing_{p_2}^1)$ is *Extraction*, Case 2.2 and Case 2.1 are applied respectively. This generates the *Extraction* vertex in Fig. 4(e) and (f) that maximizes their contribution to $MCS^s(C_{p_1}^2, C_{p_2}^2)$ (circled). But for the *Manipulation* vertex in the Fig. 4(c) and (d) circle, $ref(Manipulation_{p_1}^1)$ is *Manipulation* which is the same as $ref(Manipulation_{p_2}^1)$. Therefore, we have to reach the maximum semantic coverage by applying case 2.4. That is why the weights of both virtual vertices in Fig. 4(e) and (f) are 1. Through this transformation, the maximum $S^s(C_{p_1}^2, C_{p_2}^2)$ can be achieved.

5   Experiments

In this section, experiments are conducted using some real datasets to evaluate the proposed method. We compare our method with the traditional structure similarity method to demonstrate the effectiveness of our approach. All the experiments are performed on a machine of i7-6700 CPU @ 3.40 GHz and 16 GB RAM, running Ubuntu 16.04. All algorithms were implemented in Java using Oracle JDK 8.

Datasets. The Genome process traces[2] are generated by the USC Epigenome Centre and represent a largely pipelined application with multiple pipelines operating on distinct chunks of data. We take the workflows in the Genome dataset

---

[2] https://confluence.pegasus.isi.edu/display/pegasus/WorkflowGenerator.

as our base workflow traces and using a WTO, that is designed to accommodate several datasets in Genome domain, to transform base workflows to their multi-granularity concept traces. WTO has 5 as the maximum depth and the workflow traces generated have 25 $\sim$ 45 vertices.

Since the base workflows generated are similar from semantic perspective, to imitate heterogenous situation, the base workflow traces are transformed into the traces with various concepts involved in WTO. Basically, we first transform base workflow traces to strict concept traces with depth $depth(WTO)$, $C_p^{depth(WTO)}$. Then we use some random concepts selected from WTO to replace labels in random selected paths from base workflows. In this way, the dataset maintains some of its original semantic features but at the same time, complexities are also purposely injected.

Table 1. Structure and semantic similarity comparison

Pair ID	Structure similarity	Semantic similarity	Max s^s	Min s^s
1	0.11	0.16	0.23	0.11
2	0.17	0.52	1	0.24
3	0.4	0.59	0.86	0.4
4	0.2	0.47	0.8	0.29
5	0.17	0.49	0.8	0.25
6	0.44	0.65	0.75	0.46
7	0.24	0.31	0.43	0.15
8	0.25	0.51	1	0.3
9	0.23	0.29	0.33	0.22
10	0.12	0.51	0.93	0.12

Semantic Similarity of Workflow Traces. We apply the proposed method to analyse the semantic similarity of randomly selected workflow traces. Table 1 shows the results of 10 pairs of workflow traces. The third column is the maximum semantic similarity computed. And the fourth and fifth column represent the maximum and minimum similarity among 5 depths.

Table 1 shows that although two workflow traces may be very different from their structure perspective, their semantics at more abstract levels may be very similar. Pair 2 and 8 with very small structure similarity are such examples that at a particular depth, each pair presents the same abstract function that their maximum semantic similarity reaches 1. But for Pair 1, both the structure similarity and semantic similarity are small with no big difference. Pair 6 has a similar structure (0.44) and their semantic similarity is 0.65.

To give readers a more concrete idea what has happened during the semantic similarity computation, Fig. 5 shows an example of the maximum common subtraces computed for depth 1, 3 and 4 after applying our transformation and

Table 2. Running time (ms)

Pair ID	Semantic computation	g_1 Transformation	g_2 Transformation
1	95	87	70
2	6121	87	63
3	6521	92	69
4	432	80	64
5	228	93	72
6	29472	106	64
7	42	98	79
8	104	116	57
9	31	97	44
10	31371	85	52

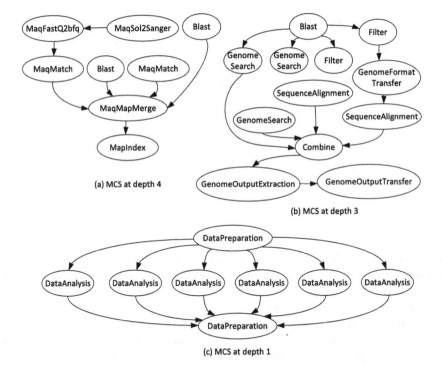

Fig. 5. Pair 10 semantic similarity structure

computation methods for Pair 10. We don't present the original two big workflow traces due to space limitation. By comparing Fig. 5(a), (b) and (c), we can see that more common abstract concepts are identified at depth 3 compared with that in depth 4. In Fig. 5(c), at depth 1, the MCS is converged to few vertices that present the most abstract common semantics that the two traces share. By the proposed method, users are able to view the common functionalities between two workflow traces at various granularities.

Running Time. Table 2 shows the semantic similarity computation time and trace transformation time for the above 10 pairs. Since it is possible that after converge procedure, the number of vertices that share the same abstract concepts may be increasing in some cases, it will take much longer time for MCS computation due to possible automorphisms. This explains the longer semantic running time for some pairs (e.g. Pair 6 and Pair 10). Otherwise, the semantic computation times are comparable with the structure MCS time. The transformation time is linear to the size of traces.

The experiment demonstrates that the proposed semantic similarity concept and its computation methods are able to identify the semantic similarity embeded in traces to be compared even they may have very different structures. This approach enables users to have a much better understanding on the conceptual level among various workflow traces.

6 Conclusion

In this paper, we propose the *Disperse* algorithm that is able to transform a workflow trace with any granularity into a concept trace with required depth. To capture the similarity of conceptual abstraction between two workflow traces, the semantic similarity concept that not only considers the structure similarity but also the semantic coverage during transformation is proposed. The maximum semantic similarity is analysed and its computation method is also presented. Our similarity method is able to capture the semantic information embedded in the workflow traces and it provides a better solution on how to evaluate workflow traces with various granularities.

References

1. Bao, Z., Cohen-Boulakia, S., Davidson, S.B., Eyal, A., Khanna, S.: Differencing provenance in scientific workflows. In: ICDE, pp. 808–819 (2009)
2. Bowers, S.: Scientific workflow, provenance, and data modeling challenges and approaches. J. Data Semant. 1(1), 19–30 (2012)
3. Chapman, A.P., Jagadish, H.V., Ramanan, P.: Efficient provenance storage. In: SIGMOD, pp. 993–1006 (2008)
4. Gotz, D., Zhou, M.X.: Characterizing users' visual analytic activity for insight provenance. Inf. Vis. 8(1), 42–55 (2009)
5. Groth, P., Moreau, L.: Prov overview. W3C Working Draft, 11 December 2012
6. Liu, Q., Zhao, X., Taylor, K., Lin, X., Squire, G., Kloppers, C., Miller, R.: Towards semantic comparison of multi-granularity process traces. Knowl. Based Syst. 52, 91–106 (2013)
7. David Allen, M., Len Seligman, B.: Provenance capture and use: a practical guide. MITRE Corporation (2010)

8. Moreau, L., Clifford, B., Freire, J., Futrelle, J., Gil, Y., Groth, P., Kwasnikowska, N., Miles, S., Missier, P., Myers, J., et al.: The open provenance model core specification (v1. 1). Future Gener. Comput. Syst. **27**(6), 743–756 (2011)

9. Scheidegger, C., Koop, D., Santos, E., Vo, H., Callahan, S., Freire, J., Silva, C.: Tackling the provenance challenge one layer at a time. Concurrency Comput. Pract. Exp. **20**(5), 473–483 (2008)

10. Stephan, E.G., Halter, T.D., Ermold, B.D.: Leveraging the open provenance model as a multi-tier model for global climate research. In: McGuinness, D.L., Michaelis, J.R., Moreau, L. (eds.) IPAW 2010. LNCS, vol. 6378, pp. 34–41. Springer, Heidelberg (2010). doi:10.1007/978-3-642-17819-1_5

11. Xie, Y., Muniswamy-Reddy, K.K., Long, D.D., Amer, A., Feng, D., Tan, Z.: Compressing provenance graphs. In: Tapp (2011)

12. Zhao, J., Wroe, C., Goble, C., Stevens, R., Quan, D., Greenwood, M.: Using semantic web technologies for representing e-Science provenance. In: McIlraith, S.A., Plexousakis, D., Harmelen, F. (eds.) ISWC 2004. LNCS, vol. 3298, pp. 92–106. Springer, Heidelberg (2004). doi:10.1007/978-3-540-30475-3_8

Intermediate Semantics Based Distance Metric Learning for Video Annotation and Similarity Measurements

Wen Qu[1(✉)], Xiangmin Zhou[2], Daling Wang[1], Shi Feng[1],
Yifei Zhang[1], and Ge Yu[1]

[1] School of Computer Science and Engineering, Northeastern University,
Shenyang, People's Republic of China
quwen@research.neu.edu.cn, {wangdaling,fengshi,
zhangyifei,yugeg}@ise.neu.edu.cn
[2] School of Science, RMIT University, Melbourne, Australia
Xiangmin.Zhou@rmit.edu.cn

Abstract. The similarity metric between videos is integral to several key tasks, including video retrieval, classification and recommendation. Since there is no standard criterion for the similarity measurement between videos except measuring manually, it is difficult to collect large training dataset for distance metric learning algorithms. Moreover, the existing distance metric learning (DML) methods for multimedia data suffer from two critical limitations: (1) they typically attempt to learn a distance function on the single label setting, in which each item is only labeled with single label; (2) they are often designed for learning distance metrics on low-level features, which ignore the semantic similarity of the multimedia data. To address these problems, in this paper, we propose a novel framework of Intermediate Semantics based Distance Learning (ISDL) for video clips, which aims to integrate semantics of multiple modals optimally for distance metric learning. In particular, the proposed framework: (1) generates the training pairs automatically; (2) defines multi-modal concepts for similarity measure among videos; (3) learns the distance metric for video clips based on the intermediate semantics. We conduct an extensive set of experiments to evaluate the performance of the proposed algorithms, and the results validate the effectiveness of our proposed approach.

Keywords: Distance metric learning · Video similarity measure

1 Introduction

The similarity metric between videos is the fundamental issue for many tasks, such as video retrieval, video classification and recommendation, etc. Despite the abundance of literature on this topic, searching for semantically similar and relevant videos remains a very challenging problem for two difficulties. First of all, video data typically exhibit multiple modalities, including textual, acoustic and visual modals. It is difficult to integrate such heterogeneous data to form a holistic similarity space. Moreover, the definition of "similarity" varies from person to person. Some methods measure the

© Springer International Publishing AG 2016
W. Cellary et al. (Eds.): WISE 2016, Part I, LNCS 10041, pp. 227–242, 2016.
DOI: 10.1007/978-3-319-48740-3_16

similarity based on the visual contents, while regard the textual contents contribute most to the similarity. Thus, the heterogeneous data and diverse definition make the problem intractable. Figure 1 shows the multi-modal contents of three videos. From the visual modal, all of these videos contain the action "shake hands". But from the text modal and audio modal, the second and the third videos are more similar, because they are similar in both text and audio modalities.

Visual Modality	Text Modality	Audio Modality
		Applause
	Knox. How are you? Joe Danburry. Nice to meet you, sir. He's the spitting image of his father. Isn't he. How is he?Come on in.	Speech
	How do you do? I am Sam. Mr. Dawson. It is a pleasure.	Speech

Fig. 1. An example of the comparison between three videos from contents of three modalities.

In recent years, researchers have noticed the limitations of conventional rigid proximity functions for multimedia similarity search, which leads to the active research on Distance Metric Learning. The Distance Metric Learning studies provide a feasible way to optimize a similarity function using the side information (e.g. relative comparisons, class labels), which makes the metric learning for multi-modal data tractable.

Researches on the distance of multi-modal data evolve through two ways: distances fusion from each modality, and heterogeneous data fusion before similarity measurement. Most existing works belong to the first paradigms. These works can be grouped into four categories according to the type of distance functions for each modality and the combination strategies: (1) linear distance metric for each modality combined with linear function; (2) linear distance metric for each modality combined with nonlinear function; (3) non-linear distance metric for each modality combined with linear function; (4) non-linear distance metric for each modality combined with nonlinear function. The first category is the simplest and the most primitive way to conduct similarity measure for multi-modal data. Nonlinear combination Attention Fusion Function was used to fuse distances of multiple modalities in [16], which belongs to the second category. Wu et al. [28] proposed a framework for multi-modal similarity learning. The distance leaned for each individual modality was nonlinear, the final similarity is the optimal linear combination of multiple modalities. Recently, the Multiple Kernel Learning (MKL) based methods [15, 29] have been used to integrate multiple sources of heterogeneous data into a single and unified data space. McFee [15] proposed multi-modal distance learning scheme based on MKL.

Despite the pioneering studies, most existing methods have the following limitations: (1) the similarity function is learned based on low-level features, which can not handle the "semantic gap" between low-level features and semantic concepts in videos; (2) the distance function learned are lack of interpretability and prone to overfit the training data.

In this paper, we obtain a similarity measure of the videos taking into account the semantics of the multi-modal data. First, the transformations from the low-level features to the semantic concepts are learned. Then the multiple concepts are combined to construct the intermediate semantics space under the MKL framework. Finally, the similarity metric is learned based on the intermediate semantic space. The proposed method is based on both the distance fusion and heterogeneous data fusion.

The rest of this paper is organized as follows. In Sect. 2, we review the related work. In Sect. 3, we present the proposed ISDL algorithm in details. Then we give the experimental results and analysis in Sect. 4. Finally, in Sect. 5, we conclude this paper and discuss the future work.

2 Related Work

Our research involves multimedia analysis and machine learning. Related work includes multimedia representation, video annotation and distance metric learning. We review the key related works in these fields respectively.

2.1 Multimedia Representation

The representation of multimedia plays important roles in many applications, such as video classification, indexing and retrieval. The early works use the low-level features as the representation of multimedia. However, low-level features suffer from the disadvantage of semantic gap with high-level semantic contents. To bridge the semantic gap, researchers have proposed methods using intermediate concept lexicon to help understand the multimedia data [11].

Concept-based representation applies discriminative classifiers to associate specific concept labels with the videos and describe the videos by the concept detection results of those detectors [25]. Ma et al. [13] proposed a classifier-specific intermediate representation for multimedia tasks. There are a variety of semantic concept detectors and a series of concept lexica that have been established. LSCOM [5] and MediaMill [22] include semantic concepts such as people (face, anchor), acoustic (speech, music) and scene, etc. 346 concepts have been defined for the TRECVID 2011 semantic indexing task. The concepts based video representations are more capable of reflecting the semantics and have witnessed encouraging results in many multimedia semantic analysis tasks [6].

2.2 Video Annotation

Automatically annotating videos at the semantic concept level has emerged as an important topic in the multimedia research community [18]. The concepts of interest include a wide range of categories such as actions (e.g., hug, kiss), scenes (e.g., urban, kitchen) and events (e.g., explosion-fire, people-marching). Video annotation is basically a classification problem, which aims to associate videos with one or multiple semantic labels (tags).

Existing researches on video annotation mainly evolve two paradigms: multi-class annotation and multi-labeling annotation. Specifically, multi-class annotation process annotates only one concept to each video clip. Most of the works, e.g., the ones being addressed in TRECVID, are multi-class annotation [24]. In contrast, the multi-labeling process annotates a video clip with multiple labels. Many researches [19] belong to this paradigm. In multi-modal labeling annotation, each modality of the video (i.e. image, text, audio) is annotated with one concept. The annotation from each modality constitutes the multi-modal labeling of the video. For example, a video can be labeled as "dialog" in audio modality, "education" in text modality and "dining room" in visual modality simultaneously. The multi-modal labeling annotates one concept to each modality of the videos and provides a comprehensive description of the video content. In this paper, we propose multi-modal labeling annotation.

2.3 Distance Metric Learning

Distance metric learning has been extensively studied in machine learning. Many kinds of settings and methodologies are explored. From the view of the side-information, the methods can be grouped into three categories: explicit class labels, similarity/dissimilarity pairwise labels [16] or relative similarity measurements [13, 21]. Representative techniques include Relevant Component Analysis (RCA) [1], Discriminative Component Analysis (DCA) [7], Large Margin Nearest Neighbor (LMNN) [27], Neighbourhood Component Analysis (NCA) [5], Information-Theoretic Metric Learning (ITML) [2], MCML [3] and so on. The reader can prefer to the survey [9, 30] for a more comprehensive review.

However, all these works only address uni-modal data, which could not effectively handle the distance measure of multi-modal data. Recently, a number of distance measure algorithms have been developed for the multi-modal data. For example, [28] propose an online deep learning algorithm for image retrieval. The similarity function of multi-modal is the weighted combination of similarity in each modality, where the weights are learned in the training phase. The multi-modal relevance between videos is fused with Attention Fusion Function in [16]. Wang et al. [26] learn the distance metric over multi-layer graph and achieve the high-order based similarity among multimodal data. In these works, the similarity in each modality is computed first and then fused together.

The works [15, 29] are similar with ours. McFee and Lanckriet [15] present a novel multiple kernel learning technique for integrating heterogeneous data into a unified similarity space. Inspired by this work, Xia et al. [29] improve the effectiveness of the MKPOE algorithm [15] and handle the efficiency and scalability issue by online learning. These two approaches use the Multiple Kernel Learning (MKL) technique to learn an optimal ensemble of kernel transformations. The algorithms focused on finding the embedding projections that conformed to perceptual similarity measurement. Our algorithm is different from them in transforming the low-level features into high-level semantic space. Then the distance metric is learned based on the intermediate semantics. In principle, our approach is capable of simultaneously accomplishing the following two learning tasks: (1) modality-specific (e.g., image-specific) concept learning; (2) intermediate semantics based distance metric learning.

In this paper, we propose a new algorithm for semantic based video distance metric learning, which has the following contributions: (1) Developing a novel video representation, named multi-modal concepts, for similarity measurement between videos. The concepts of all the modalities are combined together to describe the semantics in the videos. (2) Automatically generating the side information (triadic comparison) for training process based on multi-modal concepts. (3) Integrating concept learning and distance metric learning into a joint framework. In this way, the distance metric is obtained based on the intermediate representation, which is semantically interpretable.

3 Intermediate Semantics Based Distance Learning

In this section, we discuss the intermediate semantics based distance learning, which consists of two main components: intermediate semantic space learning and semantic based distance metric learning. Figure 2 is the overview of the proposed method. First, multi-modal features of videos are extracted separately. Then, the transformations from low-level feature to the multi-modal concept space are learned for each modal under the supervision of annotation. The multi-modal concepts construct an intermediate semantic space. Finally, the distance metric between videos is learned based on the intermediate semantic space, where the transformations are combined under the MKL framework.

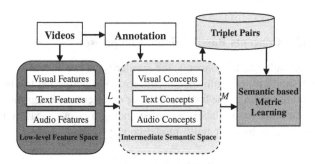

Fig. 2. Overview of the proposed ISDL algorithm.

3.1 Multi-modal Concepts

Most existing works about video annotation can be classified into two categories: single label or multiple labels. They annotate the videos according to the single modal (e.g. text contents). To describe the content of different modalities in the video, we define a multi-modal labels. Given a video v_i, the multi-modal labels of the video is denote as y_i:

$$y_i \equiv <y_i^{text}, y_i^{visual}, y_i^{audio}>$$
$$y_i^{text} \in S^{text}, y_i^{visual} \in S^{visual}, y_i^{audio} \in S^{audio} \tag{1}$$

where S is the concept set for each modality. In the following, we elaborate the concepts for each modality separately.

Textual Concepts. The textual information related to the videos includes subtitles, automated speech recognition results and Optical Character Recognition embedded in the video stream. To exploit the semantic of textual content, we make use of subtitles as the source of text information in our work. The subtitles are publicly available and provide text description of the video content. We select 6 categories to describe the textual contents with S^{text} = {Greeting, Conversation, Declarative, Imperatives, Others, and Null}. The "Null" means the video has no subtitles.

Visual Concepts. Since humans are the main objects in most videos, we analyze the human actions in the videos for the visual modality. Following the work of [12], we select several diary actions for visual concepts, where S^{visual} = {AnswerPhone, driveCar, Eat, fightPerson, HandShake, HugPerson, Kiss, Run, StandUp, SitUp, SitDown, GetOutofCar}.

Audio Concepts. Audio contents provide cues for a large number of realistic events. In order to cover the most semantic events in the dataset, we select and define seven representative events. In detail, we investigated five events, that is S^{audio} = {Dialog, Music, Mixture, Noise, Others}. Dialog and Music are the most common audio types in a video. Mixture is the mixture of speech and music. Noise contains the sounds of fight, explosion, etc. Others include environmental sounds (e.g. background noise).

3.2 Multi-modal Concepts Based Triadic Comparison

After defining the multi-modal concepts, the similarity of two videos can be measured according to the similarity of multi-modal concepts. We define the triadic comparison among videos and use them as the training data for distance metric learning. Given a set of videos $x \in V$, the multi-modal labels y corresponding to x, we measure the similarity between videos x_i, x_j, x_k by their multi-modal concepts y_i, y_j, y_k:

$$d(y_i, y_j) = \|y_i - y_j\|_F \tag{2}$$

If $d(y_i, y_j) < d(y_i, y_k)$, we generate a triplet (x_i, x_j, x_k), which indicates the triadic comparison "x_i is more similar to x_j than to x_k". Then, a collection of N triadic comparisons are generated in the form as following:

$$\left\{ (x_i, x_j, x_k)_t, t = 1, \ldots, N \right\}$$

3.3   Problem Formulation

We address two problems simultaneously, learning multi-modal labels and learning the similarity function between videos. The side information of triadic comparison is provided as training dataset. To formulate the learning task, we define the similarity function $D(x_1, x_2)$ for any two video clips x_1, x_2, and assume an embedding functions g: $R^D \rightarrow R^d$ corresponding to the projection from Euclidean space to the intermediate semantic space:

$$D(x_i, x_j) = \left\| g(x_i) - g(x_j) \right\|_F \tag{3}$$

Given a set of objects V and a set of similarity measurements $C = \{(x_i, x_j, x_k)_t, x_i, x_j, x_k \in V, t = 1, \ldots, N,\}$, where a triplet (x_i, x_j, x_k) indicates the triadic comparison "x_i is more similar to x_j than to x_k". The goal is to find an embedding function g that satisfies:

$$\forall (x_i, x_j, x_k) \in C : \quad D(x_i, x_j) + 1 < D(x_i, x_k) \tag{4}$$

The embedding function g consists of two parts: the embedding function from low-level feature space to the multi-modal concept space for each modality, which is defined as $f^m: R^D \rightarrow R^s$. And the concatenation of all the modalities: $R^s \rightarrow R^d$. Here, we restrict attention to embeddings parameterized by the projection matrix W:

$$g(x) = Wf(x) \tag{5}$$

Following the Mahalanobis distance, the distance function of the video pair can be written as:

$$D(x_i, x_j) = (f(x_i) - f(x_j))^T W^T W (f(x_i) - f(x_j)) \tag{6}$$

3.4   Intermediate Semantics Learning

In this section, we elaborate the learning approach of intermediate semantic space, based on which the metric is learned. The whole process is shown in Fig. 3. First, the transformation from low-level feature to concept space for each modality is learned under the supervised learning framework. Then, the transformations of all the modalities are combined by multiple kernel learning framework.

Suppose the distance between inputs x_i and x_j in the m-th modality is represented as:

$$D^m(x_i^m, x_j^m) = \left\| L^m(x_i^m - x_j^m) \right\|_F \tag{7}$$

where x_i^m and x_j^m are the features of videos corresponding to the m-th modality. We use the Large Margin Nearest Neighbor (LMNN) to learn the L^m, because it is the optimal distance metric for the k-nearest neighbors.

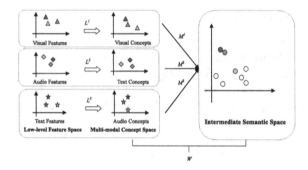

Fig. 3. The process of learning intermediate semantic space.

To extend the LMNN to support non-linear embeddings by the use of kernels, we first map the data into a reproducing kernel Hilbert space (RKHS) via a nonlinear map ϕ with corresponding kernel function $k(x_i, x_j) = \langle \phi(x_i), \phi(x_j) \rangle$. Therefore, the Eq. (7) can be rewritten as

$$D^m(x_i^m, x_j^m) = \left\| L^m(\phi(x_i^m) - \phi(x_j^m)) \right\| \tag{8}$$

According to the generalized representor theorem [20], the project L^m can be represented as a linear combination of the feature points in the form $L^m = N^m \Phi$, where $\Phi = [\phi(x_1), \ldots, \phi(x_N)]^T$. And the embedding problem can now be reformulated as an optimization over N^m rather than L^m. We adopt the method in [15] to iterate the update N^m until convergence.

After achieving the embedding function for each modality, we combine them together to form a unified intermediate semantic space, in which the distance of data points with similar concept to be near. Inspired by the works on multiple kernel technology [15], we define the transformation from low-level feature space to concept space as the concatenation of projections in each modality.

$$g(x) = (U^m(L^m(\phi(x))))_{m=1}^p \tag{9}$$

where L^m is the linear transformation of the input space for the m-th modality and p is total number of modalities.

3.5   Intermediate Semantics Based Metric Learning

In this section, we de_ne and learn the intermediate semantics based distance metric between videos. We attempt to learn the distance that keeps the videos similar if they are similar in the multi-modal concept space. Using the transformation from low-feature space to the multi-modal concept space, the distance can further be rewritten as:

$$
\begin{aligned}
D(x_i, x_j) &= \sum_{m=1}^{p} (\varphi^m(x_i) - \varphi^m(x_j))^T (U^m)^T U^m (\varphi^m(x_i) - \varphi^m(x_j)) \\
&= (\varphi(x_i) - \varphi(x_j))^T W^T W (\varphi(x_i) - \varphi(x_j)) \\
&= (\varphi(x_i) - \varphi(x_j))^T A (\varphi(x_i) - \varphi(x_j))
\end{aligned}
\tag{10}
$$

where $\varphi(x)$ is the concatenated features of all the modalities and $A = (W)^T W$. The block-matrix formulations of matrix W can be written as:

$$
\begin{bmatrix}
U^1 & 0 & \cdots & 0 \\
0 & U^2 & \cdots & 0 \\
\vdots & & \ddots & \\
0 & 0 & & U^p
\end{bmatrix}
$$

To learn the optimal similarity function on the intermediate concept space, we cast the problem into the following optimization task:

$$
\min_{A} \|A\| + \beta \sum_{t=1}^{N} h_t(x_i, x_j, x_k) \qquad s.t.\, D(x_i, x_k) - D(x_i, x_j) \geq 1, A \succ 0
\tag{11}
$$

In the constraint, we introduce a margin factor $+1$ to ensure a sufficiently large difference. The first term in Eq. (11) is a regularization term and the second term is the hinge loss:

$$
h_t(x_i, x_j, x_k) = \max(1 + D(x_i, x_j) - D(x_i, x_k), 0)
\tag{12}
$$

Here (x_i, x_j, x_k) C is a set of relative distance constraints, which are enforced through the hinge loss. Since the C may not be satisfied by a linear projection of A, we soften the constraints by introducing a slack variable for each constraint.

$$
\begin{aligned}
&\min_{A} \|A\| + \beta \sum_{C} \xi_{ijk} \\
&s.t. (\varphi(x_i) - \varphi(x_j))^T A (\varphi(x_i) - \varphi(x_j)) + 1 \\
&\leq (\varphi(x_i) - \varphi(x_k))^T A (\varphi(x_i) - \varphi(x_k)) + \xi_{ijk} \\
&\forall (x_i, x_j, x_k) \in C \\
&\xi_{ijk} \geq 0, A \geq 0
\end{aligned}
\tag{13}
$$

Denote the distance calculations in terms of Frobeninus inner products with $E_{ij} = (\varphi(x_i) - \varphi(x_j))(\varphi(x_i) - \varphi(x_j))^T$ and $D(x_i, x_j) = tr(AE_{ij})$. Let us define $\bar{C}$ as the set of all currently violated constraints, which trigger the hinge loss in the second part of Eq. (13), we can rewrite the loss function in Eq. (13) as:

$$\varepsilon(A) = tr(A) + \beta \sum_{\bar{C}} h(1 + tr(AE_{ij}) - tr(AE_{ik}))$$

(14)

The gradient of A can be computed from Eq. (14), which has two components: one for regularization and one for the hinge loss.

$$\frac{\partial \varepsilon}{\partial A} = A^T + \sum_{(v_i, v_j, v_k) \in \bar{C}} (E_{ij} - E_{ik})$$

(15)

We solve the problem based on projected gradient descent and combine the updated rules for the U_m and A together. Since the minimization of A must enforce the constraint that the matrix U_m remains positive semi-definite, the solver projects U_m onto the cone of all positive semi-definite matrices. A simplified pseudo-code implementation is shown in Algorithm 1.

Algorithm 1. The gradient projection pseudo-code implementation
Input: training data C
Ouput: matrix A
1: $t:=0$ { Initialize counter}
2: $U^m_0=I$ { Initialize U^m_0 with the identity matrix }
3: **while** (not converged) **do**
4:      **for** $m =1{:}p$
5:          Compute L_{t+1} for modality m
6:      **end for**
7:      $A_t \leftarrow$ concatenation of U^m_t
8:      $G_{t+1}=A_t^T +\text{sum}(E_{ij}\text{-}E_{ik})$ {Computeing gradient}
9:      $A_{t+1}=P(A_t - \beta G_{t+1})$ {Take gradient step and project onto SDP cone}
10:     $U^m_{t+1} \leftarrow (A_{t+1})^m$ { Update the U^m}
11:     $U^m_{t+1}=P(U^m_t)$ { Project onto SDP cone }
12:     $A=A_t,\ t:=t+1$
13: **end while**

---

4   Experiments

We conduct extensive experiments to evaluate the effectiveness of the proposed algorithm on a realistic video dataset.

4.1   Experimental Data

We use the Hollywood2-actions dataset [14] for the evaluation, which includes 1633 video clips (from 30 s to 10 min) sampled from 69 movies. The movies belong to different genres, including comedy, tragedy, horror, musical, and sci-fi, etc. The dataset consists of two parts: 810 training videos automatically collected from 33 training movies and 823 testing videos manually selected from 36 test movies. Since there are many false annotations in the training videos, we only use the testing videos (about 30 h) in our experiment. To obtain the text content in videos, we collect subtitles from the Internet. The audio and visual data are extracted from the videos. Three kinds of semantic contents are annotated manually. Specially, the multi-modal data include subtitles the videos, image frames and audios. Because the original videos are anno- tated with actions for the visual modality, we only add the annotations in text and audio, which are annotated manually by two subjects.

4.2   Experimental Setup

Since it is hard and time-consuming to manually compare similarity for videos, we generate the side-information of triplet pairs automatically in our experiments. Four kinds of triplet instances are generated, which are "Triplet 2vs1", "Triplet 2vs0", "Triplet 1vs0" and "Triplet Mix". The "Triplet 2vs1" contains triplets (x_i, x_j, x_k), in which x_i and x_j have the same labels in two modalities, while v_i and v_k are the same labels in one modality. The triplet pairs generated automatically are used as the ground truth for measuring the accuracy. Specifically, we first chose the same modalities x_i and x_j ran- domly, then randomly sample two videos that have same labels in the chose modalities. In total, we generate 10K triplet instances for each kind and 30K triplet in total.

To evaluate our method, we divide the triplet instances into training part and testing part, then adopt 10-fold cross validation on them for all the methods. The maximum iterations are set to 500 for LMNN and our method.

4.3   Multi-Modal Feature Extraction

Textual Feature. Extraction Subtitles of video clips are used as text contents, the words of which are used as the text feature. First, we conduct Log-Linear Part of Speech (POS) Tagger [23] on the sentences. Then Nouns, Adjective and Verbs are used to describe each video. The adjective and verb words are first stemmed using the Porter Stemmer. Finally, we use the Vector Space Model to describe the text modality of the videos. In the experiment, we use a 1,680-dimension vector as text feature for each video.

Visual Feature. Extraction We follow several existing works [8, 12, 14] to use the bag-of-feature framework to describe the videos. To capture the static appearance as well as motion patterns, both 2-D static point features [17] and 3-D space-time features [10] are extracted from video frames and video sequences. For 2-D features, they are described by SIFT descriptor [12]. For the 3-D features, Histograms of gradient (HoG) and histograms of optical ow (HoF) are computed to describe them. For each descriptor, we sample 80,000 descriptors and cluster them to generate a visual vocabulary (size = 2,000). Using the code-books, visual modality of each video clip can be represented by three histograms of visual words through soft-quantization [8] of local descriptors. To eliminate the influence of video length, the bag-of-features (with size 6000) are L1-normalized.

Audio Feature. Extraction Sound effects and background music in movies are essential for stimulating audience's perception. The audio features are necessary for video description. We break the audio for each clip into a sequence of mid-term windows (frames) with a 50 % overlap between successive windows [4]. For each 2 s long frame, we extract 7 features that are the most commonly used in speech processing. In detail, the features include: Zero Crossing Rate (ZCR), Energy Entropy, Spectral Rolloff, the Mel Frequency Cepstral Coefficients (MFCC), zero pitch ratios, spectrogram, chroma. We further use the statistics (e.g. standard deviation, or average value) of these feature sequences as the audio feature for each audio segment following [4].

4.4   Experimental Results

We examine the performance of the proposed algorithm on distance metric learning using the generated triplet data.

Comparison Algorithm. To extensively evaluate the effectiveness of our algorithm, we compare the proposed algorithm against existing representative distance metric learning algorithm and multi-modal similarity learning methods, including RCA [1], LMNN [27] and MKPOE [15]. Besides, we also evaluate a heuristic baseline method with Euclidean distance.

To adapt the DML methods (RCA, LMNN, Euclidean distance) for multi-modal similarity, we explore two fusion strategies for comparison, which include "Best" and "Uniform combination" following [28]. The detail of implementation for each comparison is elaborated as follows:

Eucl-B: we test the performance of Euclidean distance on each modality, and then select the modality with best performance.

Eucl-U: we test the performance of Euclidean distance on each modality, and then combine all modalities uniformly.

RCA-B: we train RCA model on the training set for all the modalities and test model on the validation set, and then select the best modality with the highest mAP for RCA.

RCA-U: we train RCA model on the training set for all the modalities, and then combine all modalities uniformly.

LMNN-B: we train LMNN model on the training set for all the modalities and test model on the validation set, and then select the best modality with the highest mAP for LMNN.

LMNN-U: we train LMNN model on the training set for all the modalities, and then combine all modalities uniformly.

MKPOE: the existing MKPOE algorithm in [15].

ISDL: The algorithm proposed in this paper.

We compare these algorithms on the accuracy, which is computed by counting the fraction of correctly predicted relative comparisons in the total number of comparisons. The higher accuracy indicates that the algorithm is more effective in measuring the similarity for triplet pairs.

Evaluation on "Triplet 1vs0". For the triplet (x_i, x_j, x_k), v_i and v_j have the same label in one modality, while v_i and v_k have different labels in all the modalities. This dataset can be viewed as two instances sampled from one same class and the third instance sampled from the different classes. This situation equals to the metric learning in single label setting. Table 1 summarizes the evaluation results on the data sets. Because little information about multi-modal concept is provided in the dataset, all the algorithms perform similarly.

Table 1. Evaluation of the mean accuracy on "Triplet 1vs0"

Algorithm	Mean accuracy	Algorithm	Mean accuracy
Eucl-B	0.5290	Eucl-U	0.5020
RCA-B	0.5020	RCA-U	0.4750
LMNN-B	0.5320	LMNN-U	0.5150
ISDL	**0.5720**	MKPOE	0.5300

Evaluation on "Triplet 2vs0". The "Triplet 2vs0" contains triplet (x_i, x_j, x_k) that v_i and v_j have same labels in two modalities while v_i and v_k are different in three modality. Compared with results on "Triplet 1vs0", the mean accuracy of all the algorithms improve in Table 2. It shows that each modality plays an important role in distance metric learning and multi-modal based metric learning performed better than uni-modal based metric learning. It is worth noting that RCA and LMNN perform better in "Best" than "Uniform combination". And the high performance in the visual modality improves the performance of Euclidean distance dramatically, which shows the Euclidean distance is easy to be influenced by performance in the single modality.

Table 2. Evaluation of the mean accuracy on "Triplet 2vs0"

Algorithm	Mean accuracy	Algorithm	Mean accuracy
Eucl-B	0.7630	Eucl-U	0.7770
RCA-B	0.5370	RCA-U	0.3810
LMNN-B	0.6250	LMNN-U	0.5250
ISDL	**0.8230**	MKPOE	0.791

Evaluation on "Triplet 2vs1". We further evaluate the algorithms on the "Triplet 2vs1" in Tabel 3, which contains triplet (x_i, x_j, x_k) that v_i and v_j have same labels in two modalities while v_i and v_k have same label in one modality. Since the instance in each pair is similar in different degree, this dataset is more challenging than "Triplet 1vs0" and "Triplet 2vs0". The performance of Euclidean distance decreases. But RCA and LMNN have better performance than "Triplet 1vs0" and "Triplet 2vs0". In different combination strategy, RCA and LMNN achieve the highest performance in the "Best" and "Uniform" combination separately. MKPOE and our ISDL have better performance than single-modal methods.

Table 3. Evaluation of the mean accuracy on "Triplet 2vs1"

Algorithm	Mean accuracy	Algorithm	Mean accuracy
Eucl-B	0.5050	Eucl-U	0.4660
RCA-B	0.5980	RCA-U	0.4340
LMNN-B	0.5030	LMNN-U	0.4740
ISDL	**0.8540**	MKPOE	0.8150

Evaluation on "Triplet Mix". We further evaluate the algorithms on the "Triplet Mix" dataset in Tabel 4, which contains triplet (x_i, x_j, x_k) that are sampled from the former three dataset, i.e. "Triplet 1vs0", "Triplet 2vs0" and "Triplet 2vs1". Table 4 summarizes the mean accuracy for each algorithm on the "Triplet Mix" dataset.

Table 4. Evaluation of the mean accuracy on "Triplet Mix"

Algorithm	Mean accuracy	Algorithm	Mean accuracy
Eucl-B	0.5890	Eucl-U	0.5910
RCA-B	0.5580	RCA-U	0.5360
LMNN-B	0.6020	LMNN-U	0.5680
ISDL	**0.7980**	MKPOE	0.7540

We can draw several observations from the results on different triplet settings. First of all, Euclidean distance performed better when the comparison pairs have large differences. Euclidean distance performs better on distinct pairs (Triplet 1vs0, Triplet 2vs0) than similar pairs (Triplet 2vs1). RCA and LMNN perform better on "Triplet 2vs1" and "Triplet 2vs0" respectively. The uniform combination of distances from three modalities seems to decrease the performance for RCA and LMNN. This is because the distances of the dissimilar modalities influence the whole performance. Besides, the proposed method achieves better results in all the evaluations especially when the differences among triplet pairs are not notable. And our method has more stable and higher performance than the other algorithms, which validate the effectiveness of the semantic for metric learning.

5   Conclusion

In this paper, we proposed a novel algorithm to learn the multi-modal concepts and intermediate semantics based metric learning for videos simultaneously. To address the difficulty of similarity measure for videos, we proposed multi-modal labels, which consist of semantic concepts from three different modalities, i.e. text, image, audio. The multi-modal concepts provide a bridge between low-level features and high-level semantic concepts. Based on the multi-modal concepts, an Intermediate Semantics based Distance Learning (ISDL) method was proposed, which integrates the learned semantic concept with distance metric leaning. In particular, ISDL explores a unified learning scheme, which learns embedding function for each modality and the optimal combination of these functions together. The experiments on the video datasets show the high effectiveness of ISDL in multi-modal similarity learning.

References

1. Bar-Hillel, A., Hertz, T., Shental, N., Weinshall, D.: Learning distance functions using equivalence relations. In: Proceedings in Conference on Machine Learning, pp. 11–18 (2003)
2. Davis, J.V., Kulis, B., Jain, P., Sra, S., Dhillon, I.S.: Information theoretical metric learning. In: Proceedings in Conference on Machine Learning, pp. 209–216 (2007)
3. Globerson, A., Roweis, S.T.: Metric learning by collapsing classes. In: Neural Information Processing Systems, pp. 451–458 (2005)
4. Giannakopoulos, T., Pikrakis, A., Theodoridis, S.: A multi-class audio classification method with respect to violent content in movies, using Bayesian networks. In: IEEE International Workshop on Multimedia Signal Processing, pp. 90–93 (2007)
5. Goldberger, J., Roweis, S., Hinton, G., Salakhutdinov, R.: Neighbourhood components analysis. In: Neural Information Processing Systems, pp. 513–520 (2004)
6. Hauptmann, A.G., Yan, R., Lin, W.H., Christel, M., Wactlar, H.: Can high-level concepts fill the semantic gap in video retrieval? A case study with broadcast news. IEEE Trans. Multimed. **9**(5), 958–966 (2007)
7. Hoi, S.C.H., Liu, W., Lyu, M.R., Ma, W.Y.: Learning distance metrics with contextual constraints for image retrieval. In: Proceedings of Computer Vision and Pattern Recognition, pp. 2072–2078 (2006)
8. Jiang, Y.G., Ngo, C.W., Yang, J.: Toward optimal bag-of-features for object categorization and semantic video retrieval. In: ACM International Conference on Image Video Retrieval, pp. 494–501 (2007)
9. Kulis, B.: Metric learning: a survey. Found. Trends Mach. Learn. **5**(4), 287–364 (2012)
10. Laptev, I.: On space-time interest points. IJCV **6**(2/3), 107–123 (2005)
11. Lin, C.Y., Tseng, B.L., Smith, J.R.: Video collaborative annotation forum: establishing ground-truth labels on large multimedia datasets. In: Proceedings of the TRECVID Workshop (2003)
12. Lowe, D.: Distinctive image features from scale invariant keypoints. IJCV **60**(2), 91–110 (2004)
13. Ma, Z., Hauptann, A.G., Yang, Y., Sebe, N.: Classifier-specific intermediate representation for multimedia tasks. In: ICMR, p. 50. ACM press, Hong Kong (2012)

14. Marszalek, M., Laptev, I.: Actions in context. In: CVPR, pp. 2929–2936. IEEE press (2009)
15. McFee, B., Lanckriet, G.R.G.: Learning multi-modal similarity. J. Mach. Learn. Res. **12**, 491–523 (2011)
16. Mei, T., Yang, B., Hua, X.S., Li, S.: Contextual video recommendation by multimodal relevance and user feedback. ACM Trans. Inf. Syst. **29**(2), 10 (2011)
17. Mikolajczyk, K., Schmid, C.: Scale and affine invariant interest point detectors. IJCV **60**(1), 63–86 (2004)
18. Naphade, M.R., Smith, J.R.: Large-scale concept ontology for multimedia. IEEE MultiMed. **13**(3), 86–91 (2006)
19. Qi, G.J., Hua, X.S., Rui, Y., Tang, J., Mei, T., Zhang, H.J.: Correlative multi-label video annotation. In: ACM MultiMedia, pp. 17–26 (2007)
20. Schölkopf, B., Herbrich, R., Smola, A.J.: A generalized representer theorem. In: Helmbold, D.P., Williamson, B. (eds.) COLT 2001 and EuroCOLT 2001. LNCS (LNAI), vol. 2111, pp. 416–426. Springer, Heidelberg (2001)
21. Schultz, M., Joachims, T.: Learning a distance metric from relative comparisons. In: NIPS, pp. 41–48 (2003)
22. Snoek, C., Worring, M., Geusebroek, J.M., Smeulders, A.W.M.: The challenge problem for automated detection of 101 semantic concepts in multimedia. In: ACM MultiMedia, pp. 421–430 (2007)
23. Toutanova, K., Klein, D., Manning, C.D., Singer, Y.: Feature-rich part-of-speech tagging with a cyclic dependency network. In: HLT-NAACL, pp. 173–180 (2003)
24. TREC video retrieval evaluation. http://www-nlpir.nist.gov/projects/trecvid
25. Wang, M., Hua, X.: Study on the combination of video concept detectors. In: ACM MultiMedia, pp. 647–650 (2008)
26. Wang, Y., Lin, X., Zhang, Q.: Towards metric fusion on multi-view data: a cross-view based graph random walk approach. In: CIKM, pp. 805–810. ACM press, San Francisco (2013)
27. Weinberger, K., Blitzer, J., Saul, L.: Distance metric learning for large margin nearest neighbor classification. In: NIPS, pp. 1473–1480 (2006)
28. Wu, P., Hoi, S.C.H., Xia, H., Zhao, P., Wang, D., Miao, C.: Online multimodal deep similarity learning with application to image retrieval. In: ACM MultiMedia, pp. 153–162 (2008)
29. Xia, H., Wu, P., Hoi, S.C.H.: Online multi-modal distance learning for scalable multimedia retrieval. In: WSDM, pp. 455–464. ACM press, Rome (2013)
30. Yang, L., Jin, R.: Distance Metric Learning: A Comprehensive Survey. Michigan State University (2006)

A Community Detection Algorithm Considering Edge Betweenness and Vertex Similarity

Hongwei Lu, Chang Liu, and Zaobin Gan[✉]

School of Computer Science and Technology,
Huazhong University of Science and Technology,
Wuhan 430074, People's Republic of China
{luhw,maggieliu,zgan}@hust.edu.cn

Abstract. Community detection is an important topic in social network analysis. It is beneficial to understand the underlying structure of the network and extract useful information from it. Most existing community detection algorithms require a prior information or neglect peripheral vertices. In this paper, we propose a divisive community detection algorithm named CDBS (Community Detection considering edge Betweenness and vertex Similarity). First, the betweenness and similarity of the connected pair of vertices are calculated for all edges in the network. Secondly, the edges with relatively high betweenness and low similarity between connected pairs of vertices are identified by two thresholds (δ and θ). Then these edges are removed from the network and the betweenness of the remaining edges are recalculated. This procedure is iterated until there is no more edge of which the betweenness is higher than δ and similarity is less than θ. Finally, the proposed algorithm is validated in both synthetic and real-world networks. Experimental results demonstrate that CDBS is effective at detecting dense community structure with high accuracy and modularity, and it is time-efficient because of low computational complexity. Besides, CDBS can cope with the isolated cluster problem.

Keywords: Community detection · Edge betweenness · Vertex similarity

1 Introduction

Complex network analysis has been the focus of much attention in the past decade. The community structure is an important property of complex networks, which can bring out much information about networks. The community structure corresponds to the groups of vertices, in which vertices are joined together in tightly-knit groups between which there are only looser connections [1].

Detecting the community structure exhibited by real networks is a crucial step toward an understanding of complex systems. Vertices belonging to the same community are more likely to share common properties or play similar roles within the network. In the World Wide Web, the pages grouped together

© Springer International Publishing AG 2016
W. Cellary et al. (Eds.): WISE 2016, Part I, LNCS 10041, pp. 243–251, 2016.
DOI: 10.1007/978-3-319-48740-3_17

may deal with the same or related topics, which is beneficial to public opinion analysis and hot topic tracking. By detecting the community in online social network sites, an efficient personalized recommendation system can be set up for customers shared similar interests or backgrounds. As a whole, community detection is of great importance in social network analysis, knowledge discovery and behavior prediction.

The preliminary knowledge is required in some existing community detection algorithms, which is hard to obtain in most cases. Besides, some peripheries connecting loosely to the other vertices tend to be classified as separated clusters, leading to an unreasonable division. To address these issues, we propose a divisive community detection algorithm named CDBS.

The remainder of this paper is organized as follows. First, the related works about community detection are briefly reviewed in Sect. 2. Section 3 describes the process of CDBS in detail. Then in Sect. 4 CDBS is tested on synthetic and real-world networks and the experimental analysis is given. The conclusion and direction for further development are drawn at the end of the paper.

2   Related Work

So far, a large number of methods have been developed to detect the community structure in networks.

Kernighan-Lin [2] is a traditional method based on graph partitioning, which aims at dividing the vertices to a predefined size such that the number of edges between communities reaches the minimum. Donath and Hoffmann [3] devised a spectral clustering method using the eigenvectors of the adjacency matrix for graph partition. This kind of method needs to specify the number of communities at the beginning of the calculation, which is typically unknown for real networks. Agglomerative algorithms merge similar clusters recursively, the typical algorithm is CNM proposed by Clauset et al. [4]. But it often scales badly in large networks. Moreover, it does not provide a stopping criterion for the iteration. To better evaluate the quality of the division, modularity is introduced in GN [5]. The partition with the maximum modularity is regarded as the most indicative community structure in the network. Louvain [6] is a multi-level technique based on local optimization of modularity. In another work, modularity is optimized via spectral bisection by replacing the Laplacian matrix with modularity matrix [7]. The main limitation of modularity-based algorithm is the fact that the true maximum is out of reach.

In addition, Reichardt et al. [8] proposed that the community structure of networks are interpreted as the spin configuration that minimizes the energy of the spin glass with the spin states being the community indices. As random walks can also be applied to community detection, Pons and Latapy [9] devised a new distance between vertices based on random walks. Lu et al. [10] introduced an improved agglomerative algorithm named ACSS. A near linear time algorithm called LPA is presented in [11]. Le et al. [12] put forward a method to solving a class of optimization problems over label assignments via projection onto low-dimensional subspace. In order to overcome the drawbacks mentioned above, a divisive community detection algorithm is proposed in the following.

3 CDBS Algorithm

The underlying assumption of CDBS is that intercommunity edges are characterized by a relatively high edge betweenness and a low similarity of the connected vertices. By removing these intercommunity edges, denser community structure therefore can be detected.

3.1 Metrics

Edge Betweenness. Betweenness is introduced by Freeman [13], which is usually interpreted as the degree to which an edge is in a position of brokerage. Let $\sigma(s,t)$ be the number of geodesic paths from vertex s to vertex t. Then, the number of geodesic paths from vertex s to vertex t that pass through intermediary edge e can be denote by $\sigma(s,t|e)$. The betweenness $c_B(e)$ of an edge e in the network can be expressed as Eq. (1)

$$c_B(e) = \sum_{s,t \subseteq V} \frac{\sigma(s,t|e)}{\sigma(s,t)} \tag{1}$$

Jaccard Similarity. Similarity is a measure of closeness between a pair of vertices. It is intuitive that a pair of vertices are likely to be closer if they share some common neighbors. Assuming that N_i, N_j are respectively defined as the neighbor set of vertex i and j. More notation, given a set N, the cardinality of N representing by $|N|$ counts how many elements are in N. The intersection and union between N_i and N_j can be denoted by $N_i \cap N_j$ and $N_i \cup N_j$, respectively. The definition of Jaccard metric can be defined as Eq. (2)

$$sim(i,j) = \frac{|N_i \cap N_j|}{|N_i \cup N_j|} \tag{2}$$

3.2 Algorithm Description

To make things more concrete, here, we give an example to illustrate the core of CDBS. The famous zachary's network [14] is showed in Fig. 1(a). There are 34 vertices linked by 78 edges in total. The network is composed of two groups, with one centered around vertex 1 and the other around vertex 34. Figure 1(b) shows the plot of similarity as a function of betweenness for each edge, where edge betweennesses and similarities are calculated by Eqs. (1) and (2), respectively. To better identify these intercommunity edges, here, we introduce two thresholds represented by two red lines, where θ and δ stand for the similarity

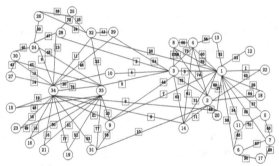

(a) Edges and vertices are marked in zachary's network.

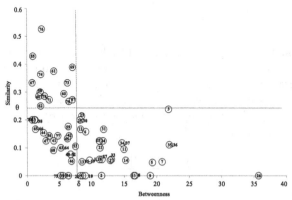

(b) Similarity is a function of betweenness for each edge in 1(a).

Fig. 1. CDBS in zachary's network.

and betweenness threshold. By tuning thresholds, we assign the value of θ and δ to be 0.246 and 8.16. Edges at the right of δ below θ will be added to the removal set on the first iteration. We mark them from 1 to 37 for simplicity. As anticipated, almost all intercommunity edges are included in the removal set.

In particular, vertex 12 has only one neighbor. It is intuitive that edge 33 will have a high betweenness score, as all shortest paths to vertex 12 must pass through edge 33. And the similarity between vertex 1 and vertex 12 is zero because of sharing no common neighbors. If edge 33 is removed from the graph, which will directly lead vertex 12 to be isolated. To solve the problem of isolated cluster, we make this rule that an edge with the minimum betweenness will be retained if all of its neighbors are included in the removal set. Analogously, edge 20 is retained on this iteration for the same reason. Then the betweennesses are recalculated until there are no more edges at the right of δ below θ.

3.3 Computational Complexity Analysis

The removing process of edges is described in Algorithm 1. Assuming that the network has n vertices and m edges. The computational complexity of

Algorithm 1 is $O(mnt)$, where t denotes the number of iterations. The preliminary division has been formed after Algorithm 1. In some cases, the partitions may contain several tiny communities. Here, we refer tiny community as the community with vertices less than 20 % of the average number of vertices within the community. By merging these tiny communities with some bigger communities, the accuracy and modularity of the division will be increased in a manner. The merger process can be done in $O(n)$ time. Thus, for a sparse network with $m \sim n$, the total running time of CDBS is $O(n^2 t)$. In practical case, the number of removing iterations appears independent of the graph size, or grows very slowly with it. CDBS is therefore a time-efficient algorithm.

Algorithm 1. Removing Edges

Input: $G(V, E)$
Output: $C = \{c_1, c_2, \dots c_n\}$
1:   $flag = true$; //there exist edges satisfying the removal condition
2: **for** each $e \in E$ **do**
3:     do calculate s_e; //denote the similarity of the pair of vertices connected by e
4: **end for**
5: **while** $flag$ **do**
6:     $D = \varnothing$; //denote the removal set
7:     **for** each $e \in E$ **do**
8:        do calculate b_e; //denote the edge betweenness of e
9:     **end for**
10:    **for** each $e \in E$ **do**
11:       **if** $e_b > \delta$ && $e_s < \theta$ **then**
12:          $D \cup e$; //add the edge to D
13:       **end if**
14:    **end for**
15:    **for** each $v \in V$ **do**
16:       calculate N_v; //denote all the edges connected to node v
17:       **if** $N_v \subseteq D$ **then**
18:          $B_{min} = \infty$; //find the edge with minimum betweenness
19:          **for** each $e \in N_v$ **do**
20:             **if** $e_b < B_{min}$ **then**
21:                $B_{min} = e_b$;
22:                $E_{min} = e$;
23:             **end if**
24:          **end for**
25:          $D = D - e$; //remove the edge from D
26:       **end if**
27:    **end for**
28:    $E = E - D$;
29:    **if** $D == \varnothing$ **then**
30:       $flag = false$; //there are no more edges satisfying the removal condition
31:    **end if**
32: **end while**
33: $G = c_1 \cup c_2 \cup \dots \cup c_n$; //$G$ is split into several communities

---

4 Experimental Results and Analysis

In this paper, NMI (Normalized Mutual Information) is adopted to evaluate the accuracy of the detected division in synthetic networks. While the quality of the community structure in real networks is measured by modularity. A higher NMI or modularity score indicates a denser community structure. CDBS is tested on synthetic and real networks. Several community detection algorithms are applied on the same set of networks for comparison. The results of Spin glass and LPA are averaged over 10 independent runs due to their unstable performance.

4.1 Synthetic Network

The LFR benchmark [15] can be considered as a proxy of a real network, and it has became a standard in the evaluation of the performance of community detection algorithm. Besides, the code to create the LFR benchmark is freely available.[1] Here, we select the network with 1000 vertices. The average degree is 15, and the exponents of the degree distribution and community size distribution are 2 and 1, respectively. The mixing parameter μ represents the ratio between external and internal degree of each vertex in its community. By increasing μ, the community structure becomes fuzzier and harder to be identified.

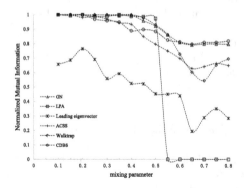

Fig. 2. Test of algorithms on the LFR benchmark.

Figure 2 illustrates the results tested on the LFR benchmark. Each curve shows the variation of NMI with the mixing parameter μ. The performance of Leading eigenvector is obviously worse than the other algorithms. Observe that the performance of ACSS and Walktrap are severely affected by the increase of the mixing parameter, where the curves go down quickly when $\mu > 0.5$. LPA can't even figure any community structure when $\mu \geq 0.55$, i.e., all vertices are joined together in a single community. The accuracy of GN is rather good,

---

[1] The software can be downloaded from https://sites.google.com/site/santofortunato/inthepress2.

and its value of NMI displays a similar pattern with CDBS, where NMI can reach nearly 0.8 even when $\mu > 0.5$. However, because of the intrinsic resolution limit of GN, there exist some separated clusters containing isolated vertex when $\mu \geq 0.4$. Moreover, the average running time of GN is up to 4051 s. The other algorithms can be completed within 10 s, which are 400 times faster than GN. The experimental result provides empirical evidence that CDBS can detect good division even if the community structure is hard to identify, and the proposed algorithm is time-efficient as well.

4.2   Real Networks

Then all the algorithms are tested on four real networks with known community structure. The results on networks are listed in Table 1. Each row presents the number of communities and modularity detected by different algorithms.

In zachary's network, we note that GN, Walktrap and Louvain have mis-classified vertices. The structure by Leading eigenvector and LPA is not so strong that Q is lower than the others. By further analyzing the division, the division of Spin glass and CDBS are very similar, the fine distinction exists in the placement of vertex 24. In Spin glass, vertex 24 belongs to the sub-group{25,26,28,29,32}. In CDBS, instead, vertex 24 belongs to a bigger sub-group{9,10,15,16,19,21,23,27,30,31,33,34}. According to a brief analysis, vertex 24 has three edges connected to the bigger subgroup and two edges to the smaller one. It seems plausible to suggest that vertex 24 should be allocated to the bigger community. As excepted, the NMI of CDBS is 0.6956, slightly higher than 0.6872 by Spin glass and 0.6021 by ACSS. In the other real-world networks, CDBS remarkably outperforms the other algorithms because of its highest Q.

Overall, the experimental results imply that the algorithm combining edge betweenness and similarity performs robustly well in both accuracy and running time. CDBS can be applied to real-world networks, and it runs extremely fast

Table 1. Comparisons in real-world networks

Network (#vertices, #edges)	Zachary (34,78)		Dolphins[16] (62,159)		Polbooks[17] (105,441)		Football [1] (115,615)	
Algorithm	#C	Q	#C	Q	#C	Q	#C	Q
GN	2	0.4012	5	0.5194	5	0.5168	11	0.6005
Walktrap	5	0.3532	7	0.5007	3	0.5069	10	0.6038
Leading eigenvector	4	0.3934	5	0.4912	4	0.4672	8	0.4877
LPA	2	0.3715	4	0.5118	3	0.4986	12	0.5769
Spin glass	4	**0.4198**	5	0.3223	6	0.3356	12	0.3222
Louvain	4	0.4188	5	0.5185	4	0.5205	9	0.6021
ACSS	4	0.4156	4	0.5238	2	0.5140	9	0.6028
CDBS	4	0.4174	4	**0.5268**	4	**0.5235**	9	**0.6054**

on the LFR benchmark. The downside of CDBS is its strongly depending on the parameters. Inappropriate thresholds may result in bad division.

5 Conclusions

The advantages of the CDBS are threefold. First, it doesn't need to know the number of communities or the other prior knowledge in advance. Second, it can detect community with high accuracy and perform rather well even when the community structure is fuzzy. Third, it can solve the isolated cluster problem. For large networks, the results of CDBS are robust with respect to the choice of θ and δ. As next step, we intend to investigate the distributions of the betweenness and similarity in the network, which may provide a solution for the automatic choice of the threshold of betweenness and similarity.

References

1. Girvan, M., Newman, M.E.J.: Community structure in social and biological networks. Proc. Natl. Acad. Sci. **99**, 7821–7826 (2002)
2. Kernighan, B.W., Lin, S.: An efficient heuristic procedure for partitioning graphs. Bell Syst. Tech. J. **49**, 291–307 (1970)
3. Donath, W.E., Hoffman, A.J.: Lower bounds for the partitioning of graphs. IBM J. Res. Dev. **17**, 420–425 (1973)
4. Clauset, A., Newman, M.E.J., Moore, C.: Finding community structure in very large networks. Phys. Rev. E. **70**, 066111 (2004)
5. Newman, M.E.J., Girvan, M.: Finding and evaluating community structure in networks. Phys. Rev. E. **69**, 026113 (2004)
6. Blondel, V.D., Guillaume, J.L., Lambiotte, R., Lefebvre, E.: Fast unfolding of communities in large networks. J. Stat. Mech. **2008**, P10008 (2008)
7. Newman, M.E.J.: Finding community structure in networks using the eigenvectors of matrices. Phys. Rev. E. **74**, 036104 (2006)
8. Reichardt, J., Bornholdt, S.: Statistical mechanics of community detection. Phys. Rev. E. **74**, 016110 (2006)
9. Pons, P., Latapy, M.: Computing communities in large networks using random walks. In: Yolum, I., Güngör, T., Gürgen, F., Özturan, C. (eds.) ISCIS 2005. LNCS, vol. 3733, pp. 284–293. Springer, Heidelberg (2005). doi:10.1007/11569596_31
10. Lu, H., Zhao, Q., Gan, Z.: A community detection algorithm based on the similarity sequence. In: Benatallah, B., Bestavros, A., Manolopoulos, Y., Vakali, A., Zhang, Y. (eds.) WISE 2014. LNCS, vol. 8786, pp. 63–78. Springer, Heidelberg (2014). doi:10.1007/978-3-319-11749-2_5
11. Raghavan, U.N., Albert, R., Kumara, S.: Near linear time algorithm to detect community structures in large-scale networks. Phys. Rev. E. **76**, 036106 (2007)
12. Le, C.M., Levina, E., Vershynin, R.: Optimization via low-rank approximation for community detection in networks. Ann. Stat. **44**(1), 373–400 (2016)
13. Freeman, L.C.: A set of measures of centrality based on betweenness. Sociometry **40**, 35–41 (1977)
14. Zachary, W.W.: An information flow model for conflict and fission in small groups. J. Anthropol. Res. **33**, 452–473 (1977)

15. Lancichinetti, A., Fortunato, S., Radicchi, F.: Benchmark graphs for testing community detection algorithms. Phys. Rev. E. **78**, 046110 (2008)
16. Lusseau, D.: The emergent properties of a dolphin social network. Proc. R. Soc. B. **270**, S186–S188 (2003)
17. Newman, M.E.J.: Modularity and community structure in networks. Proc. Natl. Acad. Sci. **103**, 8577–8582 (2006)

Measuring and Ensuring Similarity of User Interfaces: The Impact of Web Layout

Sebastian Heil[1]([⊠]), Maxim Bakaev[2], and Martin Gaedke[2]

[1] Technische Universität Chemnitz, 09107 Chemnitz, Germany
sebastian.heil@informatik.tu-chemnitz.de
[2] Novosibirsk State Technical University, 630073 Novosibirsk, Russia
bakaev@corp.nstu.ru,
martin.gaedke@informatik.tu-chemnitz.de

Abstract. Given the rapid update cycles in modern web information systems and the abundance of legacy software being migrated to the web, controlling similarity between user interfaces (UI) is an actual problem of interaction engineering. The similarity (consistency) aspect is also increasingly considered in computer-aided design, where it is included in the optimized goal function, to minimize re-learning effort for users. In this paper, we explore the impact of the proposed layout distance measure, which is calculated for different levels of hierarchy in web UIs, which we identify as: Region – Block – Group – Element. To support our approach, we conducted an experimental pilot study in the context of an ongoing medical information system (IS) web migration project. The regression analysis suggests that layout distance (particularly, its *orientation* dimension) does have effect on web UI similarity as perceived by users. The results can be used by web engineers, in particular to smoothen the transition between versions of a UI for users and IS operators.

Keywords: Similarity measure · User interface · Web migration · HCI

1 Introduction

Web information systems (WIS) increasingly supersede existing information systems (IS), even in domains where traditional desktop-based IS were common, such as medical software or in the financial sector [1]. In particular, cloud-based SaaS offerings gain more popularity. As deployment is simplified by the centralized architecture, development cycle lengths are significantly reduced. Updates in the production versions of SaaS WIS can be applied up to several times a day. Given these rapid update cycles and the abundance of legacy software being migrated to the web, controlling similarity between user interfaces (UI) is an actual problem of interaction engineering. Being able to determine the similarity between different versions of a user interface allows to control the amount of change during update cycles and to ensure smoother transitions and reduced learning efforts when migrating user interfaces to the web.

Migrating a legacy desktop IS into a web IS implies a fundamental paradigm shift which involves changes in all parts of the application: presentation (user interface), application logic and persistence (database) [2]. However, existing approaches like [1, 3]

© Springer International Publishing AG 2016
W. Cellary et al. (Eds.): WISE 2016, Part I, LNCS 10041, pp. 252–260, 2016.
DOI: 10.1007/978-3-319-48740-3_18

mainly focus on code and data migration and often disregard problems and costs associated with changes to the user interface and user interaction. The transformation or re-development of a user interface as a combination of HTML, CSS and Javascript source code does not only change the internal structure, but also impacts the visual appearance and user interaction.

As a first step towards supporting developers to measure and control user interface similarity, in this paper we explore the impact of layout, which is arguably the top consideration when migrating a UI to the web. In Sect. 2, some related works are examined and our approach for measuring distances between interfaces based on counting and visual area measurements is outlined. In Sect. 3, we describe the pilot experimental study that we have undertaken to support our approach, while the analysis of the results is presented in Sect. 4.

2   Related Work

Direct applications of UI similarity seem to be scarce in HCI, where it is overshadowed by the concept of consistency, whose importance is widely recognized, but which conceptually remains rather vague [4]. Consistency is, in a way, similarity within a single interface – between different screens, elements, minor and major conventions, etc. The reason we chose to speak of not consistency but similarity is that the latter is already extensively used in the AI field, in particular for case-based reasoning (CBR) that is gaining increased popularity in intelligent and recommender systems. The main idea is that a new problem is identified and the search for similar but already resolved problems is undertaken, in the assumption that similar problems have similar solutions.

In HCI, practical application of CBR methods remains somehow limited, in particular because interaction problems' formal identification remains unresolved, although in one of our research works we proposed to employ concepts of dedicated web design support ontology to describe the interaction context [5]. An important milestone in the field was the development of the SUPPLE system that is capable of auto-generating user interfaces with a model-driven approach. The authors introduced an *interface dissimilarity* metric that was included in the optimized goal function, so that new interfaces produced by the system resemble the old ones, familiar to the users [6]. The metric was determined as linear combination of factors {0/1} reflecting whether or not the two considered interface widgets are similar according to a certain criterion. The authors also put forward a list of widget features: language, orientation of data presentation, primary manipulation method, widget geometry, etc.; but their groping and layout don't seem to get further adequate consideration. We'd like to note that with the use of the totally same widgets designers would be able to create radically different interfaces, although it may be less of a problem in case of interfaces auto-generated by an intelligent system from the same interface model.

Thus it so far remains unclear whether robust quantitative identification of web interfaces is feasible, but we believe such an undertaking will be of considerable potential use for AI methods advance in HCI.

3 Method

As we mentioned above, we first decided to consider the layout similarity of user interfaces, for which end we propose approaches to determining the *distances* between several interfaces in terms of their visual layout. By analogy with AI-based CBR algorithms, we could try to infer weights for the several potentially meaningful dimensions of the distance, which we identify and quantify as the following:

1. **Orientation:** the share of interface items that have different visual orientation – horizontal, vertical, or other.
2. **Order:** the share of interface items that have different order relative to both their neighbors (start and end are virtual neighbors). This dimension is especially relevant for migration, since it implies that the interfaces in comparison (legacy and web ones) have the same or comparable items.
3. **Density:** the share of interface items that have different visual density of sub-items. It shouldn't matter whether the density is increased or decreased – we shall consider the effect on similarity in the same way.

The interface items that we mention here are somehow close to widgets introduced in interface dissimilarity metric in [6]. They can belong to either level in the modern web interfaces organizational hierarchy, which we see as Region – Block – Group – Element. The relations between the levels may .be described with more complex models, but we will so far consider all the levels to be equal in importance and assume that changes in any of them have comparable effects on perceived similarity.

In order to support our approach and provide a first evaluation, we conducted an experimental pilot study, which took place in the context of an ongoing medical IS (patient management system) web migration project.

4 Experiment Description

The experiment scenario is adapted from a research collaboration project with an industrial partner – the migration of medical software system to the web. The migration of existing user interfaces used by doctors and nurses introduces changes both in layout and interaction, and in the health sector there is generally not much time to conduct extensive user training, so similarity becomes especially prominent.

4.1 Experimental Design and the Hypothesis

The experiment had within-subjects design, with main independent variables being the layout distances and dependent variables being the similarity of old and new user interfaces, as perceived by users. We also added additional dependent variables that we outline in more detail below, to more fully capture the users' experience with interfaces. Our main hypothesis, related to the approach we propose for expressing the layout distance measure, is thus the following:

H_0: there's no effect of distance measures on perceived interface similarity.

To evaluate the validity of our approach and the experimental design, as well as the diversity of the subjects' evaluations, we'll also explore the differences in the calculated dimensions of the distance, as well as correlations between the evaluation scales.

4.2   The User Interfaces

We chose three legacy user interfaces representing different levels of complexity:

- User interface screen A is a simple graphical shift schedule (complexity: 1)
- User interface screen B is a calendar for appointment scheduling (complexity: 2)
- User interface screen C is an extensive patient data form (complexity: 3)

For each of them we created web versions, implemented in HTML, CSS and Javascript using Bootstrap and jQuery. In terms of layout, they are copies of their original desktop counterparts with no intentional changes apart from those changes introduced by the migration to the web. We assigned identifiers A0, B0 and C0 to these web user interface versions of A, B and C respectively. Then, for each of the three web interfaces, maintaining their original functionalities, we created three variations by varying one of the three main aspects per variation and assigned identifiers 1, 2 and 3: Orientation (A1, B1, C1), Order (A2, B2, C2), Density (A3, B3, C3).

For the *orientation* variations, we changed the layout from horizontal to vertical and vice versa by repositioning groups of UI elements. To vary *order*, we changed the positions of elements like buttons or text fields (along with their labels) within regions. We did not mix them between different regions like by re-ordering days in the calendar or by moving patient data inputs into the medical billing region, as this would result in an unrealistic user interface that intendedly confuses users. To achieve *density* variations, we replaced color fills in the shift schedule (A) by letters and adjusted position and spacing of elements for the other two interfaces (B, C). Obviously, these changes cannot be regarded as completely independent – e.g. changing the orientation for instance may also results in a change of density. The interfaces can be found at https://vsr.informatik.tu-chemnitz.de/demos/LayoutSimilarity.

4.3   Calculation of Distances

The values for the main independent variables in our experiment, the distances, were determined using the method we proposed above. So, we first determined the total number of items on each level of hierarchy for the web interfaces (results are shown in Table 1). Then, we asked an expert to determine the number of items in the web interfaces that altered relatively to respected desktop versions in regard to orientation, order or density, so that we could determine the distances. Table 2 shows the numbers of items that changed in regions (R), blocks (B), groups (G), and elements (E), as well as the calculated distances for each of the dimension. It should be noted that for *order,* changes in a single region are impossible, so this hierarchy level was not included in the calculation. For *density*, changes within an element are not possible, so the elements level was not included in the calculation.

Table 1. The total numbers of items per hierarchy levels in web interfaces

Interfaces	Regions	Blocks	Groups	Elements	Notes
A0-A3	1	2	4	7	We consider the calendar inner table as one element.
B0-B3	1	3	7	22	Interiors of small and large calendars are one element each.
C0-C3	1	3	7	48	Each input field and its label are one element.

Table 2. Numbers of changed items and distances (Dist.) per similarity dimensions

Interface	Orientation		Order		Density	
	Changed	Dist.	Changed	Dist.	Changed	Dist.
A0	1B	0.125	none	0	1G	0.083
A1	1R, 2B, 2G	0.625	1B	0.167	1R, 1G	0.417
A2	none	0	1B	0.167	1B, 1G	0.250
A3	1B	0.125	none	0	2G[a]	0.167
B0	none	0	none	0	1B, 1G	0.159
B1	1R, 1B	0.333	none	0	1R, 2B, 1G	0.603
B2	none	0	3B	0.333	1B, 1G	0.159
B3	1B	0.083	none	0	1R, 2B, 1G	0.603
C0	none	0	none	0	1R[b]	0.333
C1	1R, 2B	0.417	none	0	1R	0.333
C2	none	0	4G, 24E (50 %)	0.357	1R	0.333
C3	none	0	none	0	1R, 2B	0.556

[a] Although A3 was supposed to be dedicated version with changes in density (1R would be expected), visually the density didn't change with the removal of color fills, according to the evaluating expert.
[b] For the interface screen C, all web versions visually had different density compared to the desktop one, so 1R was assigned for the each version.

4.4   Subjects and Procedure

In our pilot experimental study we employed 7 subjects (of which 2 were female), all of them students majoring in Informatics or staff of Chemnitz Technical University, Germany. Their ages ranged from 21 to 50, average being 28.4 and SD = 9.83. All but one participant were of German nationality, and all of the subjects were proficient in English. Before the experiment informed consent to take part in the study was obtained. The subjects didn't have previous experience with the medical WIS, but have good background in software development, however, not related to HCI. As such, they can be rather considered experts than representatives of the system's target users.

We would show the participants how to achieve the specially designed tasks in the legacy UI, and they were then asked to achieve them in one of the web UIs. Our experiment environment would randomly select one of the four web interfaces and display it to participant. When the task list was completed, the participants answered several questions, assessing *difficulty*, *Like* and *similarity* impressions. Then, the entire process was repeated, showing a new task list on another legacy UI and then having the participant replicate it in the four web versions, for the remaining interfaces, overall three times. In order to avoid participants being biased from recognizing A0, B0 and C0 as "basic" versions, we re-numbered all versions in what was visible to the participants. The experimental sessions with each participant were scheduled and conducted for the duration of about one week. Each session lasted about one hour and was performed on the same desktop PC and screen, for the sake of consistency in the interfaces representation.

5 Results

5.1 Descriptive Statistics

In Table 3 we show the values of the distance factors together with the evaluations provided by the participants, per the three scales.

Table 3. Values for the factors and the subjects' evaluations

Interface	Distances			Evaluations		
	Orientation	Order	Density	Difficulty	Like	Similarity
A0	0.125	0.000	0.083	1.714	3.429	4.000
A1	0.625	0.167	0.417	2.000	2.571	2.857
A2	**0.000**	**0.167**	**0.250**	2.000	3.571	**4.286**
A3	0.125	0.000	0.167	1.571	2.571	3.714
B0	**0.000**	**0.000**	**0.159**	1.857	3.286	**4.143**
B1	0.333	0.000	0.603	2.143	2.000	3.000
B2	0.000	0.333	0.159	1.714	3.714	3.286
B3	0.083	0.000	0.603	1.857	3.143	3.571
C0	0.000	0.000	0.333	2.143	3.143	3.714
C1	0.417	0.000	0.333	2.857	2.000	2.714
C2	0.000	0.357	0.333	2.143	3.000	4.000
C3	**0.000**	**0.000**	**0.556**	1.286	3.571	**4.143**
Avg.	0.142	0.085	0.333	1.940	3.000	3.619
(SD)	(0.207)	(0.137)	(0.181)	(1.057)	(1.299)	(1.029)

The *difficulty* evaluation had the lowest absolute value (1.940), which is understandable since the employed interfaces were relatively simple, especially given the subjects' proficiency in computers. The greatest standard deviation of the *Like* evaluation (1.299) was also to be expected, as these answers have the highest degree of subjectivity. We detected highly significant positive correlation (Pearson's $\rho = 0.734$, $p = 0.007$) between *Like* and *similarity* evaluations, which may imply that people prefer familiar interfaces, although the experimental environment may have affected this judgement, hinting that similar equals good. The significant negative correlation

(ρ = -0.632, p = 0.027) between *difficulty* and *Like* should have been expected, as it's well known that in interaction perceived difficulty invokes negative feelings. The negative correlation between *difficulty* and *similarity* was significant at α = 0.06 (ρ = -0.563, p = 0.057), which supports the assumption that familiar interfaces have lower perceived difficulty.

5.2   Regression Analysis

Regression analysis for the three factors and *difficulty* evaluation did not find any significant effects (p = 0.549, R^2 = 0.221). In regressions for *Like* and *similarity*, only the *orientation* factor was significant, so the other factors were removed from the models. The resulting model for *Like* (1) was highly significant: p = 0.005, R^2 = 0.561. The model for *similarity* (2) had even greater significance and better fit: p = 0.001, R^2 = 0.670.

$$Like = 3.3 - 2.14 * D_{ORIENT} \tag{1}$$

$$Similarity = 3.92 - 2.14 * D_{ORIENT} \tag{2}$$

5.3   The Layout Distance Measures

The obvious way to calculate the final distance measure for layout is to take the average of *orientation*, *order* and *density* distances, which would result in what we'll call the D_{basic} layout distance. However, we can also take normalized coefficients for all factors in regression for similarity (even though order and density distance factors were not found to be significant) as weights in calculating the overall distance measure, in which case the formula will be the following:

$$D_{LAYOUT} = 0.726 * D_{ORIENT} + 0.162 * D_{ORDER} + 0.112 * D_{DENSITY} \tag{3}$$

The correlation between the two measures was highly significant (ρ = 0.862, p < 0.001). However, correlation between D_{layout} and *similarity* (-0.831, p = 0.001) was found to be higher than for D_{basic} and *similarity* (−0.716, p = 0.009). The only another significant correlation at α = 0.05 was the negative one between D_{layout} and *Like* evaluation (ρ = -0.727, p = 0.007). We further compared the two distance measures by attempting regressions for *similarity* evaluation. The model for D_{basic} was significant (p = 0.009, R^2 = 0.513), but the regression for D_{layout} (4) showed even higher significance and considerably better fit (p = 0.001, R^2 = 0.691).

$$Similarity = 4.07 - 2.89 * D_{LAYOUT} \tag{4}$$

6   Conclusions and Future Work

In our research work we sought to explain why we consider interface similarity to be an important and potentially useful metric. Our assumption was that familiar interfaces, other things being equal, are more usable to users, and this should be considered in interface design and re-design activities. In our current work we focused on interface layout and proposed an approach for quantitative expression of distances between two interfaces in this regard. The considered dimensions included *orientation*, *order* and *density*, and we also sought to determine their relative importance for users and thus contribution to the overall layout distance measure. To support our approach, we designed and conducted a pilot study with 7 subjects and 12 interface screens constructed from 3 legacy interfaces.

The results of the analysis suggest that our hypothesis H_0 could be rejected, and the proposed distance measures do have effect on web interface similarity as perceived by users. The regression model (2) was highly significant ($p = 0.001$) and had reasonably fair $R^2 = 0.670$. The only significant layout dimension was *orientation* distance, which predictably had a negative coefficient in the equation. Based on the model, we calculated normalized weights for the three dimensions and determined the overall layout distance metric, D_{layout} (3). Compared to the simple average (D_{basic}), this metric had higher correlation with similarity ($\rho = -0.831$) and produced considerably better regression (4): $p = 0.001$, $R^2 = 0.691$.

Among the limitations of our current study we'd like to note, first of all, the small sample of users and low diversity of interfaces. Although the experiment participants performed specially developed realistic tasks, they were not quite representative of the target user group, had no previous experience with the employed WIS, and there was little interaction with the interfaces. All in all, we are far from asserting that the results of our pilot study can be used directly, but the proposed approach may be still sound. Our plans for future work include further exploration of interface complexity factor and coverage of other aspects of interface similarity.

Acknowledgements. This research was funded by RFBR, according to the research project No. 16-37-60060 mol_a_dk, and supported by the eHealth Research Laboratory funded by medatixx GmbH & Co. KG.

References

1. Aversano, L., Canfora, G., Cimitile, A., De Lucia, A.: Migrating legacy systems to the Web: an experience report. In: Proceedings Fifth European Conference on Software Maintenance and Reengineering, pp. 148–157 (2001)
2. Canfora, G., Cimitile, A., De Lucia, A., Di Lucca, G.A.: Decomposing legacy programs: a first step towards migrating to client–server platforms. J. Syst. Softw. **54**(2), 99–110 (2000)
3. Colosimo, M., de Lucia, A., Francese, R., Scanniello, G.: Assessing legacy system migration technologies through controlled experiments. In: IEEE International Conference on Software Maintenance, pp. 365–374 (2007)

4. Molich, Rolf, Jeffries, Robin, Dumas, Joseph S.: Making usability recommendations useful and usable. J. Usability Studies **2**(4), 162–179 (2007)
5. Bakaev, M., Avdeenko, T.: Indexing and comparison of multi-dimensional entities in a recommender system based on ontological approach. Computación y Sistemas **17**(1), 5–13 (2013)
6. Gajos, K., Wu, A., Weld, D.S.: Cross-device consistency in automatically generated user interfaces. In: Proceedings 2nd Workshop on Multi-User and Ubiquitous UIs, pp. 7–8 (2005)

Context-Aware Recommendation

Semantic Context-Aware Recommendation via Topic Models Leveraging Linked Open Data

Mehdi Allahyari[✉] and Krys Kochut

Computer Science Department, University of Georgia, Athens, GA, USA
{mehdi,kochut}@cs.uga.edu

Abstract. Context aware recommendation systems are used to provide personalized recommendations by exploiting contextual situation. They take into account not only user preferences, but also additional relevant information (context). Statistical topic models such as Latent Dirichlet Allocation (LDA) have been extensively used for discovering latent semantic topics in text documents. In this paper, we propose a probabilistic topic model that incorporates user interests, item representation and context information in a single framework. In our approach, the contextual information is represented as a subset of the items feature space which is acquired from the knowledge available in the Linked Open Data (LOD). We use DBpedia, a well-known knowledge base in LOD, to utilize the context information in recommendation. Our proposed recommendation framework computes the conditional probability of each item given the user preferences and the additional context. We use these probabilities as recommendation scores to find *top-n* items for recommendations. The performed experiments demonstrate the effectiveness of our proposed method and shows that leveraging semantic context from the Linked Open Data can improve the quality of the recommendations.

1 Introduction

Since its introduction, the amount of data published on the Web has grown dramatically. As a result of this information explosion, it has become very difficult for the users to find appropriate items relevant to their needs. Recommender Systems (RS) are widely used as some of the most essential techniques for *information filtering*. They help users in making decisions and finding what is relevant to them in a personalized manner. Recommender systems have also been successfully employed in the industry, for example, for product recommendations at Amazon and movie recommendations at Netflix. Recently, there has been a substantial amount of research on various recommendation techniques [3,9,24,26,27,33,35].

Although recommender systems are broadly used in multiple domains, they mostly do not consider the contextual situation in which the item is evaluated or used. Incorporating additional contextual information, such as time, location and other factors into the recommendation process can significantly increase the quality of the recommendations in many cases. For example, taking temporal context

© Springer International Publishing AG 2016
W. Cellary et al. (Eds.): WISE 2016, Part I, LNCS 10041, pp. 263–277, 2016.
DOI: 10.1007/978-3-319-48740-3_19

into consideration, a movie recommender system is able to provide movie recommendations for weekends that can be entirely different from the recommendations for the week days. In order to address this issue, *context-aware recommender systems* (CARS) have been introduced in recent years. CARS have shown to improve the accuracy and provide more relevant recommendations [6–8,25].

One of the most commonly used definitions of context, proposed by Abowd et al. [1], states that it is *"any information that can be used to characterize the situation of an entity. An entity is a person, place, or object that is considered relevant to the interaction between a user and an application, including the user and applications themselves"*. Apart from this general definition of context, some more specific definitions were proposed in recent years. For example, Cantador and Castells [15] focused on semantic contextualization and represent context as *"the background topics under which activities of a user occur within a given unit of time"*.

Incorporation of context into the recommender systems can be done in a variety of ways. In general, the representation and integration of context rely on the available contextual information and the way context is defined. Typically, there are two approaches to acquire the context. In the first, users specify the context *explicitly* each time they interact with the system [5,17]. For example, in a music recommender system, users can describe their current interests in a particular genre of music by giving that information as a query to the system. The recommender system then, recommends songs that best match the user's needs considering the previously established user preferences as well as the queried context. Although, explicit context assumption simplifies the system, it does not hold for many applications where the context is hidden and should be inferred. In the second approach, contextual information is *implicitly* inferred from the user's behavior [16,20]. As an example, a restaurant recommender may produce recommendations considering implicitly variables such as time of the day and user's location along with past user's interests. In this paper, we consider a setting where context is not pre-specified, but rather is learned from the knowledge existing in the knowledge bases.

Within the Semantic Web, numerous data sources have been published as ontologies. Many of them are interconnected which have created a huge decentralized knowledge base commonly known as **Linked Open Data** (LOD) [10]. For example, DBpedia [11] (as part of LOD) is a publicly available knowledge base extracted from Wikipedia in the form of an ontology of concepts and relationships, making this vast amount of information programmatically accessible on the Web. This freely available knowledge has been used in several works to improve the quality of the recommender systems [4,14,18,29–31,34].

We propose a single probabilistic topic model that integrates user preferences, item descriptions and contextual information based on sound principles. We represent the context information as a subset of features representing the items. In our model, the context is acquired from the semantic descriptions of the items obtained through DBpedia. We should point out that there exist several other knowledge bases such as YAGO [22], and Freebase [13] that could be exploited as the prior knowledge in our work. For this research, we selected DBpedia as arguably more frequently used for Semantic Web tasks, but our approach could be used with other knowledge bases, as well.

In our semantic context-aware recommendation system (SCRM) each user profile is represented as a multinomial distribution over a set of latent topics, while topics are distributions over items and item features. The main difference between our work with all prior works is that we propose a probabilistic model that both *infers* the semantic context and *models* this context in a systematic way. For a given user's profile u and context c, we compute the recommendation score for each item v as $p(v|c, u)$, rank the items based on these scores and select the *top-n* recommendations for the user.

The paper is organized as follows. In Sect. 2, we discuss the prior work on context-aware recommendation systems. In Sect. 3, we present a brief overview of Latent Dirichlet Allocation (LDA), the state-of-art probabilistic topic modeling technique. We formally define our semantic context-aware recommendation model in Sect. 4. In Sect. 5, we demonstrate the effectiveness of our method on a real-world dataset. Finally, we present our conclusions and future work in Sect. 6.

2 Related Work

Several approaches have been recently proposed that make use of contextual information in recommender systems. [7] proposes a context-aware matrix factorization method for rating prediction. The system proposed in [6] uses contextual factors such as "temperature" or "weather" to recommend places of interest. [28] introduces a context-aware system for recommending playlists to the users according to their moods. There is also prior works [36–39] that propose spatiotemporal recommendation models exploiting location and time as context for recommendations. Our method is different with these works, since in our system the context is not pre-specified. Ma et al. [27] propose matrix factorization methods that exploit social information to improve the prediction accuracy of recommender systems. The main distinction of our system is that we do not incorporate social-context.

Extracting contextual information from unstructured text is relatively recent and has not been widely addressed. Aciar [2] presents a classification method to identify review sentences containing contextual information. Our work is different from this work as they do not incorporate the retrieved information in the recommendation system. [21] proposes a context-aware system for hotel recommendation that obtains contextual information by mining user reviews and combining it with user rating history to compute a utility function over a set of items. They represent the context as a distribution function over the set of "trip types" which are pre-determined. Then, using Labeled-LDA topic model as a multi-class supervised classifier, they find context distributions over trip types. Eventually, they define a context score for each context and combine these scores with a item-based kNN recommender system for recommendation. Our method is different from this approach in two ways. First, unlike [21], we learn the contextual information through DBpedia knowledge about items. Second, we propose a single topic model that combines the user, items and the context in a systematic manner. Hariri et al. [20] introduce a context-aware recommender system that is more similar to our system. However, their item features are known and available (e.g., item tags) whereas we model and learn them via existing knowledge in DBpedia.

3   Latent Dirichlet Allocation (LDA)

Probabilistic topic models are a set of algorithms that are used to uncover the hidden thematic structure from a collection of documents. The main idea of topic modeling is to create a probabilistic generative model for the corpus of text documents. In topic models, documents are mixture of topics, where a topic is a probability distribution over words. The two main topic models are Probabilistic Latent Semantic Analysis (pLSA) [23] and Latent Dirichlet Allocation (LDA) [12]. Hofmann (1999) introduced pLSA for document modeling. pLSA model does not provide any probabilistic model at the document level which makes it difficult to generalize it to model new unseen documents. Blei et al. [12] extended this model by introducing a Dirichlet prior on mixture weights of topics per documents, and called the model Latent Dirichlet Allocation (LDA). In this section we describe the LDA method.

The Latent Dirichlet Allocation (LDA) [12] is a generative probabilistic model for extracting thematic information (topics) of a collection of documents. LDA assumes that each document is made up of various topics, where each topic is a probability distribution over words.

Let $\mathcal{D} = \{d_1, d_2, \ldots, d_{|\mathcal{D}|}\}$ is the corpus and $\mathcal{V} = \{w_1, w_2, \ldots, w_{|\mathcal{V}|}\}$ is the vocabulary of the corpus. A topic $z_j, 1 \leq j \leq K$ is represented as a multinomial probability distribution over the $|\mathcal{V}|$ words, $p(w_i|z_j), \sum_i^{|\mathcal{V}|} p(w_i|z_j) = 1$. LDA generates the words in a two-stage process: words are generated from topics and topics are generated by documents. More formally, the distribution of words given the document is calculated as follows:

$$p(w_i|d) = \sum_{j=1}^{K} p(w_i|z_j)p(z_j|d) \tag{1}$$

The graphical model of LDA is shown in Fig. 1(a) and the generative process for the corpus $\mathcal{D}$ is as follows:

1. For each topic $k \in \{1, 2, \ldots, K\}$, sample a word distribution $\phi_k \sim \mathrm{Dir}(\beta)$
2. For each document $d \in \{1, 2, \ldots, \mathcal{D}\}$,
   (a) Sample a topic distribution $\theta_d \sim \mathrm{Dir}(\alpha)$
   (b) For each word w_n, where $n \in \{1, 2, \ldots, N\}$, in document d,
      i. Sample a topic $z_i \sim \mathrm{Mult}(\theta_d)$
      ii. Sample a word $w_n \sim \mathrm{Mult}(\phi_{z_i})$

The joint distribution of the model (hidden and observed variables) is:

$$P(\phi_{1:K}, \theta_{1:\mathcal{D}}, z_{1:\mathcal{D}}, w_{1:\mathcal{D}}) = \prod_{j=1}^{K} P(\phi_j|\beta) \prod_{d=1}^{|\mathcal{D}|} P(\theta_d|\alpha) \left(\prod_{n=1}^{N} P(z_{d,n}|\theta_d) P(w_{d,n}|\phi_{1:K}, z_{d,n}) \right)$$

4   Semantic Context-Aware Recommendation Model

In this section, we formally describe our model. We propose a probabilistic topic model, which integrates the users, items and the contextual information in a unified framework. The underlying idea is that ontological knowledge existing in the LOD forms a *semantic context* which can be incorporated with user preferences to improve recommendations in collaborative filtering recommendation systems. Note, that our semantic context-aware recommendation model (SCRM) resembles the model introduced by Hariri et al. [20], with users being analogous to the corpus and user profiles being analogous to documents. Yet, our model differs from [20] in a way that we acquire the context from the information available in DBpedia. We describe items by using their related entities in the DBpedia ontology. Similarly for each item, the features are represented over its description. We run the standard LDA on the collection of items and then for each item, we extract a subset of its feature space (latent factors) and obtain the context. Details are explained in Sect. 4.2.

The graphical representation of the SCRM is shown in Fig. 1(b) and the generative process is defined as follows:

1. For each topic $k \in \{1, 2, \ldots, |K|\}$, draw an item distribution $\phi_k \sim \text{Dir}(\beta)$
2. For each topic $k \in \{1, 2, \ldots, |K|\}$, draw a feature distribution $\vartheta_k \sim \text{Dir}(\gamma)$
3. For each user $u \in \{1, 2, \ldots, |U|\}$,
   (a) Draw a topic distribution $\theta_u \sim \text{Dir}(\alpha)$
   (b) For each of the M_u items v_i of user's profile u,
      i. Draw a topic $z_i \sim \text{Mult}(\theta_u)$
      ii. Draw an item $v_i \sim \text{Mult}(\phi_{z_i})$
      iii. For each of the N_{v_i} features f_i of item v_i, draw $f_i \sim \text{Mult}(\vartheta_{z_i})$

The generative process for the SCRM model corresponds to the following joint distribution of the hidden and observed variables:

$$P(\phi, \vartheta, \theta, z, v, f) =$$
$$\prod_{k=1}^{|K|} p(\phi_k|\beta) \prod_{k=1}^{|K|} p(\vartheta_k|\gamma) \prod_{u=1}^{|U|} p(\theta_u|\alpha) \prod_{i=1}^{|M|} \left(p(z_i|\theta_u) p(v_i|z_i, \phi) \prod_{j=1}^{|N|} p(f_j|z_i, \vartheta) \right) \quad (2)$$

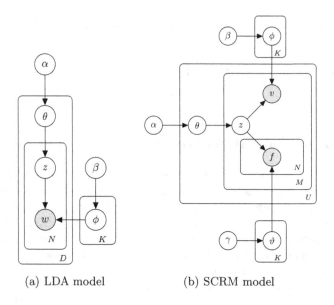

(a) LDA model   (b) SCRM model

Fig. 1. Graphical representation of different models

4.1 Inference and Estimation

Since the posterior inference of SCRM is intractable, we need to find an algorithm for estimating this posterior inference. A variety of algorithms have been used to estimate the parameters of topic models, such as variational EM [12] and Gibbs sampling [19]. In our SCRM topic model presented in this paper, we use the collapsed Gibbs sampling procedure. Collapsed Gibbs sampling [19] is a Markov Chain Monte Carlo (MCMC) algorithm, which constructs a Markov chain over the latent variables in the model and converges to the posterior distribution after a number of iterations. In our case, we aim to construct a Markov chain that converges to the posterior distribution over z conditioned on the observed items v, features f and hyperparameters α, β and γ.

$$P(z|v, f, \alpha, \beta, \gamma) = \frac{P(z, v, f|\alpha, \beta, \gamma)}{P(v, f|\alpha, \beta, \gamma)} \propto P(z, v, f|\alpha, \beta, \gamma)$$

$$\propto P(z)P(v|z) \prod_{i=1}^{N} P(f_i|z)$$

Let $c = \{f_1, f_2, \ldots, f_{|N|}\}$. Subsequently, the update equation for the hidden variable can be derived as:

$$P(z_i = k|v_i = v, c_i = c, z_{-i}, v_{-i}, c_{-i}, \alpha, \beta, \gamma) \propto$$

$$\frac{n_{k,-i}^{(u)} + \alpha}{\sum_{k'} (n_{k',-i}^{(u)} + \alpha)} \times \frac{n_{v,-i}^{(k)} + \beta}{\sum_{v'} (n_{v',-i}^{(k)} + \beta)} \times \prod_{f \in c} \frac{n_{f,-i}^{(k)} + \gamma}{\sum_{f'} (n_{f',-i}^{(k)} + \gamma)} \tag{3}$$

After Gibbs sampling, we can easily estimate the topic-item distributions ϕ, topic-feature distributions ϑ, and user-topic distributions θ by:

$$\phi_{kv} = \frac{n_v^{(k)} + \beta}{\sum_{v'} (n_{v'}^{(k)} + \beta)} \quad \vartheta_{kf} = \frac{n_{f,-i}^{(k)} + \gamma}{\sum_{f'} (n_{f',-i}^{(k)} + \gamma)} \quad \theta_{uk} = \frac{n_k^{(u)} + \alpha}{\sum_{k'} (n_{k'}^{(u)} + \alpha)} \quad (4)$$

where ϕ_{kv} is the probability of an item given a topic, ϑ_{kf} is the probability of a feature given a topic and θ_{uk} is the probability of a topic given a user.

4.2   Context Extraction from DBpedia

In this section, we describe how to extract the contextual information about items from the DBpedia ontology. The intuition is that items have a set of hidden features (e.g. topics) that can be learned or uncovered. We assume that knowledge bases such as DBpedia can be utilized to discover the features. Subsequently, the contextual information can be extracted as a subset of these features, and be used in context-aware recommendation systems to improve personalized recommendations. Since in this paper the application of the recommendation is for the movie domain, we regard movies as items. However, it should be noted that our overall approach is domain independent and applicable to other knowledge domains.

DBpedia is a publicly available knowledge base belonging to the LOD cloud which covers a diverse range of domains. In DBpedia, the knowledge about an entity, for example, a movie, a person, or a song is represented by linking different entities (vertices) to each other via semantic relationships (edges). Given an entity representing an item, its *description* includes the knowledge associated with it, i.e., related entities. In order to create an item description, in addition to DBpedia's *object properties*, we exploit other properties, including the `dcterms:subject` property, since these properties convey rich ontological knowledge about the entities. On the other hand, we exclude the properties `dbpedia-owl:thumbnail` and `dbpedia-owl:wikiPageExternalLink` since they do not give useful semantic information about the entities. Figure 2 shows a snippet of the description for the entity *"back_to_the_future"*.

For each movie, we extract all the related entities such as movies, actors, genres, directors, etc. from DBpedia and create a bag of entities. In our approach, each item corresponds to a document where entity labels represent the words of the document. Moreover, we consider the set of item features as latent variable $T = \{f_1, f_2, \ldots, f_{|T|}\}$ that needs to be discovered. We assume each item is a distribution over the set of features, while each feature is a distribution over the words. We run the LDA model and extract the set of features T for the collection of items. The probability of each feature f_i under item v, $p(f_i|v)$, indicates the significance of the f_i for item v.

We assume that the contextual information is represented as a subset of the items feature space. Therefore, for each item v, $c_v = \{f_{1v}, f_{2v}, \ldots, f_{|N|v}\}$ where $|N| \leq |T|$, i.e. the context c_v consists of the top-N features having the highest marginal probability under v.

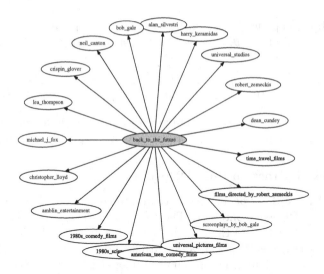

Fig. 2. Snippet of the description for the entity *"back_to_the_future"*

4.3 Semantic Context-Aware Ranking Score

In this section, we detail the computation of the ranking score used in our SCRM model. Let $c = \{f_1, f_2, \ldots, f_{|N|}\}$ be a given context for user u with n items. We define a scoring function to rank the items and provide the personalized recommendations for user u considering the given context.

Given our model, for an item v, context c and user u, $p(v|c, u)$ is computed and used as the *ranking score*. Thus, according to Bayes rule:

$$p(v|c, u) \propto p(v|u) \cdot p(c|v, u) \tag{5}$$

where $p(v|u)$ and $p(c|v, u)$ are computed as follows:

$$p(v|u) = \sum_{k=1}^{K} p(v|z_k) \cdot p(z_k|u) = \sum_{k=1}^{K} \phi_{kv} \cdot \theta_{uk} \tag{6}$$

$$p(c|v, u) = \prod_{f \in c} p(f|v, u) = \prod_{f \in c} \sum_{k=1}^{K} p(f|z_k) \cdot p(z_k|v, u)$$

$$= \prod_{f \in c} \sum_{k=1}^{K} \vartheta_{kf} \cdot p(z_k|v, u) \tag{7}$$

$$p(z_k|v, u) = \frac{p(z_k, v, u)}{p(v, u)} = \frac{p(v|z_k) \cdot p(z_k|u)}{\sum_{i=1}^{K} p(v|z_i) \cdot p(z_i|u)}$$

$$= \frac{\phi_{kv} \cdot \theta_{uk}}{\sum_{i=1}^{K} \phi_{kv} \cdot \theta_{uk}} \tag{8}$$

Combining Eqs. 6, 7 and 8, Eq. 5 can ber simplified as:

$$p(v|c,u) \propto \left(\sum_{k=1}^{K} \phi_{kv} \cdot \theta_{uk} \right) \cdot \prod_{f \in c} \sum_{k=1}^{K} \frac{\vartheta_{kf} \cdot \phi_{kv} \cdot \theta_{uk}}{\sum_{i=1}^{K} \phi_{kv} \cdot \theta_{uk}} \qquad (9)$$

5 Experiments and Results

In this section, we evaluate our method and compare it with several baselines. The evaluation has been carried out on a real-world data set `MovieLens` from the movie domain with ratings in $\{1, 2, 3, 4, 5\}$.

5.1 Dataset

We performed the evaluation on `MovieLens` 1M dataset[1], one of the most commonly used datasets for movie recommender systems. This dataset contains $1,000,209$ ratings of roughly $3,900$ movies made by $6,040$ users. Since our method is based on semantic context-aware recommendation, in order to use this dataset, we need to link each movie in `MovieLens` to the corresponding entities in DBpedia. However, Noia et al. [18] have carried out the mappings and created a mapping file which is publicly available[2].

We represented the user profile based on a binary rating such as *like/dislike*. Nonetheless, in `MovieLens` user u rates a movie based on a five-value scale: $r(u, v) \in \{1, 2, 3, 4, 5\}$ where a rate 1 indicates a *terrible* movie whereas a rate 5 implies an *excellent* movie. In order to map the five-scale ratings to a binary one, we considered the ratings above 3 as *like* and the others as *dislike*. Therefore, we model the user profile as:

$$profile(u) = \{v_i | r(u, v_i) > 3\} \qquad (10)$$

5.2 Experimental Setup

For evaluation, we performed a 5-fold cross validation. In our setting, 80 % of the items in each user's profile used for training the model, while the remaining 20 % were put aside for testing. We also used the method explained in Sect. 4.2 to obtain the contextual information for the movies in the dataset. For each user u, held-out item v and context $c_v = \{f_{1v}, f_{2v}, \ldots, f_{|N|v}\}$ (consisting of $|N|$ features f_i), we compute a recommendation score from each of the compared methods. As described in Sect. 4.3, our SCRM model computes the ranking score for each item v_i as $p(v_i | c_{v_i}, u)$ and sorts the items in descending order based on these scores. We compared our proposed SCRM model with the following baselines:

---

[1] http://grouplens.org/datasets/movielens/1m/.
[2] http://sisinflab.poliba.it/semanticweb/lod/recsys/datasets.

1. **User-based kNN:** This approach is one of the most commonly used collaborative filtering methods where items are recommended to a user based on
   similar user profiles using the k-nearest neighbor approach.
2. **Item-based kNN:** This approach applies the same idea as **User-based**
   kNN, but exploits the similarity between the items instead of the users.
3. **BPRMF:** This is a matrix factorization method that is optimized based on
   Bayesian Personalized Ranking (BPR) criterion for personalized rankings of
   items [32].

For our SCRM model, we set the number of topics $K = 10$, and assumed
the symmetric Dirichlet prior and set $\alpha = 50/K$, $\beta = 0.01$ and $\gamma = 0.01$, respectively. We ran the Gibbs sampling algorithm for 1000 iterations and computed
the posterior inference after the last sampling iteration. For extracting the set
of features from the items, we implemented the LDA model with the Mallet
toolkit[3]. We set the number of features to $|T| = 10$, and all other settings the
same as in the SCRM model. We also restricted the size of contextual features
to $|N| = 1$, meaning for each item v, the context c_v consists of the top-1 feature
having the highest marginal probability under v. For both user-based kNN and
item-based kNN, we set the number of neighbors to $k = 10$.

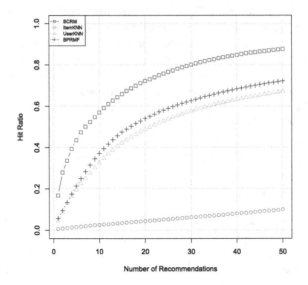

Fig. 3. Hit ratio for different number of recommendations for MovieLens dataset

5.3   Experimental Results

In this section, in order to show the the effectiveness of our proposed approach, we
compare different algorithms in terms of their capability to provide personalized

---

[3] http://mallet.cs.umass.edu/.

recommendation based on the given context. Thus, we measure how well they discover the movies in the test data. Since top recommendations are more important to the user, for each held-out movie, we find its rank in the overall recommendation list. We run a cross validation and evaluate the results by measuring the **Hit Ratio** metric. The hit ratio computes the probability that a removed movie is recommended as part of the *top-n* recommendations. If a removed movie is part of the *top-n* recommendations, we consider it a *hit*. We represent the *top-n* recommendations for a given user u as $R_n(u)$. For the held-out movie m_u, we define an indicator function:

$$\mathbb{1}_n(u, m_u) = \begin{cases} 1 & \text{if } m_u \in R_n(u) \\ 0 & \text{otherwise.} \end{cases}$$

For any given rank n, the hit ratio of the recommendation algorithm is defined as follows:

$$h(n) = \frac{\sum_{i=1}^{|U_{test}|} \mathbb{1}_n(u_i, m_{u_i})}{|U_{test}|} \tag{11}$$

where U_{test} is the set of users in the test set. Figure 3 illustrates the average hit ratio results (i.e. average of five-fold cross validation) of our approach as well as all other baselines for the first 50 recommendations. It shows that our SCRM model significantly outperforms the other methods at all ranks. Similarly, Table 1 presents the hit ratio measurement values of *top-n* where $n = \{1, 5, 10, 20, 30, 40, 50\}$, which shows that our approach achieves considerably better results. For example, if we consider the top-10 recommendations, the hit ratio of our approach is 20 times better than item-based kNN and 1.6 times higher than user-based kNN.

Table 1. Average hit ratio for *top-n* recommendations for various methods.

	n = 1	n = 5	n = 10	n = 20	n = 30	n = 40	n = 50
Item-based kNN	0.006	0.016	0.026	0.044	0.063	0.081	0.100
User-based kNN	0.046	0.195	0.327	0.488	0.578	0.634	0.673
BPRMF	0.058	0.214	0.372	0.541	0.627	0.684	0.722
SCRM	**0.168**	**0.436**	**0.570**	**0.722**	**0.801**	**0.848**	**0.876**

5.4 Context Analysis

As mentioned in Sect. 4.2, context is represented as a subset of the features of items. In our setting, features are analogous to categories (topics) of the items and are learned through LDA model (i.e. items are grouped per item category). Thus, each item has a distribution over all the categories with different probabilities, and a category with a high probability under an item indicates that the item is classified under that category.

In our experiments, we set the size of the context $|N| = 1$, i.e. the topmost (top-1) feature having the highest marginal probability under a given item was considered as the context. In other words, for each item, we selected the topmost category. Figure 4 illustrates the hit ratio of our semantic context-aware model with varying sizes of the context. As can be seen, when we increase the size of the features in the context, the hit ratio drops. The reason is that growing the size of the context adds noise to the model, which impacts the accuracy of the recommendations. Baltrunas et al. [7] propose several context-aware matrix factorization methods and consider the interaction of context with items at different levels of granularity (i.e. contextual information with different sizes). They demonstrate that the model having one contextual parameter for each item category outperforms the other models with larger number of contextual factors. Additionally, they explain that the best number of contextual features with respect to the items, depends on the domain and the amount of data available. It should be noted that this is consistent with our assumptions that (1) the context is a subset of features for a given item and (2) setting the size of the context to $|N| = 1$ means that we select one contextual factor for each item category (i.e. the top feature as the context).

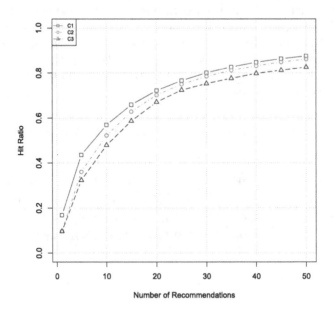

Fig. 4. Hit ratio with different sizes of the context. C_i shows that the size of the context is $|N| = i$

6   Conclusion and Future Work

In this paper, we proposed SCRM, a semantic context-aware recommendation system that integrates user profiles, item descriptions and contextual information in a unified probabilistic topic model. The system learns the context using the knowledge from the DBpedia ontology and utilizes this additional information to compute the ranking score for each item and provides personalized recommendations for users. We demonstrated that exploiting *semantic context* from ontologies can improve the recommendations, by conducting thorough experiments.

There are many interesting future directions of this work. We evaluated our approach on one data set in the domain of movies. As a future work, we plan to perform comprehensive evaluations on multiple datasets and extend our model to other domains such as music and scientific articles. Also, in this work, we did not use the social context, i.e. interactions between the users, in the recommender system. It would be interesting to investigate how to integrate social information network among users in the model. In the work presented here, we set the number of features to $T = |10|$. Hence, an interesting direction for future research is to explore approaches to determine the best number of features to consider. Furthermore, exploring a much richer set of semantic context-aware recommendation models that combine the contextual information, items and users would be a promising direction for future work.

References

1. Abowd, G.D., Dey, A.K., Brown, P.J., Davies, N., Smith, M., Steggles, P.: Towards a better understanding of context and context-awareness. In: Gellersen, H.-W. (ed.) HUC 1999. LNCS, vol. 1707, pp. 304–307. Springer, Heidelberg (1999). doi:10.1007/3-540-48157-5_29
2. Aciar, S.: Mining context information from consumers reviews. In: Proceedings of Workshop on Context-Aware Recommender System, vol. 201. ACM (2010)
3. Agarwal, D., Chen, B.C.: fLDA: matrix factorization through latent Dirichlet allocation. In: Proceedings of the Third ACM International Conference on Web Search and Data Mining, pp. 91–100. ACM (2010)
4. Anand, S.S., Kearney, P., Shapcott, M.: Generating semantically enriched user profiles for web personalization. ACM Trans. Internet Technol. (TOIT) **7**(4), 22 (2007)
5. Ardissono, L., Goy, A., Petrone, G., Segnan, M., Torasso, P.: Intrigue: personalized recommendation of tourist attractions for desktop and hand held devices. Appl. Artif. Intell. **17**(8–9), 687–714 (2003)
6. Baltrunas, L., Ludwig, B., Peer, S., Ricci, F.: Context relevance assessment and exploitation in mobile recommender systems. Pers. Ubiquit. Comput. **16**(5), 507–526 (2012)
7. Baltrunas, L., Ludwig, B., Ricci, F.: Matrix factorization techniques for context aware recommendation. In: Proceedings of the Fifth ACM Conference on Recommender Systems, pp. 301–304. ACM (2011)

8. Baltrunas, L., Ricci, F.: Context-based splitting of item ratings in collaborative filtering. In: Proceedings of the Third ACM Conference on Recommender Systems, pp. 245–248. ACM (2009)
9. Beutel, A., Murray, K., Faloutsos, C., Smola, A.J.: CoBaFi: collaborative bayesian filtering. In: Proceedings of the 23rd International Conference on World Wide Web, pp. 97–108. ACM (2014)
10. Bizer, C., Heath, T., Berners-Lee, T.: Linked data-the story so far. Semant. Serv. Interoperability Web Appl. Emerg. Concepts 5, 205–227 (2009)
11. Bizer, C., Lehmann, J., Kobilarov, G., Auer, S., Becker, C., Cyganiak, R., Hellmann, S.: Dbpedia-a crystallization point for the web of data. Web Semant. Sci. Serv. Agents World Wide Web 7(3), 154–165 (2009)
12. Blei, D.M., Ng, A.Y., Jordan, M.I.: Latent Dirichlet allocation. J. Mach. Learn. Res. 3, 993–1022 (2003)
13. Bollacker, K., Evans, C., Paritosh, P., Sturge, T., Taylor, J.: Freebase: a collaboratively created graph database for structuring human knowledge. In: Proceedings of the 2008 ACM SIGMOD International Conference on Management of Data, pp. 1247–1250. ACM (2008)
14. Cantador, I., Bellogín, A., Castells, P.: A multilayer ontology-based hybrid recommendation model. AI Commun. 21, 203–210 (2008)
15. Cantador, I., Castells, P.: Semantic contextualisation in a news recommender system. In: Workshop on Context-Aware Recommender Systems (CARS 2009) (2009)
16. Chen, G., Chen, L.: Recommendation based on contextual opinions. In: Dimitrova, V., Kuflik, T., Chin, D., Ricci, F., Dolog, P., Houben, G.-J. (eds.) UMAP 2014. LNCS, vol. 8538, pp. 61–73. Springer, Heidelberg (2014). doi:10. 1007/978-3-319-08786-3_6
17. Cheverst, K., Davies, N., Mitchell, K., Friday, A., Efstratiou, C.: Developing a context-aware electronic tourist guide: some issues and experiences. In: Proceedings of the SIGCHI Conference on Human Factors in Computing Systems, pp. 17–24. ACM (2000)
18. Di Noia, T., Mirizzi, R., Ostuni, V.C., Romito, D., Zanker, M.: Linked open data to support content-based recommender systems. In: Proceedings of the 8th International Conference on Semantic Systems, pp. 1–8. ACM (2012)
19. Griffiths, T.L., Steyvers, M.: Finding scientific topics. Proc. Natl. Acad. Sci. USA 101(Suppl 1), 5228–5235 (2004)
20. Hariri, N., Mobasher, B., Burke, R.: Query-driven context aware recommendation. In: Proceedings of the 7th ACM Conference on Recommender Systems, pp. 9–16. ACM (2013)
21. Hariri, N., Zheng, Y., Mobasher, B., Burke, R.: Context-aware recommendation based on review mining. General Co-Chairs, p. 27 (2011)
22. Hoffart, J., Suchanek, F.M., Berberich, K., Weikum, G.: Yago2: a spatially and temporally enhanced knowledge base from Wikipedia. In: Proceedings of the Twenty-Third international Joint Conference on Artificial Intelligence, pp. 3161–3165. AAAI Press (2013)
23. Hofmann, T.: Probabilistic latent semantic indexing. In: Proceedings of the 22nd Annual International ACM SIGIR Conference on Research and Development in Information Retrieval, pp. 50–57. ACM (1999)
24. Jahrer, M., Töscher, A., Legenstein, R.: Combining predictions for accurate recommender systems. In: Proceedings of the 16th ACM SIGKDD International Conference on Knowledge Discovery and Data Mining, pp. 693–702. ACM (2010)

25. Karatzoglou, A., Amatriain, X., Baltrunas, L., Oliver, N.: Multiverse recommendation: n-dimensional tensor factorization for context-aware collaborative filtering. In: Proceedings of the Fourth ACM Conference on Recommender Systems, pp. 79–86. ACM (2010)

26. Leskovec, J., McAuley, J.: Hidden factors and hidden topics: understanding rating dimensions with review text. Department of Computer Science, Stanford University (2013)

27. Ma, H., Zhou, D., Liu, C., Lyu, M.R., King, I.: Recommender systems with social regularization. In: Proceedings of the Fourth ACM International Conference on Web Search and Data Mining, pp. 287–296. ACM (2011)

28. Meyers, O.C.: A mood-based music classification and exploration system. Ph.D. thesis, Massachusetts Institute of Technology (2007)

29. Mobasher, B., Jin, X., Zhou, Y.: Semantically enhanced collaborative filtering on the web. In: Berendt, B., Hotho, A., Mladenič, D., Someren, M., Spiliopoulou, M., Stumme, G. (eds.) EWMF 2003. LNCS (LNAI), vol. 3209, pp. 57–76. Springer, Heidelberg (2004). doi:10.1007/978-3-540-30123-3_4

30. Ostuni, V.C., Di Noia, T., Di Sciascio, E., Mirizzi, R.: Top-n recommendations from implicit feedback leveraging linked open data. In: Proceedings of the 7th ACM Conference on Recommender Systems, pp. 85–92. ACM (2013)

31. Passant, A.: dbrec — music recommendations using DBpedia. In: Patel-Schneider, P.F., Pan, Y., Hitzler, P., Mika, P., Zhang, L., Pan, J.Z., Horrocks, I., Glimm, B. (eds.) ISWC 2010. LNCS, vol. 6497, pp. 209–224. Springer, Heidelberg (2010). doi:10.1007/978-3-642-17749-1_14

32. Rendle, S., Freudenthaler, C., Gantner, Z., Schmidt-Thieme, L.: BPR: Bayesian personalized ranking from implicit feedback. In: Proceedings of the Twenty-Fifth Conference on Uncertainty in Artificial Intelligence, pp. 452–461. AUAI Press (2009)

33. Salakhutdinov, R., Mnih, A.: Bayesian probabilistic matrix factorization using Markov chain Monte Carlo. In: Proceedings of the 25th International Conference on Machine Learning, pp. 880–887. ACM (2008)

34. Semeraro, G., Lops, P., Basile, P., de Gemmis, M.: Knowledge infusion into content-based recommender systems. In: Proceedings of the Third ACM Conference on Recommender Systems, pp. 301–304. ACM (2009)

35. Wang, C., Blei, D.M.: Collaborative topic modeling for recommending scientific articles. In: Proceedings of the 17th ACM SIGKDD International Conference on Knowledge Discovery and Data Mining, pp. 448–456. ACM (2011)

36. Yin, H., Cui, B., Chen, L., Hu, Z., Huang, Z.: A temporal context-aware model for user behavior modeling in social media systems. In: Proceedings of the 2014 ACM SIGMOD International Conference on Management of Data, pp. 1543–1554. ACM (2014)

37. Yin, H., Cui, B., Chen, L., Hu, Z., Zhou, X.: Dynamic user modeling in social media systems. ACM Trans. Inf. Syst. (TOIS) 33(3), 10 (2015)

38. Yin, H., Cui, B., Sun, Y., Hu, Z., Chen, L.: LCARS: a spatial item recommender system. ACM Trans. Inf. Syst. (TOIS) 32(3), 11 (2014)

39. Yin, H., Zhou, X., Shao, Y., Wang, H., Sadiq, S.: Joint modeling of user check-in behaviors for point-of-interest recommendation. In: Proceedings of the 24th ACM International on Conference on Information and Knowledge Management, pp. 1631–1640. ACM (2015)

Optimizing Factorization Machines for Top-N Context-Aware Recommendations

Fajie Yuan[1(✉)], Guibing Guo[2], Joemon M. Jose[1], Long Chen[1], Haitao Yu[3],
and Weinan Zhang[4]

[1] University of Glasgow, Glasgow, UK
`f.yuan.1@research.gla.ac.uk`, {`Joemon.Jose,Long.Chen`}`@glasgow.ac.uk`
[2] Northeastern University, Shenyang, China
`guogb@swc.neu.edu.cn`
[3] University of Tsukuba, Tsukuba, Japan
`yuhaitao@slis.tsukuba.ac.jp`
[4] Shanghai Jiao Tong University, Shanghai, China
`wnzhang@sjtu.edu.cn`

Abstract. Context-aware Collaborative Filtering (CF) techniques such as Factorization Machines (FM) have been proven to yield high precision for rating prediction. However, the goal of recommender systems is often referred to as a top-N item recommendation task, and item ranking is a better formulation for the recommendation problem. In this paper, we present two collaborative rankers, namely, Ranking Factorization Machines (RankingFM) and Lambda Factorization Machines (LambdaFM), which optimize the FM model for the item recommendation task. Specifically, instead of fitting the preference of individual items, we first propose a RankingFM algorithm that applies the cross-entropy loss function to the FM model to estimate the pairwise preference between individual item pairs. Second, by considering the ranking bias in the item recommendation task, we design two effective *lambda*-motivated learning schemes for RankingFM to optimize desired ranking metrics, referred to as LambdaFM. The two models we propose can work with any types of context, and are capable of estimating latent interactions between the context features under sparsity. Experimental results show its superiority over several state-of-the-art methods on three public CF datasets in terms of two standard ranking metrics.

Keywords: Context-aware · Learning to rank · Factorization machines · RankingFM · LambdaFM

1 Introduction

Commonly used recommendation techniques such as collaborative filtering (CF) have gained much attention in recent years. However, typical collaborative filtering (CF) methods mainly focus on mining interactions between users and items without considering the additional context which the users or items are associated with [21]. For example, in a music recommender system, the location of the

© Springer International Publishing AG 2016
W. Cellary et al. (Eds.): WISE 2016, Part I, LNCS 10041, pp. 278–293, 2016.
DOI: 10.1007/978-3-319-48740-3_20

user and the time of the user-item interaction may be important contextual factors when the user listened to a song. Ignoring the contextual information may result in considerable degradation in recommendation performance. Currently, there are some hybrid approaches performing pre- or post-filtering of the input data to make standard methods context-aware. Although such ad-hoc strategies may work in practice, they suffer from two drawbacks [10,21]: (1) pre- or post-filtering the data based on the context can potentially lead to information loss about the interactions between different contextual variables; (2) all steps in the process need supervision and manual tuning. On the other hand, a variety of specialized models designed for specific tasks, such as TimeSVD [11] and Tensor Factorization [18], are able to leverage contextual information, but they rely on very strict assumptions, which make them cumbersome to incorporate different types of context and usually require complicated inference algorithms. Therefore, the models capable of integrating any types of context are more practical, as well as more elegant in theory. So far, two of the most flexible and effective methods for context modelling are Multiverse Recommendation [10] and Factorization Machines (FM) [17]. Unfortunately, Multiverse Recommendation relies on Tucker decomposition, which leads to $O(k^m)$ computational complexity, where k is the dimensionality of factorization and m is the number of predictor variables involved [21]. In contrast, FM enjoys linear complexity (both in k and m), which gives fast learning and prediction with contextual features.

It has been recognized that both Multiverse Recommendation and FM were originally designed for the rating prediction task based on explicit user feedback [6,21]. However, it is a commonplace that in real-world scenarios most observed feedback is not explicit but implicit [20]. Typical implicit feedback includes the number of purchases, clicks, played songs, etc., and thus it is much more accessible, because the user does not have to express his feelings explicitly [19]. As a result, implicit feedback is often one-class, i.e., only positive class is available. In addition, for item recommendation task, the recommendation accuracy near the top of the ranked list is usually more important than that at the end of the list, known as the top-N (item) ranking task. Some recent work has shown that rating prediction algorithms optimized for error metrics such as RMSE (root mean squared error) empirically do not guarantee accuracy in terms of top-N item recommendations [6].

To address the above drawbacks, we propose to optimize FM for the item recommendation task based on implicit feedback, which is also knowns as One-Class Collaborative Filtering (OCCF). More specifically: Firstly, we present RankingFM, which adopts FM as a ranking function to model the interactions between context features, and apply it to the Learning-to-Rank (LtR) approach by using pairwise cross-entropy (CE) loss. We propose to optimize the RankingFM by widely used stochastic gradient descent method. Secondly, inspired by LambdaRank [15], we explore to further improve the top-N recommendation performance of RankingFM by adapting the original lambda weighting function with two alternative sampling schemes, referred to as LambdaFM[1]. Lastly, we

---

[1] A full version of LambdaFM has been published at CIKM'16 [26].

carry out a set of experiments on three public datasets. The results indicate that our proposed methods (i.e., RankingFM and LambdaFM) achieve superior recommendation quality in terms of two standard ranking metrics. In particular, LambdaFM largely outperforms a bunch of strong baselines for top-N recommendations.

2   Related Work

Learning-to-Rank. Recently, Learning-to-Rank (LtR) has been attracting broad attention due to its effectiveness and importance in machine learning community. There are two major approaches, namely, pairwise [1,18] and listwise approaches [3,15]. Specifically, the pairwise ranking usually treats an objective pair as an 'instance' in learning. For example, Herbrich et al. [8] employed the approach and utilized the SVM technology to build a classifier, referred to as Ranking SVM; Burges et al. [1] adopted cross-entropy and gradient descent to train a Neural Network model, known as RankNet. Empirically, pairwise methods perform better than traditional pointwise methods. However, typical pairwise objective functions are devised to maximize the AUC metric, which is clearly position-independent. But for item recommendation, the recommendation quality is highly position-biased because the accuracy near the top of the ranked list is usually more important. In this regard, pairwise loss functions might still be a suboptimal scheme for the top-N item ranking task. In contrast, listwise approaches address the problem more directly because the models are usually formalized to optimize a specific ranking measure. Generally, it is difficult to directly optimize the ranking metrics because they are either flat or non-differentiable. One way to solve this problem is to propose smooth approximations of the target measures. For example, Shi et al. proposed smooth variants of MAP [24] and Mean Reciprocal Rank (MRR) [25] to optimize ranking performance. The other way is the lambda-based approach, such as LambdaRank [15] and LambdaMart [2], which is designed to add listwise information into pairwise implementation to bypass the major challenges of traditional listwise methods.

Factorization Models. Recommender systems (RS) have two characteristics that distinguish themselves from conventional LtR (e.g., Ranking SVM, RankNet) in web search: (1) The user-item matrix is usually highly sparse e.g., $\geq 95\%$ in most scenarios where the conventional LtR is likely to fail [16,21]; (2) RS aim at personalization, which means each user should receive one personalized ranking, whereas the conventional LtR learns only one ranking for a query, which, in effect, is non-personalization [20]. To tackle the above problems, researchers have proposed factorization models for recommendation tasks. Specifically, a series of matrix factorization (MF) based algorithms have been devised in the literature, e.g., Singular Value Decomposition (SVD) [11], Tensor Factorization (TF) [22], Probability Matrix Factorization (PMF) [23]. In particular, Rendle [16] unified factorization based models by developing a general predictor called Factorization Machines (FM), and showed that FM worked in linear time and can mimic several state-of-the-art MF models by feature engineering.

Furthermore, FM demonstrates high recommendation accuracy for rating prediction by mining the latent interactions between pairwise features in sparse settings [16]. However, it has been pointed out that the least square based loss for rating prediction is suboptimal for item recommendation task [6,18]. Accordingly, a variety of ranking-based MF models have been proposed, e.g., WRMF [9] (pointwise), PITF [22] & RTF [18] (pairwise) and CLiMF [25] (listwise), which, however, were designed for specific tasks (e.g., tag recommendation) and cannot handle general scenarios of context-aware recommendations.

In our work, we adapt FM to RankingFM by applying the pairwise cross-entropy loss, and then explore to improve the way of pairwise learning by optimizing a rank biased performance measure. In contrast to the previous work, our proposed method is a general context-aware algorithm that is capable of effectively optimizing item ranking performance.

3   Ranking Factorization Machines

In this section, we first briefly review Factorization Machines (FM), and then elaborate our RankingFM algorithm. Lastly, the stochastic gradient descent (SGD) is applied to train the RankingFM model.

3.1   Factorization Machines

FM is a state-of-the-art pointwise prediction model, which is capable of capturing all nested interactions up to order d among n input variables in $\boldsymbol{x}$ with a factorized representation. For a detailed description, please refer to Rendle [17]. The FM model of order $d = 2$ is defined as:

$$\hat{y}(\boldsymbol{x}) = \underbrace{w_0 + \sum_{i=1}^{n} w_i x_i}_{\text{linear}} + \underbrace{\sum_{i=1}^{n} \sum_{j=i+1}^{n} \langle \boldsymbol{v}_i, \boldsymbol{v}_j \rangle x_i x_j}_{\text{polynomial}} \tag{1}$$

where the model parameters $\Theta = \{w_0, w_1, ..., w_n, v_{1,1}, ..., v_{n,k}\}$ to be estimated are: $w_0 \in \mathbb{R}, \boldsymbol{w} \in \mathbb{R}^n, \boldsymbol{V} \in \mathbb{R}^{n \times k}$, and $\langle \cdot, \cdot \rangle$ denotes the dot product of two vectors of size k:

$$\tau_{i,j} \approx \langle \boldsymbol{v}_i, \boldsymbol{v}_j \rangle = \sum_{f=1}^{k} v_{i,f} \cdot v_{j,f} \tag{2}$$

A row vector $\boldsymbol{v}_i$ of V is the i-th variable with k factors. The linear term of the FM model is identical to a linear regression model. The polynomial term models the interaction between the i-th and j-th variables by using a factorized parametrization $\langle \boldsymbol{v}_i, \boldsymbol{v}_j \rangle$ instead of an independent parameter $\tau_{i,j}$. In [16], it shows that FM can be computed in linear runtime $O(kn)$ because Eq. (1) can be reformulated as:

$$\hat{y}(\boldsymbol{x}) = w_0 + \sum_{i=1}^{n} w_i x_i + \frac{1}{2} \sum_{f=1}^{k} \left(\left(\sum_{i=1}^{n} v_{i,f} x_i \right)^2 - \sum_{i=1}^{n} v_{i,f}^2 x_i^2 \right) \tag{3}$$

3.2 RankingFM Framework

FM is recognized as being very successful for a variety of prediction problems with variables each of which may have interactions with one another. Specifically, pointwise error loss functions are adopted in the latent factor model for rating prediction. However, as previously mentioned, the pointwise optimization results in a suboptimal solution for the item recommendation task, which is known as a ranking task. Thus we aim to extend FM to a Ranking FM approach by applying pairwise LtR techniques.

Consider that the learning algorithm is given a set of pairs of samples (a, b), with known probabilities $\overline{P}_{ab}$ that sample a will be ranked higher than sample b. Also, there exists an input vector $x \in \mathbb{R}^n$, where n is the number of features. Meanwhile there is an output space of ranks represented by label $y = \{y_1, y_2, ..., y_L\}$ with the number of ranks L. We denote the modeled posterior $P(x^a \rhd x^b)$ by P_{ab}. Then $s_a = \hat{y}(x^a)$ (i.e., Eq. (1)) and $s_{ab} = s_a - s_b = \hat{y}(x^a) - \hat{y}(x^b)$. Finally, the deviation between P_{ab} and $\overline{P}_{ab}$ can be formulated by cross-entropy (CE) loss [1]:

$$C_{ab} = C(s_{ab}) = -\overline{P}_{ab} \log P_{ab} - (1 - \overline{P}_{ab}) \log(1 - P_{ab}) \qquad (4)$$

where the two outputs s_a, s_b of the models are mapped into a probability using sigmoid function, i.e., $P_{ab} = \frac{1}{1+e^{-(s_a-s_b)}}$.

In the case of Recommender Systems (RS)[2], for a given user $u \in \mathcal{U}$, let $S_{ab} \in \{0, \pm 1\}$ be defined as 1 if u prefers item a over item b, -1 if opposite, and 0 if u has the same preference of them. $\overline{P}_{ab}$ is assumed to be deterministically known from the ground truth, so that $\overline{P}_{ab} = \frac{1}{2}(1 + S_{ab})$. By combining the above equations, C_{ab} becomes:

$$C_{ab} = \frac{1}{2}(1 - S_{ab})(s_a - s_b) + \log\left(1 + e^{-(s_a-s_b)}\right) \qquad (5)$$

The difference of s_a and s_b can be computed with the computational complexity of $O(kn)$ by applying Eq. (3):

$$s_{ab} = \sum_{i=1}^{n} w_i(x_i^a - x_i^b) - \frac{1}{2}\sum_{f=1}^{k}\left(\sum_{i=1}^{n} v_{i,f}^2 x_i^{a\,2} - \sum_{i=1}^{n} v_{i,f}^2 x_i^{b\,2}\right)$$

$$+ \frac{1}{2}\sum_{f=1}^{k}\left(\left(\sum_{i=1}^{n} v_{i,f} x_i^a\right)^2 - \left(\sum_{i=1}^{n} v_{i,f} x_i^b\right)^2\right) \qquad (6)$$

The objective of CE is to train the scoring function (i.e., FM) so that the loss of ordering probability estimation can be minimized by:

$$C = \sum_{a,b \in D_s} C_{a,b} + \sum_{\theta \in \Theta} \gamma_\theta ||\theta||^2 \qquad (7)$$

where D_s represents all the pair collections, $||\cdot||^2$ is the Frobernius norm and γ_θ is a hyper-parameter for the L2 regularization term.

---

[2] User-item pairs function similarly as query-url pairs in the conventional LtR task.

3.3   Optimization Methods

We adopt the stochastic gradient descent (SGD) to optimize the loss function. By differentiating Eq. (7), the parameter θ can be updated:

$$\theta \leftarrow \theta - \eta \left(\frac{\partial C_{ab}}{\partial \theta} + \gamma_\theta \theta \right) \tag{8}$$

where

$$\frac{\partial C_{ab}}{\partial \theta} = \frac{\partial C_{ab}}{\partial s_a} \frac{\partial s_a}{\partial \theta} + \frac{\partial C_{ab}}{\partial s_b} \frac{\partial s_b}{\partial \theta} \tag{9}$$

$\frac{\partial C_{ab}}{\partial s_a}$ and $\frac{\partial C_{ab}}{\partial s_b}$ are the learning weights (i.e., the strength for updating θ), defined as:

$$\frac{\partial C_{ab}}{\partial s_a} = \left(\frac{1 - S_{ab}}{2} - \frac{1}{1 + e^{(s_a - s_b)}} \right) = -\frac{\partial C_{ab}}{\partial s_b} \tag{10}$$

According to Eqs. (8)-(10), we obtain:

$$\theta \leftarrow \theta - \eta \left(\left(\frac{1 - S_{ab}}{2} - \frac{1}{1 + e^{(s_a - s_b)}} \right) \left(\frac{\partial (s_a - s_b)}{\partial \theta} \right) + \gamma_\theta \theta \right) \tag{11}$$

According to the property of Multilinearity [17], the gradient of FM can be derived:

$$\frac{\partial \hat{y}(x^a)}{\partial \theta} = \begin{cases} 1 & \text{if } \theta \text{ is } w_0 \\ x_i^a & \text{if } \theta \text{ is } w_i \\ x_i^a \sum_{j=1}^{n} v_{j,f} x_j^a - v_{i,f} x_i^{a2} & \text{if } \theta \text{ is } v_{i,f} \end{cases} \tag{12}$$

By combining Eqs. (11)-(12), we have:

$$v_{i,f} \leftarrow v_{i,f} - \eta \left(\left(\frac{1 - S_{ab}}{2} - \frac{1}{1 + e^{(s_a - s_b)}} \right) \right.$$
$$\left. \left(\sum_{j=1}^{n} v_{j,f} (x_i^a x_j^a - x_i^b x_j^b) - v_{i,f} (x_i^{a2} - x_i^{b2}) \right) + \gamma_{v_{i,f}} v_{i,f} \right) \tag{13}$$

$$w_i \leftarrow w_i - \eta \left(\left(\frac{1 - S_{ab}}{2} - \frac{1}{1 + e^{(s_a - s_b)}} \right) (x_i^a - x_i^b) + \gamma_{w_i} w_i \right) \tag{14}$$

We show the general learning process of RankingFM in Algorithm 1, which can handle multi-class ranking tasks, e.g., Trec Contextual Suggestion[3] and conventional LtR scenarios. Nevertheless, as explained in Sect. 1, in most real-world scenarios of CF, negative examples and unknown positive examples are mixed together and hardly to be distinguished [14], known as one-class collaborative filtering (OCCF). It can be seen the algorithm has $O(|\mathcal{U}||\mathcal{A}||\mathcal{B}|)$ training triples, where $|\mathcal{A}|$ and $|\mathcal{B}|$ represent the size of observed and unobserved actions by the user $u \in \mathcal{U}$, so we have $\mathcal{A} \cup \mathcal{B} = \mathcal{I}, \mathcal{A} \cap \mathcal{B} = \varnothing$, where $|\mathcal{I}|$ represents the number of items. That is, we need to compute all the full gradient in each update step, which is infeasible because $|\mathcal{B}|$ is usually huge in practice. To solve this problem,

---

[3] https://sites.google.com/site/treccontext/.

Algorithm 1. RankingFM Learning

1: **Input:** Training dataset, regularization parameters γ, learning rate η
2: **Output:** $\Theta = (w, V)$
3: Initialize Θ: $w \leftarrow (0, ..., 0)$; $V \sim \mathcal{N}(0, 0.1)$;
4: **repeat**
5:      **for** a, b with different labels given by u **do**
6:          $s_a = \hat{y}(x^a)$, $s_b = \hat{y}(x^b)$
7:          **for** $f \in \{1, ..., k\}$ **do**
8:              **for** $i \in \{1, ..., n\} \wedge x_i \neq 0$ **do**
9:                  Update $v_{i,f}$ according to Eq. (13)
10:             **end for**
11:         **end for**
12:         **for** $i \in \{1, ..., n\} \wedge x_i \neq 0$ **do**
13:             Update w_i according to Eq. (14)
14:         **end for**
15:     **end for**
16: **until** convergence
17: return Θ

Algorithm 2. RankingFM Learning for OCCF

1: Uniformly draw u from $\mathcal{U}$
2: Uniformly draw a from $\mathcal{A}$
3: Uniformly draw b from $\mathcal{I} \backslash \mathcal{A}$

it is natural to propose a sampling scheme (e.g., bootstrapping [20]), which, on one hand, can make the best use of unobserved feedback for the learning; on the other hand, helps to reduce the runtime of the algorithm. The slightly revised RankingFM for OCCF is shown in Algorithm 2, i.e., Line 5 in Algorithm 1 is replaced by Line 1–3 of Algorithm 2. In this case, x^a denotes the observed positive sample vector, while x^b denotes the unobserved sample vector.

In terms of the computational complexity, it can be seen that the complexity of Eqs. (13) and (14) is $O(kn)$ and $O(n)$ respectively. Thus RankingFM also has a linear computational complexity for each training pair. Moreover, for a CF scenario, most elements x_i in a vector x are zero. For example, let $N(x)$ be the number of non-zero elements in the feature vector x and $\overline{N(x)}$ be the average number of non-zero elements in all vectors. We can see that $\overline{N(x)} \ll n$ under huge sparsity, i.e., the complexity becomes $O(k\overline{N(x)})$ in the CF settings.

4 Efficient Lambda Samplers

4.1 Sampling Analysis

RankingFM is made to work quite well due to the design of the pairwise CE loss function, which is fine if that is the desired loss. However, typical pairwise loss functions are devised to maximize the AUC metric, which is clearly position-independent. For item recommendations, the recommendation quality is highly position-biased because high accuracy near the top of a ranked list is more important to users. To solve this challenge, lambda-based approaches (e.g., LambdaRank [7,15]) have been presented by incorporating ranking bias into pairwise comparison. Inspired by this idea, we may design a similar weighting term $\xi_{a,b}$ for further optimization of RankingFM, which is hereafter called

Lambda Factorization Machines (LambdaFM for short). $\xi_{a,b}{}^4$ is designed to incorporate the size of change of a specific ranking measure by swapping two items (i.e., a and b) of this pair with different relevance levels, the way of which is called lambda (or λ). The new learning weight is defined as:

$$\lambda_{ab} = \left(\frac{1 - S_{ab}}{2} - \frac{1}{1 + e^{(s_a - s_b)}}\right)\xi_{ab} \tag{15}$$

where ξ_{ab} can be the difference of any ranking measure, e.g., NDCG, Reciprocal Rank (RR), or Average Precision (AP), computed by:

$$\xi_{ab} = \begin{cases} |N(2^{l_a} - 2^{l_b})\left(\frac{1}{\log(1+r_a)} - \frac{1}{\log(1+r_b)}\right)| & \text{if } \xi_{a,b} \text{ is } |\triangle NDCG_{ab}| \\ |\frac{1}{R}\left[\frac{n+1}{r_b} - \frac{m}{r_a}\right] + \sum_{k=r_b+1}^{r_a-1}\frac{l_k}{k}| & \text{if } \xi_{a,b} \text{ is } |\triangle AP_{ab}| \\ |\frac{1}{r_b} - \frac{1}{r}| & \text{if } \xi_{a,b} \text{ is } |\triangle RR_{ab}| \text{ and } r_b < r \leq r_a \end{cases} \tag{16}$$

where N is the reciprocal of maximum DCG for a user; l_a and l_b are levels of relevance for item a and b, respectively; r_a and r_b are the rank positions of a and b, respectively; n and m are the number of relevant items at the top r_b and the top r_a positions, respectively; l_k is the binary relevance label of the item at rank position k, i.e., 1 for relevance and 0 otherwise, R is the number of relevant items; r is the rank of the top relevant item in the ranking list. Note that the above equation of RR_{ab} holds only when $r_b < r \leq r_a$, otherwise there is no RR gain. We find that the above implementation is reasonable for multi-class scenarios in typical LtR tasks but impractical in OCCF settings. The reason is that to calculate ξ_{ab} it requires to compute scores of all items using Eq. (3) to obtain the rank, i.e., r_a and r_b in Eq. (16). For typical IR tasks, the candidate documents for a query in training datasets have usually been limited to a small size (e.g., 1000) because of query filtering [27]. However, for recommendation with implicit feedback, the size of candidate items is usually very huge (e.g., 10 million) as all unobserved items should be considered as candidates. Thus, the computational complexity before the update of each training pair has becomes $O(kn|\mathcal{I}|)$. In other words, the original lambda implementation for LambdaRank is not suitable for OCCF settings [26].

To bypass this complexity issue, we devise two efficient lambda-based sampling schemes in the followings. Assume we have an ideal lambda function λ_{ab}, if we have a sampling scheme that generates the training item pairs with the probability proportional to $\lambda_{ab}/(\frac{1-S_{ab}}{2} - \frac{1}{1+e^{(s_a-s_b)}})$ (just like ξ_{ab}), then we can have almost equivalent training models. Further, we give an example of a ranked list (with implicit feedback) to show which item pairs should be assigned with higher sampling weights, where $+1$ and -1 are positive and unobserved items, respectively.

$$\text{Rank Order} : \overbrace{-1, -1, +1, -1, -1, \underbrace{-1, -1, +1}_{\xi_{86}}, -1, -1}^{\xi_{81}}$$

---

^{4} The work in [26] only adopted NDCG for the analysis of lambda whereas we here consider multiple measures.

According to Eq. (16), we calculate that ξ_{81} is 0.42, 0.54 and 0.67 when ξ_{81} is $\triangle NDCG$, $\triangle AP$ and $\triangle RR$ respectively, and that ξ_{86} is 0.02, 0.04 and 0 when ξ_{81} is $\triangle NDCG$, $\triangle AP$ and $\triangle RR$ respectively. Obviously, ξ_{81} is always larger than ξ_{86} regardless of the ranking of the positive item and which ranking measure we employ. This implies ξ_{81} is likely to be a more informative[5] training pair (compared with ξ_{86}) if the unobserved item b has a higher ranking position. Based on this insightful finding, we believe the item pairs whose unobserved item has a higher rank should be drawn with higher probability. This is because the top ranked unobserved items hurt the ranking performance more than those with lower ranked positions [27,28]. With the intuitive observation and above analysis, we devise two simple yet effective sampling schemes to further optimize RankingFM for top-N item ranking.

4.2 Lambda-Based Learning Schemes

Scheme I. According to the above analysis, we argue that the item pair $(8, 1)$ is supposed to be sampled with higher probability than the $(8, 6)$ pair. In addition, we observe that the value of s_{81} is smaller than that of s_{86} because

$$s_{81} = \hat{y}(x^8) - \hat{y}(x^1)$$
$$s_{86} = \hat{y}(x^8) - \hat{y}(x^6) \tag{17}$$
$$\hat{y}(x^1) > \hat{y}(x^6)$$

The above observation suggests that we should sample more item pairs with small preference difference, such as s_{81}. We thus propose an intuitive learning scheme with a dynamic utility function $\rho(u, a, b)$ to judge whether a (u, a, b) triple contains an informative training pair (or a good negative item) such that swapping the positions of a and b could lead to a larger change of a desired ranking loss. The lambda-motivated learning scheme is shown in Algorithm 3, where ρ is a sigmoid function. Hereafter we denote the new algorithm (replacing Line 5 of Algorithm 1 with Algorithm 3) as LFM-I. It can be clearly seen that a smaller s_{ab} will contribute to a larger utility $\rho(u, a, b)$. In terms of the computational complexity, the complexity to calculate the original ξ_{ab} is $O(k\overline{N(x)}|\mathcal{I}|)$, while the complexity of Algorithm 3 is $O(k\overline{N(x)}T)$, where T is the size of sampling trials. In general, we have $T \ll |\mathcal{I}|$ in the beginning of the training and $T < |\mathcal{I}|$ when the training reaches convergence. The reason is because in the beginning, the elements of matrix V are initialized by a standard normal distribution with mean 0 and variance 0.1 (see Algorithm 1), and thus the distribution of s_{ab} also follows an approximate normal distribution with mean 0. In this case, it is quick to find an unobserved item that meets the condition (i.e., $\rho_{\mathbf{rand}} \le \rho(u, a, b)$) as most possible values of $\rho(u, a, b)$ are around 0.5. After several training round, most observed items are likely to ranked higher than the unobserved items

---

[5] In the followings, we refer to a training pair (a,b) as an informative pair if ξ_{ab} is larger after swapping a and b, the unobserved item b is called a good or informative item.

Algorithm 3. Lambda Learning Scheme I (LFM-I)

1: Uniformly draw u from $\mathcal{U}$
2: Uniformly draw a from $\mathcal{A}$
3: **repeat**
4:     Uniformly draw b from $\mathcal{I}\backslash\mathcal{A}$
5:     Generate a random variable $\rho_{\mathbf{rand}}(u,a,b) \in [0,1]$
6:     Calculate the utility function $\rho(u,a,b) = \frac{e^{-s_{ab}}}{1+e^{-s_{ab}}}$
7: **until** $\rho_{\mathbf{rand}} \leq \rho(u,a,b)$

Algorithm 4. Lambda Learning Scheme II (LFM-II)

1: **Require:** Unobserved item set $\mathcal{I}\backslash\mathcal{A}$, parameter ρ and m, scoring function $\hat{y}(\cdot)$
2: Sample a rank r from the power law distribution $pr(r(b)) \propto (\frac{1}{r(b)+1})^{2\rho}$, $\rho \in [0,1]$, where $r(b)$
   (starting from 0) is the rank of item b, and ρ is a coefficient that can be tuned for the
   optimal results. Note that LFM-II will reduce to RankingFM when $\rho = 0$.
3: Uniformly draw $b_1,...,b_m$ from $\mathcal{I}\backslash\mathcal{A}$
4: Compute $\hat{y}(\boldsymbol{x}^{b_1}),...,\hat{y}(\boldsymbol{x}^{b_m})$, and then sort $b_1,...,b_m$ by descending order of $\hat{y}(\boldsymbol{x}^{b_1}),...,\hat{y}(\boldsymbol{x}^{b_m})$
5: Return one item b, which is currently ranked on the r-th position.

(i.e., $s_{ab} > 0$), and thus $\rho(u,a,b)$ is likely to be smaller than 0.5, which will lead to a bit larger T. However, it is impossible that all positive items are ranked higher than unobserved items, so in general we still have $T < |\mathcal{I}|$ with $\rho_{\mathbf{rand}}(u,a,b) \in [0,1]$.

Scheme II. According to Sect. 4.1, a straightforward sampling scheme with the same training effect of the original lambda can be implemented by calculating scores of all items to obtain the possible ranks, and then oversample higher ranked unobserved items. Unfortunately, this learning scheme has the same computational complexity with the original lambda strategy, which is clearly infeasible in practice. To overcome this issue, we first employ a uniform sampling to select m candidates. Then we compute the scores of these candidate items to achieve possible rank orders, and sample the rank by a power-law distribution $pr(r)$ (In practice, $pr(r)$ can be replaced with other distributions, such as exponential and linear distributions as long as $pr(r)$ meets the condition that assigns larger sampling weight to top ranked unobserved items.). Because the first sampling is uniform, the sampling probability density for each item has almost equivalent effect with that from the original (expensive) global sampling. The proposed sampler is shown in Algorithm 4. We refer to RankingFM with the seconding learning scheme as LFM-II. The complexity before performing each pairwise comparison reduces to $O(mkn + m\log m)$, where m is often set to a small value (e.g., $m = 20, 50$). Therefore, by implementing scheme II, we are also able to find an efficient way to bypasses the expensive computational complexity.

5 Experiments

We conduct a set of experiments to evaluate the top-N recommendation accuracy of RankingFM and LambdaFM, compared to several state-of-the-art methods.

5.1 Settings

We use three real-world CF datasets to verify the performance of our proposed methods, namely Libimseti.cz[6] (user-user pairs, where the users recommended as daters are regarded as items here), Lastfm[7] (user-music-artist tuples) and Yahoo[8] (user-music-artist-album tuples). In order to speed up the experiments, we follow the common practice as in [5] by randomly sampling a subset of users from the Libimseti and Yahoo datasets, and a subset of items from the Lastfm dataset. Table 1 summarizes the statistics of the three datasets used in this work. We evaluate the results of top-N item recommendations by two standard metrics, namely, Precision@N and Recall@N (denoted by Pre@N and Rec@N, respectively) [12], where N is the number of recommended items. Details about the two metrics are omitted for saving space.

In our experiments, we compare our methods with four powerful baseline methods: Most Popular(MP) [20], Factorization Machines(FM) [17], Bayesian Personalized Ranking with matrix factorization (BPR) [20], Pairwise Interaction Tensor Factorization (PITF)[9] [22]. Note we adapt FM for the top-N recommendation task by binarizing rating values[10] (denoted as FMB). Since the frequency of a user listening to a song (i.e., relevance feedback) can be obtained from the Lastfm dataset, we also recommend songs by leveraging such information (denoted as FMF[11]). Note that the frequency information has a large range compared with ratings (e.g., $[1,5]$ interval). For example, a user may listen to a song in hundreds of times. In this paper, we employ a trivial function $\frac{1}{1+f^{-1}}$ to map the frequency into $[0.5,1)$, where f represents the frequency. Besides, for a fair comparison, we also exploit the same bootstrap sampling as in BPR to make use of the large number of unobserved items.

All factorization models use a factorization dimension of k = 30. Results for k = 10, 50, 100 give consistent conclusion but are omitted due to space limitations. In terms of η and γ_θ, we apply the 5-fold cross-validation to find optimal

Table 1. Basic statistics of datasets.

DataSets	#Users	#Items	#Records	Density	Rsize	Csize	#Artists	#Albums
Libimseti	5000	82444	642454	0.16 %	128.49	7.79	-	-
Lastfm	983	60000	246853	0.42 %	251.12	4.11	25147	-
Yahoo	2450	124346	911466	0.29 %	372.03	7.33	9040	19851

The "Rsize" and "Csize" columns are the average number of records (e.g., ratings) for each user and for each item respectively.

---

[6] http://www.occamslab.com/petricek/data/.
[7] http://dtic.upf.edu/~ocelma/MusicRecommendationDataset/lastfm-1K.html.
[8] http://webscope.sandbox.yahoo.com/catalog.php?datatype=r\&did=2.
[9] Due to lack of contexts, PITF is not applicable to the Libimseti dataset.
[10] It is a standard way to solve the one-class problem in CF [14].
[11] FMF is identical to FMB in the other datasets, since the frequency of all observed actions is 1.

values for BPR. For PITF, our results show that it performs best with the same η and γ_θ of BPR. For FM (FMB and FMF), we apply the same method to tune η and γ_θ individually; For RankingFM and LambdaFM, we use the same η and γ_θ with BPR for comparison. Specifically, η is set to 0.01 on the Libimseti and Yahoo datasets, and 0.08 on the Lastfm dataset; γ_θ (including γ_{w_i}, $\gamma_{v_{i,f}}$) is set to 0.01 on Libimseti dataset, and 0.05 on Lastfm and Yahoo datasets. Note that we find that all FM based models perform well enough by just using polynomial term (see Eq. (1)). $\rho \in [0,1]$ is specific for LFM-II, which is discussed later.

5.2   Results

Accuracy Analysis. Figure 1(a–f) shows the top-N recommendation accuracy of all algorithms on the three datasets. First, we clearly observe that our proposed RankingFM (RFM) largely outperforms the original FM model (i.e., FMB), which empirically indicates the pairwise approach outperforms the pointwise approach with the common 0/1 interpretation [20]. The reason is because the two algorithms have the same scoring function but only differ in loss functions: FMB applies the pointwise square loss, while RFM uses the pairwise CE loss for optimization. Second, the proposed LambdaFM (LFM-I, LFM-II) consistently outperforms other methods in terms of both ranking metrics. This is because LambdaFM (1) directly optimizes the ranking metrics by the design of two lambda-based sampling schemes (vs. BPR, PITF, FMB, FMF and RFM); (2) estimates more accurate ordering relations between candidate items by

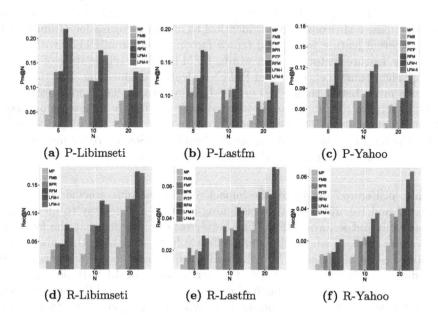

Fig. 1. Performance comparison w.r.t. top-N values, i.e., Pre@N (P) and Rec@N (R). ρ is fixed to 0.8 for LFM-II, and m is fixed to 50.

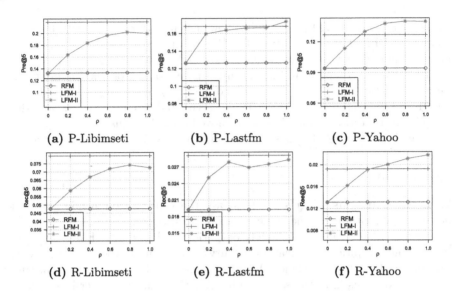

Fig. 2. Parameter tuning for LFM-II w.r.t. Pre@5 (P) and Rec@5 (R). $\rho \in \{0, 0.2, 0.4, 0.6, 0.8, 1.0\}$, $m = 50$.

incorporating additional contextual variables (e.g., artists and albums) (vs. BPR and PITF). Third, we find that RFM achieves almost the same results with BPR and PITF on the Libimseti and Lastfm datasets. The reason is because all the three approaches exploit the pairwise loss function but with different prediction functions. FM (from RFM) is identical to matrix factorization (from BPR) with user-item feature vector and tensor factorization (from PITF) with user-item-artist feature vector. In other words, RFM is able to mimic state-of-the-art ranking algorithms (i.e., BPR and PITF) by feature engineering. Fourth, FMF performs much better than FMB on the Lastfm dataset. This indicates that a user's preference to a song can be inferred more accurately by leveraging playing times information. The intuition is that the more times she played a music track, the higher preference she expresses implicitly. Several other insights can be obtained in Fig. 1 but are omitted for space reasons.

Tuning ρ. Figure 2 illustrates the impact of ρ for learning scheme II in terms of Pre@5 and Rec@5[12]. First, by assigning a larger ρ, it is easy to find LFM-II noticeably outperforms RFM. The better results indicate that the lambda-motivated sampler (scheme II) works effectively to deal with the suboptimal results of pairwise ranking. Note LFM-II is equivalent to RFM when $\rho = 0$ according to Algorithm 4. Second, the performance of LFM-II on all three datasets increases with the growth of ρ. In particular, on the Libimseti dataset, the performance achieves the optimal value when $\rho = 0.8$, and starts to decrease when $\rho = 1.0$. The reason is because only several top ranked items have the

---

[12] The results w.r.t. other top-N values are consistent, but are omitted for saving space.

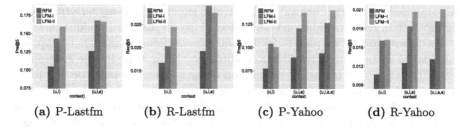

Fig. 3. Performance comparison w.r.t. Pre@5 (P) & Rec@5 (R) by adding context. ρ is fixed to 0.8, and m is fixed to 50 for LFM-II. (u, i) is a user-item (i.e., music) pair and (u, i, a) is a user-item-artist tuple in (a-d); (u, i, a, a) is a user-item-artist-album tuple in (c) and (d).

chance to be selected as candidates for the pairwise comparison when $\rho = 1.0$, and in this case, many unobserved items will not be seen by the learning algorithm, which probably leads to relatively worse performance because of insufficient training samples. On the other side, its performance has not obviously decreased when ρ is set to 1.0 on the Lastfm and Yahoo datasets, which suggests that (1) LFM-II may work well even by picking the top from m randomly selected items; (2) the performance is expected to be improved further by setting a larger sampling size (i.e., m)[13].

Impact of Context. We compare the performance changes of RankingFM and LambdaFM by gradually adding additional contextual variables. First, Fig. 3 (a-d) indicates that both RankingFM and LambdaFM with (u, i, a) noticeably outperform that with (u, i) tuples on both datasets. Second, we can see RankingFM and LambdaFM with (u, i, a, a) tuples outperform that with (u, i, a) tuples from Fig. 3(c-d). The intuition behind is that a use's preference to a music track can be inferred more accurately by taking into account of the artist and album information. Hence, we argue that in general by adding useful context features, our models are able to obtain significant recommendation improvements.

6 Conclusion and Future Work

In this paper, we have introduced two ranking predictors, namely RankingFM and LambdaFM. In contrast to other CF algorithms, RankingFM and LambdaFM are general context-aware recommendation algorithms that are able to incorporate any types of context information. Besides, we design two intuitive sampling schemes for LambdaFM, with which LambdaFM is made more reasonable for optimizing item ranking in OCCF settings. Our experiments on three public CF datasets show that RankingFM and LambdaFM performs better than several state-of-the-art CF methods. In particular, LambdaFM (with

---

[13] Note that a larger sampling size m will result in a larger computational complexity.

two proposed sampling schemes) demonstrates superior ranking performance in the top-N item recommendation task, reflected in two standard ranking metrics.

For future work[14], we plan to (i) develop more advanced samplers to improve LambdaFM without negative effects on efficiency; (ii) investigate the generalization of the suggested lambda strategies on other well-known pairwise loss functions, e.g., hinge Loss [8], exponential loss [4] as well as fidelity loss [13] (iii) investigate the performance of both RankingFM and LambdaFM for traditional multi-class ranking tasks, such as web search and Trec Contextual Suggestion.

References

1. Burges, C., Shaked, T., Renshaw, E., Lazier, A., Deeds, M., Hamilton, N., Hullender, G.: Learning to rank using gradient descent. In: ICML, pp. 89–96 (2005)
2. Burges, C.J.: From ranknet to lambdarank to lambdamart: An overview
3. Cao, Z., Qin, T., Liu, T., Tsai, M., Li, H.: Learning to rank: from pairwise approach to listwise approach. In: ICML, pp. 129–136 (2007)
4. Chen, W., Liu, T.-Y., Lan, Y., Ma, Z.-M., Li, H.: Ranking measures and loss functions in learning to rank. In: NIPS, pp. 315–323 (2009)
5. Christakopoulou, K., Banerjee, A.: Collaborative ranking with a push at the top. In: WWW, pp. 205–215 (2015)
6. Cremonesi, P., Koren, Y., Turrin, R.: Performance of recommender algorithms on top-n recommendation tasks. In: RecSys, pp. 39–46 (2010)
7. Donmez, P., Svore, K.M., Burges, C.J.: On the local optimality of lambdarank. In: SIGIR, pp. 460–467 (2009)
8. Herbrich, R., Graepel, T., Obermayer, K.: Support vector learning for ordinal regression (1999)
9. Hu, Y., Koren, Y., Volinsky, C.: Collaborative filtering for implicit feedback datasets. In: ICDM, pp. 263–272 (2008)
10. Karatzoglou, A., Amatriain, X., Baltrunas, L., Oliver, N.: Multiverse recommendation: n-dimensional tensor factorization for context-aware collaborative filtering. In: RecSys, pp. 79–86 (2010)
11. Koren, Y.: Collaborative filtering with temporal dynamics, pp. 89–97 (2010)
12. Li, X., Cong, G., Li, X.-L., Pham, T.-AN., Krishnaswamy, S.: Rank-GeoFM: a ranking based geographical factorization method for point of interest recommendation. In: SIGIR, pp. 433–442 (2015)
13. Tsai, M., Liu, T., Qin, T., Chen, H., Ma, W.: Frank: a ranking method with fidelity loss. In: SIGIR, pp. 383–390 (2007)
14. Pan, R., Zhou, Y., Cao, B., Liu, N.N., Lukose, R., Scholz, M., Yang, Q.: One-class collaborative filtering. In: ICDM, pp. 502–511 (2008)
15. Quoc, C., Le, V.: Learning to rank with nonsmooth cost functions. In: Advances in Neural Information Processing Systems 19, pp. 193–200 (2007)
16. Rendle, S.: Factorization machines. In: ICDM, pp. 995–1000 (2010)
17. Rendle, S.: Factorization machines with libFM. TIST **3**, 57:1–57:22 (2012)
18. Rendle, S., Balby Marinho, L., Nanopoulos, A., Schmidt-Thieme, L.: Learning optimal ranking with tensor factorization for tag recommendation. In: SIGKDD, pp. 727–736 (2009)

---

[14] We would refer the interested reader to [26] for a detailed analysis about LambdaFM.

19. Rendle, S., Freudenthaler, C.: Improving pairwise learning for item recommendation from implicit feedback. In: WSDM, pp. 273–282 (2014)
20. Rendle, S., Freudenthaler, C., Gantner, Z., Schmidt-Thieme, L.: BPR: Bayesian personalized ranking from implicit feedback. In: UAI, pp. 452–461 (2009)
21. Rendle, S., Gantner, Z., Freudenthaler, C., Schmidt-Thieme, L.: Fast context-aware recommendations with factorization machines. In: SIGIR, pp. 635–644 (2011)
22. Rendle, S., Schmidt-Thieme, L.: Pairwise interaction tensor factorization for personalized tag recommendation. In: WSDM, pp. 81–90 (2010)
23. Salakhutdinov, R., Mnih, A.: Probabilistic matrix factorization. In: NIPS 20, pp. 1257–1264 (2008)
24. Shi, Y., Karatzoglou, A., Baltrunas, L., Larson, M., Hanjalic, A., Oliver, N.: TFMAP: optimizing map for top-n context-aware recommendation. In: SIGIR, pp. 155–164 (2012)
25. Shi, Y., Karatzoglou, A., Baltrunas, L., Larson, M., Oliver, N., Hanjalic, A.: CLiMF: learning to maximize reciprocal rank with collaborative less-is-more filtering. In: RecSys, pp. 139–146 (2012)
26. Yuan, F., Guo, G., Jose, J., Chen, L., Yu, H., Zhang, W.: Lambdafm: learning optimal ranking with factorization machines using lambda surrogates. In: CIKM (2016)
27. Zhang, W., Chen, T., Wang, J., Yu, Y.: Optimizing top-n collaborative filtering via dynamic negative item sampling. In: SIGIR, pp. 785–788 (2013)
28. Zhong, H., Pan, W., Xu, C., Yin, Z., Ming, Z.: Adaptive pairwise preference learning for collaborative recommendation with implicit feedbacks. In: CIKM, pp. 1999–2002 (2014)

Taxonomy Tree Based Similarity Measurement of Textual Attributes of Items for Recommender Systems

Longquan Tao[(⊠)], Fei Liu, and Jinli Cao

Department of Computer Science and Information Technology,
La Trobe University, Bundoora, VIC 3083, Australia
{C.Tao,F.Liu,J.Cao}@latrobe.edu.au

Abstract. Recommender systems have become indispensable tools for numerous industries and individual who utilize e-commerce. Although recommender systems rely on the similarities between the items to be recommended, most current research projects in this area utilize traditional algorithms for similarity measurement such as cosine distance or derivatives, etc. However, the most challenging problems occur due to the difficulties of quantification for those non-numeric values that are quite intractable and cannot be solved by using regular similarity measurement algorithms. This paper proposes a novel and effective method which utilizes a taxonomy tree to measure similarities between textual attributes based on their natural characteristics. Furthermore, a 7-type-rule to cleanse the textual terms a is implemented for improving the recommender system. Finally, we evaluate our methods by implementing a recipe recommender system. The system achieves a 74.4 % overall satisfaction rate as evaluated by its users.

Keywords: Recommender systems · Taxonomy tree · Item similarity · Textual attribute similarity

1 Introduction

With the development of information systems, the volume of information online has grown enormously. Consequently, it has become very difficult to find information of interest from the Internet. Additionally, it is desirable if items of interest are actively recommended to users when they start a search.

Recommender systems are a type of information systems that can make intelligent decisions, based on some known information associated to each user, such as user's profile, preferences, buying/browsing history and even context. The items to be recommended are the general terms in the system, which are usually information of interest, consumptions or any types of services [12]. Moreover, personalization is always a desirable feature to a recommender system. Within this large-scale research area, most research projects focus on the accuracy of recommendations, which is based on measuring the similarity between the items utilizing traditional algorithms or derivatives, such as Euclidean distance and cosine similarity. This paper will propose a

© Springer International Publishing AG 2016
W. Cellary et al. (Eds.): WISE 2016, Part I, LNCS 10041, pp. 294–301, 2016.
DOI: 10.1007/978-3-319-48740-3_21

new way of measurement that the similarity measurement of items can be improved by taking semantic taxonomies into account utilizing their natural characteristics.

In this paper, we propose a new similarity measurement for items based on the structure of the taxonomy tree. We then apply the algorithm to our recommender system to test its effectiveness. Experiments reveals the users' satisfaction rate was improved to 74.4 %.

The rest of this paper is organized as follows. In Sect. 2, several current projects are discussed and their advantages and disadvantages are analyzed. In Sect. 3, using the methodologies will be proposed. Firstly, the dimensions of these items are analyzed as the theoretical basis of these methodologies. Then, the 7-rule-type is introduced for data cleansing the dimension type "group textual attributes" for these general items in any recommender systems. After this, the construction of a knowledge-tree based on natural taxonomies is proposed to measure the similarity between the items. The proposed recommender system is evaluated in Sect. 4, utilizing recipes as the items, and the results show that a high level of satisfaction was reached as evaluated by the volunteers.

2 Related Work

Since the 1990s, recommender systems have been applied in numerous areas, such as e-commerce, and industries such as the tourism, film, music industries.

There are two major approaches in terms of collecting users' preferences, namely, explicit and implicit [1]. The former refers to explicitly collect users' ratings on items, which is the most commonly utilized method, whereas the latter refers to strategies that do not depend on users' ratings, such as monitoring their behaviors, or context-aware methods that take the circumstances of the users as parameters and criteria for recommendations [15].

It is commonly believed that the most critical component of recommender systems is the filtering method that rates the items to be recommended to users. The most widely accepted classifications are presented in a survey paper [5] as follows:

- collaborative filtering that pairs of users are correlated based on their preferences [5],
- content-based filtering which relies on the similarity between items [9],
- knowledge-based filtering which verifies item features and user demands, based on a knowledge base which is usually based on ontology [2],
- demographic filtering that analyses the attributes of users in their profiles, such as gender and age [13],
- matrix factorization which is a sub-category of collaborative filtering but utilizes matrix decomposition [14],
- hybrid filtering [10] that combines any of the above to produce more accurate results.

In order to conduct item content-based filtering, the similarity measurement between items is the most critical technique. [4] suggested utilizing LOD (Linked Open Data) to analyse the content of the items, so that the dimensions obtained will be formed by the VSM (Vector Space Model) to calculate their cosine distances. Although the approaches that discover the dimensions are various, the similarity algorithms are still quite common and conservative.

On the basis of these conservative similarity measurements, [3] argued that these traditional methods implicitly assign equal weights to all the features of the items, i.e. the dimensions. Nevertheless, the latent criteria by which users judge whether items are preferable or not is usually at different importance levels. For example, the price of a camera is subjectively more important than the colour of it for most customers.

However, the inherent issue of these algorithms is that all the dimensions need to be quantified and mapped into numbers. Additionally, string distance measurement is still not able to reflect the nature of the features of items, since they only measure the appearance of these words rather than their real meanings. Therefore, only semantic-based methods are able to solve this issue.

[6] proposed a semantic-based news recommender system, who utilize a knowledge base to extract concepts that are mentioned in the news, as well as the preferred concepts of the users. Specifically, text snippets from the news are formed into vectors.

Therefore, semantics relatedness is calculated based on the ontology concept vectors. Similarly, [8] utilized an ontology-based similarity measurement in terms of topic matching.

However, the ontologies used in these research projects all extract the main topic or concepts from free text, rather than entities. This paper proposes an approach that the similarities of the textual attributes should be measured using numeric values after they have been quantified. The next section introduces the proposed methodology to resolve this problem.

3   Methodology

In order to cope with the text-based elements contained in a field, semantic concepts are necessarily used to capture these elements and map them into the numeric domain. Therefore, it will be feasible to compare the similarities.

To begin with, the dimensions of any types of items are analyzed. Then, the 7-type-rule is proposed as a methodology to cleanse the raw data that is collected from general sources, such as websites. The core algorithm based on the taxonomic knowledge tree is introduced to determine the relationships between these textual terms. At the end, those recommendation items have been classified into the categories with their particular rules.

4   Dimension Classification

In order to solve the aforementioned problem, it is necessary to classify the dimension types of the items to be recommended. The following is a classification based on the nature and format of the fields, which covers all the existing items that could appear in any recommender system.

- **Single continuous numeric**: e.g. the <u>price</u> of products in the supermarket
- **Group continuous numeric**: e.g. the <u>flavor</u> of a recipe can be measured by several sub-dimensions (bitter, sour, sweet, piquant, meaty, etc.) which are all numeric continuous values within an interval
- **Single numeric discrete value**: e.g. users' <u>ratings</u> of the items
- **Group of numeric discrete values**: e.g. users' <u>ratings of multiple aspects</u> of a movie, such as sound effects, visual effects, plot and average actor skills
- **Single textual attribute**: e.g. the <u>name</u> of the movie's director, cuisine type of a recipe
- **Group of textual attributes**: e.g. actors' <u>names</u> in a movie, ingredients of a recipe
- **Free text**: e.g. User's <u>comments</u>

Where types 1–4 in the above are numeric values, similarities are relatively easy to measure as they have a natural deviation from each other. It is also feasible to calculate the group numeric values with distances utilizing common approaches, such as cosine distance, so that the fields are considered as vectors by themselves.

Compared with numerical fields, similarities in text fields are more challenging to calculate. It is necessary, therefore, to take semantics into account to resolve this issue. The main challenge, is the ambiguity of human languages. Fortunately, types 5 and 6 are restricted within a certain scope, which means that we can organize them based on domain-specific criteria or existing taxonomies.

4.1   7-Type-Rule

The first difficulty in relation to the ambiguity of human language relates to synonyms and flexibilities of nature languages. Especially, it is even more challenging when coping with group of textual attributes. Although there are some effective similarity metrics for strings, such as Jaccard distance [11], they are not appropriate for recommender systems, since the distance between textual terms within the context is not about the removal of letters or substitutions, but meanings. Therefore, it is necessary to generalize nouns based on their actual meaning in order to ensure that different terms with same meaning are matched with a high similarity rather than being considered to be completely different. Therefore, 7-type-rule is organized and proposed for generalizing, cleansing the textual terms in order to facilitate the pattern matching for later stages.

We suggest that 7 types of rules should be applied to every single term. The rules are matched by pre-defined keywords. For example, a rule will be applied on a term if the terms contain keyword x, and then the modifications will be executed if the terms also satisfy the condition of the rule. These rules are firstly filtered by the containing terms. If any attributes contain the filtering terms, all the rules will be looped and applied when applicable. The 7 rules are presented as follow:

1. $h_1(f, c, other) = (f, c)$
2. $h_2(f, \not\exists c(c\varepsilon\{c_o, c_1, \ldots, c_i\} \subseteq other)) = (f, c)$
3. $h_3(f, c, other) = (f, other)$

4. $h_4(f, c, other) = (f, r, other)$
5. $h_5(f, c, other) = (r)$
6. $h_6(f) = (r)$
7. $h_7(f, other) = (f)$.

4.2   Taxonomy Tree Based Similarity Measurement

To begin with, some common criteria need to be identified and extracted from these items in order to compare them. It is not too difficult to find these common fields. For instance, most of the ingredients in recipes are either animals or plants, and they have natural similarities in terms of biology.

For those terms that can be categorized and ranked based on either natural characteristics or human knowledge, it is simple to build a tree-like hierarchy structure of them. For example, the zoology taxonomy methodologies are able to satisfy similarity requirements, since they are tested and proved with a large amount of scientific biological research. Also, it is feasible to find "taxonomies" for other realms, such as the products of a supermarket which also have natural classifications.

Therefore, all the items are ranked and classified based on their taxonomies. Suppose the total number of levels is denoted by N, lv represents the nearest common ancestor's level number and μ is the predefined minimum similarity when the pair of terms have a common ancestor at the highest level. Then the similarity $fsim(t_i, t_j)$ is calculated as follows:

$$
fsim(t_i, t_j) = \begin{cases} 0, & no\ common\ ancestor \\ \dfrac{N\mu - \mu}{(1 - \mu)lv_k + N\mu - 1}, & have\ common\ ancestor\ on\ level\ lv_k \end{cases}
$$

Regarding the nearest common ancestor level lv, the level of the target term is considered as 1, therefore the maximum similarity between two different terms will occur when $lv = 2$.

The reason why hyperbolas are utilized rather than linear functions is that the further nearest common ancestors result in much longer distance between two items in the experiments. One of the reasons for this could be that the number of atomic elements contained in a certain category level is much higher than the ones in its offspring categories.

However, there are more challenges in practice, as not all the terms can be considered as simple nodes in taxonomy trees. Therefore, based on the position and function of the nodes, they can be classified as follows:

- **ROOT** – like traditional data trees, taxonomic tree must also have a root, which represent the cornerstone of the domain for a particular item textual attribute.
- **CATEGORY** – these categories are general terms that must contain offspring. They might be the terms that appears as item textual attributes, but more probably will only find their meanings by classifying offspring nodes.

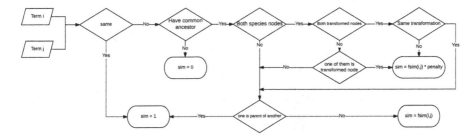

Fig. 1. Flow chart of the algorithm: Taxonomy tree based similarity measurement

- **TRANSFORMATION** – occasionally, some terms are transformed from other terms within this domain. They normally inherit less traits from their ancestors in this case, although they do share little similarities with their ancestors. The process of transforming terms into other terms is called transformation in taxonomic tree.
- **TRANSFORMED NODE** – these are the nodes that are transformed from their ancestors. Transformed nodes must be indirect nodes, since they at least have one ancestor which may be direct node or indirect node that they transformed from.
- **SPECIES NODE** – the nodes in the tree that are not categories.

Basically, this algorithm (Fig. 1) utilizes the taxonomy tree, as well as the position of the terms as nodes in the tree, to determine the similarity between these terms. It maps terms with their semantic meanings in the tree, so that the similarities calculated are also firmly based on their semantics.

5 Evaluation

In order to validate the aforementioned algorithms, they implemented in a recipe recommender system. There are several reasons why this particular recommender system is chosen. Firstly, recipe recommenders are popular with the general public, but unfortunately there are too many of them, making it difficult for users to choose their preferred ones. Secondly, the most significant dimension of the recipes is the ingredients, which are classified as grouped textual terms, so it is suitable for evaluating the effectiveness of the algorithms. Thirdly, there is already an Integrated Taxonomic Information System [7] which is reliable that can be utilized to generate the taxonomic tree of those animal and plants types of ingredients.

A recipe recommender system is implemented with PHP as the backend server and presented as web UI. This enables users to easily access the system via web browsers. The evaluation activities are conducted on 30 volunteers from different backgrounds.

The results show that the recipe recommender system achieved a 74.4 % satisfaction rate based on the statistics in Fig. 2, calculated as follows:

$$ s = \frac{\sum_{u \in U} \sum_{r_i \in R_U} r_i}{\sum_{u \in U} \sum_{n=i}^{|R_U|} r_{max}} $$

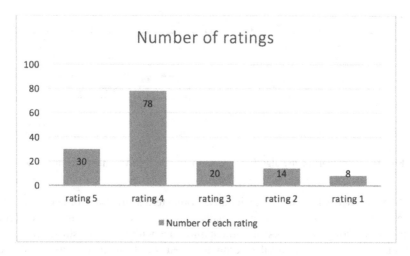

Fig. 2. Statistics of user satisfaction rates.

where U is the set of all users, U is the individual user, R_U is all the feedback ratings of the corresponding user U, r_i is the ith rating in the current rating set, and r_{max} is the max rating in principle, which equals 5 in this case.

6 Conclusion

Recommender systems have become important tools to a large number of online services for providing useful information to users on items of interest. This paper addressed the current weaknesses in this research area and proposed a viable solution. Since the core activity of recommender systems is to measure the similarity between items, the pivotal activity to be undertaken in order to improve the current performance level is to enhance the similarity measurements. Therefore, this paper suggested that there are different types of dimensions for all items, so different algorithms must be used. The most significant contribution of this paper is the taxonomy tree based similarity measurement, which resolved the similarity measurement of the textual attributes of items by measuring natural semantic distance rather than simply counting co-occurrences. Consequently, a novel and effective similarity measurement approach was developed which integrated data cleansing techniques (7-type-rule) which are able to amalgamate terms with similar meanings.

References

1. Bobadilla, J., Ortega, F., Hernando, A., Glez-de-Rivera, G.: A similarity metric designed to speed up, using hardware, the recommender systems k-nearest neighbors algorithm. Knowl. Based Syst. **51**, 27–34 (2013)

2. Carrer-Neto, W., Hernández-Alcaraz, M.L., Valencia-García, R., García-Sánchez, F.: Social knowledge-based recommender system. Application to the movies domain. Expert Syst. Appl. **39**(12), 10990–11000 (2012)
3. Debnath, S., Ganguly, N., Mitra, P.: Feature weighting in content based recommendation system using social network analysis. In: Proceedings of the 17th International Conference on World Wide Web, pp. 1041–1042. ACM, April 2008
4. Di Noia, T., Mirizzi, R., Ostuni, V.C., Romito, D., Zanker, M.: Linked open data to support content-based recommender systems. In: Proceedings of the 8th International Conference on Semantic Systems, pp. 1–8. ACM, September 2012
5. Gavalas, D., Konstantopoulos, C., Mastakas, K., Pantziou, G.: Mobile recommender systems in tourism. J. Netw. Comput. Appl. **39**, 319–333 (2014)
6. IJntema, W., Goossen, F., Frasincar, F., Hogenboom, F.: Ontology-based news recommendation. In: Proceedings of the 2010 EDBT/ICDT Workshops, p. 16. ACM, March 2010
7. Itis.gov. Integrated Taxonomic Information System (2015). Retrieved 9 November 2015. http://www.itis.gov/
8. Middleton, S.E., De Roure, D., Shadbolt, N.R.: Ontology-based recommender systems. In: Staab, S., Studer, R. (eds.) Handbook on Ontologies. International Handbooks on Information Systems, pp. 779–796. Springer, Heidelberg (2009). doi:10.1007/978-3-540-92673-3_35
9. Nguyen, T.T., Hui, P.M., Harper, F.M., Terveen, L., Konstan, J.A.: Exploring the filter bubble: the effect of using recommender systems on content diversity. In: Proceedings of the 23rd International Conference on World Wide Web, pp. 677–686. ACM, April 2014
10. Porcel, C., Tejeda-Lorente, A., Martínez, M.A., Herrera-Viedma, E.: A hybrid recommender system for the selective dissemination of research resources in a technology transfer office. Inf. Sci. **184**(1), 1–19 (2012)
11. Resnick, P., Varian, H.R.: Recommender systems. Commun. ACM **40**(3), 56–58 (1997)
12. Ricci, F., Rokach, L., Shapira, B.: Introduction to recommender systems handbook. In: Ricci, F., Rokach, L., Shapira, B., Kantor, P.B. (eds.) Recommender Systems Handbook. International Handbooks on Information Systems, pp. 1–35. Springer, Heidelberg (2011). doi:10.1007/978-0-387-85820-3_1
13. Wang, Y., Chan, S.C.F., Ngai, G.: Applicability of demographic recommender system to tourist attractions: a case study on trip advisor. In: Proceedings of the The 2012 IEEE/WIC/ACM International Joint Conferences on Web Intelligence and Intelligent Agent Technology vol. 3, pp. 97–101. IEEE Computer Society, December 2012
14. Yu, H.F., Hsieh, C.J., Si, S., Dhillon, I.: Scalable coordinate descent approaches to parallel matrix factorization for recommender systems. In: 2012 IEEE 12th International Conference on Data Mining, pp. 765–774. IEEE, December 2012
15. Zheng, Y., Xie, X.: Learning travel recommendations from user-generated GPS traces. ACM Trans. Intell. Syst. Technol. (TIST) **2**(1), 2 (2011)

A Personalized Recommendation Algorithm for User-Preference Similarity Through the Semantic Analysis

Haolin Zhang[✉] and Feiyue Ye

Shanghai University, Shanghai, China
zhllyyyl@126.com, yefy@shu.edu.cn

Abstract. Traditional personalized recommendation algorithms do not involve the analysis of semantic information, so the recommendation results are less accurate. Aiming at this problem, based on the semantic analysis, genres similarity and content features similarity of the projects rated by users are used to measure user-preference and therefore calculate users similarity. Moreover, the number of projects in the same genres is applied to measure project-relevancy and thereby project similarity. Based on these, this study puts forward a personalized recommendation algorithm for user-preference similarity through the semantic analysis. The contrast experiment results based on Movielens data set show that the recommendation accuracy and quality of the proposed algorithm are significantly improved.

Keywords: Personalized recommendation · Semantic analysis · User-Preference · Project-Relevancy

1 Introduction

Personalized services are a service mode that provides various service strategies and contents for different users [1]. At present, the main form of the network personalized services is personalized recommendation systems. However, sometimes only a little information in the returned lists of personalized recommendation results is in line with user-preferences. Therefore, this study introduces the idea of semantic analysis into the traditional personalized recommendation algorithm, so as to provide users with the personalized information accorded with their preferences.

On the basis of summarizing existing researches, this study proposes a personalized recommendation algorithm for user-preference similarity from the perspective of semantic analysis. This algorithm is employed to carry out semantic analysis from users and projects separately to solve the problem of similarity measure. Therefore, this algorithm firstly measures users similarity based on user-preference and obtains similar neighbor sets. Then by combining the calculation of projects similarity based on project-relevancy, the nearest neighbor set is selected. Finally the recommendation results are obtained. By doing so, the recommendation quality and accuracy are ensured.

© Springer International Publishing AG 2016
W. Cellary et al. (Eds.): WISE 2016, Part I, LNCS 10041, pp. 302–310, 2016.
DOI: 10.1007/978-3-319-48740-3_22

2 A Personalized Recommendation Algorithm for User-Preference Similarity Through the Semantic Analysis

2.1 User-Project Semantic Analysis and Relevant Definitions

In fact, users similarity is not only associated with their ratings for projects but also with the preference for genres of projects. For example, in Movielens, each film has been marked with genres, accompanied by a text summary of the film content. Therefore, if two users evaluate comedy or films relating to family life, it indicates that the two users are interested in the types and contents of the film, so the two users have high similarity. Hence this study introduces the idea of semantic analysis into the calculation for users similarity, defined as follows:

Definition 1: User-preference. According to semantic analysis, projects can belong to one or more genres and possess one or more content features. Therefore, this study divides user-preference into that for genres or content features.

Definition 2: User-preference for genres of project. Suppose that the set $G = \{Genre_1, Genre_2, \ldots, Genre_k\}$ is used to represent genres of projects. The more times that users evaluate a certain genre, the higher the user-preference for this genre is. User-preference for project genres is expressed by $PG_{u,j}$:

Definition 3: User-preference for content features of project. Based on the bag-of-words model [2], the feature words are selected from the content descriptions for all n projects in the project set I to form a bag-of-words set W comprised of the h unordered feature words. Hence in the content descriptions for all projects evaluated by a user, the more frequently a certain feature word appears in W, the higher the user-preference for the content features is. $PC_{u,j}$ represents user-preference for the content features of project.

Similarly, for instance, two films in different languages probably belong to comedy and romance simultaneously. While they are possibly not rated by same users due to users preference for the language, resulting in poor recommendation effects. Therefore, this study also introduces the idea of semantic analysis into the calculation of projects similarity, as follow:

Definition 4: Project-relevancy. According to semantic analysis, project-relevancy which represents the correlation strength of projects, indicates the probability that any two projects in the project genre set $G = \{Genre_1, Genre_2, \ldots, Genre_k\}$ belong to the same genres. It is used to measure the similarity of projects. Moreover, C_{ij} stands for project-relevancy.

According to the above definitions, this study calculates the user-preference and project-relevancy through the semantic analysis. In this way, the authors obtain users and projects similarities and thereby the final recommendation result lists.

2.2   Measure of User-Preference and Selection of the Similar Neighbor Set

Suppose that the matrix $R(m,n)$ is the rating matrix of users for projects. Here, the sets U and I represent all the users and projects, respectively. In addition, the m rows and n columns denote the m users and n projects, separately [3]. This study simplifies score values as binary scores. This is because if users do not rate a project after knowing its genres and content features, it indicates that the users are not interested in this project, and therefore will not watch it and rate it.

The matrix of genres of projects $G(n, k)$ is a binary matrix. When the value of $G_{i,j}$ is 1, it means that an project i belongs to the genre j if it is 0, it indicates that the project i does not belong to this genre. The preference $PG_{u,j}$ of user u for a genre $Genre_j$ is expressed by formula (1):

$$PG_{u,j} = \frac{N1_{u,j}}{N1_u} \tag{1}$$

Where $N1_{u,j}$ represents the total rating number of the user u for projects in j genre, and $N1_u$ refers to the number of rated projects by the user u. Besides, $PG_{u,j}$ indicates that user-preference for a certain genre of project is calculated based on the overall user-preference.

In semantic analysis based user-preference, the user-preference for content features in the content description for projects is another main factor considered in this study. For example, nouns including kids, home, love and life generally appear in the content description for the films relating to family life.

In this study, based on the idea of the bag-of-words model, feature words are extracted from content description for all n projects in the project set I. Firstly, these words are preprocessed through word segmentation, part-of-speech tagging, reserving nouns, and removing stop words, and then the preprocessed nouns are simplified by based on word frequency to remove the nouns with minimal word frequency [4]. Finally, the duplicated words in the sets are deleted, so as to obtain the bag-of -words set $W = \{W_1, W_2, \ldots, W_h\}$ based on the content descriptions for all projects in which h feature words are disordered. Due to short content description and few feature words for each project and as the overall preference for each user needs to be measured, this study uses content description for all projects rated by each user as a whole to represent the user-preference for content features of projects.

For the rated project set I_u of a user u, the user-preference for the content features of projects can be expressed as a h-dimensional vector, $PC_u = (PC_{u,1}, PC_{u,2}, \ldots, PC_{u,h})$, $PC_{u,h}$ is expressed by formula (2):

$$PC_{u,h} = f(u, w_h) \tag{2}$$

Where, $f(u, w_h)$ represents the times of a feature word $w_h \in W$ appearing in the rated project set I_u, so the vector PC_u indicates the preference of the user u from the semantic features.

Sets $PG_u = (PG_{u,1}, PG_{u,2}, \ldots, PG_{u,k})$ and $PC_u = (PC_{u,1}, PC_{u,2}, \ldots, PC_{u,h})$ obtained through the formula (1) and (2), which represent the preference of the user u for genres and content features of projects, respectively. Therefore, the user-preference similarities $Sim_{PG}(u, v)$ and $Sim_{PC}(u, v)$ based on the semantic analysis of any two users u and v are substituted into the formula (3), (4) for calculation:

$$Sim_{PG}(u, v) = \frac{\sum_{j=1}^{q}(PG_{u,j} - \overline{PG_u})(PG_{v,j} - \overline{PG_v})}{\sqrt{\sum_{j=1}^{q}(PG_{u,j} - \overline{PG_u})^2}\sqrt{\sum_{j=1}^{q}(PG_{u,j} - \overline{PG_v})^2}} \tag{3}$$

$$Sim_{PC}(u, v) = \frac{\sum_{j=1}^{p}(PG_{u,j} - \overline{PG_u})(PG_{v,j} - \overline{PG_v})}{\sqrt{\sum_{j=1}^{p}(PG_{u,j} - \overline{PG_u})^2}\sqrt{\sum_{j=1}^{p}(PG_{u,j} - \overline{PG_v})^2}} \tag{4}$$

Where, $\overline{PG}$ and $\overline{PC}$ represent the average preferences of the users u and v for genres and content features of all projects, while q and p indicate the numbers of common genres and common content features of the users u and v, respectively. Both $Sim_{PG}(u, v)$ and $Sim_{PG}(u, v)$ compute the similarity of user-preference based on the semantic analysis by calculating genres similarity and content features similarity of the projects rated by users using the modified cosine similarity algorithm, thus reflecting the users similarity.

In this study, the Pearson user similarity [5] $Sim_{PR}(u, v)$ of traditional collaborative filtering algorithms is combined with the similarities $Sim_{PG}(u, v)$ and $Sim_{PC}(u, v)$ of the user-preference based on semantic analysis. Therefore, the improved user similarity algorithm $Sim_{user}(u, v)$ can comprehensively measure user similarity. In addition, three impact factors α, β, and γ are employed to control the importance of the three similarities, as shown in Formula (5):

$$Sim_{user}(u, v) = \alpha \cdot Sim_{PR}(u, v) + \beta \cdot Sim_{PG}(u, v) + \gamma \cdot Sim_{PC}(u, v) \tag{5}$$

Where, the three impact factors α, β, and γ are valued in the range of (0,1) and can be adjustable. Moreover, they meet relations $\alpha + \beta + \gamma = 1$ and $\alpha > \beta \geq \gamma$. The impact factors are adjusted through the experiment in the following paragraphs. $Sim_{user}(u, v)$ reflects that the overall users similarity is measured by combining the three factors through semantic analysis.

The similarity between the target user u and v can be calculated using formula (5), and then supposing that $Sim_{user}(u, v) \geq \omega$, $\forall u \in U. \forall v \in U, u \neq v$, then the user v is placed in the set of similar neighbors $U1_u$. The threshold ω is set as $\overline{Sim_{user}(u)}$, which is the average similarity of the target user u, as expressed in formula (6):

$$\omega = \overline{Sim_{user}(u)} = \frac{\sum_{v=1}^{m-1} Sim_{user}(u, v)}{m - 1} \tag{6}$$

Where $Sim_{user}(u, v)$ denotes the similarity between the target user u and v, $\forall v \in U$, and $(m - 1)$ refers to the number of users excluding the target user. Formula (6) demonstrates that the selection range is set while preliminarily selecting the set of similar neighbors, which makes the double neighbor choosing more accurate.

2.3   Measure of the Project-Relevancy and Recommendation

The similar neighbor set obtained from the above procedures are the results of the semantic analysis of users. However, some users possibly have a wide range of interests and evaluate many projects, and if the recommendations are only produced based on the similar neighbor set, inaccurate recommendation results can be obtained. Therefore, by conducting semantic analysis of projects similarity, this study reselects the most suitable nearest neighbors to produce recommendation results.

The formula (7) for calculating the project-relevancy C_{ij} of projects i and j is defined as:

$$C_{ij} = \frac{N2_i \cap N2_j}{N2_i \cup N2_j} \tag{7}$$

Where, $N2_i \cap N2_j$ represents the intersection of the genres of projects i and j, that is, the number of common genres of the two projects, while $N2_i \cup N2_j$ indicates the union of projects of the projects i and j. For example, if the project i belong to genres $Genre_1$, $Genre_2$ and $Genre_3$, and the project j belong to genres $Genre_2$, $Genre_3$ and $Genre_4$, thus $C_{ij} = \frac{2}{4} = 50\%$, which indicates that C_{ij} comprehensively takes the semantics based correlation between wholeness and individual of projects into account.

Measuring strong relevance of projects through the number of projects in common genres, given a threshold σ, for $\forall i, j \in I$, if the project-relevancy $C_{ij} > \sigma$, then the two projects i and j are strong relevance and they constitute a pair of strongly relevancy 2-frequent project set Re, this strong relevancy reflects the similarity of projects. Thereinto, the threshold σ is adjustable and set as 20 % as that in pervious study [6].

According to the strong relevance of projects, if all the projects rated by a user in the similar neighbor set show weak relevance, the user is eliminated from the set. The prediction rating $P_{u,i}$ of the user u for the non-rated project i can be calculated by using the formula (8):

$$P_{u,i} = \overline{R_u} + \frac{\sum_{v \in U2_u} Sim_{user}(u, v) \times (R_{v,i} - \overline{R_v})}{\sum_{v \in U2_u} |Sim_{user}(u, v)|} \tag{8}$$

Where $\overline{R_u}$ is the average rating of the target user u, $\overline{R_v}$ refers to the average rating of the nearest neighbor user v, and $R_{v,i}$ represents the rating of a user v for the target project i.

2.4   Integrated Algorithm Description

This study firstly calculates the preferences $PG_{u,j}$ and $PC_{u,j}$ of users, and the similarities of $Sim_{PG}(u,v)$ and $Sim_{PC}(u,v)$ based on the user-preferences. By combining the Pearson similarity $Sim_{PR}(u,v)$, the impact factors α, β, γ are selected, and the $Sim_{user}(u,v)$ is obtained through combining and assigning weights to the three factors. In addition, after selecting the similar neighbor set $U1_u$, the project-relevancy C_{ij} is calculated. The study determine whether the 2-frequent project set Re exists in the set I of projects rated by each user in the set $U1_u$, so as to dually select the nearest neighbor set $U2_u$ that meets conditions. Finally, the TOP-N recommendation set of the target user u in a descending order.

Algorithm. A Personalized Recommendation Algorithm for User-Preference Similarity through the Semantic Analysis

Input: The rating matrix $R(m,n)$, The equilibrium factor α、β、γ, The target user u

Output: The set of recommendations of the target user u

Step1 For$\forall j \in I$, $\forall u \in U$ calculate each $PG_{u,j}$ using formula (1)

Step2 For $\forall j \in I$, $\forall u \in U$ calculate each $PC_{u,j}$ using formula (2)

Step3 For $\forall u,v \in U, u \neq v$, $Sim_{PG}(u,v)$ and $Sim_{PC}(u,v)$ using formula (3)(4)

Step4 Select the value of α,β,γ, calculate $Sim_{user}(u,v)$ using formula (5)(6) and select $U1_u$

Step5 For $\forall i,j \in I, i \neq j$, calculate C_{ij} using formula (7)

Step6 According to Step6, select all the 2-frequent project sets Re of I

Step7 For $\forall v \in U1_u$

        If $\exists i,j \in I, i \neq j, Re_u \neq 0$ then v is retained in $U1_u$ Else v is rejected

Step8 Based on sorted double choosing $U1_u$, select the set $U2_u$

Step9 For $\forall i \in I1_u$, calculate $P_{u,i}$ using formula (8) and TOP-N recommendations of u

---

3   Experimental and Analysis

3.1   Dataset

The Movielens dataset [7] used by the classical CF recommendation model is applied. In the study, the dataset ML_100 K containing 23 documents data is utilized. The dataset is composed of 100,000 pieces of rating data from 943 users for 1,682 movies, and we add the text set of each film's content description is added to the dataset. We consider the rating was simplified into 1 and no rating was 0 in this paper. Each user rates at least 20 movies, and each movie is found to be one or more of 19 genres. The rating sparsity of the dataset is 93.695 %, indicating that the rating matrix of the dataset is highly sparse. The experimental data are divided into a training set and a test set. Likewise the previous research [8], 80 % and 20 % of the data are included in the training set and the test set, respectively, for the convenience of comparison in following sections.

3.2   Evaluation Criterion

To evaluate the recommendation quality of the algorithm, the most commonly used measurements Mean Absolute Error (MAE) and Precision in CF recommender systems are adopted. Assume that the predicted rating set of the target user u is $\{p_{u1}, p_{u2}, \ldots, p_{uN}\}$ and actual rating set is $\{r_{u1}, r_{u2}, \ldots, r_{uN}\}$, then the formula (9) for computing the MAE is:

$$MAE_u = \frac{\sum_{i=1}^{N} |p_{ui} - r_{ui}|}{N} \tag{9}$$

Precision in the TOP-N recommendation results produced by the system is the proportion of the number of correct projects in the recommendation result set, and the formula (10) for computing precision is shown as follows:

$$Precision = \frac{|test \cap TOPN|}{N} \tag{10}$$

Where, test represents the test sample set, TOP-N represents the recommendation result set produced by using the recommendation algorithm. Therefore, the greater the values of the accuracy, the larger the proportion of the projects recommended correctly and the higher the accuracy of the recommended results are.

3.3   Experimental Results and Analysis

Experiment 1: In this study, several experiments are conducted by valuing the combination of the three factors, α, β, γ, and due to limited space, Fig. 1 only presents value examples of 6 sets of parameters. The experimental results show that when the value of α. is 0.5, the sum of the values of β and γ is 0.5. In the condition, the recommended absolute deviation is small. Moreover, superior recommendation results can be obtained when $\beta > \gamma$ compared with $\beta = \gamma$, so the final impact factor is $\alpha = 0.5$, $\beta = 0.3$, $\gamma = 0.2$ and the value is used in the contrast experiments 2 and 3.

Experiment 2: To verify the effectiveness of the proposed personalized recommendation algorithm for user-preference similarity through the semantic analysis (ours), the MAE and Precision of the algorithm are compared with those of other algorithms which without considering the semantic information using the same data in the same environment. These algorithms include the traditional CF algorithms based on cosine similarity (UCBCF) and Pearson correlation coefficient (UPBCF) [9], as well as IRPCF that proposed in previous study and has been widely used for comparison. In the experiments, the number N of the nearest neighbor users grows from 5 to 60 by 5 in each time. The experimental results are shown in Figs. 2 and 3, it can be seen from the figure that with any a K, the algorithm proposed ours in the research shows the minimum value of MAE and the maximum value of Precision. In addition, with the increase of the number of the nearest neighbors, the MAE decreases and the Precision increases, so that the accuracy of the recommender system is improved and stabilized after N = 25.

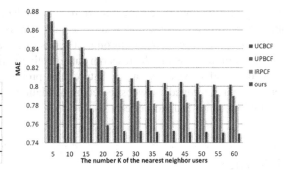

No.	α	β	γ
1	0.4	0.3	0.3
2	0.5	0.3	0.2
3	0.5	0.25	0.25
4	0.6	0.2	0.2
5	0.6	0.3	0.1
6	0.7	0.2	0.1

Fig. 1. The sample values of the impact factors

Fig. 2. The comparison in MAE of contrast experiments

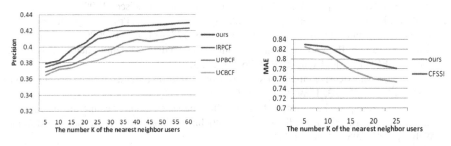

Fig. 3. The comparison in Precision of contrast experiments

Fig. 4. The comparison in MAE of experiment

Experiment 3: Furthermore, this study conducts the contrast experiments between the proposed ours algorithm and CFSSI algorithm based on semantics proposed in the previous research [10] that is widely acknowledged. In addition to this, the mean absolute errors (MAE) of the two algorithms are compared in the same conditions with the same experimental data. In this experiment, the number K of the nearest neighbors increases from 5 to 25 with an interval of 5. As shown in Fig. 4, for the arbitrary K value, the proposed new algorithm ours obtains smaller MAE value compared with algorithm CFSSI.

4   Conclusions

Traditional personalized recommendation algorithms present shortcomings such as sparsity of rating data and without considering the semantic information. As a result, the application of traditional methods for measuring similarities brings about less accurate calculation results and therefore causes undesirable recommendations. To solve this problem, the authors comprehensively consider the influences of user-preference for genres of projects and the relevancy of projects on the calculation of similarities. By doing so, a personalized recommendation algorithm for user-preference similarity through the semantic analysis is put forth. By comparing the accuracy of the algorithm with those of others through a series of comparison, it is demonstrated that the algorithm proposed in the research can effectively improve the accuracy and quality of recommender systems. In the future, the proposed algorithm is expected to be verified and improved in website recommendations, so as to make it present more favorable recommendation effects.

References

1. Breese, J., Hecherman, D., Kadie, C.: Empirical analysis of predictive algorithms for collaborative filtering. In: Proceedings of the 14th Confon Uncertainty in Artificial Intelligence (UAI98), pp. 43–52. Morgan Kaufmann, San Francisco (1998)
2. Wallach, H.M.: Topic modeling: beyond bag-of-words. In: Proceedings of the 23rd International Conference on Machine Learning, pp. 977–984. ACM (2006)
3. Linden, G., Smith, B., York, J.: Amazon. com recommendations: Item-to-item collaborative filtering. IEEE Internet Comput. 7(1), 76–80 (2003)
4. Forman, G.: An extensive empirical study of feature selection metrics for text classification. J. Mach. Learn. Res. 3, 1289–1305 (2003)
5. Sarwar, B., Karypis, G., Konstan, J., et al.: Item-based collaborative filtering recommendation algorithms. In: Proceedings of the 10th International Conference on World Wide Web, pp. 285–295. ACM (2001)
6. You-shi, H., Cui-li, S.: Improved collaborative filtering recommendation based user's purchase records mining. Comput. Eng. Des. 35(9) (2014)
7. Miller, B.N., Albert, I., Lam, S.K., et al.: MovieLens unplugged: experiences with an occasionally connected recommender system. In: Proceedings of the 8th International Conference on Intelligent User Interfaces, pp. 263–266. ACM (2003)
8. Ai-Lin, D., Yang-Yong, Z., Bai-Le, S.: A collaborative filtering recommendation algorithm based on item rating prediction. J. Softw. 14(9) (2003)
9. Herlocker, J.L., Konstan, J.A., Terveen, L.G., et al.: Evaluating collaborative filtering recommender systems. ACM Trans. Inf. Syst. (TOIS) 22(1), 5–53 (2004)
10. Ming, X., Qian-Xing, X., Bai-Le, S.: Collaborative filtering recommendation algorithm based on semantic similarity between items. J. Wuhan Univ. Technol. 31(3) (2009)

Prediction

Learning-Based SPARQL Query
Performance Prediction

Wei Emma Zhang[1(✉)], Quan Z. Sheng[1], Kerry Taylor[2], Yongrui Qin[3],
and Lina Yao[4]

[1] School of Computer Science, The University of Adelaide, Adelaide , Australia
wei.zhang01@adelaide.edu.au
[2] Research School of Computer Science, Australian National University,
Canberra, Australia
[3] School of Computing and Engineering, University of Huddersfield, Huddersfield, UK
[4] School of Computer Science and Engineering, UNSW Australia, Sydney, Australia

Abstract. According to the predictive results of query performance,
queries can be rewritten to reduce time cost or rescheduled to the time
when the resource is not in contention. As more large RDF datasets
appear on the Web recently, predicting performance of SPARQL query
processing is one major challenge in managing a large RDF dataset efficiently. In this paper, we focus on representing SPARQL queries with
feature vectors and using these feature vectors to train predictive models
that are used to predict the performance of SPARQL queries. The evaluations performed on real world SPARQL queries demonstrate that the
proposed approach can effectively predict SPARQL query performance
and outperforms state-of-the-art approaches.

Keywords: SPARQL · Feature modeling · Prediction

1   Introduction

The Semantic Web, with its underlying data model RDF and its query language
SPARQL, has received increasing attention from researchers and data consumers
in both academia and industry. RDF essentially represents data as a set of three-
attribute tuples, i.e., triples. The attributes are *subject*, *predicate* and *object*,
where *predicate* is the relationship between *subject* and *object*. Over the recent
years, RDF has been increasingly used as a general data model for conceptual
description and information modeling. Since the number of publicly available
RDF datasets and their volume grows dramatically, it becomes essential to provide efficient querying of large scale RDF datasets. This is an important issue
in the sense that whether to obtain knowledge efficiently affects the adoption of
RDF data as well as the underlying Semantic Web technologies.

Substantial works focus on the prediction of query performance (e.g., execution time) [1,9,16]. Prediction of query execution performance can benefit many
system management decisions, including workload management, query scheduling, system sizing and capacity planning. Studies show that cost model based

© Springer International Publishing AG 2016
W. Cellary et al. (Eds.): WISE 2016, Part I, LNCS 10041, pp. 313–327, 2016.
DOI: 10.1007/978-3-319-48740-3_23

query optimizers are insufficient for query performance prediction [2,6]. Therefore, approaches that exploit the machine learning techniques to build predictive models have been proposed [2,6]. These approaches treat the database system as a black box and focus on learning a query performance predictive model, which are evaluated as feasible and effective [2]. These works extract the features of queries by exploring the query plan with which they can provide estimations for execution time and row count.

However very few efforts have been made to predict the performance of SPARQL queries. SPARQL query engines can be grouped into two categories: RDBMS-based triple stores and RDF native triple stores. RDBMS-based triple stores rely on optimization techniques provided by relational databases. However, due to the absence of schematic structure in RDF, cost-based approaches show problematic query estimation and cannot effectively predict the query performance [15]. RDF native query engines typically use heuristics and statistics about the data for selecting efficient query execution plans [14]. Heuristics and statistics based optimization techniques generally work without any knowledge of the underlying data, but in many cases, statistics are often missing [15]. Hassan [8] proposes the first work on predicting SPARQL query execution time by utilizing machine learning techniques. The key contribution of the work is to model a SPARQL query to a feature vector, which can be used in machine learning algorithms. However, in practice, we observe that modeling approach is very time consuming.

To address this issue, we leverage both syntactical and structural information of the graph-based SPARQL queries and propose to use the hybrid features to represent a SPARQL query. Specifically, we transform the algebra and BGPs of a SPARQL query into two feature vectors respectively and perform a feature selection process based on heuristic to build hybrid features from these two feature vectors. Our approach reduces the computation time of feature modeling in orders of magnitude. Once the features are built, we use machine learning algorithms to train the prediction model. The input of the algorithm is the feature matrix of the training queries (we concatenate the feature vectors of individual queries into a matrix) and the query performance of these queries (here we only consider the elapsed time used to perform a query and get the result). The output is the trained prediction model. When a new query q is issued, we obtain its feature vector using our feature modeling approach. Then we use the trained prediction model to predict the performance of q. K-Nearest Neighbors (k-NN) regression and Support Vector Regression (SVR) are both considered as the predictive model. We develop a two-step prediction process to improve the prediction result compared to one-step prediction. Moreover, we evaluate our approach on both cold (i.e., fresh queries) and warm (i.e., repeated queries) stages of the system. In the cold stage, elapsed time consists of both compile and execution time while in the warm stage, elapsed time equals to the execution time. The reason we can ignore the compile time is because that our work only considers static querying data. Thus a repeated query has the same execution plan each time it is issued and the system only compiles once.

The consideration of cold stage is useful as knowing execution performance for unseen queries is more important for system management than to previously seen queries.

Our approach can be applied in the situation that no estimation of query execution performance is provided, or such estimations are implicit or inaccurate. In practice, this applies to most triple stores that are publicly accessible. Moreover, no domain expertise is required. In a nutshell, the main contributions of this work are summarized as follows:

- We adopt machine learning techniques to predict the SPARQL query performance before their execution. We transform the SPARQL queries to feature vectors that are required by the machine learning algorithms. Hybrid feature modeling is proposed based on the features that can be obtained without the information of the underlying systems.
- We consider both warm and cold stage prediction, and the latter one has not been discussed in the state-of-the-art works, but is important to the examination of execution performance of a query.
- We perform extensive experiments on real-world queries obtained from widely accessed SPARQL endpoints. The triple store we used is one of the most widely used systems in the Semantic Web community. Thus our work can benefit a large population of users. Moreover, our approach is system independent that can be applied to other triple stores.

The remainder of this paper is structured as follows. Existing research efforts on the related topics are discussed in Sect. 2. In Sect. 3, the background knowledge is briefly introduced. Section 4 describes our prediction approaches in detail. Section 5 reports the experimental results. Finally, we discuss some issues we observed and conclude the paper in Sect. 6.

2   Related Work

There are limited previous works that pertain to predicting query performance via machine learning algorithms in the context of SPARQL queries. We introduce here the works of predicting SQL queries performances that we draw ideas from and discuss the work in [8], which focuses on SPARQL queries.

Akdere et al. [2] propose to predict the execution time using Support Vector Machine (SVM). They build predictors by leveraging query plans provided by the PostgreSQL optimizer. The authors also choose operator-level predictors and then combine the two with heuristic techniques. The work studies the effectiveness of machine learning techniques for predicting query latency of both static and dynamic workload scenarios. Ganapathi et al. [6] consider the problem of predicting multiple performance metrics at the same time. The authors also choose query plan to build the feature matrix. Kernel Canonical Correlation Analysis (KCCA) is leveraged to build the predictive model as it is able to correlate two high-dimension datasets. As addressed by the authors, it is hard to find a reverse mapping from feature space back to the input space and they consider the performance metric of k-NN to estimate the performance of target query. Hassan [8] proposes the first work on predicting SPARQL query execution time by utilizing machine learning techniques. In the work, multiple regression using SVR is adopted. The evaluation is performed using benchmark queries on an open source triple store Jena TDB[1]. The feature models are extracted based on *Graph Edit Distances* (GED) between each of training queries. However, in practice, we observe that the calculation of GED is very time consuming, which is not a desirable method when the training dataset is large. Our work draws idea from this work and improves it by largely reducing the computation time.

3   Preliminaries

3.1   SPARQL Query

A SPARQL query can be represented as a graph structure, the SPARQL graph [7]. Given the notation B for blank nodes, I for Internationalized Resource Identifier (IRIs), L for literals and V for variables, a SPARQL graph pattern expression is defined recursively (bottom-up) as follows [11]:

(i)   A valid triple pattern $T \in (IVB) \times (IV) \times (IVLB)$ is a *Basic Graph Pattern* (BGP), where a triple pattern is the triple that any of its attributes is replaced by a variable (A BGP is a graph pattern represented by the conjunction of multiple triple patterns. It models the SPARQL conjunctive queries and is the most widely used subset of SPARQL queries [4]).

(ii) For BGP_i and BGP_j, the conjunction (BGP_i and BGP_j) is a BGP.

(iii) If P_i and P_j are graph patterns, then (P_i AND P_j), (P_i UNION P_j) and (P_i OPTIONAL P_j) are graph patterns.

(iv) If P_i is a graph pattern and R_i is a SPARQL build-in condition, then the expression (P_i FILTER R_i) is a graph pattern.

3.2   Multiple Regression

Multiple regression focuses on finding the relationship between a dependent variable and multiple independent variables (i.e., predictors). It estimates the expectation of the dependent variable given the predictors. Given a training set $(\mathbf{x}_i, y_i), i = 1, ...n$, where $\mathbf{x}_i \in \mathbb{R}^m$ is a m-dimensional feature vector (i.e., m predictors), the objective of multiple regression is to discover a function $y_i = f(\mathbf{x}_i)$ that best predicts the value of y_i associated with each $\mathbf{x}_i$ [12].

Support Vector Regression is to find the best regression function by selecting the particular hyperplane that maximizes the margin, i.e., the distance between the hyperplane and the nearest point [13]. The error is defined to be zero when the difference between actual and predicted values are within a certain amount ξ. The problem is formulated as an optimization problem:

$$\min \mathbf{w}^T \mathbf{w}, \quad s.t. \quad y_i(\mathbf{w}^T \mathbf{x}_i + b) \geq 1 - \xi, \xi \geq 0 \qquad (1)$$

---

[1] https://jena.apache.org/documentation/tdb/.

where parameter $\frac{b}{\|\mathbf{w}\|}$ determines the offset of the hyper-plane from the origin along the normal vector $\mathbf{w}$. If we extend the dot product of $\mathbf{x}_i \cdot \mathbf{x}_j$ to a different space of larger dimensions through a functional mapping $\Theta(\mathbf{x}_i)$, then SVR can be used in non-linear regression. $\Theta(\mathbf{x}_i) \cdot \Theta(\mathbf{x}_j)$ is called kernel function. An advantage of SVR is its insensitivity to outliers [17].

K-Nearest Neighbors is a non-parametric classification and regression method [3]. The k-NN regression predicts based on k nearest training data. It is often successful in the cases where the decision boundary is irregular, which applies to SPARQL queries [8]. By training the k-NN model, the predicted query time can be calculated by the average time of its k nearest neighbors.

$$t_Q = \frac{1}{k} \sum_{i=1}^{k} (t_i) \tag{2}$$

where t_i is the elapsed time of the ith nearest query.

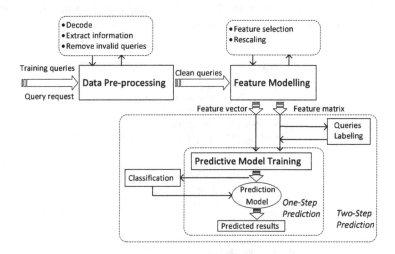

Fig. 1. Steps for query performance prediction

4   SPARQL Query Performance Prediction

Our prediction process consists of four main phases, namely *Data Pre-Processing*, *Feature Modeling*, *Predictive Model Training* and *Prediction* (one-step and two-step) (Fig. 1). Both training and new requested queries are cleaned in the *Data Pre-Processing* phase, valid queries are extracted during this phase. In the *Feature Modeling* phase, queries are represented as a set of features. In the *Predictive Model Training* phase, predictive models are derived from the training queries with observed query performance metrics. In the *Prediction* phase, trained predictive models are used to predict the performance of a new issued query. Compared to one-step prediction, the two-step prediction has labeling before predictive model training and classification step before prediction. We focus on discussion of feature modeling in Sect. 4.1 and describe the predictive models training and two-step prediction in Sect. 4.2. We ignore the description of data pre-processing due to the space constraint.

4.1 Feature Modeling

In order to utilize machine learning algorithms for SPARQL query performance prediction, we transform the SPARQL query into vector representation where each value in a vector is regarded as a feature instance of a query. The performance of prediction highly depends on how much information the features can represent the data. In this study, we use only static, compile time features that are extracted prior to execution. The algebra and BGP features are obtained by parsing the query text (Sects. 4.1.1 and 4.1.2). The hybrid features are generated by applying a selection algorithm on the algebra and BGP features (Sect. 4.1.3). We concatenate the feature vectors of a set of training queries and form a feature matrix as the input of learning algorithms.

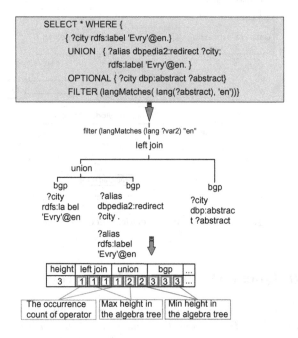

Fig. 2. Algebra feature modelling on example query

4.1.1   Algebra Feature

The algebra of a SPARQL query can be presented as a tree where the leaves are BGPs and nodes are operators presented hierarchically. The parent of each node is the parent operator of current operator. We traverse the tree to construct a set of tuples $\{(opt_i, c_i, maxh_i, minh_i)\}$, where opt_i is the operator name, c_i is the occurrence count of opt_i in the algebra tree, $maxh_i$ and and $minh_i$ are opt_i's maximum height and minimum height in the algebra tree, respectively. We then concatenate all the tuples sequentially to form a vector. We further insert the tree's height at the beginning of the vector. Figure 2 illustrates an example of algebra feature modelling.

Fig. 3. Example triple patterns

4.1.2   BGP Feature

Algebra features take occurrences and some hierarchical information of operators into consideration, but fail to represent BGPs, the most widely used subset of SPARQL queries [4]. To represent BGPs, we propose to build BGP features. We examine that BGPs consist of sets of triple patterns (Sect. 3) thus can also be represented as tree structure. But we choose not to use similar transformation approach (i.e., record occurrences and maximum/minimum heights) as in algebra feature modelling. Instead, we propose a new way to transform BPGs to vector representation.

Specifically, we leverage the edit distance between graphs to build BGP features because it can capture complete information of a BGP graph. Figure 3 illustrates the graph representation of two triple patterns *(?s, p, o)* and *(?s, p, ?o)* where the *subject(s)* and *object(o)* are nodes and *predicate(p)* are edges. The question mark indicates that the corresponding component is a variable. However, it is hard to tell the differences between the two graphs, as they are structurally identical. To address this problem, we propose to map the eight types of triple patterns to eight structurally different graphs, as shown in Fig. 4(left). The black circles are inner conjunction nodes. To exemplify, we model the triple patterns of BGPs in the example query in Fig. 2, to a graph, which is depicted in Fig. 4(right). The black rectangles are outer conjunction nodes.

We then calculate the Graph Edit Distance (GED)[2] between the graph of a query q and graphs of some representative queries and regard each distance as an instance of a feature. Thus we obtain a n-dimensional feature vector for q, where n is the number of representative queries. We choose to use the 18 valid out of

---

[2] Graph edit distance is the minimum amount of edit operations (i.e., deletion, insertion and substitutions of nodes and edges) needed to transform one graph to the other.

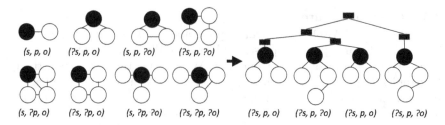

Fig. 4. Mapping triple patterns to graphs. Left: eight types of triple patterns are mapped to eight structurally different graphs. Right: mapping example query in Fig. 2 to a graph.

22 templates from DBPSB benchmark [10] to generate representative queries. We build the graph for each of the 18 queries and compute the GED between q and these graphs. Thus we obtain a 18-dimension feature vector for q.

4.1.3 Hybrid Feature

We build hybrid feature vector for q by selecting the most predictive features based on the algebra and BGP features. Most feature selection approaches rank the candidate features (often based on their correlations) and use this ranking to guide a heuristic search to identify the most predictive features. In this paper, we use a similar forward feature selection algorithm, but we choose the contribution to overall prediction performance as the heuristic. The algorithm starts with building predictive model (we use k-NN as the predictive model here) using a small number of features and iteratively build more complex and accurate model by using more features. In each iteration, a new feature is tested and added to the feature set. If it improves the overall prediction performance, the feature is selected. Otherwise, it is removed from the feature set. Finally, we simply consider the completion of traversing all features as the stopping condition. In this work, we considers BGP features first and then select features from algebra features. The output of the algorithm is the list of selected features that form the feature vector for each query.

4.2 Prediction

We propose two prediction processes, namely one-step prediction and two-step prediction. In the one-step prediction, feature vector of a new query is input into the trained predictive model obtained in the predictive model training phase. The output is the predicted value of the query performance metrics. The two-step prediction differs with one-step prediction by adding classification step. We present the predictive models used in this work in Sect. 4.2.1 and describe how we do two-step prediction in Sect. 4.2.2.

4.2.1   Predictive Models

We choose two regression approaches SVR and k-NN regression in this work (Sect. 3.2). The models are trained with the actual query performance of training queries and then be used to estimate the performance of a new issued query. Both models require the features vector-represented. We compare several variations of these two models in this work. The description is as follows.

SVR. Four commonly used kernels are considered in our model: namely *Linear*, *Polynomial*, *Radial Basis* and *Sigmoid*, with different kernel parameters γ and r:

– Linear: $K(\mathbf{x}_i, \mathbf{x}_j) = \mathbf{x}_i^{\mathbf{T}} \mathbf{x}_j$
– Polynomial: $K(\mathbf{x}_i, \mathbf{x}_j) = (\mathbf{x}_i^{\mathbf{T}} \mathbf{x}_j)^r$
– Radial Basis: $K(\mathbf{x}_i, \mathbf{x}_j) = exp(-\gamma ||\mathbf{x}_i - \mathbf{x}_j||^2), \gamma > 0$
– Sigmoid: $K(\mathbf{x}_i, \mathbf{x}_j) = tanh(\gamma \mathbf{x}_i^{\mathbf{T}} \mathbf{x}_j) + r$

k-NN. We apply three variations of k-NN regression by considering different weighting methods to the neighbors.

– *Average.* We assign equal weights to each of the k nearest neighbors and get the average of their elapsed time as the predicted time:

$$t_Q = \frac{1}{k} \sum_{i=1}^{k} (t_i) \qquad (3)$$

where t_i is the elapsed time of the ith nearest query.
– *Power.* The weights in *Power* is the power value of weighting scale α. The predicted query time is calculated as follows:

$$t_Q = \frac{1}{k} \sum_{i=1}^{k} (\alpha^i * t_i) \qquad (4)$$

where α^i is the weight of the ith nearest query.
– *Exponential.* We apply an exponential decay function with decay scale β to assign weights to neighbors with different distance.

$$t_Q = \frac{1}{k} \sum_{i=1}^{k} (e^{-d_i * \beta} * t_i) \qquad (5)$$

where d_i is the distance between target query and its ith nearest neighbor.

All the scaling parameters are chosen through 5-fold cross-validation.

4.2.2   Two-Step Prediction

We observe that the one-step prediction, where all the training data are fed into a single predictive model, gives undesirable performance. A possible reason is the fact that our training dataset has queries with various different elapsed time ranges. Fitting a curve for such irregular data points is often inaccurate. Then we propose a two-step prediction process, where we split queries according to their elapsed time and train different predictive models. Specifically, we firstly put the training queries in four bins, namely *short, medium short, medium*, and *long*. The time ranges in these four bins are <0.1 s, 0.1 to 10 s, 10 to 3,600 s, and >3,600 s respectively. We correspondingly label all the training queries with these four labels. Then we train four predictive models and one for each bin (or class). When a new query q arrives, we perform classification for q and obtain its label (or class). Here we use Support Vector Machine (SVM) as classification algorithm as it is the mostly used classification algorithm. Then we use the trained predictive model for the class that q is labelled to predict q's performance.

5   Experiments

5.1   Setup

Data. We used real world queries gathered from USEWOD challenge[3], which provides query logs from DBPedia's SPARQL endpoint[4] (DBpedia3.9). We randomly chose 10,000 valid queries in our prediction evaluation. Then these queries were executed 11 times as suggested in [7], including the first time as cold stage, and the remaining 10 times as the warm stage. Finally, we split the collection to training and test sets according to the 4:1 tradition. We set up a local mirror of DBpedia3.9 English dataset to execute the queries.

System. The backing system of our local triple store is Virtuoso 7.2, installed on 64-bit Ubuntu 14.04 Linux operating system with 32 GB RAM and 16 CPU. All the machine learning algorithms are performed on a PC with 64-bit Windows 7, 8 GB RAM and 2.40 GHZ Intel i7-3630QM CPU.

Implementation. We used SVR for kernel and linear regression available from LIBSVM [5]. k-NN and weighted k-NN regression was designed and implemented using Matlab. The algebra tree used for extracting algebra features was parsed using Apache Jena-2.11.2 library, Java API. Graph edit distance was calculated using the Graph Matching Toolkit[5].

Evaluation Metric. We followed the suggestion in [2] and used the *mean relative error* as our prediction metric:

$$relativeerror = \frac{1}{N} \sum_{i=1}^{N} \frac{|actual_i - estimate_i|}{actual_{mean}} \tag{6}$$

The difference with the calculation in [2] is that we divide $actual_{mean}$ instead of $actual_i$ because we observe there are zero values for $actual_i$.

---

[3] http://usewod.org/.

[4] http://dbpedia.org/sparql/.

[5] http://www.fhnw.ch/wirtschaft/iwi/gmt.

Table 1. Relative error of elapsed time prediction (one-step)

	Elapsed time (cold)	Elapsed time (warm)
SVR-Linear	99.69 %	97.59 %
SVR-Polynomial	99.46 %	97.33 %
SVR-RadialBasis	99.74 %	97.86 %
SVR-Sigmoid	99.68 %	97.57 %
k-NN ($k = 1$)	21.94 %	20.89 %

5.2 Models Comparison

We compared the Linear SVR and SVR with three kernels, namely Polynomial, Radial Basis and Sigmoid with k-NN when $k = 1$. The feature model used in the experiments was the hybrid feature. Table 1 shows the performance of the four models in one-step prediction. SVR models perform poorer than k-NN. We investigate this phenomenon and find two possible reasons. One is that the elapsed time has a broad range and SVR considers all the data points in the training set to fit the real value, whereas k-NN only considers the points close to the test point. The other reason is that we use mean of actual values in Eq. 6, and the values that are far from average will lead to distortion of mean value. Given this result, we chose to use k-NN model by default in the following evaluations.

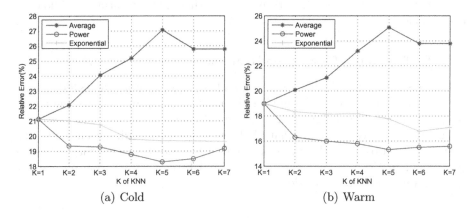

(a) Cold   (b) Warm

Fig. 5. Performance comparison of different weighted k-NN model (one-step)

We evaluated three weighting schemes for k-NN regression discussed in Sect. 4.2.1, namely *Average*, *Power* and *Exponential*. From Fig. 5 we observe that the power weighting gives the best performance. In the warm stage, the 15.32 % relative error is achieved when $k = 5$. The trend of relative error returns to upward after $k = 5$. Average weighting is the worst weighting method for our data. Exponential weighting does not perform as well as we expected although it

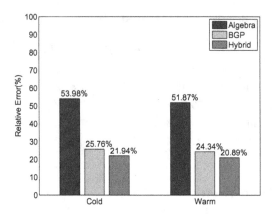

Fig. 6. Feature model selection (one-step)

is better than average weighting. Weighting schemes show similar performances when the query execution is in the cold stage, i.e., when $k = 5$, the power weighting achieves the lowest relative error of 18.29 %. We therefore used $k = 5$ power weighting in following evaluations.

We also compared the three feature models: *Algebra*, *BGP* and *Hybrid*. Figure 6 shows the prediction performance of elapse time on both warm and cold stages. The hybrid feature performs the best and the BGP feature performs better than the algebra feature. Thus we chose hybrid features in following evaluations.

5.3   Performance of Two-Step Regression

We used SVM for the classification task and achieved accuracy of 98.36 %, indicating that we can accurately predict the time range. Table 2 shows the result of two-step prediction comparison between k-NN and SVR-Polynomial on both warm and cold stage. It shows that SVR regression model still does not perform desirably. It also shows the two-step prediction performs better than one-step prediction. In Fig. 7 we compare one-step and two-step prediction on elapsed time in warm stage using log-log plotting.

Table 2. Relative errors (%) of two-step prediction with k-NN and SVR. In the parentheses are the values from one-step prediction

Predictive model	Elapsed time (cold)	Elapsed time (warm)
5-NN ($\alpha = 0.3$)	11.06 (21.94)	9.81 (20.89)
SVR-Polynomial	22.39	20.30

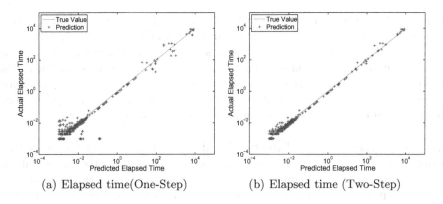

(a) Elapsed time(One-Step)   (b) Elapsed time (Two-Step)

Fig. 7. Elapsed time prediction fitting in warm stage (log-log)

5.4   Comparison to the State-of-the-Art

We compare the approach in the work [8] with our approach, as it is the only work that exploits machine learning algorithms to predict SPARQL query. Table 3 shows the result of comparison on warm stage querying. The training time includes feature modelling, clustering and classification for work in [8]. The first part takes the most time because the calculation of GED for all training queries is time-consuming. In our approach (Sect. 4.1.2), we reduce the GED calculation drastically. But this calculation still takes most time in the prediction process. The time gap of training process between ours and the approach in [8] will be enlarged when more training queries are involved because their approach takes squared time. We do not have clustering process, which further reduces the time used. Our approach also shows better prediction performance with lower relative error for the prediction metric.

Table 3. Comparison to the state-of-the-art work. Training times for 1000 queries (Time1k) are compared as well as the relative errors for elapsed time.

	Models	Time1k	Relative error
Ours	SVM + Weighted k-NN	51.36 s	9.81 %
[8]	X-means + SVM + SVR	1548.45 s	14.39 %

6   Discussions and Conclusion

In this section, we first discuss some observations and issues of this work. Then we conclude this paper.

Plan Features. There are two obstacles for using query plan as features in our work. Firstly, this information is based on the cost model estimation, which has been proven ineffective [2,6]. Secondly, most of the open source triple stores fail to provide explicit query plans. Thus we turn to choose structure-based features that can be obtained directly from query texts. From our practical experience in this work, we observed that although it leads to distortion of the prediction, the error rate is acceptable based on limited features we can acquire.

Training Size. Larger size of the training data would lead to better prediction performance. The reason is that more data variety is seen and the model will be less sensitive to unforeseen queries. However, in practice, it is time consuming to obtain the query elapsed time of a large collection of queries. That is the possible reason why many other works only use small size of queries in their evaluation. This fact will cause the bias of the prediction result and makes similar works hard to compare. Although our experimental query set is larger than theirs, we will consider to further enlarge the size of our query set to cover more various queries in the future.

Dynamic vs Static Data. In dynamic query workloads, the queried data is updated. Therefore, the prediction might perform poorly due to lack of update of the training data. Our work focuses on prediction on static data and we expect training to be done in a periodical manner. In the future we plan to investigate the techniques to make prediction more available for continuous retraining which reflects recently executed queries.

To conclude, in this paper, we build feature vectors for SPARQL queries by exploiting the syntactic and structural characteristics of the queries. We observe that k-NN performs better than SVR on predicting the elapsed time of real-world SPARQL queries. The proposed two-step prediction performs better than one-step prediction because it considers the broad range of observed elapsed time. The prediction in the warm stage is generally better than in the cold stage. We identify the reason comes from same structured queries because many queries are issued by programmatic users, who tend to issue queries using query templates. Our work is on static data and we will consider dynamic workload in the future. Techniques that can incorporate new training data into an existing model will also be considered.

References

1. Ahmad, M., Duan, S., Aboulnaga, A., Babu, S.: Predicting completion times of batch query workloads using interaction-aware models and simulation. In: Proceedings of the 14th International Conference on Extending Database Technology (EDBT 2011), Uppsala, pp. 449–460, March 2011
2. Akdere, M., Çetintemel, U., Riondato, M., Upfal, E., Zdonik, S.B.: Learning-based query performance modeling and prediction. In: Proceedings of the 28th International Conference on Data Engineering (ICDE 2012), Washington, DC, pp. 390–401, April 2012

3. Altman, N.S.: An introduction to kernel and nearest-neighbor nonparametric regression. Am. Stat. **46**(3), 175–185 (1992)
4. Bursztyn, D., Goasdoué, F., Manolescu, I.: Optimizing reformulation-based query answering in RDF. In: Proceedings of the 18th International Conference on Extending Database Technology (EDBT 2015), Brussels, pp. 265–276, March 2015
5. Chang, C., Lin, C.: LIBSVM: a library for support vector machines. ACM Trans. Intell. Syst. Technol. **2**(3), 27 (2011)
6. Ganapathi, A., Kuno, H.A., Dayal, U., Wiener, J.L., Fox, A., Jordan, M.I., Patterson, D.A.: Predicting multiple metrics for queries: better decisions enabled by machine learning. In: Proceedings of the 25th International Conference on Data Engineering (ICDE 2009), Shanghai, pp. 592–603, March 2009
7. Gubichev, A., Neumann, T.: Exploiting the query structure for efficient join ordering in SPARQL queries. In: Proceedings of the 17th International Conference on Extending Database Technology (EDBT 2014), Athens, pp. 439–450, March 2014
8. Hasan, R.: Predicting SPARQL query performance and explaining linked data. In: Presutti, V., d'Amato, C., Gandon, F., d'Aquin, M., Staab, S., Tordai, A. (eds.) ESWC 2014. LNCS, vol. 8465, pp. 795–805. Springer, Heidelberg (2014). doi:10. 1007/978-3-319-07443-6_53
9. Li, J., König, A.C., Narasayya, V.R., Chaudhuri, S.: Robust estimation of resource consumption for SQL queries using statistical techniques. VLDB Endow. (PVLDB) **5**(11), 1555–1566 (2012)
10. Morsey, M., Lehmann, J., Auer, S., Ngomo, A.N.: Usage-centric benchmarking of RDF triple stores. In: Proceedings of the 26th AAAI Conference on Artificial Intelligence, Toronto, July 2012
11. Pérez, J., Arenas, M., Gutierrez, C.: Semantics and complexity of SPARQL. ACM Trans. Database Syst. **34**(3), 16 (2009)
12. Rajaraman, A., Ullman, J.D.: Mining of Massive Datasets. Cambridge University Press, Cambridge (2011)
13. Smola, A., Vapnik, V.: Support vector regression machines. Adv. Neural Inf. Process. Syst. **9**, 155–161 (1997)
14. Stocker, M., Seaborne, A., Bernstein, A., Kiefer, C., Reynolds, D.: SPARQL basic graph pattern optimization using selectivity estimation. In: Proceedings of the 17th International World Wide Web Conference (WWW 2008), Beijing, pp. 595–604, April 2008
15. Tsialiamanis, P., Sidirourgos, L., Fundulaki, I., Christophides, V., Boncz, P. A.: Heuristics-based query optimisation for SPARQL. In: Proceedings of the 15th International Conference on Extending Database Technology (EDBT 2012), Uppsala, pp. 324–335, March 2012
16. Wu, W., Chi, Y., Zhu, S., Tatemura, J., Hacigümüs, H., Naughton, J.F.: Predicting query execution time: are optimizer cost models really unusable? In: Proceedings of the 29th International Conference on Data Engineering (ICDE 2013), Brisbane, pp. 1081–1092, April 2013
17. Wu, X., Kumar, V., Quinlan, J.R., Ghosh, J., Yang, Q., Motoda, H., McLachlan, G.J., Ng, A.F.M., Liu, B., Yu, P.S., Zhou, Z., Steinbach, M., Hand, D.J., Steinberg, D.: Top 10 algorithms in data mining. Knowl. Inf. Syst. **14**(1), 1–37 (2008)

Can Online Emotions Predict the Stock Market in China?

Zhenkun Zhou[1], Jichang Zhao[2(✉)], and Ke Xu[1]

[1] State Key Laboratory of Software Development Environment,
Beihang University, Beijing, China
[2] School of Economics and Management, Beihang University, Beijing, China
jichang@buaa.edu.cn

Abstract. Whether the online social media, like Twitter or its variant Weibo, can be a convincing proxy to predict the stock market has been debated for years, especially for China. However, as the traditional theory in behavioral finance states, the individual emotions can influence decision-makings of investors, so it is reasonable to further explore this controversial topic from the perspective of online emotions, which is richly carried by massive tweets in social media. Surprisingly, through thorough study on over 10 million stock-relevant tweets from Weibo, both correlation analysis and causality test demonstrate that five attributes of the stock market in China can be competently predicted by various online emotions, like disgust, joy, sadness and fear. Specifically, the presented model significantly outperforms the baseline solutions on predicting five attributes of the stock market under the K-means discretization. We also employ this model in the scenario of realistic online application and its performance is further testified.

Keywords: Social media · Stock market · Sentiment analysis · Causality test

1 Introduction

With explosive development of online social media, tremendous amounts of tweets are posted and reposted in popular platforms like Twitter and Weibo. These tweets, spreading in terms of word-of-mouth, not only convey the factual information, but also reflect the emotional statuses of the authors. Taking Weibo as an example, around 100 million Chinese tweets are posted every day and from which we can not only sense what happens in China, but also how 500 million users feel about their lives. In fact, the online social media indeed provide us an unprecedented opportunity to study the detailed human behavior from many new views. The investment decision in the stock market, as one of the most important issues, attracts much attention in recent decades.

However, whether online social media like Twitter can be excellent predictors is still controversial, especially for the stock market in China [2,9,16]. Different from the west, the marketing policy intervention in China will introduce

© Springer International Publishing AG 2016
W. Cellary et al. (Eds.): WISE 2016, Part I, LNCS 10041, pp. 328–342, 2016.
DOI: 10.1007/978-3-319-48740-3_24

more non-market factors that might disturb the fluctuation of the stock market. And moreover, those possible interventions could be leaked through the social media and then greatly influence the investors' emotions and decisions. In the mean time, considering the irrationality of huge amount of individual investors in China (which is also rare in the west), their actions might be more easily affected by online news and other investors' feelings about the market. Then the messages about the stock market and the sentiments they convey could be good indicators for the market prediction. Thus, like the conventional behavioral finance theory claims, which the emotion can influence the decision-process of the investors, it is necessary to investigate the following important issues:

- Is there indeed significant correlation between online emotions and attributes of Chinese stock market?
- Can online emotions predict the attributes of the stock market in China?
- Which emotion does play the critical role in predicting various attributes of the Chinese stock market?

In the present study, we collect over 10 million Chinese stock-relevant tweets from Weibo and classify them into five emotions, including anger, disgust, joy, sadness and fear. Besides the daily closing index of Shanghai Stock Exchange[1], we consider the daily opening index, the intra-day highest index, the intra-day lowest index and the daily trading volume of the stock market. By both correlation analysis and Granger causality test, it is revealed that disgust has a Granger causal relation with the closing index, joy, fear and disgust have Granger causal relations with the opening index, joy, sadness and disgust have Granger causal relations with the intra-day highest and lowest index, and correlation between trading volume and sadness is unexpectedly strong. It's also surprising to find that anger in online social media possesses the weakest correlation or even is no relation with the Chinese stock market.

Based on the findings, we develop classification-oriented predictors, in which different emotions are selected as features, to predict five daily attributes of the stock market in China. The comparison with other baseline methods show that our model can outperform them according to K-means discretization. And the model is also deployed in a realistic application and achieves the accuracy of 64.15 % for the intra-day highest index (3-categories) and the accuracy of 60.38 % for the trading volume (3-categories). Our explorations demonstrate that the online emotions, specially disgust, joy, sadness and fear, in Weibo indeed can predict the stock market in China.

2   Related Works

Emotion expression and stock fluctuation are usually bonded together in the traditional theory and even in social media. Behavioral economics studies the

---

[1] In the paper, index refers in particular to Shanghai Stock Exchange Composite Index and the trading volume refers in particular to the daily volume of the Shanghai Stock Exchange.

effects of social, emotional and psychological factors on the economic decisions of individuals and institutions and the consequences for market prices. It demonstrates that mood can affect individual behavior and decision-makings of investors [6,10].

Owing to lack of effective measurement method of sentiment, stock prediction using emotions had been in dispute [1,4]. However, with the recent widespread presence of computers and Internet, public emotions can be extracted from data on online platforms. Using Twitter as a corpus, some researchers built sentiment classifiers, which are able to determine different sentiments for a tweet [13] [12] [8]. Specially on Sina Weibo platform, Zhao et al. trained a fast Naive Bayes classifier for Chinese emotion classification, which is now available online for temporal and spatial sentiment pattern discovery [18].

In addition, there have long been controversies on predictive power of social media aiming at different fields [7,14]. In the field of finance, Bollen et al. found that public mood on Twitter could predict the Dow Jones Industrial Average [2]. The public mood dimensions of Calm and Happiness seemed to have a predictive effect. However, the tweets they collected were associated with whole social status, not just the stock market in America, which could not represent online investors' sentiment. Oh et al. also showed stock micro-blog sentiments did have predictive power for market-adjusted returns. Instead of emotion on social media [11], some researchers examined textual representations in financial news articles for stock prediction [5,15]. Ding et al. proposed a deep learning method for event-driven stock market prediction on large-scale financial news dataset [17]. Besides, Bordino et al. showed that daily trading volumes of stocks traded in NASDAQ100 were correlated with daily volumes of queries related to the same stocks [3].

However, to the best of our knowledge, existing studies referring to the stock market in China are relatively few. Mao et al. pointed out that Twitter did not have a predictive effect, as regards to predicting developments in Chinese stock markets [9]. They advised adopting the tweets on Weibo platform to research Chinese stock market. Based on 66,317 tweets of Weibo with one year and two emotion categories, Cheng and Lin found that the investors' bullish sentiment of social media can help to predict trading volume of the stock market, but still does not work for the market returns [16]. Because of less collected data set and simple emotion classification, it is not easy to generalize their conclusions to other realistic scenarios.

While in this paper, we focus purely on the stock market in China and try to understand the predictive ability of multiple online emotions in Weibo. Different from the previous study, we hope to develop predictors from more data sets and more sentiments and to predict more attributes of the real market.

3   Data Sets

3.1   Online Stock Market Emotions

The feelings of investors can be collected through many different approaches, like questionnaires in previous study. While with the explosive development of the social media in China, more and more investors express their seeings, hearings and feelings on Weibo. Therefore, we choose and utilize the characteristic of Weibo to obtain online emotion referring to Chinese stock market.

From December 1st 2014 to December 7th 2015, the massive public tweets on Weibo are collected through its open APIs. However, only a fraction of the tweets are semantically related with Chinese stock market. Filtering out the irrelevant tweets and remaining the data that truly represents the stock market emotion is a very significant step. Therefore, we manually select six Chinese keywords, including Stock, Stock Market, Security, The Shenzhen Composite Index, The Shanghai Composite Index and Component Index with help of expertise from the background of finance. These manually selected keywords are supposed to depict the overall status of Chinese stock market sufficiently. We postulate that if the text of tweet contains one or more of the six selected keywords, the tweet describes the news, opinions or sentiments about Chinese stock market. In our database, the number of tweets related to stock market, involving one or more keywords, is a total of 10,550,525 from December 1st 2014 to December 7th 2015.

In the paper, the emotions are divided into five categories, including anger, sadness, joy, disgust and fear. In our previous work [18], a fast Naive Bayes classifier is trained on Weibo data for emotion classification. The system named MoodLens whose vital part is the emotion classifier, is now available online for temporal and spatial sentiment pattern discovery. We arrange the tweets related to stock market, with one day as the time unit, and employ the system to label them with the emotions. There are five online emotion time series: $X_{anger}, X_{sadness}, X_{joy}, X_{disgust}$ and X_{fear}. Online emotions are represented by $X = (X_{anger}, X_{sadness}, X_{joy}, X_{disgust}, X_{fear})$.

Observing the time series, the volume of tweets reduces significantly on non-trading days. Figure 1 shows the volume of the tweets related to the stock market from September 1st to 16th in 2015. There are separately Memorial Day between September 3rd and 5th and a weekend between September 12th and 13th, which are both non-trading days. We consider that online stock market emotion on non-trading days could not help us analyze and predict Chinese stock market. Hence, removing the data items on non-trading days from the time series, the results retain significant emotion data. It also partly reflects that tweets selected by the keywords could represent the stock market.

For the sake of stability of online emotion data, we measure the relative value (proportion) of each mood on one day as the final online stock market emotion X. Figure 2 shows online stock market emotion time series X from September 1st to 16th in 2015. We observe the spike in X_{anger} on May 12th in 2015, when there is a heated argument between CEOs of listed companies. On June 19th in 2015, there was a plunge in Chinese stock market, with a fall of the index with

Fig. 1. Volume of the tweets related to the stock market from September 1st to 16th in 2015. There are respectively Memorial Day day between 9–3 and 9–5 and a weekend between 9–12 and 9–13, which are also non-trading days.

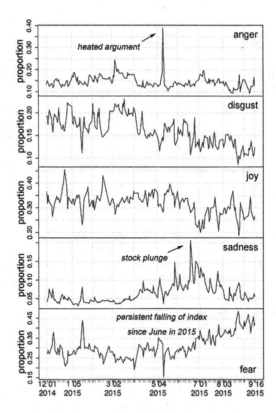

Fig. 2. Time series of each online stock market emotion from December 1st 2014 to September 16th 2015.

6.41 % and $X_{sadness}$ arrived the maximum. Since June in 2015, persistent falling of the index caused inward fears of investors, which can be seen from the sharp growth of X_{fear} in Fig. 2.

From the above observations, it can be concluded that the fluctuation of the sentiments can be connected with remarkable events in the stock market. It further inspires us to investigate the correlation and even causality between emotions and the market, which will provide the foundation for the predicting models.

3.2 Stock Market Data

In China, the economists and traders regard the Shanghai Stock Exchange Composite Index as reflecting the overall status of the Chinese stock market. Therefore, the index is selected as price attribute of the stock market to analyze and predict. In particular, there are four values in candlestick charts of the index, which are respectively the closing index, the opening index, the intra-day highest index, the intra-day lowest index. We transform the values of the index into *Close*, *Open*, *High* and *Low* (to express rate of change on i-th day), and they can be written as

$$
\begin{aligned}
Close_i &= \frac{Index_{close,i} - Index_{close,i-1}}{Index_{close,i}} \times 100, \\
Open_i &= \frac{Index_{open,i} - Index_{close,i-1}}{Index_{close,i}} \times 100, \\
High_i &= \frac{Index_{high,i} - Index_{close,i-1}}{Index_{close,i}} \times 100, \\
Low_i &= \frac{Index_{low,i} - Index_{close,i-1}}{Index_{close,i}} \times 100.
\end{aligned}
\tag{1}
$$

In addition to these four attributes, the trading volume of Shanghai Stock Exchange is also a key target used to reflect the status of the Chinese stock market. The time series of trading volume on each day is not transformed at all.

We crawl historical data of the index and trading volume from December 1st 2014 to December 7th 2015. In this period, the number of trading days is totally 249 in our research. As a result, we obtain five time series which depict stock market's state on each day including $Y_{close}, Y_{open}, Y_{high}, Y_{low}$ and Y_{volume}. Each time series is a column vector of Y (shown in Fig. 4), i.e., $Y = (Y_{close}, Y_{open}, Y_{high}, Y_{low}, Y_{volume})$.

The dataset (X and Y) is divided into two parts according to the date: the 80 % data for training (from December 1st 2014 to September 16th 2015) and the 20 % data for testing (from September 17th to December 7th in 2015). The training set is used to not only analyze the relation between online emotions and the stock market but also fit and estimate the prediction model. The testing set is kept in a vault and brought out only at the end of evaluation in realistic application.

4   Correlation Between Online Emotions and the Stock Market

The preceding part of the paper describes the two groups of time series (in the training set): X (represents online stock market emotions) and Y (represents the stock market), which contribute to discuss the correlation between online emotions and the stock market. However, the purpose of the paper is to find out whether online emotions can predict the stock market in China. Supposing that online emotions ahead of 1 to 5 days are available for stock prediction, we shifted emotion series to an earlier date: 1 to 5 days. Hence, each emotion corresponds to 5 time series according to shifted time. Each category of online emotions can be defined as (the categories of emotions are represented by e, $e = anger, sadness, joy, disgust,$ or $fear$) $X_e = (X_{e,1}, X_{e,2}, X_{e,3}, X_{e,4}, X_{e,5})$.

For the analysis the relation of X and Y (T represents one certain time series of X or Y), we normalize all the time series, of which data items are transformed to the values from 0 to 1 as

$$T_i = \frac{T_i - T_{min}}{T_{max} - T_{min}}, \tag{2}$$

T_i is the i-th item in time series T, T_{max} is the maximal value of T, and T_{min} is the minimal value of T. Then, by using Pearson correlation analysis, we measure the linear dependence between x ($X_{e,t}$, the emotion e ahead of t days in X) and y (one target of Y) as Eq. (3). ρ is the Pearson correlation coefficient of time series x and y defined as

$$\rho = \frac{\Sigma(x_i - \bar{x})(y_i - \bar{y})}{\sqrt{\Sigma(x_i - \bar{x})^2 \Sigma(y_i - \bar{y})^2}}. \tag{3}$$

For observing whether there are distinct differences of correlation coefficient between online emotions and the stock market, the emotion time series associated with stock market time series are sampled 100 times. In one time, we sample randomly 150 pairs (from 191 pairs in the training set) of data items respectively from emotion time series and stock market time series. We calculate 100 sampling results' correlation coefficient, and then obtain the mean values and standard deviations. Figure 3 shows the means and error bars, which depicts sampling results' correlation coefficient. It can be seen that there are significant differences between different emotions.

In addition, we randomly shuffle the time series 100 times and calculate the Pearson correlation coefficient of them. Comparing the mean coefficients with that of non-shuffled time series, we find that all the correlation coefficients (shuffled) are near 0, and it suggests that most of correlation coefficients (not shuffled) are relatively higher than random results, indicating the significance of the correlation we find.

Through the correlation coefficients above (shown in Fig. 3), we set the threshold of correlation coefficient ρ as 0.2 (the absolute value) and find some interesting and valuable results. The correlation between all online emotion time

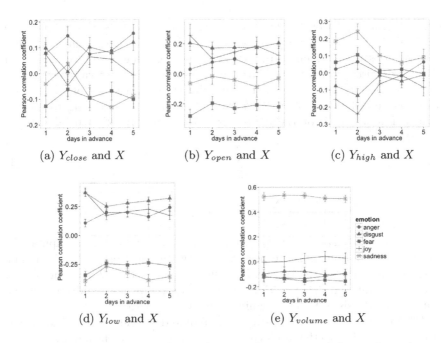

Fig. 3. Pearson correlation coefficient between five targets of the stock market and online emotion time series.

series (in X) and Y_{close} is very low ($\rho < 0.2$), which indicates little linear dependence between them. As to Y_{open}, the correlation coefficients with X_{fear} (ahead of 1, 3, 4 and 5 days), X_{joy} (ahead of 1 day) and $X_{disgust}$ (ahead of 1 and 5 days) are more than 0.2. Y_{open} is negatively correlated with X_{fear}, positively correlated with X_{joy} and $X_{disgust}$. As to Y_{high}, the correlation coefficients with X_{joy} (ahead of 2 days) and $X_{sadness}$ (ahead of 2 days) are more than the threshold. Y_{high} is negatively correlated with X_{joy}, positively correlated with $X_{sadness}$. Y_{low} and 5 types of emotion time series have relatively high correlation, and the correlation coefficients between Y_{low} and $X_{sadness}$ (ahead of 1 and 4 days) is the highest ($|\rho| > 0.4$). Y_{low} is negatively correlated with X_{anger}, $X_{disgust}$ and X_{joy}, positively correlated with X_{sad} and X_{fear}. An interesting finding is that the correlation between Y_{volume} and $X_{sadness}$ (no matter ahead of how many days) is unexpectedly high, correlation coefficients ρ of which is more than 0.5. Besides, Y_{volume} and other online emotion time series don't have a comparatively strong ($\rho > 0.2$) correlation.

5   Granger Causality Test of Online Emotions and the Stock Market

Despite the correlation analysis, we also preform the causality test further on the training data. Here we apply the econometric approach named Granger causality test to study the relation between online emotions and the stock market. The Granger Causality Test is a statistical hypothesis test for determining whether one time series is functioning in forecasting another. One time series x is said to Granger-cause y if it can be shown that x provides statistically significant information about future values of y, usually through a series of t-tests and F-tests on lagged values of x. We perform the analysis according to models shown in Eqs. 4 and 5 for the period from December 1st 2014 to September 16th 2015.

$$y_t = \alpha + \sum_{i=1}^{n} \beta_i y_{t-i} + \epsilon_t, \tag{4}$$

$$y_t = \alpha + \sum_{i=1}^{n} \beta_i y_{t-i} + \sum_{i=1}^{n} \gamma_i x_{t-i} + \epsilon_t. \tag{5}$$

The Granger causality test could select only two time series as inputs. We apply Granger causality test respectively on two groups: online emotion time series and stock market time series. Delaying time is set to 1, 2, 3, 4 and 5 days. According to different delaying time, we calculate the p-value to determine the results of hypothesis test. Here, significance level is set to 5 %.

We list testing results whose p-values are required to different significant levels in Table 1. According to the results of Granger causality test, the null hypothesis, $X_{disgust}(lag = 1, 2)$ series do not predict Y_{close}, with a high level of confidence (p-value < 0.01) can be rejected. However, the other emotions do not have causal relations with Y_{close}. Y_{open} and X_{joy} (p-value < 0.001), X_{fear} (p-value < 0.001) and $X_{disgust}$ (p-value < 0.05 or even 0.01) have causal relations. X_{joy}, $X_{sadness}$ and $X_{disgust}$ have causal relations with Y_{high} and Y_{low} (p-value < 0.05 or even 0.01). At last, the results also suggest trading volume in stock market time series do not have significant causal relation with any emotion time series (p-value ≥ 0.05). It's surprising to find that X_{anger} in online emotion time series does not have causal relation with any attribute of stock market in China.

The above analysis shows that $X_{disgust}$, X_{joy}, $X_{sadness}$ and X_{fear} can be promising features for the stock prediction models, except for X_{anger}.

6   Predict the Stock Market

Firstly, in this section, based on discretization methods, regression problems of predicting the stock market are converted to corresponding classification problems. Next, we perform linear and non-linear methods to solve the classification problems of stock market prediction. Eventually, the classification models are validated by 5-fold cross-validation on training set and we obtain a group of high-performance prediction models named SVM-ES.

For the prediction issue, we make use of the online emotion time series set (composed by shifted time series with different lags ranging from 1 to 5 for five emotions) or its subsets within the period from December 1st 2014 to September 9th 2015. Setting the longest lag to 5 trading days, the actual stock market time series are Y from December 8th 2014 to September 16th 2015.

Table 1. Results of Granger causality test of online emotion and stock market time series. Only significant results are listed because of the limited space. p-value < 0.05: *, p-value < 0.01: **, p-value < 0.001.

emotion	lag (days)	Close	Open	High	Low	Volume
anger	1					
	2					
	3					
	4					
	5					
disgust	1	0.0057**			0.0322*	
	2	0.0062**				
	3			0.0067**		
	4			0.0190*		
	5				0.0280*	
joy	1			0.0005***	0.0234*	
	2			6.e − 5***	0.0304*	
	3			2.e − 5***	0.0087**	0.0058**
	4			7.e − 5***	0.0385*	0.0352*
	5			0.0006***		
sadness	1				0.0115*	0.0272*
	2				0.0224*	
	3				0.0303*	
	4					
	5					
fear	1			0.0001***		
	2			2.e − 5***		
	3			3.e − 6***		
	4			6.e − 6***		
	5			0.0002***		

6.1 Discretization of Stock Market Data

As illustrated in the previous sections, $Y_{close}, Y_{open}, Y_{high}, Y_{low}$ and Y_{volume} are our targets of prediction in the stock market. Investors always just care for whether $Y_{close,i}$ (the element on i-th day in Y_{close}) are positive or negative, which will help investors make decisions to conduct stock transactions, and the binary classification (positive or negative) of Y_{close} and Y_{open} are also the part of our targets for prediction.

Besides, we convert regression problems of predicting five attributes in the stock market to classification problems by discretization methods through which we classify each of attributes to three categories. Specifically, $Y_{close}, Y_{open}, Y_{high}$, and Y_{low} are divided into three categories: **bearish**(-1), **stable**(0) and **bullish**(1) represented by CLOSE, OPEN, HIGH and LOW below. Y_{volume} are divide into three categories: **low**(-1), **normal**(0) and **high**(1) represented by VOLUME below.

The discretization of five attributes in the stock market is conducted by two methods: equal frequency and K-means clustering. Equal frequency discretization is a simple but effective method that we sort items from large to small then cut them into 3 clusters of even size. K-means clustering, another method we use, is popular for cluster analysis in data mining. In this paper, K-means clustering aims to partition observations of the stock market into 3 clusters in which each observation belongs to the cluster with the nearest distance. The results of three categories discretization by K-means are shown in Fig. 4 with 3 different grey levels.

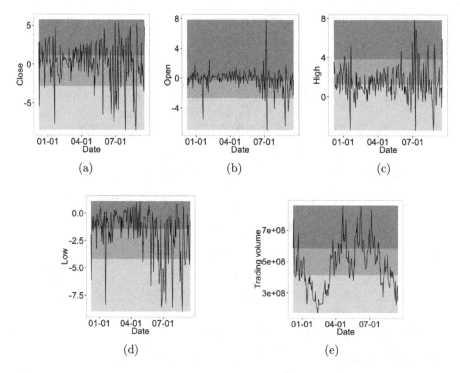

(a)   (b)   (c)

(d)   (e)

Fig. 4. Stock market time series and discretization results (by K-means) of $Y_{close}, Y_{open}, Y_{high}, Y_{low}$ and Y_{volume}.

Intuitively, as compared to the approach of equal frequency, the discretization based on K-means is more flexible and adjustable to the dynamic of the market. The categories it generates can better reflect the actual market status and thus can offer us a better benchmark to test the prediction results.

6.2   Classification Model for Stock Prediction

In this paper, we perform machine learning methods, Logistic Regression (linear) and Support Vector Machine (non-linear), to solve the classification problems for stock prediction. These methods are both popular for training binary or multiple classification. To predict the categories $(-1, 0, 1)$ or $(0, 1)$ (just for CLOSE and OPEN) of Y on i-th day, the input attributes of our Logistic Regression model (LR) and Support Vector Machine model (SVM) include only online emotion values of the past 5 days or a subset of them, except for other variables in the field of finance. We adapt 5-fold cross-validation to examine the accuracies of models.

At the outset, we consider all five emotions of the past 5 days as the input attributes of LR and SVM. The accuracies of models by 5-fold cross-validation are shown in Table 2 (3-categories and 2-categories). For the classification problem in this paper, the performance of SVM is always better than that of LR. Therefore, we conjecture that, relation between online emotions and the stock market is not simply linear, and the relation is more likely complicated and nonlinear.

While 3-categories discretization $(-1, 0, 1)$ results in the stock market as predicted targets, K-means clustering is always better than equal frequency discretization. In other words, the accuracies of models by 5-fold cross-validation, of which predicted targets are the results by K-means clustering, are relatively higher. Considering the categories generated by K-means discretization better represent the market status, we can conclude that our models indeed capture the essence of the stock fluctuation.

However, recalling the correlation analysis and Granger causality test of online emotions and the stock market, not all the emotions play roles on predicting the stock market and the analysis results should be used for the feature selection. Consequently, we build support vector machine model based emotions selected (SVM-ES) for stock prediction (discretized by K-means). The input attributes are based on analysis results of Granger causality test and Pearson correlation. We select $X_{disgust}$ (ahead of 1, 2 days) for the SVM-ES to predict CLOSE, X_{fear} (ahead of

Table 2. Accuracies of 5-fold cross-validation for 3-categories and 2-categories prediction models.

Target (3)	equal frequency		K-means		
	LR	SVM	LR	**SVM**	**SVM-ES**
CLOSE	34.0%	43.5%	52.9%	**58.1%**	57.6%
OPEN	37.7%	44.0%	53.4%	61.3%	**64.4%**
HIGH	36.7%	39.3%	48.7%	53.4%	**54.5%**
LOW	42.4%	49.2%	57.0%	63.4%	**64.4%**
VOLUME	50.8%	63.9%	53.4%	**67.0%**	66.5%

Target (2)	LR	SVM	SVM-ES
CLOSE	58.1%	**61.3%**	60.2%
OPEN	58.1%	**66.0%**	64.9%

1–5 days), X_{joy} (ahead of 1–5 day) and $X_{disgust}$ (ahead of 3 and 4 days) as the input attributes for predicting OPEN, X_{joy} (ahead of 1-4 days), $X_{sadness}$ (ahead of 1–3 day) and $X_{disgust}$ (ahead of 5 days) as the input attributes for predicting HIGH, and $X_{sadness}$ (ahead of 1 day), X_{joy} (ahead of 1–3 day) and $X_{disgust}$ (ahead of 5 days) as the input attributes for predicting LOW. Correlation analysis of Y_{volume} indicates that Y_{volume} and $X_{sadness}$ (ahead of 1–5 days) have the strongest correlation ($\rho > 0.5$) among all online emotions, however, just using sadness as the learning feature surprisingly can not guarantee the expected performance. Thus, we try to select $X_{sadness}$ (ahead 1–5 days) and X_{fear} (ahead 1–5 days) which is the second strongest relation with Y_{volume} as the input attributes to predict VOLUME.

After adjusting and fixing the input attributes of SVM-ES, we train the models for stock prediction. The last column of Table 2 shows the accuracy of 5-fold cross-validation, respectively for 3-categories and 2-categories classification models. There are slight differences in performance between SVM-ES and the SVM trained using all the emotions, indicating emotions selected are playing dominant roles in forecasting the market. It is noteworthy that input attributes of all the SVM-ES don't include anger and it's surprising that anger shown in online social media possesses the weakest correlation or even no relation with the Chinese stock market.

From Table 2 it should be also noted that, emotions selected can boost the classification results attributes like OPEN, HIGH and LOW, while for CLOSE and VOLUME, SVM with all emotions as features is still the most competent solution, with slight increment (around 1 %) to SVM-ES (few attributes of input). This result explains that emotions except for input attributes of SVM-ES have very weak effects on the stock market prediction.

6.3   Evaluation in Realistic Application

For further evaluating our prediction models, we sustain collecting stock-relevant tweets on Weibo with APIs and process them so as to obtain online emotion time series as our testing set from September 17th to December 7th in 2015. Then we apply our classification models SVM-ES for stock prediction in the realistic Chinese stock market and we can get the daily predictions of five attributes before the market open. Framework of realistic application based on SVM-ES is demonstrated in Fig. 5. We evaluate the stock market prediction application and the accuracies are shown in Table 3. It turns out that the model achieves the high prediction performance, especially with accuracy of 64.15 % for the intraday highest index (3-categories) and the accuracy of 60.38 % for the trading volume (3-categories).

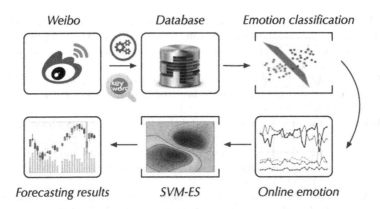

Fig. 5. Framework of realistic application for stock prediction based on SVM-ES.

Table 3. Accuracies of SVM-ES on realistic application.

CLOSE (3)	OPEN (3)	HIGH (3)	LOW (3)	VOLUME (3)	CLOSE (2)	OPEN (2)
56.60 %	43.40 %	64.15 %	56.60 %	60.38 %	60.38 %	56.60 %

7   Conclusion

In this paper, we collect massive tweets in Weibo with five categories of sentiments and focus on the stock market in China. The correlation analysis and Granger causality test are performed, which suggest that several emotions can be directly used to predict the market. Based on this, we establish several models to predict the closing index, the opening index, the intra-day highest index, the intra-day lowest index and trading volume. The results show that our model SVM-ES can outperform baseline solutions. Finally, we also testify its performance in the realistic application. In conclusion, our findings in this paper confirm that the stock market in China can be predicted by various online emotions including disgust, joy, sadness and fear.

This study has inevitable limitations, which might be interesting directions in the future work. For example, the detailed connection between the emotion and the market still remains unclear and how it evolves with time is also not discussed, however, which could help to design incremental learning schemes.

References

1. Baker, M., Wurgler, J.: Investor sentiment in the stock market. Working Paper 13189, National Bureau of Economic Research, June 2007
2. Bollen, J., Mao, H., Zeng, X.: Twitter mood predicts the stock market. J. Comput. Sci. **2**(1), 1–8 (2011)

3. Bordino, I., Battiston, S., Caldarelli, G., Cristelli, M., Ukkonen, A.: Web search queries can predict stock market volumes. PLoS ONE **7**(7), e40014 (2012)
4. Brown, G.W., Cliff, M.T.: Investor sentiment and the near-term stock market. J. Empir. Financ. **11**(1), 1–27 (2004)
5. Cohen-Charash, Y., Scherbaum, C.A., Kammeyer-Mueller, J.D., Staw, B.M.: Mood and the market: can press reports of investors' mood predict stock prices? PLoS ONE **8**(8), e72031 (2013)
6. Dolan, R.J.: Emotion, cognition, and behavior. Science **298**(5596), 1191–1194 (2002)
7. Gayo-Avello, D.: "I wanted to predict elections with twitter and all i got was this lousy paper"-a balanced survey on election prediction using twitter data. arXiv preprint arXiv:1204.6441 (2012)
8. Go, A., Bhayani, R., Huang, L.: Twitter sentiment classification using distant supervision. Technical report, Stanford Digital Library Technologies Project (2011)
9. Mao, H., Counts, S., Bollen, J.: Quantifying the effects of online bullishness on international financial markets. In: ECB Workshop on Using Big Data for Forecasting and Statistics, Frankfurt (2014)
10. Nofsinger, J.R.: Social mood and financial economics. J. Behav. Financ. **6**(3), 144–160 (2005)
11. Oh, C., Sheng, O.: Investigating predictive power of stock micro blog sentiment in forecasting future stock price directional movement. In: Galletta, D.F., Liang, T.P. (eds.) ICIS. Association for Information Systems (2011). http://dblp.uni-trier.de/db/conf/icis/icis2011.html#OhS11
12. Pak, A., Paroubek, P.: Twitter as a corpus for sentiment analysis and opinion mining. In: LREC, vol. 10, pp. 1320–1326 (2010)
13. Parikh, R., Movassate, M.: Sentiment analysis of user-generated twitter updates using various classification techniques. Technical report (2009)
14. Sakaki, T., Okazaki, M., Matsuo, Y.: Earthquake shakes twitter users: real-time event detection by social sensors. In: Proceedings of the 19th International Conference on World Wide Web, pp. 851–860. ACM (2010)
15. Schumaker, R.P., Chen, H.: Textual analysis of stock market prediction using breaking financial news: the AZFin text system. ACM Trans. Inf. Syst. **27**(2), 1–19 (2009)
16. Wanyun, C., Jie, L.: Investors' bullish sentiment of social media and stock market indices. J. Manag. **5**, 012 (2013)
17. Ding X., Zhang Y., Liu T., Duan J.: Deep learning for event-driven stock prediction. In: IJCAI, pp. 1–7, July 2015
18. Zhao, J., Dong, L., Wu, J., Xu, K.: Moodlens: an emoticon-based sentiment analysis system for Chinese tweets. In: Proceedings of the 18th ACM SIGKDD International Conference on Knowledge Discovery and Data Mining, pp. 1528–1531. ACM (2012)

Predicting Replacement of Smartphones with Mobile App Usage

Dun Yang[1], Zhiang Wu[1(✉)], Xiaopeng Wang[2], Jie Cao[1], and Guandong Xu[3]

[1] School of Info. Engineering, Nanjing University of Finance and Economics,
Nanjing, China
zawuster@gmail.com
[2] Jiangsu Posts & Telecommunications Planning and Designing Institute,
Nanjing, China
[3] Advanced Analytics Institute, University of Technology, Sydney, Australia

Abstract. To identify right customers who intend to replace the smartphone can help to perform precision marketing and thus bring significant financial gains to cellphone retailers. In this paper, we provide a study of exploiting mobile app usage for predicting users who will change the phone in the future. We first analyze the characteristics of mobile log data and develop the temporal bag-of-apps model, which can transform the raw data to the app usage vectors. We then formularize the prediction problem, present the hazard based prediction model, and derive the inference procedure. Finally, we evaluate both data model and prediction model on real-world data. The experimental results show that the temporal usage data model can effectively capture the unique characteristics of mobile log data, and the hazard based prediction model is thus much more effective than traditional classification methods. Furthermore, the hazard model is explainable, that is, it can easily show how the replacement of smartphones relate to mobile app usage over time.

Keywords: App usage · Smartphone replacement · Hazard model · Mobile log data

1 Introduction

Recent few years have witnessed a smartphone surge, due in large part to the rapid advances in mobile Internet technologies. Such high-speed yet ubiquitous access to the mobile Internet has given birth to a variety of applications (apps) to facilitate people's daily life. For example, as of the end of June 2015, there are more than 3 million apps and 594 million smartphones in China.

The study on the usage of mobile apps can help to understand users' behavior and preferences, which not only motivates the development of many intelligent services or adaptive user interfaces but also provides new marketing opportunities [6,14]. There are mainly two aspects of this interesting problem that the existing research has been done. On the one hand, the usage prediction and classification of mobile apps themselves [8,10,14], which can help users to search

© Springer International Publishing AG 2016
W. Cellary et al. (Eds.): WISE 2016, Part I, LNCS 10041, pp. 343–351, 2016.
DOI: 10.1007/978-3-319-48740-3_25

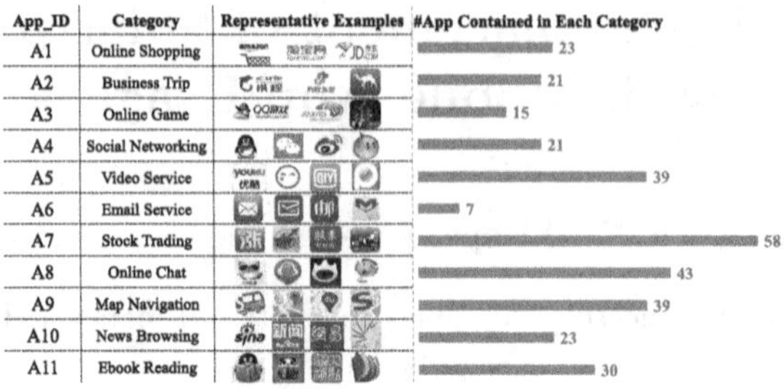

App_ID	Category	Representative Examples	#App Contained in Each Category
A1	Online Shopping		23
A2	Business Trip		21
A3	Online Game		15
A4	Social Networking		21
A5	Video Service		39
A6	Email Service		7
A7	Stock Trading		58
A8	Online Chat		43
A9	Map Navigation		39
A10	News Browsing		23
A11	Ebook Reading		30

Fig. 1. Selected applications and categorization.

and launch apps efficiently. On the other hand, to exploit the app usage for developing other business intelligence services, such as recommendation, churn controlling, target advertising [13]. The topic of this paper falls into the second aspect, but we address another business problem. That is, we target at predicting the replacement of smartphones by mining mobile app usage data[1]. Since mobile Internet service providers (e.g., China Telecom) usually sell mobile phones simultaneously, pinpointing a set of users who intend to change their phones will improve marketing strategies and thus bring significant financial gains. There is therefore a clear need to apply data mining techniques on mobile Internet log data for the phone-replacement prediction.

This paper investigates the phone-replacement prediction problem by applying a unique data set that consists of individual-level log records. We present the temporal bag-of-apps model for representing these raw data in a convenient format. Then, we propose a hazard model from survival analysis to predict the replacement of phones. There is a varying time interval between a user changes his/her phone and the time point that we censor the app usage of this user. Hence, the hazard based models are preferred over both standard regression models and classification models for this problem, due to their ability to model particular factors of *duration data*. Furthermore, the hazard based prediction model is very simple and explainable, which are the important merits for real-world applications. In other words, with the hazard model, we can easily examine how the replacement of cell-phones relate to mobile app usage over time.

2   Data Description

Our sample consists of 411,331 mobile users who used 3G and 4G Internet service between January 12, and January 14, 2016. There are over 130 million log records, each of which denotes a user has used a specific application. Every record contains the information with the following attributes: user_ID, MEID (Mobile Equipment IDentifier), app_ID, start_time and end_time. MEID is a globally unique number identifying a physical mobile phone, and it is used to judge a

---

[1] We use the terms "replace phone" and "change phone" interchangeably in this paper. Both terms mean a user changes his/her *physical* mobile device.

user whether replace his/her phone. That is, if we observe a user with same user_ID yet different MEID, we say this user has changed his/her cell-phone.

We extract a total of 319 typical applications and manually divide them into 11 categories, including online shopping, business trip, online game, social networking, video service, email service, stock trading, online chat, map navigation, news browsing, and ebook reading, denoted by A1 to A11 respectively. Figure 1 shows several representative apps of each category, and the number of apps contained in every category. As can be seen, video service and social networking are two most popular categories, including 58 and 43 apps respectively.

3   Data Model

As introduced above, our raw data contains information about at what time and how long a user has used an application. For instance, a user spends averagely one hour on Wechat in evening. To analyze these temporal frequency data, we need to adopt a specific data model for transforming these raw data into a convenient format. To address the behavioral changes with respect to the time of a day, we discretize 24 h into 4 timeslots [5]: *Morning (6 am–12 am)*, *Afternoon (12 am–6 pm)*, *Evening (6 pm–0 am)* and *Night (0 am–6 am)*. These timeslots are denoted by their initial letters (i.e., $m - a - e - n$).

We then quantify the app usage for each timeslot on our observed data. Let T denote the observed time period (e.g., 3 days) and $s \in \{m, a, e, n\}$ denote one of timeslots. Denote the usage time of ith user on jth app in the timeslot s as t_{ij}^s. Thus, the average usage on the observed data is defined as $U_{ij}^s = \frac{1}{|T|} \sum_{s \in T} \frac{t_{ij}^s}{|s|}$. We use the time bucket rather than the frequency to model the app usage. With the temporal bag-of-apps (TBoA) representation, we can represent a user as a vector, each element is a triplet $\langle timeslot, app, usage \rangle$. Since we have categorized all apps into 11 classes and a day into 4 timeslots, a total of 44 features are generated for each user.

4   Prediction Model

In this section, we first formularize the prediction problem mathematically. Then we present the hazard based prediction model and derive its inference procedure.

4.1   Problem Statement

Assume the observed period is $[t_0, t_0 + T]$, where t_0 is the starting time point and T is the interval of observation. Besides the averaged app usage described by TBoA model, we can also observe whether a user has replaced his/her smartphone. It provides class labels denoted as $y_l, l \in \{1, 0\}$ indicating the user has changed his/her phone or not. In the meanwhile, the event that a replacement of phone happens also has a timestamp. Let S be a non-negative random variable representing the waiting time until the occurrence of this event. In what follows, we call S as *survival time* [4]. For example, we might observe a user replaces his/her phone at $t_0 + \Delta t, \Delta t \leq T$, where this user is labeled as y_1 with survival time $S = \Delta t$. In contrast, if a user did not change his phone during the observation period, the user is labeled as y_0 with survival time $S = T$.

We aim to learn a prediction model based on the observation data during $[t_0, t_0 + T]$, including the usage vector, class label and survival time. Then, given any user represented by a usage vector, we want to predict whether he/she will replace his/her mobilephone in a pre-defined future period denoted by t.

4.2   Hazard Model

The nature of our problem is to predict the probability that an event is going to happen after t units of time. In the literature [4,7], a statistical approach called survival analysis has provided a rich set of methods for handling the time of occurrence of events. Here, we formularize the problem of predicting the replacement of phones by using the survival analysis model.

Let $\mathbf{U}_i = [\mathbf{U}_i^m, \mathbf{U}_i^a, \mathbf{U}_i^e, \mathbf{U}_i^n]$ be the usage vector of ith user, where each element $U_{ij}^s, s \in \{m, a, e, n\}$ is the average usage vector. The vector $\mathbf{U}_i$ can be interpreted as a set of covariates. We then define the *hazard* function $\lambda(t|\mathbf{U}_i)$ to measure the instantaneous rate of occurrence of the event.

$$\lambda(t|\mathbf{U}_i) = \lim_{dt \to 0} \frac{\Pr(t \leq S < t + dt)}{dt}. \tag{1}$$

According to Eq. (1), $\lambda(t|\mathbf{U}_i)$ means the probability that the replacement of phones to be occurred at time t, given an individual with covariates $\mathbf{U}_i$. A large family of models introduced by [4] focus directly on the hazard function. Among them, the *proportional hazard* model is the simplest and most widely-used one. Denote a set of regression coefficients corresponding to covariates as $\boldsymbol{\beta}$. The Cox's proportional hazard model is expressed as

$$\lambda(t|\mathbf{U}_i) = \lambda_0(t)\exp(\mathbf{U}_i\boldsymbol{\beta}^T), \tag{2}$$

where $\lambda_0(t)$ is the baseline hazard function describing the risk for individuals with $\mathbf{U}_i = \mathbf{0}$. Note that $\lambda_0(t)$ does not depend on $\mathbf{U}_i$ but only on t. In contrast, $\exp(\mathbf{U}_i\boldsymbol{\beta}^T)$ is the relative risk, a proportionate increase or decrease in risk, associated with the set of features $\mathbf{U}_i$.

4.3   Model Estimation

The parameter estimation of Cox's proportional hazard model generally consists of two parts: (1) the parameter to determine the baseline hazard function $\lambda_0(t)$, and (2) the set of regression coefficients $\boldsymbol{\beta}$. In the literature [4], various kinds of distributions for modeling $\lambda_0(t)$ are introduced, among which the *Weibull* distribution has gained the particular attention. This is because the Weibull distribution can provide a flexible model to depict the baseline risk varying with time. Meanwhile, the Weibull distribution remains in the exponential family that coincides the Cox's proportional hazard function. The baseline hazard function in Weibull model is defined as $\lambda_0(t) = \frac{p}{k^p}t^{p-1}e^{-(t/k)^p}$, where $p > 0$ is the shape parameter and $k > 0$ is the scale parameter. We can fit the training data to $\lambda_0(t)$, i.e., to determine the parameter p and k.

As for the estimation of $\boldsymbol{\beta}$, the *partial likelihood* is often employed as the objective function. We define an indicator $\delta_i = 1$ to signify the ith user has changed the phone, otherwise for $\delta_i = 0$. The objective function is

$$L(\boldsymbol{\beta}) = \prod_{i=1}^{N} \left(\frac{\exp(\mathbf{U}_i\boldsymbol{\beta}^T)}{\sum_{j \in R(t_i)} \exp(\mathbf{U}_j\boldsymbol{\beta}^T)} \right)^{\delta_i}, \tag{3}$$

where N is the total number of observed users, t_i is the time unit that ith user changed the phone and $R(t_i)$ is the set of users observed at time t_i. Taking the logarithm, we have

$$\log L(\boldsymbol{\beta}) = \sum_{i=1}^{N} \delta_i (\mathbf{U}_i\boldsymbol{\beta}^T - \log \sum_{j \in R(t_i)} \mathbf{U}_j\boldsymbol{\beta}^T). \tag{4}$$

Similar to the Logistic regression, we can solve the $\log L(\boldsymbol{\beta})$ maximization problem by using the Newton-Raphson method [2]. In fact, we focus on estimation of the regression coefficients $\boldsymbol{\beta}$, yet regarding the estimation of $\lambda_0(t)$ as another independent procedure. This approach is also known as the *non-parametric* strategy. A handsful of standard packages are available for estimating $\boldsymbol{\beta}$, such as R, SAS and Python, among which we choose the Python package to improve the computational efficiency on the large-scale data.

5 Experimental Results

In this section, we evaluate the performance of the hazard model for solving our proposed smartphone replacement prediction problem. To obtain the ground-truth label indicating whether changing the phone, we gather the MEID attribute of every user in the log data in the next three weeks. Figure 2 shows statistics of the collected ground-truth information. As can be seen, there are totally 87,346 (21.2%) users who have replaced the phone in three weeks, where nearly half users changed the phone in the first week.

We select three classification models, including Logistic Regression (LR), Naïve Bayesian (NB) and Random Forest (RF). The usage vector of every users (see Sect. 3) are employed as the input of all competitive classifiers. As the ground-truth is available, we adopted the widely-used precision (P), recall (R) and F-measure (F) as evaluation measures. The precision is the ratio of truly identified users changing the phone and the total users who are predicted to change the phone. The recall is the ratio of truly identified users changing the phone and the number of users who have changed the phone.

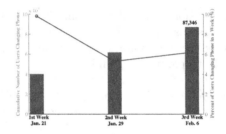

Fig. 2. Statistics of users who change the phone in three weeks.

5.1   Prediction Performance

Here, we evaluate the effectiveness of Cox's proportional hazard model for predicting the smartphone replacement. Firstly, we compare our hazard model with three baseline algorithms via the 10-fold cross validation. The comparative results are shown in Table 1, where the maximal value of each metric is bolded. As can be seen, except the NB model performs poorly, the precision of other three models are very close to each other, i.e., around 0.6. However, the recall value of our hazard model is much higher than those of other models, which leads to the its best overall performance indicated by F. These results state that the traditional classifiers are very strict, that is, they tend to identify a small fraction of users who truely change the phone. By the control of the survival time, our hazard model can seize more potential users with changing behaviors. Interestingly, the RF model was regarded as one of best algorithms for chrun prediction [7,11], and also RF performed best among three baseline methods. Since the RF model used TBoA for modeling mobile log data, this result demonstrate our data model, i.e., TBoA to represent the usage, can effectively fit the prediction for changing phones.

Table 1. Performance comparison (10-fold cross validation)

Algorithm	P	R	F
Logistic Regression	0.614	0.253	0.358
Naïve Bayesian	0.484	0.167	0.248
Random Forest	**0.646**	0.567	0.604
Cox's Prop. Hazard	0.581	**0.762**	**0.660**

5.2   Analysis on Covariates

Besides the satisfactory prediction accuracy, the explainability of a prediction model is crucial to decision makers, who except to know whether there is a positive or negative interdependence between any factor and a given event. With the hazard regression, we can easily observe the importance of every covariate in

Table 2. Several important positive and negative covariates

	No	ID	Coefficient	Standard error	Semantics
Positive	1	n_5	4.93e–02	8.96e–03	Night, Video Service
	2	n_4	4.82e–02	5.26e–03	Night, Social Networking
	3	n_8	3.62e–02	3.18e–03	Night, Online Chat
	4	m_5	1.39e–03	8.96e–03	Morning, Video Service
	5	m_8	1.37e–04	1.78e–04	Morning, Online Chat
Negative	6	m_{10}	−1.47e–03	8.12e–05	Morning, News Browsing
	7	m_7	−8.65e–03	9.26e–05	Morning, Stock Trading
	8	a_7	−3.81e–03	8.81e–05	Afternoon, Stock Trading
	9	a_2	−4.95e–03	2.45e–05	Afternoon, Business Trip
	10	a_{11}	−5.97e–03	1.68e–05	Afternoon, Ebook Reading

terms of their coefficients, i.e., β. We select a case from 10-fold cross validation and show several covariates with their coefficients. Table 2 shows ten important covariates exerting both positive and negative impacts to the event of replacing smartphones. All covariates listed in Table 2 are highly significant with p-value < 10^{-4}, using two-tailed t-test. The regression coefficient tells us how much a unit change in the value of the covariate impacts the user's rate of changing the phone.

Three observations are noteworthy from Table 2. First, none app category in evening has been included in the important covariates, because the usage in evening is much higher than that of other timeslots. Thus, the usage vector in evening has weak discriminative power for different users. Second, as indicated by #1, #2 and #3 covariates, users playing their phones at night tend to change their phones frequently, and the launched apps are very trendy, e.g., video service, social networking and online chat. We can bold guess most users changing their phones are probably teeny-bopper. Third, the negative covariates are in sharp contrast to the positive ones. In detail, many traditional apps exert negative affect to the smartphone replacement behavior, such as news browsing, stock trading, business trip and ebook reading.

6 Related Work

Our work is related to a group of literature about mobile apps usage analysis and churn prediction. Firstly, the studies on the usage analysis focus on understanding when and where apps are used in mobile phones [1]. The contextual usage patterns can then be leveraged for apps prediction (or recommendation) [8,10], which usually guides the development of adaptive user interfaces [5]. Moreover, a number of mobile recommender systems and target advertising engines have been designed based on the app usage analysis [9,13].

Secondly, a related line of research has studied the problem of churn prediction. This problem is often defined as a binary classification problem where users are categorized based on a set of behavioral features into two classes: future churners or non-churners. A lot of classification models have been utilized for churn prediction including logistic regression, neural networks and support vector machines, though the random forest is found to be better in performance [3,11]. Furthermore, the survival analysis is widely used for analyzing or predicting user behavior in online environment, including how user participation patterns affect the lifetime on online knowledge sharing communities [12], and predicting return time for web services [7]. The above work inspires our research a lot when handling this new problem of predicting the replacement of smartphones.

7   Concluding Remarks

This paper provided a study of exploiting mobile app usage for predicting users who will change the phone in the future. In particular, we first analyzed the characteristics of mobile log data that we obtained from a large telecommunications service company in China. We designed the temporal bag-of-apps (TBoA) model for data representation, and presented the hazard based prediction model. The experimental results demonstrated that the hazard base prediction model was thus much more effective than traditional classification methods. Furthermore, we analyzed the important positive and negative covariates to show the good explainability of the proposed hazard prediction model.

Acknowledgments. This research was partially supported by National Natural Science Foundation of China under Grants 71571093, 71372188 and 61502222, National Center for International Joint Research on E-Business Information Processing under Grant 2013B0135, National Key Research and Development Program of China under Grant 2016YFB1000901, and Industry Projects in Jiangsu S&T Pillar Program under Grant BE2014141.

References

1. Böhmer, M., Hecht, B., et al.: Falling asleep with angry birds, Facebook and kindle: a large scale study on mobile application usage. In: MobileHCI, pp. 47–56 (2011)
2. Böhning, D.: Multinomial logistic regression algorithm. Ann. Inst. Stat. Math. **44**(1), 197–200 (1992)
3. Buckinx, W., Van den Poel, D.: Customer base analysis: partial defection of behaviourally loyal clients in a non-contractual FMCG retail setting. EJOR **164**(1), 252–268 (2005)
4. Cox, D.R.: Regression models and life-tables. In: Kotz, S., Johnson, N.L. (eds.) Breakthroughs in Statistics. Springer, New York (1992)
5. Do, T.M.T., Gatica-Perez, D.: By their apps you shall understand them: mining large-scale patterns of mobile phone usage. In: MUM (2010)
6. Ghose, A., Han, S.P.: An empirical analysis of user content generation and usage behavior on the mobile internet. Manag. Sci. **57**(9), 1671–1691 (2011)

7. Kapoor, K., Sun, M., Srivastava, J., Ye, T.: A hazard based approach to user return time prediction. In: KDD, pp. 1719–1728 (2014)
8. Parate, A., Böhmer, M., Chu, D., et al.: Practical prediction and prefetch for faster access to applications on mobile phones. In: UbiComp, pp. 275–284 (2013)
9. Shi, Y., Karatzoglou, A., Baltrunas, L., Larson et al.: TFMAP: optimizing map for top-n context-aware recommendation. In: SIGIR, pp. 155–164 (2012)
10. Shin, C., Hong, J.H., Dey, A.K.: Understanding and prediction of mobile application usage for smart phones. In: UbiComp, pp. 173–182 (2012)
11. Xie, Y., Li, X., Ngai, E., Ying, W.: Customer churn prediction using improved balanced random forests. Expert Syst. Appl. **36**(3), 5445–5449 (2009)
12. Yang, J., Wei, X., et al.: Activity lifespan: an analysis of user survival patterns in online knowledge sharing communities. ICWSM **10**, 186–193 (2010)
13. Yuan, B., Xu, B., Chung, T., Shuai, K., Liu, Y.: Mobile phone recommendation based on phone interest. In: Benatallah, B., Bestavros, A., Manolopoulos, Y., Vakali, A., Zhang, Y. (eds.) WISE 2014. LNCS, vol. 8786, pp. 308–323. Springer, Heidelberg (2014). doi:10.1007/978-3-319-11749-2_24
14. Zhu, H., Chen, E., Xiong, H., Cao, H., Tian, J.: Mobile app. classification with enriched contextual information. IEEE Trans. Mob. Comput. **13**(7), 1550–1563 (2014)

Real Time Prediction on Revisitation Behaviors of Short-Term Type Commodities

Xiangzhen Xu[2], Jinghua Fu[2], Yuliang Shi[1], Shijun Liu[1], and Lizhen Cui[1,2,3(✉)]

[1] School of Computer Science and Technology, Shandong University, Jinan 250101,
Shandong, People's Republic of China
clz@sdu.edu.cn
[2] School of Software Engineering, Shandong University, Jinan 250101,
Shandong, People's Republic of China
[3] Electronic Commerce Research Center of Shandong University, Jinan 250101,
Shandong, People's Republic of China

Abstract. With the advent of Internet we have entered a new era, more and more enterprises take the network as the new profit growth point of the marketing channel. The target of network trading platform is to attract consumers firstly. In this paper, we first extract the five common features in customers' visitation behaviors, i.e. commodity's type popularity, commodity's type revisitation ratio, user revisitation ratio, window repeat ratio and brand popularity. These features constitute the major factors that affect people's short-term types of commodity revisitation behaviors. Based on features of revisitation behaviors, we propose method that can quickly predict consumers will have a types of commodity revisitation behavior. The method is based on Fuzzy Comprehensive Evaluation Method, namely FCEM-II. The experiment of our method adopts a large scale real and reliable data set, and the experimental results show that our proposed method are more accurate in the prediction tasks compared with the baselines.

Keywords: Revisitation · Feature extraction · Fuzzy comprehensive evaluation method

1 Introduction

In the 21st century, with the advent of the era of Internet, it has subtly changed our way of production and consumption. Online shopping as a new social and cultural phenomenon has a great impact on the way of people's life, behavior and consumption patterns. At the same time, with the development of material culture society, online shopping consumption is gradually becoming an important part of resident's living consumption in our country. So we research the consumer's short-term revisitation behavior is of great importance on attract and keep the customers [1–3].

In the real life, people's behavior habits show the difference in thousands ways restricted by the influence of various social factors and personal internal factors. However, some of one's behaviors for each individual have shown some certain regularities and repeatability at a certain time and under the background of the social environment [4–6].

© Springer International Publishing AG 2016
W. Cellary et al. (Eds.): WISE 2016, Part I, LNCS 10041, pp. 352–360, 2016.
DOI: 10.1007/978-3-319-48740-3_26

We use the consumer's visitation history for goods on the internet platform as a data source. Each consumer's visitation history at every time that contains some basic properties, such as the user ID, commodity category ID, brand ID and the visitation time. The information of these properties include the side of consumer, commodity, as well as the side of consumers and commodity interact with each other.

For enterprises, it is important to predict consumers' revisitation, which is able to find potential customers and provide personalized recommendation for them, attracting more consumers. The study of predicts short-term revisitation behaviors are facing the following challenges: (1) the short-term revisitation behavior is dynamic, it is difficult to describe its dynamics through simulation model. (2) The short-term revisitation behavior is random and it is easily affected by the outside factors. (3) It is difficult to extract the features of the general to represent the user's revisitation behavior. (4) Because of the lack of source control, it lead to all sorts of dirty data.

In this paper, firstly, we extract five features of revisitation behavior from a large number of data source: commodity's type popularity, commodity's type revisitation ratio, user revisitation ratio, window repeat ratio and brand popularity. On the basis of the five features, using the improved fuzzy comprehensive evaluation method [7, 8]: FCEM-II, we can predict the customers' short-term revisitation behaviors of types of commodity and calculated its precision. The prediction results as an important basis for the identification of potential and loyal customers to help shopping trading platform. We summarize our main contributions as follows:

- Based on the factors that may affect people's visitation behaviors of types of commodity, we can infer the five general features of the impact of short-term revisitation behaviors: commodity's type popularity, commodity's type revisitation ratio, user revisitation ratio, window repeat ratio, brand popularity.
- We propose our own method FCEM-II based on Fuzzy Comprehensive Evaluation Method as a fast prediction method. Using the five features of the revisitation behavior, we analyze and predict the short-term revisitation behaviors of types of commodity. The experimental results show that our method is effective and efficient.

The paper is organized as follows. In Sect. 2, we give an overview of related research work in prediction on revisitation behaviors of short-term. In Sect. 3, we propose binary short-term revisitation behavior prediction. Section 4 introduces the experimental results and analysis. Finally, we conclude this paper in Sect. 5.

2   Related Work and Preliminaries

In this paper, on the basis of the existing literature, we draw lessons from the relevant theory, the study predicts consumer short-term revisitation behavior. We introduce several studies on people's revisitation behaviors in this section.

- Research of consumers repeated visitation behavior. In the study of repeat queries [9–12], through the data information of digital network trading platform, it can be obtained the user information in the information excavation work.

- Analysis of consumer consumption psychology and behavior research. At the beginning of twentieth century, the research of consumer psychology has been enriched, developed and improved. The study pointed out that the process of consumers' behavior, but also the consumers' psychology and demand to meet the process [13, 14].
- Research on the relationship between consumers' demand and market. Research of the relationship between consumers' demand and the market, can make the market supply on macroscopic overall balance [15].
- Analysis of reconsumption. In [16, 17], they study the short-term reconsumption behavior, the method begins to extract multiple features, on the basis of the extracted features using the linear method and the quadratic method to predict short-term reconsumption behavior.

However, in this paper, we predict whether a user will revisitated commodity's types at a specific time. According to the above browse and buy records of possible revisitation consumer groups. We can recommend to the user with the more appropriate commodity and make the shopping website management more intelligent, provides the necessary reference for the maker's decision. It can reduce the operating costs of the site greatly, but also improve the management efficiency significantly, and promote the increase in the volume of transactions. To the best of our knowledge, this problem has not been solved in the research literature in the field of consumption.

On the whole, there are several basic concepts in our work:

Definition 1. We define a k-length sliding window as W_k, which is a queue which maintains the k most recent visitation of a user till now.

Definition 2. Assuming the k-length sliding window $W_k^{u,t}$ of user u right before performing visitation commodity's type t the problem of the binary prediction of short-term revisitation behaviors is to predict whether or not $t \in W_k^{u,t}$ where t, is unknown.

3 Short-Term Revisitation Behavior Prediction

In this section, we introduce the five common features in detail firstly which influence the consumers' revisitation behaviors in the short term, and on the basis of it, using the improved fuzzy comprehensive evaluation method to predict consumers' revisitation behavior in a short time.

3.1 Feature Extraction

In order to simulate and predict the willing of the consumers' behavior of revisitation, we propose the following features: (1) commodity's type popularity, (2) commodity's type revisitation ratio, (3) user revisitation ratio, (4) window repeat ratio, (5) brand popularity.

Commodity's type Popularity. In the given data set, we define the freq(x) to be the frequency of commodity's type x. We measure the popularity of commodity's type x is its fraction of the maximum commodity's type frequency in the data set

$$h_{IP}(x) = \frac{\log(1 + \text{freq}(x))}{\max_{y \in X} \log(1 + \text{freq}(y))} \tag{1}$$

Where X is the type of commodities set. The *log* operator is used to adjust the skewed distribution of commodity's type popularity. We use the average popularity to represent the given sliding window W_k:

$$h_{IP}(W_k) = \frac{1}{|W_k|} \sum_{x \in W_k} h_{IP}(x) \tag{2}$$

Commodity's type Revisitation Ratio. The absolute commodity's type revisitation ratio is defined as its probability to be observed as a revisitation along the user visitation sequences,

$$h_{AIRR}(x) = \log(1 + \frac{\sum_{u \in U} \sum_{t \in T_u} 1_{t = x \wedge t \in C_k^{u,t}}}{\sum_{u \in U} \sum_{t \in T_u} 1_{t = x}}) \tag{3}$$

Where the user set is U, T_u represent the visitation sequence of user u, 1_{cond} represent the indicator function which returns 1 when *cond* is satisfied, and or else returns 0. The value of the absolute commodity's type revisitation ratio is similar to the definition of the commodity's type popularity.

In the data set, the relative commodity's type revisitation ratio is defined as the fraction of the maximum absolute commodity's type revisitation ratio,

$$h_{IRR}(x) = \frac{h_{AIRR}(x)}{\max_{y \in X} h_{AIRR}(y)} \tag{4}$$

We use the average commodity's type revisitation ratio (abbr. IRR) of a sliding window as commodity's type revisitation ratio,

$$h_{IRR}(W_k) = \frac{1}{|W_k|} \sum_{x \in W_k} h_{IRR}(x) \tag{5}$$

User Revisitation Ratio. The user revisitation ratio is defined as the probability that a user performs a visitation along her visitation transaction sequence,

$$h_{URR}(u) = \frac{\sum_{t \in T_u} 1_{t \in W_k^{u,t}}}{|T_u|} \tag{6}$$

Window Repeat Ratio. Set of distinct commodity's type in W_k is expressed as $DS(W_k)$, where $1 \leq |DS(W_k)| \leq k$. We use the proportion of the revisitation in the current sliding window to measure the window repeat ratio (abbr. WRR),

$$h_{WRR}(W_k) = 1 - \frac{|DS(W_k)|}{k} \tag{7}$$

It is obvious that $0 \leq h_{WRR}(W_k) \leq 1$.

Brand Popularity. In the given data set, we define the freq(x) to be the frequency of brand x. The fraction of the maximum commodity's type frequency in the data set is the popularity of brand x which we measured.

$$h_{BP}(x) = \frac{\log(1 + \text{freq}(x))}{\max_{y \in X} \log(1 + \text{freq}(y))} \tag{8}$$

Where X is the brand set. The log operator is used to adjust the skewed distribution of brand popularity. We used the average popularity to represent brand popularity given sliding window W_k:

$$h_{BP}(W_k) = \frac{1}{|W_k|} \sum_{x \in W_k} h_{BP}(x) \tag{9}$$

3.2 Fast Prediction Methods

We propose an improved fuzzy comprehensive evaluation method to predict the short-term revisitation behavior.

(1) Determining factor set. All these features constitutes the evaluation index system of collection, namely the factor set, remember to

$$U = \left\{ h_{IP}(W_k^{u,t}), h_{IRR}(W_k^{u,t}), h_{URR}(u), h_{WRR}(W_k^{u,t}), h_{WBE}(W_k^{u,t}) \right\} \tag{10}$$

(2) Determine the weight of each factor. In this paper, we use the Delphi method to determine the weight of each factor, remember to

$$A = [a_1, a_2, a_3, a_4, a_5] \tag{11}$$

Among them: a_i for the weight of the i-th factor, and meet the $\sum_{i=1}^{5} a_i = 1$.

(3) Determine the fuzzy comprehensive evaluation matrix. The fuzzy comprehensive evaluation matrix of each index is

$$W = \begin{bmatrix} w_{11} & w_{12} & w_{13} & w_{14} & w_{15} \\ w_{21} & w_{22} & w_{23} & w_{24} & w_{25} \end{bmatrix}^T \qquad (12)$$

(4) Comprehensive evaluation. By transformation, the comprehensive evaluation results can be obtained.

$$B = A \cdot W \qquad (13)$$

(5) Single feature impact. We believe that if the value of a particular feature is particularly large, then the probability of short-term visitation is very large.

$$\text{for } i = 1:5 \quad \text{if } w1j > 0.5 \text{ then } B_1 = (w1j - 0.5) * 0.5 + B_1 \qquad (14)$$

4 Experimental Results

4.1 Data Sets

Tmall has a large visitation of a large number of customers and the types of goods. Consumers often visit the goods on the Tmall, and repeatedly visit the same type of goods.

We use the consumer's visitation history on the internet platform (Tmall) as a data source. The data source includes a detailed record of 428 consumers on the Tmall to visit commodities within six months. The statistics of this data set is shown in Table 1.

Table 1. Statistics of the data sets

Data Set	#. User	#. Types of commodity	#. Transactions
Tmall	428	725	38409

4.2 Experimental Settings

To the best of our knowledge, at present, method linear can solve this problem. So we use the linear method and our improved fuzzy comprehensive evaluation method: FCEM-II for comparison. Linear method use a linear hyperplane to separate the points in the 5-dimensional feature space. Linear method is the state-of-the-art classifiers and it is quite suitable to compare with our method (abbr. ST-L).

4.3 Short-Term Revisitation Behavior Prediction

As shown in Table 2, the prediction of whether a new view is a short-term revisitation behavior results when the window length k = 20. TP, TN, FN and FP represent True-Positive, True-Negative, False-Negative and False-Positive, respectively. Thus, the values of $\frac{TP}{TP+TN}$ and $\frac{FN}{FN+FP}$ can be used to measure the ability of the methods to identify the revisitation behaviors and the novel visitation behaviors, respectively. Furthermore, the overall prediction accuracy of these methods can be evaluated by $\dfrac{TP + FN}{TP + TN + FP + FN}$.

Table 2. Results of predicting revisitation behaviors.

Method	TP/(TP + TN)	FN/(FN + FP)	(TP + FN)/(TP + TN + FP + FN)
FCEM-II	0.82	0.39	0.625
ST-L	0.15	0.94	0.51

Table 2 shows the FCEM-II have the dominating overall prediction accuracy on data set and their performance is outstanding and stable. FCEM-II has the highest prediction accuracy 0.625. We can see that our methods could balance the prediction results better.

4.4 Impacts of Window Length

Figure 1 shows the overall revisitation rate for different window lengths for Tmall. The changes of short-term revisitation behavior ratio of Tmall is not smooth as the window length increases. By contrast, the short-term revisitation rate of Tmall increases quickly from 0.37 (k = 5) to 0.44 (k = 25). This means that most people often revisit goods that are far from now.

Figure 2 illustrates the accuracy of short-term revisitation behaviors prediction under different settings of window length of Tmall, which shows that FCEM-II has a similar trend with the linear method. But our method has higher accuracy in predicting short-term revisitation behavior. With the increase of the window length, the accuracy of the improved fuzzy comprehensive evaluation method is not very large.

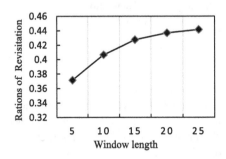
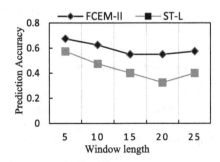

Fig. 1. The ratios of the revisitation under different sliding window length.

Fig. 2. The prediction accuracy under different settings of window length.

5 Conclusion and Future Work

In this paper, we studied whether consumers would have a revisitation on a commodity's type in a short-term and the probability of a revisitation behavior. According to the revisitation history of consumers in online shopping platform, five common features were extracted, value of which were analyzed by improved fuzzy comprehensive evaluation method. Based on the analytic results, we predicted the consumers' revisitation behavior in the short term. Comparing the predicted results with the actual behavior of the consumers, the accuracy of our prediction method was obtained, which showed that our method was efficient and stable.

In the future, our research focus will go deep into studying correlation degree between the commodities that consumers visit, the personalized recommendation service and information re-finding.

Acknowledgments. The research work was supported by the National Natural Science Foundation of China under Grant No. 61303005, No. 61572295, the Innovation Method Fund of China No. 2015IM010200, the Natural Science Foundation of Shandong Province of China under Grant No. ZR2014FM031, ZR2013FQ014, the Science and Technology Development Plan Project of Shandong Province No. 2015GGX101015, and the Shandong Province Independent Innovation Major Special Project No. 2015ZDJQ01002, No. 2015ZDXX0201B03.

References

1. Ahmed, B.S., Maati, M.L.B., Al, Mohajir B.: Improve intelligence of E-CRM applications and customer behavior in online shopping. Int. J. Bus. Intell. Res. (IJBIR) **6**(1), 1–10 (2015)
2. Bilgihan, A., Kandampully, J., Zhang, T.: Towards a unified customer experience in online shopping environments: antecedents and outcomes. Int. J. Qual. Serv. Sci. **8**(1), 102–119 (2016)
3. Moon, Y.J.: Online consumer's shopping motives of personality and shopping values in the lodging industries. **120**, 105–108 (2015)
4. Yu, C., Balakrishnan, R., Hinckley, K., Moscovich, T., Shi, Y.: Implicit bookmarking: improving support for revisitation in within-document reading tasks. Int. J. Hum. Comput. Stud. **71**, 303–320 (2013)
5. Zhang, H., Zhao, S.: Measuring web page revisitation in tabbed browsing. In: CHI, pp. 1831–1834 (2011)
6. Liu, J., Yu, C., Xu, W., Shi, Y.: Clustering web pages to facilitate revisitation on mobile devices. In: IUI, pp. 249–252 (2012)
7. Shiguang, Z.: The fuzzy comprehensive evaluation method of civil servant in the application of evaluation. Nat. Hazards **77**, 1243–1259 (2012)
8. Lzhang, M., Wp, Y.: Fuzzy comprehensive evaluation method applied in the real estate investment risks research. Phys. Procedia **24**, 1815–1821 (2012)
9. Park, Y.H., Fader, P.S.: Modeling browsing behavior at multiple websites. Mark. Sci. **23**(3), 280–303 (2004)
10. Awad, M.A., Khalil, I.: Prediction of user's web-browsing behavior: application of markov model. IEEE Trans. Syst. Man Cybern. B: Cybern **42**(4), 1131–1142 (2012)
11. Adar, E., Teevan, J., Dumais, S.T.: Largescale analysis of web revisitation patterns. In: CHI, pp. 1197–1206 (2008)
12. Kawase, R., Herder, E., Nejdl, W.: Beyond the usual suspects: context-aware revisitation support. In: ACM Conference on Hypertext and Hypermedia, pp. 27–36 (2011)
13. Kasser, T.E., Kanner, A.D.: Psychology and Consumer Culture: The Struggle for a Good Life in a Materialistic World. American Psychological Association, Washington (2004)

14. Gourville, J., Soman, D.: Pricing and the psychology of consumption. Harv. Bus. Rev. **80**(9), 90–96 (2002). 126
15. Al-Gahaifi, T.H., Světlík, J.: Factors influencing consumer behaviour in market vegetables in Yemen. Acta Universitatis Agriculturae et Silviculturae Mendelianae Brunensis **59**(7), 17–28 (2014)
16. Chen, J., Wang, C., Wang, J.: Will you "reconsume" the near past? fast prediction on short-term reconsumption behaviors. In: AAAI Conference on Artificial Intelligence (2015)
17. Anderson, A., Kumar, R., Tomkins, A., Vassilvitskii, S.: The dynamics of repeat consumption. In: WWW, pp. 419–430 (2014)

Big Data Processing

Parallel Materialization of Datalog Programs with Spark for Scalable Reasoning

Haijiang Wu[1,2], Jie Liu[1,3]($\boxtimes$), Tao Wang[1], Dan Ye[1], Jun Wei[1,3], and Hua Zhong[1]

[1] Institute of Software, Chinese Academy of Sciences, Beijing, China
{wuhaijiang12,ljie,wangtao,yedan,wj,zhongh}@otcaix.iscas.ac.cn
[2] University of Chinese Academy of Sciences, Beijing, China
[3] State Key Laboratory of Computer Science, Beijing, China

Abstract. As the volume of semantic data increases rapidly, semantic reasoning becomes a very challenging task. Existing scalable reasoners focus on fragments of OWL 2 RL (eg. RDFS, OWL Horst), and cannot support Semantic Web Rules Language (SWRL) rules, which are widely used in real-world knowledge-based applications. As reasoning of OWL 2 RL ontology extended with SWRL rules can be implemented by materialization of Datalog programs, we propose an approach on parallel materialization of Datalog programs with Spark for scalable reasoning. Since existing scalable reasoners aimed for deterministic rule sets, they used rule-specific strategies for translation of rule execution to parallel jobs and performance optimization techniques. Thus, they cannot be easily extended to support application-specific semantics. In this paper, we propose a rule-independent automatic translation strategy, and several optimization techniques including a dynamic data partition model, a duplication removing strategy and a dependency-aware rule scheduling strategy. These techniques can generalize to vast application-specific semantic rules. Finally, we evaluate our approach with both synthetic and real knowledge bases. The experimental results show our implementation is scalable and the reasoning speed is comparable with that of CiChild, the state-of-the-art scalable reasoner for RDFS/OWL Horst semantics using rule-specific optimizations.

Keywords: Semantic web · Datalog · In-memory computing · Scalable reasoning

1 Introduction

Semantic data used in knowledge-based systems often implies significant information, which can be revealed by reasoning tasks. With the development of knowledge construction techniques, the volume of semantic data grows rapidly, and some big knowledge bases even evolve to contain billions of RDF triples [1]. Such large scale semantic data brings new challenges to semantic reasoning. Aiming for scalable reasoning, the World Wide Web Consortium (W3C) proposed

© Springer International Publishing AG 2016
W. Cellary et al. (Eds.): WISE 2016, Part I, LNCS 10041, pp. 363–379, 2016.
DOI: 10.1007/978-3-319-48740-3_27

the OWL 2 RL specification, which enables the implementation of polynomial time reasoning algorithms using rule-extended technologies operating directly on RDF triples [2].

Recently, many research efforts have been devoted to scalable reasoning [7,8]. While most of them focus on the RDFS/OWL Horst semantics [5], which are fragments of the OWL 2 RL semantics. Moreover, they do not support SWRL rules, which are widely used in real-world knowledge-based applications [3]. Since the RDFS/OWL Horst semantics are deterministic rule sets, existing scalable reasoners propose rule-specific strategies for the translation of rule execution to parallel jobs and performance optimization techniques, such that these strategies and techniques cannot generalize to support the OWL 2 RL and SWRL semantic rules. As Datalog can capture OWL 2 RL ontologies extended with SWRL rules, modern Semantic Web systems often use Datalog "behind the scenes" [18]. Although many single-node reasoners have adopted materialization of Datalog programs to implement semantic reasoning [10,12,14], they are not economic viable for large scale data sets due to hardware resources limitations.

In this paper, we propose an approach on parallel materialization of Datalog programs with Spark. The Datalog programs can be used to express more semantics than standard RDFS/OWL Horst, and a Spark-based architecture makes the scalable reasoner more efficient than traditional MapReduce-based ones. The major contributions and novelties of our work are as follows:

We first introduce an automatic method to translate rule execution into Spark RDD transformations, and use this method to materialize Datalog programs with Spark. In order to speed up the rule execution process, we proposed two optimizations, including a dynamic data partition model and a duplication removing strategy.

As the rule execution order is an significant factor of reasoning efficiency, which is determined by the dependencies among rules. While the rule dependencies varies with the knowledge base, and false dependencies will conduct unnecessary job running. We then propose a sample-based approach to capture the true rule dependency, and schedule rule executions with a dependency-aware strategy.

We have implemented a prototype with Spark, and evaluate our approach on real and synthetic data sets, with standard RDFS/OWL 2 RL semantic rules and a Datalog program defined with LUBM ontology. The experimental results show that our approach has good performance in scalability and efficiency.

The rest of this paper is organized as follows. Section 2 introduces the preliminary of Datalog for semantic reasoning and the Spark programming model. Section 3 presents how we materialize Datalog programs with spark and proposes two optimizations for rule execution. Section 4 introduces a sampling-based method to capture rule dependencies, and a dependency-aware strategy to schedule the rule executions. Section 5 evaluates our approach on real and synthetic data sets with three Datalog programs. Section 6 discusses the related works. Finally, we conclude our work and give the future work in Sect. 7.

2   Preliminaries

2.1   Materialization of Datalog Program for Semantic Reasoning

A knowledge base is usually treated as a finite set of RDF facts, which are triples of resources $<s, p, o>$, called subject, predicate and object. OWL 2 RL extended with SWRL not only includes the RDFS/OWL ter Horst semantics, but also can be used to express application-specific semantics. Reasoning such rules on a knowledge base can be implemented by Datalog techniques. One can do reasoning by materialising all consequences of the rules and the knowledge base, which is PTIME-complete in data complexity [18]. In classic Datalog, a program is defined as a set of recursive rules, and the termination of recursions is implicitly determined by a fixed point strategy [21]. Although the recursion expression is clean and simple, it is not easy to encode sophisticated optimization strategies using parallel programming model [20]. To encode optimization strategies, we use an iteration statement and define the termination condition as the fixed point that no new result is derived.

A Datalog rule r has the form (1), where $B = B_1 \wedge ... \wedge B_n$ is the body of the rule, and H is the head of the rule. Both B_i and H are RDF atoms, which are triples of resources or variables. B_i is a body atom, and H is the head atom. To make a rule safe, we constrain that a variable in the head must occur at less one time in the body.

$$B_1 \wedge B_2 \wedge ... \wedge B_n \to H \tag{1}$$

A variable substitution [19] $\varphi_t^{?x}$ is an operation that apply to an atom to replace the occurrences of the variable $?x$ with a resource t. A substitution application (SA) $A[\varphi]$ is a triple obtained by applying a substitution φ to an atom A. Composition of substitutions φ_1 and φ_2 is also a substitution and defined as usual in [21]. Given a knowledge base I and a rule r, the inference result $r(I)$ is the smallest set containing $H(r)[\varphi]$ for each substitution φ such that $B_i(r)[\varphi] \in I$ for each $1 \le i \le len(r)$.

Example 1. Given a rule r_1, "B_1 : $\langle ?x,\ lubm\ :\ worksFor,\ ?y_1 \rangle \wedge B_2$: $\langle ?y_1,\ lubm\ :\ subOrganizationOf,\ ?y_2 \rangle \to H$: $\langle ?x,\ lubm\ :\ worksFor,\ ?y_2 \rangle$", and a knowledge base I with three triples: tr_1 : $\langle p_1,\ lubm\ :\ worksFor,\ o_1 \rangle$, tr_2 : $\langle p_2,\ lubm\ :\ worksFor,\ o_1 \rangle$, tr_3 : $\langle o_1,\ lubm\ :\ subOrganizationOf,\ o_2 \rangle$, all the possible SAs of B_1 and B_2 in I are $SA_{B_1}^I = \{tr_1, tr_2\}$, and $SA_{B_2}^I = \{tr_3\}$, and then SAs corresponding to the rule body $B = \{B_1 \cdots B_n\}$ is $SA_B^I = \{tr_1, tr_2\} \bowtie_{SA_{B_1}^I.3 = SA_{B_2}^I.1} \{tr_3\}$. SA_B^I entails two composite substitutions $\varphi_{p_1}^{?x} \cdot \varphi_{o_1}^{?y_1} \cdot \varphi_{o_2}^{?y_2}$ and $\varphi_{p_2}^{?x} \cdot \varphi_{o_1}^{?y_1} \cdot \varphi_{o_2}^{?y_2}$. The inference result $r(I)$ can be computed by applying these substitutions to H, $r(I) = \{tr_4 :< p_1,\ lubm\ :\ worksFor,\ o_2 >,\ tr_5 :< p_2,\ lubm\ :\ worksFor,\ o_2 >\}$.

A Datalog program is a finite set of rules. Given a Datalog program p, $p(I) = \cup_{r \in p} r(I)$ is the result of evaluating p on I in one iteration. Such inference process will be executed iteratively, and will not terminate until no new result is derived. The final inference result of p on I is $p^\infty(I) = \cup_i p(I)$, where $i > 0$ is the iterations that p has been evaluated for. The latter set is the materialization of I with P. In this paper, we concentrate on parallel materialization of Datalog programs for scalable reasoning.

2.2 Spark Computing Framework

As a popular in-memory and distributed cluster computing framework, Spark has two key features to make scalable reasoning more efficient. The first is its in-memory parallel execution model, in which the intermediate results can be persisted into memory and be reused as input in the subsequent rule execution. The second is that, Spark can provide more flexible DAG-based (directed acyclic graph) data flow. Because materialization of a Datalog program is an iterative process, such data flow is more suitable for semantic reasoning than that of MapReduce [9].

The programming model of Spark is upon a fault-tolerant distributed memory abstraction called Resilient Distributed Dataset (RDD). RDDs are fault-tolerant, parallel data structures that let users explicitly persist intermediate results in memory, control their partitioning to optimize data placement, and manipulate them using a rich set of transformations (e.g., map, filter, groupBy, join). RDDs and transformations sequences are defined according to the computing tasks in the driver program, and some actions are invoked on these RDDs to return results. With RDDs and transformations, Spark implements distributed data set operations through the methods of manipulating local data set. Besides, spark provides broadcast variables which allow the programmer to keep a read-only variable cached on each machine rather than shipping a copy of it with tasks. For a more detailed description of Spark, we refer to its official web site [1].

3 Materialization of Datalog Program with Spark

3.1 Datalog Rule Execution with Spark

Essentially, semantic reasoning is an iterative process. In each iteration, the semantic rules in a Datalog program are applied over the knowledge base. Existing scalable reasoners aimed for RDFS/OWL Horst semantics, they used a rule-specific strategy to translate rule execution to parallel jobs. Such strategy is not generic and can not easily extended to execute OWL 2 RL and SWRL rules. In this section, we introduce an automatic method to translate Datalog rule execution into Spark RDD transformations, which can be further used to evaluate a Datalog program.

As described in Sect. 2.1, we divide the Datalog rule execution procedure into three stages: (1) finding SAs for each body atoms (FindSA), (2) joining SA sets of all the body atoms of the rule (JoinSA) (3) finding out the variable substitutions of the rule body and applying them to the rule head (ApplySH).

---

[1] http://spark.apache.org/docs/latest/index.html.

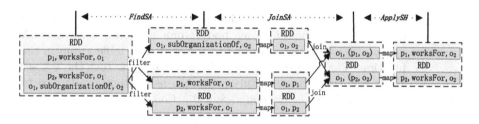

Fig. 1. The execution workflow of r_1 on Spark

Initially, the input data set is used to create a Spark RDD, and then the rule execution procedure can be expressed by Spark RDD transformations to achieve parallelization.

A rule with one shared variable need only one join transformation in the JoinSA stage. For simplicity, we use a one-shared-variable rule to illustrate the rule execution procedure. Here, we take the rule r_1 given in Sect. 2.1 as an example. As shown in Fig. 1, the FindSA stage can be implemented with a *filter* transformation. In this transformation, we find out the SA sets of the body atoms. If an input fact matches a body atom of the rule, it will be taken as a SA of the body atom. More specifically, we constrain that a SA of B_1 has a predicate "lubm:worksFor", and that of B_2 has a predicate "lubm:subOrganizationOf". Then in the JoinSA stage, the SAs are mappd into key/value pairs in the *map* transformation. A key is a set of elements corresponding to the shared variable $?y_1$, and a value is a set of elements corresponding to the no-shared variables $?x$ and y_2. Further, the rule body's SA set is generated by a *join* transformation over the SA sets of B_1 and B_2. In the ApplySH stage, we find out the substitution for the head atom H, and produce a reasoning result. From the execution procedure of rule r_1, we can see that the important thing is to find out the SA set for each body atom and then apply the rule by performing *map* and *join* transformations over these SA sets.

However, there are two problems for a rule with multiple shared variables. First, only one *join* transformation is not sufficient. Second, the conditions for different join transformations are not the same. Taking a rule "$cls - svf_2$" in the OWL 2 RL semantics as an example, "$B_1 : \langle ?v, owl : someValuesFrom, ?w \rangle \wedge B_2 : \langle ?x, rdf : type, ?w \rangle \wedge B_3 : \langle ?v, owl : onProperty, ?p \rangle \wedge B_4 : \langle ?u, ?p, ?x? \rangle \rightarrow H : \langle ?u, rdf : type, ?v \rangle$", To join the SA sets of B_1 and B_2, the elements corresponding to the variable $?w$ should be equal. The *join* transformation on the SA sets of B_2 and B_3 requires that the elements corresponding to $?v$ are equal. As a result, we need three *join* transformations in a cascade way to find out the SA set of the body of the rule, $J_1 : ASset_{B_1} \bowtie ASset_{B_2}$, $J_2 : ASset_{B_1,B_2} \bowtie ASset_{B_3}$ and $J_3 : ASset_{B_1,B_2,B_3} \bowtie ASset_{B_4}$. Noticing that J_3 has two shared variables $?p$ and $?x$ in the join condition, then we need to generate a key with two elements when doing the *map* transformations.

The strategy of translating a rule execution into Spark RDD transformations is shown in Algorithm 1. We first resolve the rule and generate the logic plan of the rule execution. Each step of the plan corresponds to a group of shared variables. And then the SA sets for each body atoms are found out, each SA is mapped into a key/value pair according to the shared variable group in the *mapTriple* function. The mapped SA sets are joined to construct the SA set of the rule body, and then the substitutions for the head atom are derived to generate reasoning result in the *generateResult* function.

Algorithm 1. Evaluate a Datalog Rule with Spark.

Require: knowledge base kb (key/value pairs), Datalog rule r.;
Ensure: reasoning result (key/value pairs).;
  1: **function** REASONING(kb, r)
  2:    $sharedVGroups$ = logicPlan(r); ▷ a logic plan corresponds to a shared variable group
  3:    **for** each $bodyAtom \in r.bodyAtoms()$ **do**   ▷ FindSA stage
  4:      $saSet = kb.$**filter**($t \Rightarrow$ isSubstitutionApplication($t._2, bodyAtom$));
  5:      $SASetList.$add($saSet$);
  6:    **end for**
  7:    **for** each $svg \in sharedVGroups$ **do**   ▷ JoinSA stage
  8:      $saSetOfSV$ = getSAsetsforSVG($r, svg, SASetList$);
  9:      **for** each $saSet \in saSetOfSV$ **do**
 10:        $mappedSASet = saSet.$**map**($t \Rightarrow$ MAPTRIPLE($t._1, t._2, bodyAtom, svg$));
 11:        $joinResult = joinResult.$**join**($mappedSASetList[i]$)
 12:      **end for**
 13:    **end for**
 14:    $result = joinResult.$**map**($t \Rightarrow$ generateResult($t, r.head$))   ▷ ApplySH stage
 15: **end function**
 16: **function** MAPTRIPLE($key, value, bodyAtom, svg$)
 17:    $outKey$=getKeyElems($key, bodyAtom, svg$)   ▷ elements corresponding to shared variables
 18:    $outValue$=getValueElems($value, bodyAtom, svg$)   ▷ elements to unshared variables*
 19:    **return** ($outKey, outValue$);
 20: **end function**

3.2   Evaluation of Datalog Program

Evaluation of a Datalog program is an iterative process that repeatedly applies all the rules. We adopt the Spark RDD model to describe the workflow of materializing a Datalog program in Fig. 2. Each round starts with a rule. It performs *join* transformation on the SA sets of the body atoms. If the rule needs another *join* transformation, the join results will be used as the input of the next *join* transformation. Otherwise, they will be used for producing new results of the rule. After a rule execution, the derived results will be merged into the original input facts and are used as the input of the next rule. All the rules are iteratively applied to the knowledge base until no new data is generated.

3.3   Optimizations for Rule Execution

A naive translation of rule execution into RDD transformations is inefficient. As most of the body atoms can match a small part of the knowledge base, applying

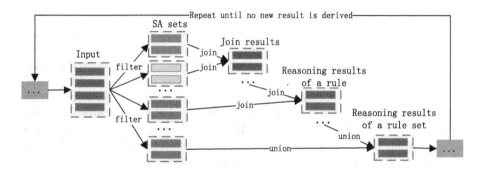

Fig. 2. The execution workflow of Materializing a Datalog program with Spark

rules to the whole knowledge base brings unnecessary computation. Furthermore, new derived facts can duplicate the original facts in the knowledge base, and reasoning with duplicated facts generates more duplication, vicious circle like this again and again. In this section, we then introduce two optimizations for rule execution including a data partition model and a duplication removing strategy.

Dynamic Data Partition Model. If a rule is applied by doing join operations on the whole knowledge base, too many self-join operations on a large scale data set result in a large amount of computation. This is one of the key factors why existing reasoners cannot scale to large data set. Some works notice that most of the body atoms can match a small partition of facts [8], they partition the knowledge base into small pieces, and then most of the facts can be filtered before taken to match the body atoms. However, partitioning the knowledge base before reasoning is not suitable for materialization of Datalog programs, because the rules are unknown in advance. Thus, we partition the knowledge base dynamically according to the rules during the reasoning process. We first find out all the constant elements in the body atoms. And then in the FindSA stage of a rule execution, if the body atom contains constants, and the size of the SA set is less than a upper bound u and a lower bound l, they will be put into a new partition. In our practice, we choose a half of the triples in the whole knowledge base as u, and select 10000 as l. Taking r_1 as an example, we put a triple with a predicate "lubm:worksFor" into a new partition after the FindSA stage of the rule execution, so that these triples can be used in the subsequent rule executions. At the beginning of a reasoning process, we separate the schema and type data from the whole knowledge base, because the schema data usually has a small size and is more prone to match the body atom.

Eliminating Duplicated Facts. It is inevitable that some results of the rules duplicate the facts in the knowledge base during the rule execution process. The duplicated facts not only degrade system performance but also increase management overhead. However, removing duplication also bring additional computation to the reasoner. Existing systems adopt a series of duplication elimination strategies, authors in [7] eliminated duplicated data from the derived results, authors in [8] remove all the duplicated data. While the experimental results show that none of them is the best for all the data sets and rule sets, when there are dependency circles between rules [8]. Since we aim to reason an indeterministic rule set with complex dependencies, we provide three strategies to remove all the duplicated facts as follows: (1) removing duplication after a rule execution, (2) removing duplication till the materialization of a Datalog program finished, and (3) removing duplication when the ratio of the derived data to the original data exceeds a defined threshold. The last strategy is to avoid that too many duplicated facts bring extra disk I/O overhead.

4   Dependency-Aware Rule Scheduling

4.1   Dependency Between Rules

The materialization of a Datalog program is a process of continuously applying each rule to the knowledge base, until no new result is derived. Then a rule dependency exists when the evaluation results of a rule trigger another rule execution. Thus improper rule evaluation order will conduct unnecessary job running and bring significant performance degradation [7,8]. However, analyzing the dependencies among rules is not an easy task such that existing works achieve it in a manual way. This section introduces how we automatically capture the rule dependencies with a sampling-based method.

---

rdfs2:?p rdfs:domain ?x; ?s ?p ?o → ?s rdf:type ?x
rdfs3:?p rdfs:range ?x; ?s ?p ?o → ?o rdf:type ?x
rdfs5:?p rdfs:subPropertyOf ?q; ?q rdfs:subPropertyOf ?r → ?p rdfs:subPropertyOf ?r
rdfs7:?s ?p ?o; ?p rdfs:subPropertyOf ?q → ?s ?q ?o
rdfs9:?s rdf:type ?x; ?x rdfs:subClassOf ?y → ?s rdf:type ?y
rdfs11:?x rdfs:subClassOf ?y; ?y rdfs:subClassOf ?z → ?x rdfs:subClassOf ?z

---

Fig. 3. RDFSM rule set

Given a Datalog program contains only two rules $rdfs_9$ and $rdfs_{11}$ in RDFSM (listed in Fig. 3), if we execute the rules in the order "$rdfs_9 \rightarrow rdfs_{11}$", then the execution results of $rdfs_{11}$ can match the second body atom of $rdfs_9$, and $rdfs_9$ will be executed two times. However, if the reasoning process start with $rdfs_{11}$ instead of $rdfs_9$, the execution results of $rdfs_9$ can not match any body atom of $rdfs_{11}$, and then the reasoning task can be implemented by executing $rdfs_9$ and $rdfs_{11}$ only one time respectively.

To a certain extent, the literals of the rules indicate some information about rule dependency. For example, all the facts in the knowledge base can match the second body of $rdfs_2$, then it depends on all the all rules. However, the rule dependencies analyzed by literal could be false. Taking two rules $rdfs_2$ and $rdfs_{11}$ in RDFSM as an example, let a none-empty set $ER_{rdfs_{11}}$ be the execution result of $rdfs_{11}$, then $ER_{rdfs_{11}}$ can match the second body atom of $rdfs_2$, and is a incremental SA set of it. As a result, it will trigger a new execution of $rdfs_2$ according to rule dependency analyzed by literal. However, there might be no fact that describes the domain of "$rdfs : subClassOf$" in a knowledge base. That means the join operation on the SA set of "$<?p, rdfs : domain, ?x >$" and $ER_{rdfs_{11}}$ produces an empty set, and then no new result is generated. As shown in Fig. 4(a), when analyzing RDFSM literally, $rdfs_2$ depends on all the rules. Actually, when executing RDFSM on the LUBM knowledge base [17], $rdfs_2$ depends only on $rdfs_7$, and the left rule dependencies do not exist all.

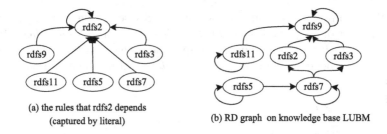

(a) the rules that rdfs2 depends
(captured by literal)

(b) RD graph on knowledge base LUBM

Fig. 4. False rule dependency vs. True rule dependency

Since the rule dependency varies with the knowledge base, it can not be captured unless we execute the rules on the knowledge base. For better illustration, we define *rule dependency* as follow:

Definition 1. [Rule Dependency] Let I be a knowledge base. Given two Datalog rules r_1 and r_2, ER_1 is the execution result of r_1, SAS_2 is the SA set of r_2's body, and ER_2 is the execution result of r_2. As a subset of SAS_2, each element in SAS_{2-r_1} contains triples from ER_{r_1}. $ER_{r_2-r_1}$ is a subset of ER_2 and derived from SAS_{2-r_1}. r_1 depends r_2 iff $ER_{r_2-r_1} \neq \phi$. The degree how r_1 depends on r_2, when executing them on I, is measured as follows:

$$rd_I(r_2, r_1) = (|ER_{r_2-r_1}|/|ER_{r_2}|) * |ER_{r_1}| \qquad (2)$$

In practice, in order to identify the elements in $ER_{r_2-r_1}$, all the execution results of a rule are associated with the rule identifier.

4.2 Capture Rule Dependency by Sampling

For a large knowledge base, it is costly to capture dependencies among rules by applying them over the whole knowledge base. A natural choice is to execute it on a little sample knowledge base. As a widely used sampling method, Simple

Random Sampling (SRS) ensures that each sample has the same probability of being chosen at any stage during the sampling process. While the SA sets of the body atoms isn't distributed uniformly in the knowledge base, then a body atom with less SAs on the original knowledge base is more likely to have an empty SA set on the sample knowledge base.

When the SA set of a rule body atom is empty, we will lose the dependencies about this rule. In a worse case, the "empty SA set" problem transmits, then we will lose more rule dependencies. As described in Sect. 2.1, if any of the body atoms has a empty SA set on the sample knowledge base, no result is generated. As a result, the degrees that other rules depend on this rule are 0. That means we fail to capture these rule dependencies. Moreover, if the reasoning result of a rule is empty, rules that depend on this rule may also have a empty SA set with the their body. Then we will lose more rule dependencies. Taking $rdfs_7$ and $rdfs_2$ in RDFSM as an example, as shown in Fig. 5, an arrow indicates that an empty-SA-set atom causes the another atom's SA set to be empty. Assuming that the SA set of $bodyAtom_1$ or $bodyAtom_2$ is empty, the SA set of $headAtom_1$ will be empty if there is no fact in the sample knowledge that match $headAtom_1$, and then the dependency among $rdfs_7$ and $rdfs_2$ will not be captured.

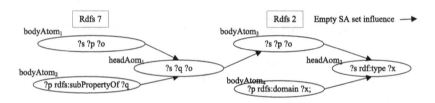

Fig. 5. Transmission of "empty SA set" problem

When applying the rules on the sample knowledge base, a better sampling method should ensure that each body atom's SA set has the same size. In this paper, we adopt a stratified sampling method and consider the structure of the knowledge base. At the beginning of the sample stage, we identify the knowledge base files of each body atom of a rule, and count the number of the facts in them. Then, we sample the same size of facts for each body atom of the rules. As we only need to ensure that, on the sampled knowledge base SI, $rd_{SI}(r_2, r_1) > 0$ if $rd_I(r_2, r_1) > 0$. In practice, we can capture the rule dependencies of RDFS and OWL rules on LUBM [17] by sampling no more than 0.1 % of the whole data set.

Compared to SRS, our approach can capture the rule dependency more precisely for the following two reasons. First, since different types of facts are not uniformly distributed in a knowledge base, the SRS method make the body atoms with a small SA set which has less probability to be match. Second, most of the rules in a Datalog program are applied to a small proportion of the knowledge base, then sampling the facts that a rule is not be applied to, increases the size of the sample set unnecessarily.

4.3   Dependency-Aware Rule Scheduling

Considering the dependencies among rules, we schedule the rule execution dynamically as shown in Algorithm 2. In order to identify all the rules that a rule depends, each rule should be executed at less one more time after the whole rule set has been executed. Then we execute the rule set for two rounds to construct the rule dependency graph. While such graph often contains cycles, which brings the reasoner into an endless loop. We remove the cycles in the rule dependency graph by deleting the edge with the minimum value, and then it becomes a directed acyclic graph (DAG). The rule evaluation order is a topological sort of all the nodes in the DAG. Note that the minimum rule dependency in a cycle is removed, but all the rules in the cycle will be executed, which ensures the correctness of reason results. In runtime, the rule dependency graph is updated according to the rule execution results.

A transitive rule is in the form "$B_1 : \langle ?x,\ p,\ ?y \rangle \wedge B_2 : \langle ?y,\ p,\ ?z \rangle \rightarrow H : \langle ?x,\ p,\ ?z \rangle$". Conducting transitive rule reasoning is essentially to compute the transitive closure on the knowledge base. A common method is to perform the rule execution as other rules. While reasoning such rules usually need much more iterations. Scheduling such rules together with other rules makes the reasoner inefficient. Take a transitive rule $rdfs_5$ in RDFSM as an example, it is depended by $rdfs_7$. Assuming that reasoning of $rdfs_5$ terminates after three iterations, then in the first two iterations, the execution result of $rdfs_5$ will trigger $rdfs_7$'s execution. Actually, if $rdfs_5$ is executed repeatedly until no new triple is produced, execution of $rdfs_7$ will be triggered only once. Thus, we sperate the execution of transitive rules from other rules, and we don't step to execute the next rule, until its execution derives no result. To improve the reasoning performance, we adopt the smart transitive closure algorithm [15] to conduct transitive rule reasoning as described in Algorithm 2.

5   Evaluation

We constructed a prototype system PLogSpark with Spark, and implemented all the optimizations introduced in this paper. In this section, we evaluate the performance of this system in efficiency and scalability aspects.

5.1   Experiment Setup

The experiments are conducted on a local cluster with 9 nodes. One of them is reserved as master and the left are used for slave nodes. Each node has 8 cores and 16 G RAM. The prototype system is built on Spark v 1.4 with Java 1.6 installed.

Data Set. We used two datasets for the experiments. LUBM [17] is a benchmark that can generate semi-realistic data sets of arbitrary size. As a widely used benchmark for semantic reasoning, it gives a good reference

Algorithm 2. Dependency-Awared Rule Scheduling.

Require: knowledge base kb, sample knowledge base skb, rule set rs
Ensure: rule evaluation order reo

 1: **function** SCHEDULERULES(kb, skb, rs) ▷ scheduling the rule executions according rdg
 2:     rdg = constructRDGraph(skb, rs)
 3:     **do**
 4:         reo = decyclizeAndTopologicalSort(rdg, rs)▷ traverse rdg to do decyclization
 5:         **for** $r \in reo$ **do**
 6:             **If** isTransitive(r) **Then** kb = transExecute(r, kb)
 7:                 **Else** evaluate r on kb and update rdg
 8:         **end for**
 9:     **while** result(rs) > 0
10: **end function**
11: **function** TRANSEXECUTE(r, kb)
12:     **do**
13:         $incrKB = kb$;
14:         $SAset_1$ = SASet($r.ba_1, incrKB$); $SAset_2$ = SASet($r.ba_2, incrKB$);
15:         $tempKB_1$ = GenerateResult($SAset_1, SAset_2$); $incrKB = tempKB_1 - kb$;
16:         $SAset_2$ = SASet($r.ba_1, kb$); $tempKB_2$ = GenerateResult($SAset_1, SAset_3$);
17:         $kb = tempKB_2$.union(kb).union($incrKB$)
18:     **while** kb has changed
19:     **return** kb
20: **end function**

---

for a comparison. In our experiment, we generate 5 data sets with different size: LUBM400, LUBM800, LUBM1200, LUBM1600 and LUBM2000, each including semantic data of 50 million, 100 million, 150 million, 200 million and 250 million triples respectively. DBpedia [4] is a multi-lingual knowledge base from Wikipedia using knowledge extraction technologies and is freely available on the Web. Here, we used five data sets of it: DBpedia40M(40 million), DBpedia80M(80 million), DBpedia120M(120 million), DBpedia160M(160 million) and DBpedia200M(160 million). Each element of a triple is expressed by integer.

Comparison and Rule Set. CiChild is the state-of-the-art scalable reasoning system, which is established on the Spark computing framework, and is specific to the RDFS and the OWL ter Host semantics. Since our approach aims to execute Datalog rules, which are indeterministic. We need to associate rule ID with the rule execution results to support generic rule execution. For comparison purposes, we create two Datalog programs RDFSM and OWLM which include all the RDFS/OWL semantics rules supported in Cichild. In addition, we created a Datalog program (called plogRules) with the LUBM ontology, by choosing the rules that contain "lubm:Organization" and "lubm:Course" from the Datalog program provided in [18].

5.2 Effectiveness of the Optimizations

We evaluate the effectiveness of our proposed optimizations: dynamic data partition model (DAPM) and dependency-aware rule scheduling (DAS) on LUBM2000 and DBpedia200M with above Datalog Programs. As DRS highly depends on DAPM, we evaluate DRS together with DAPM. All the experiments run five times and the average execution time is shown in Fig. 6. We can see that the proposed optimizations have significant effectiveness on the data sets with all the rules, they can even make the reasoner 10 times faster than the naive approach with OWL rules.

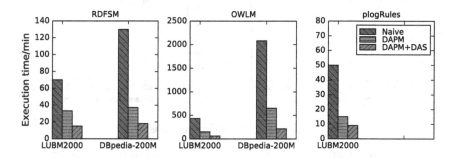

Fig. 6. Effectiveness of the optimizations

5.3 Scalability Performance Analysis

Data Scalability. We evaluate the data scalability of PLogSpark, compared with CiChild. The experiments are performed with RDFSM and OWLM on different scale of data sets including five sets of LUBM benchmark and five sets of DBpedia data. Experimental results are shown in Fig. 7. We can see that PLogSpark's performance is comparable with CiChild. There are two reasons for this performance gap. First, we associate the rule ID to the results to construct the rule dependency graph. Second, we use a more complex schedule strategy. Although both of them need more computation than that of CiChild, they are necessary for indeterministic rule reasoning.

Node Scalability. We reason both RDFSM and OWLM on the LUBM400 and DBpedia40M data sets to evaluate the node scalability of PLogSpark. Figure 8 shows the execution performance of these two systems when the number of computing nodes varies. We can see that the execution time of PLogSpark decreases more than 5 times as cluster nodes increase from one to 8 nodes in all of these four experiments. It achieves close node scalability to Cichild.

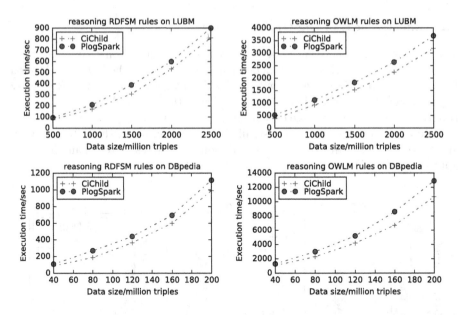

Fig. 7. Data scalability analysis

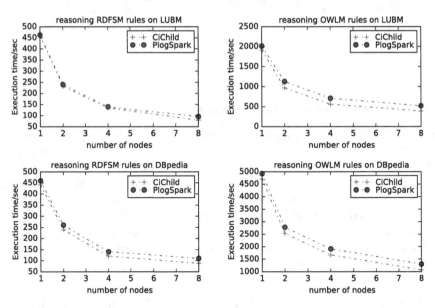

Fig. 8. Node scalability analysis

6　Related Work

As the one of the most popular reasoners, Jena is a Datalog engine that supports
OWL 2 RL semantics extended with SWRL rules [14]. However, it is not viable

to reason on billions of triples due to the limitation of hardware resources. Peters et al. [10] introduced an efficient implementation of RETE algorithm, which can apply RDFS rules on 1 billion triples on a Laptop. However, both these two systems run on single-node environments, then they can hardly cope with complex reasoning on large-scale semantic data. Consequently, some multi-core reasoning systems have been proposed in the past few years. Authors in [12] presented an approach to parallel materialization of Datalog programs in centralized main-memory, multi-core RDFS systems. Subercaze et al. [23] implemented better performance than [12] by using fine-gained data structure. However, the scalability of these systems is constrained by the volume of main memory.

Due to the excellent scalability and fault tolerance features of the MapReduce and the Spark parallel programming models, many research efforts have been devoted to implement highly scalable reasoning systems with them. Urbani et al. introduce WePIE, a MapReduce based inference engine [8], which can scale to reason large knowledge bases on a cluster. Liu et al. [13] then extended it to reason over fuzzy RDF data. Rong et al. proposed Cichild, a reasoning engine on top of Spark [7]. While all of these systems handle only OWL 2 RL fragments. Since the authors aim to infer deterministic rule sets, they use a static data partition model and manually analyze the rule dependencies to determine the rule evaluation order. Thus, the optimizations introduced in these works are not applicable for more generic rule execution. Existing Datalog engines were mainly built for database analysis. Existing Datalog engines are mainly used for database analysis, and most of them are built with MapReduce. Afrati et al. introduced an algorithm of transitive and recursive Datalog rule evaluation on clusters, but given no experimental analysis. J. Gao et al. proposes a high level graph analysis system using MapReduce [20]. As these systems don't materialize rule execution results, they didn't remove duplication and didn't analyze rule dependencies.

7   Conclusion and Future Work

In this paper, we have shown an approach of parallel materialization of Datalog programs with Spark for semantic reasoning. Compared to existing scalable reasoners for RDFS/OWL Horst semantics, our approach can express OWL 2 RL semantics extended with SWRL rules, which is widely used in real semantic-based applications. To do so, we proposed an algorithm to automate the translation from a rule execution to Spark RDD transformations and make several optimizations to speed up the rule evaluation process. More importantly, we introduced an automated method to establish the dependency graph of rules to produce an efficient rule evaluation order, which is manually determined in existing scalable reasoners. Compared with existing single-node reasoners, our approach is based on the Spark programming framework and can scale out to reason large data sets. In future work, we intend to evaluate our system with more large data set, and extend this approach to infer semantic data incrementally.

Acknowledgement. This work was partially supported by National Key research and Development Plan (2016YFB1000103), Chinese Academy of Sciences STS Project (KFJ-SW-STS-155)and National Science Technology Support Plan (2015BAF23B03).

References

1. https://datahub.io/
2. https://www.w3.org/TR/owl2-profiles/
3. https://www.w3.org/Submission/SWRL/
4. Lehmann, J., Isele, R., Jakob, M., et al.: DBpedia - a large-scale, multilingual knowledge base extracted from Wikipedia. J. Semant. Web **6**(2), 167–195 (2015)
5. ter Horst, H.J., et al.: Completeness, decidability and complexity of entailment for RDF Schema and a semantic extension involving the OWL vocabulary. Web Semant. J. **3**, 79–115 (2005)
6. Suchanek, F., Kasneci, G., Weikum, G.: YAGO: a core of semantic knowledge. In: Proceedings of the 16th International Conference on World Wide Web (WWW) (2007)
7. Gu, R., Wang, S., Wang, F., et al.: Cichlid: efficient large scale RDFS/OWL reasoning with Spark. In: IPDPS, pp. 700–709 (2015)
8. Urbani, J., Kotoulas, S., Maassen, J., et al.: WebPIE: a web-scale parallel inference engine using MapReduce. J. Web Semant. **17**(44), 59–75 (2012)
9. Dean, J., Ghemawat, S.: MapReduce: simplied data processing on large clusters. In: OSDI, pp. 137–147 (2004)
10. Peters, M., Sachweh, S., Zündorf, A.: Large scale rule-based reasoning using a laptop. In: Gandon, F., Sabou, M., Sack, H., d'Amato, C., Cudré-Mauroux, P., Zimmermann, A. (eds.) ESWC 2015. LNCS, vol. 9088, pp. 104–118. Springer, Heidelberg (2015). doi:10.1007/978-3-319-18818-8_7
11. Xu, J., Zhang, W., Zhang, Z., et al.: Clustering-based acceleration for virtual machine image deduplication in the cloud environment. J. Syst. Softw. **121**, 144–156 (2016)
12. Motik, B., Nenov, Y., Piro, R.E.F., et al.: Incremental update of Datalog materialisation: the backward/forward algorithm. In: AAAI, pp. 1560–1568 (2015)
13. Liu, C., Qi, G., Wang, H., Yu, Y.: Large scale fuzzy pD^* reasoning using MapReduce. In: Aroyo, L., Welty, C., Alani, H., Taylor, J., Bernstein, A., Kagal, L., Noy, N., Blomqvist, E. (eds.) ISWC 2011. LNCS, vol. 7031, pp. 405–420. Springer, Heidelberg (2011). doi:10.1007/978-3-642-25073-6_26
14. Carroll, J., Dickinson, I., et al.: Jena: implementing the semantic web recommendations. In: Proceedings of the 13th International Conference on World Wide Web, pp. 74–83 (2004)
15. Rajaraman, A., Ullman, J.D.: Mining of Massive Datasets. Cambridge University Press, New York (2011)
16. Urbani, J., et al.: DynamiTE: parallel materialization of dynamic RDF data. In: Alani, H., et al. (eds.) ISWC 2013. LNCS, vol. 8218, pp. 657–672. Springer, Heidelberg (2013). doi:10.1007/978-3-642-41335-3_41
17. Guo, Y., Pan, Z.X., Heflin, J.: LUBM: a benchmark for OWL knowledge base systems. J. Web Semant. **3**(2–3), 158–182 (2005)
18. Motik, B., Nenov, Y., Piro, R., et al.: Parallel materialisation of Datalog programs in centralised, main-memory RDF systems. In: Proceedings of the Twenty-Eighth Conference on Artificial Intelligence (AAAI), pp. 129–137 (2014)

19. Abiteboul, S., Hull, R., Vianu, V.: Foundations of Databases. Addison Wesley, Boston (1995)
20. Gao, J., Zhou, J.S., Zhou, C., et al.: GLog: a high level graph analysis system using MapReduce. In: ICDE, pp. 544–555 (2014)
21. Ullman, J.D.: Principles of Database and Knowledge-base Systems, vol. I. Computer Science Press, New York (1988)
22. Afrati, F.N., Ullman, J.D.: Transitive closure and recursive Datalog implemented on clusters. In: ICDT, pp. 132–143 (2012)
23. Subercaze, J., et al.: Inferray: fast in-memory RDF inference. VLDB **9**, 468–479 (2016)

A Data Type-Driven Property Alignment Framework for Product Duplicate Detection on the Web

Gijs van Rooij, Ravi Sewnarain, Martin Skogholt, Tim van der Zaan,
Flavius Frasincar$^{(\boxtimes)}$, and Kim Schouten

Erasmus University Rotterdam, PO Box 1738, 3000 DR Rotterdam, The Netherlands
g.vanrooij1@gmail.com, r.sewnarain1@gmail.com, m.skogholt@gmail.com,
tvanderzaan1@gmail.com, {frasincar,schouten}@ese.eur.nl

Abstract. During the last decade daily life has morphed into a world of broadband ubiquity, where devices facilitate constant engagement. As a consequence of this, the area of e-commerce has seen an immense growth. Despite the market opportunities for retailers and the ease for customers to acquire products through webshops, the shift to digital retail has its drawbacks. For example, it leads to cluttered and incomparable information among different webshops, which calls for an automated method to regain homogeneity in product representations. This paper presents a product duplicate detection solution, which exploits a data type-driven property alignment framework. Based on the performed experiment, we show a statistically significant improvement of the F_1-score from 47.91 % to 78.13 % compared to an existing state-of-the-art approach.

Keywords: Web products · Data types · Property matching · Duplicate detection

1 Introduction

The evidence for e-commerce popularity surrounds us, whether it is commuters immersed in their tablet to order groceries or a shopper buying a book on her phone to avoid the hassle of carrying it home and save time spent on shopping. Online shopping has become widely used and a common way of acquiring products. Digital retail sales has hit an all time high and keeps growing according to recent projections by eMarketer [5]. As stated by Jeff Jordan of Andreessen Horowitz: 'We're in the midst of a profound structural shift from physical to digital retail'. The prosperity in e-commerce leads to more laborious product comparisons among different webshops, since characteristics of products are represented by different lexical representations and webshops provide different characteristics on products. This calls for an automated framework on product duplicate detection to regain homogeneity in product representations. The basic requirement of this framework involves enabling a fair comparison among products of different webshops. Although a lot of methods have been proposed

W. Cellary et al. (Eds.): WISE 2016, Part I, LNCS 10041, pp. 380–395, 2016.
DOI: 10.1007/978-3-319-48740-3_28

on duplicate detection [3,15], the idea of pre-processing data first, apart from data cleaning [4], is not a popular one. One of the few methods related to property alignment is [9], which takes the characteristics of products (which will be referred to as properties from now on) into account, as well as the use of different measurement units. Those ideas are both adopted and further extended upon in our research. Many methods on duplicate detection use either TF-IDF [11] or use some lexical similarity measure [13], but lack usage of a semantic similarity between words. In addition, besides some of the existing approaches on duplicate detection, such as [13], most approaches do not take typographical errors into account. To solve these drawbacks of existing methods our method exploits the semantic similarities between words and tackles the problem of typographical errors. In addition, we propose to use an elaborate pre-processing part to determine matching properties to be used in the remaining steps of our solution.

A typical product duplicate detection framework consists of three steps. The first step involves reducing the number of comparisons between products by so-called blocking. This ensures that only the products within the same block or partition of the dataset are compared. Next, the similarity scores between product-pairs within the same block are calculated. In the last step product duplicates are determined based on these scores. Our solution for product duplicate detection includes an additional step which precedes the above-mentioned steps. This step consists of an extensive data type-driven property alignment (sub)framework. Products are characterized by several properties, each property consists of a property Key and a property Value (from now on referred to as simply Key and Value). For example, for the 'aspect ratio' of a television the property Key is 'aspect ratio' and the property Value could be '4:3'. These together form a property or Key-Value pair (KVP). First, a data type is determined for a property Value. Next, the similarity of a pair of properties is evaluated based on the similarity of the two Keys and the similarity of the corresponding Values. After the extensive data type-driven property alignment (sub)framework is used to find matching properties, the product duplicate detection framework is applied, which makes use of these matched properties. The product duplicate detection framework is built upon the state-of-the-art Multi-component Similarity Method (MSM) [2]. This framework determines the product duplicates among different webshops, using a novel similarity function and an adapted single linkage clustering algorithm.

The evaluation of the proposed framework is done on a dataset that describes 1446 televisions accompanied by their characteristics or so-called properties. The data is a collection of products from two webshops, i.e., bestbuy.com, and newegg.com. The dataset contains 774 and 672 products for these webshops, respectively. Each unique television has a 'modelID', 'title', 'url', and 'webshop' along with its list of properties. In total 200 unique Keys are present in the dataset. The product with the most comprehensive characteristics contains 61 properties, whereas the product with least exhaustive characteristics only contain one property. The characteristics of bestbuy.com consists of 37 properties on average, whereas the characteristics of newegg.com only contains 22 properties on average.

The remainder of this paper is organized as follows. Section 2 elaborates upon existing techniques on product duplicate detection and property alignment. We describe the proposed methods in Sect. 3 and in Sect. 4 we evaluate the results of our proposed methods and in Sect. 5 we discuss our results and give some suggestions for further research.

2   Related Work

The aim of our research is to better detect product duplicates by using an extensive property alignment framework. In order to do so, we use existing methods and introduce new ones. As will be discussed in both Subsects. 2.1 and 2.2, there are existing methods that already give a reasonably good performance, however, they have shortcomings that we will address by means of extensions and new methods. In Subsect. 2.1 we present an overview of current methods that address property alignment. Subsect. 2.2 shows some of the current methods that deal with duplicate detection.

2.1   Property Alignment

The idea of intelligently pre-processing data, before pairwise product comparisons are conducted, is not popular. Most of the research concerns data cleaning, see for example [4], which provides an overview of methods on data preparation. Many are concerned with the Values corresponding to a Key to determine the data type of the Key, whereas [9] formalizes product information in an ontology in order to compare products. Data can be represented in different ways, for example as a string or an integer. One can imagine that comparing strings with integers makes no sense. Using such information should make the property alignment more accurate. [9] introduced the use of measurement units. '2m' is not the same as '2lbs.' and therefore they should not be considered as a match. In our method we introduce Block Classification where the insight gained from [9] is used. Furthermore, we add Value Classification and Property Classification in order to align properties and detect property matches, making use of the Block Classification.

2.2   Duplicate Detection

The first method we discuss for duplicate detection is the one proposed by [11], which is based on the popular Term Frequency–Inverse Document Frequency method (TF-IDF). First, it is determined how many times a unique word occurs in a document, which is the Term Frequency. Then, the Inverse Document Frequency is calculated by taking the logarithm of the number of documents minus the logarithm of the number of documents containing that word. The final TF-IDF score for each word is calculated by dividing the term frequency by the inverse document frequency. A cosine similarity matrix is then constructed for all combinations of products. A shortcoming of this method is that it makes no use of the semantic similarities between words.

[14] proposes the Title Model Words Method (TMWM). This method determines if two products are duplicates by considering a similarity score. At first, an initial product title similarity score is calculated. This is done by calculating the cosine similarity of two product titles (interpreted as a bag of words). Then, the product title similarity score is compared with a threshold α to determine if the two products are duplicates. If the products are not duplicates based on the product title similarity score, *model words* are extracted for the two products. Model words are words consisting of both numeric and non-numeric tokens. The two products are considered to be different if the normalized Levenshtein distance of the non-numeric part is smaller than the threshold 0.5, while the normalized Levenshtein distance of the numeric part is larger than the threshold of 0.5. If the model words match, i.e., the numeric part is equal and non-numeric part is approximately the same, then it could be an indication of duplicate products. Therefore, the authors update the initial similarity score by taking a weighted average of the initial product similarity title score and the average Levenshtein distance of the model words. Based on the final similarity score and a threshold δ, it is determined if the products are duplicates. This method does not consider different representations of data. Therefore, products that are in fact duplicates, can be labelled as different.

[1] introduces the Hybrid Similarity Method (HSM), which extends the TMWM method. Instead of only comparing the title, [1] uses information from the product properties in order to construct a similarity measure. This measure is based on two methods. The first method checks for each combination of properties from both products if the corresponding keys are matching. Then, the similarity between the corresponding Values is calculated. The authors use the cosine similarity as the first part of the similarity measure. The second part concerns properties with no matching Key. Model words are extracted from the Values of these properties and combined in two sets, one for each product. Then, the percentage of matching model words between the two sets is calculated, which is the second part of the similarity measure. At last, the authors of [1] introduce a weight θ which is based on the number of Key matches and take a weighted average of both parts to construct the similarity measure. A shortcoming of this method is that it is only applicable for comparing two shops. Differently, [2] proposes a Multi-Component Similarity Method(MSM), which is applicable to more than two shops.

MSM builds on HSM and makes use of TMWM. It introduces a hierarchical clustering approach, which allows for comparing multiple shops. Its similarity measure consists of three parts. At first, a list of brands is introduced to detect if products have different brands. If they are not different, the Keys are given a score by making use of q-grams [12]. The authors use q-grams, in order to deal with typographical errors and abbreviations at both the Key and Value level. The second part is the same as in HSM, where the percentage of matching model words is calculated. The third part consists of calculating a similarity measure based on TMWM, where TMWM is adjusted to return a similarity score instead of a boolean result.

We extended the model of [2] by introducing a data-type property alignment framework as a pre-processing. Therefore, our approach in finding the similarity measure between properties differs. Furthermore, instead of using a hierarchical clustering approach, we use a trained threshold, which we compare with the similarity score, in order to determine if two products are duplicates. In order to overcome typographical errors and abbreviations, we will make use of q-grams similarity as well, which is actually the same as the Jaccard Similarity with q-shingles [10], later on called k-shingles.

3   Method

To determine property matches and detect duplicates of products over different webshops several methods have been incorporated in two frameworks. This section describes the two frameworks and the various methods that have been implemented. The first framework is the property alignment framework, which is described in Subsect. 3.1 and the second framework is the product duplicate detection framework, which is described in Subsect. 3.2.

3.1   Property Alignment Framework

This subsection explains the property alignment framework and all incorporated methods in detail. Subsubsections 3.1.1, 3.1.2, and 3.1.3 elaborate upon the *Block-*, *Value-*, and *Property Classification* methods, respectively. In Subsubsect. 3.1.4 the property matching is explained. Subsubsections 3.1.1, 3.1.2, and 3.1.3 build upon each other's results in order to classify each property to a certain type. These property types are last used in Subsubsect. 3.1.4 to match properties.

3.1.1   Block Classification

Our goal is to align or match properties of TVs from different webshops. The data from different webshops is not easily comparable. In order to tackle this problem, we introduce *Block Classification*. Recall that a property of a product consists of a Key and a corresponding Value. Keys are directly comparable, since each Key is a string. However, Values are not always a string and therefore comparing Values is not as straightforward. The Value representation can differ quite significantly between webshops and even within a webshop. For example, the Key 'Brightness' is described with the Value '450 cd/m^2'. This Value consists of two separated parts (split by a space), namely an integer and a string. The same key is also described with the Values '350 cd/m2' and '450Nit' for other products. The latter consists of two parts as well, viz. an integer and a string. However, the representation of the Value is not separated by a space. Clearly, this shows that Values are not easily comparable.

In order to deal with the previously mentioned issues, we propose *Block Classification*. The idea is to partition a Value based on white spaces, where the resulting partitions of the Value are called blocks. Now that the Values are split into blocks, one can compare the blocks of the Values. This makes sense if one compares integers with integers, strings with strings, and so on. Therefore, we introduce *Type Classification* of the blocks. First it is checked whether a block is strictly Numerical. In this case, it is sequentially checked if the block is of the type: Integer, Float, Fraction, Ratio, Percentage, or Dash.

If a block strictly consists of alphabetical characters, it is checked whether the block belongs to the block type Boolean, Measure, or Compound. If Boolean, Measure, and Compound are not suitable as types, the block will be classified as a Word block. A Boolean block can have either 'Yes' or 'No' as Value. The block type Measure represents a measurement unit listed in Table 1. The expressions of the measurement units vary over webshops, for example, the weight of a product might be given in Kilogram(s), Kilo(s), Kg(s), Pound(s), or Lb(s). The block type Compound represents the symbol 'x' or the word 'times' to connect two or more blocks, for example: '52 × 6.8'. If a block is neither Numerical nor Alphabetical, the block type Universal will be assigned.

Table 1. Measuring units expressed by different unit symbols

Measuring unit	Unit symbol
Weight	Kilogram(s), Kilo(s), Kg(s), Pound(s), Lb(s)
Frequency	Hertz, Hz
Energy & Electricity	Watt(s), W, kWh, Joule(s), kiloJoule(s), kJ, J, Volt(s), V
Sound intensity	Bel(s), Decibel(s), B, dB
Length	Inch(es), cm, mm, "
Brightness	cd/m2, cd/m^2, cd/m, Nit

Even though *Type Classification* is introduced, one can still find cases where a comparison is not optimal. Consider for example the Values '180Hz' and '180 Hz', which describe the same information. The first block is classified as Universal, while for the latter Value the blocks are classified as Integer and Measure. Clearly, both Values are equal. In order to overcome this problem, an Universal block, which contains an integer and a measurement unit (in this case Hz), will be split into 2 blocks, we use the measurement units form Table 1. Namely, an Integer block and a Measure block. In this way, one is able to compare 180Hz and 180 Hz. Table 2 shows how several Values are split into block and how these blocks are subsequently classified. Also it is indicated how Values containing a measurement unit are classified, such as '180Hz'.

Table 2. Examples of *Block Classification*

Property value	Set of blocks	Types of blocks
180Hz	'180' + 'Hz'	Integer + Measure
180 Hz	'180' + 'Hz'	Integer + Measure
4:3 and 16:9	'4:3' + 'and' + '16:9'	Ratio + Word + Ratio
B007B9PMCO	'B007B9PMCO'	Universal
122 W	'122' + 'W'	Integer + Measure
Yes	'Yes'	Boolean
52 × 6.8	'52' + '×' + '6.8'	Integer + Compound + Float
52 % humidity	'52 %' + 'humidity'	Percentage + Word
55-2/3 mm	'55-2/3' + 'mm'	Dash + Measure

3.1.2   Value Classification

The goal is to compare properties. Recall that a property consists of a Key-Value pair. As mentioned before, comparing Values is not straightforward. In the previous section we introduced *Block Classification* and we have shown how to classify parts (blocks) of Values. However, we do not want to compare parts of the Values, but the Values themselves. Therefore, we introduce *Value Classification*, where the results of the *Block Classification* is used to determine the type of the corresponding Value.

The classification of a Value depends on the number of blocks. Three options are considered: one block, two blocks, and three or more blocks. For each of these options the *Value Classification* differs. In case a Value consists of one block, three types are considered: Boolean, Quantitative, and Qualitative. A Value is of type Boolean if the block is a Boolean block. A Value is classified as Quantitative if the block is strictly Numerical. And lastly, a Value is of type Qualitative if the block is either a Word block or a Universal block.

In case a Value consists of two blocks, the following three types are considered: Measure, Quantitative, and Qualitative. A Value is of type Measure if the first block is Numerical followed by a Measure block. A Value is of type Quantitative if the first block is of type Numerical followed by a Word block. If the two blocks are of any other combination of types, the Value is classified as Qualitative.

In case a Value consists of three or more blocks, two types are considered: Compound, and Qualitative. At first it is checked whether one or more blocks are of type Compound. A Value is classified as a Compound Value, if the following three restrictions are satisfied. First, the one or two blocks before a Compound block must be either Numerical followed by a Word block or Numerical followed by a Measure block or simply a Numerical block. Secondly, between two Compound blocks, the one or two blocks must be Numerical followed by a Word or a Measure block or simply a single Numerical block. At last, the first block after the last compound block must be Numerical. For example, '52 mm × 45 mm × 20' would be classified as a compound Value. This is visualized in Eq. 1, where a Compound Value is considered. A Value is Qualitative, if there is no Compound block present.

$$Value: \overbrace{\underbrace{Block_i + Block_j}_{\text{Numerical + Word or Measure}} \quad x \quad \underbrace{Block_k + Block_l}_{\text{Numerical + Word or Measure}} \quad x \quad \underbrace{Block_m}_{\text{Numerical Block}}}^{\text{Compound Type Value}} \quad (1)$$

3.1.3 Property Classification

In order to compare two properties we also classify the type of the properties, in other words *Property Classification* is introduced. To determine the type of a property the previously determined Value types are used. First, for each unique Key the Values belonging to this Key are aggregated in a list. For example, the key 'Brightness' will have multiple observations in the dataset all with different Values. For example, '450 Nit' could be found or '350 cd/m' or other Values with different representations. Obviously, all these Values together indicate the type of the property. It can also happen that certain Values have been misspelled or otherwise incorrectly entered. For example, it could be that for the Key 'Brightness' there is a Value which is simply '250'. This Value would be classified as a Quantitative Value, while the property 'Brightness' should still be classified as a Measure type. This is why the type of the property is equal to the type that has been assigned most frequently to the Values belonging to the unique Key. This way certain errors in the Values are circumvented.

The whole process of *Block Classification*, *Value Classification*, and *Property Classification* is shown below in Algorithm 1.

As can be seen in this Algorithm 1, the different values are first split into a list of Blocks after which the type of the Blocks is determined. When this is done the type of the value can be determined based on the Blocks. Finally, a list of aggregated properties is filled and each aggregate property is classified as being of a certain type based on the Values.

3.1.4 Property Matching

In the previous subsections we introduced *Block Classification*, *Value Classification*, and *Property Classification*. These are the building blocks for the property matching method. In order to match properties, a quantification of the similarity between properties is needed. To compare properties a *Property Score* is calculated to represent the similarity between two properties. Property matching is based on the similarity between two Keys (Key Score) and the similarity between two Values (Value Score). The Property Score is a weighted average between these two scores. If the Property Score exceeds a certain trained threshold, then the two compared properties are considered as a match.

The Key Score is a combination of two similarity measures; the lexical similarity and the semantic similarity. To find the lexical similarity, the Jaccard similarity with k-shingles [10], with k representing the number of considered characters, is used. For this metric we vary k between 2 and 8. For the semantic similarity

---
Algorithm 1. Property Classification

Initialize list of products with properties, X
Initialize list of aggregated properties, A
for Unique key $k \in X$ **do**
    Initialize list of values, V
    for $prop \in X$ with key k **do**
        Split value $v \in prop$ by space
        Initialize list of blocks, B
        for substring $s \in v$ **do**
            Create block $b(s)$
            Determine type of $b(s)$
            Add block $b(s)$ to B
        end for
        Add list B to value v
        Determine type of value v based on B
        Add value $v \in prop$ to list of values V
    end for
    Create aggregated property AP_k with key k and list of values, V
    Determine type of aggregated property AP_k
    Add aggregated property AP_k to A
end for
---

between two Keys the meaning of the words are considered, where WordNet [8] is used as a semantic lexicon. One can understand that Keys which have the same meaning, but are represented with different words, should have sets of synonyms (Synsets) in common and thus a high (semantic) similarity. First, the lemmas (the root of a word) of all the words in both Keys are retrieved. These lemmas can then be disambiguated and assigned a certain sense from WordNet using Lesk [6], where the context is defined as the words in the other keys of the current type of product, i.e., televisions in our case. A word sense is simply the meaning of a word with a set of synonyms and a certain gloss. When comparing two Keys with certain synsets assigned to each individual word or compound words, if this is the case, the Jaccard similarity is used on the glosses of the synsets. Basically, two sets are created with the combined glosses and the Jaccard similarity is taken over these two sets of glosses after stopwords[1] have been removed. The Key Score is a weighted average of the lexical similarity score and the semantic similarity score. This is summarized in Eq. 2.

$$Key\ score = (1 - \delta) \times Lexical\ Sim.\ score + \delta \times Semantic\ Sim.\ score, \qquad (2)$$

where δ is a trained weight.

The Value Score depends on the type of the property. First, we iterate through all the Values belonging to the unique Keys of the compared properties. Only the Values that have the same data type as the property are used for the comparison. In case both properties are of type Qualitative, a lexical

---

[1] The stopwords used can be found at http://www.ranks.nl/stopwords.

similarity metric is used. Here, the same lexical similarity is used as in the Key Score, i.e., k-shingles Jaccard Similarity. In case both Values are Boolean, we first calculate the fraction of the number of times 'Yes' occurs in the Values. Then, one minus the absolute value of the difference between the fractions is calculated and this is the Value Score. In case both of the Values are either of type Quantitative, Measure, or Compound, the Mann-Whitney U test is used [7]. This test is used to check whether the null hypothesis of a equal distribution of Values is statistically significant. Whenever both Values are Quantitative, the Numerical blocks are used in the Mann-Whitney U test. Whenever both Values have been classified as Measure Values and the Measure blocks are of the same unit, the Mann-Whitney U test is performed on the Numerical blocks of both Values. When both Values are Compound, the Numerical blocks are used in the Mann-Whitney U test. The Value score is then equal to the p-value of the Mann-Whitney U test. Now that the Key and Value Score have been obtained, the Property Score is a weighted average of the Key and Value score. This is shown in Eq. 3.

$$Property\ score = (1 - \theta) \times Key\ score + \theta \times Value\ score, \tag{3}$$

where θ is a trained weight. If this Property Score is higher than a certain trained threshold, the properties are classified as match.

3.2   Product Duplicate Detection Framework

The product duplicate detection framework consists of 3 heuristics to decrease the number of product comparisons before doing the actual duplicate detection computations. After heuristics-based preprocessing, the framework contains 2 steps to match products that are considered duplicates. First, the so-called shop heuristic assumes that there are no product duplicates present within the same shop. With this heuristic a lot of unnecessary comparisons are avoided. The second heuristic is the brand heuristic. This assumption presumes there are no product duplicates with the same brand. The third heuristic is the screen size heuristic. The assumption is that there are no product duplicates with different screen sizes. The second and third heuristics will be explained in more detail in Subsubsects. 3.2.1 and 3.2.2, respectively. To test whether products need to be indicated as being duplicates or not, the similarity between product titles and a list of properties are measured by means of a title-based score (i.e., title score) and properties-based score (i.e., property score). The first step consists of a score calculated based on the two product titles of the products that are being compared, as can be found in Subsubsect. 3.2.3. The second step consists of a score that is calculated based on the properties of the two products, as can be found in Sect. 3.2.4. Last, these steps will be combined into a so-called 'product score' as will be explained in Subsubsect. 3.2.5.

3.2.1   Finding the Brand for All Products

To find the brand of all the products in our dataset we make use of a list of well-known TV brands and manufacturers from Wikipedia[2]. This can easily be extended for other types of products as there are many such lists of manufacturers. First we scan all the products with their properties and count how often a property Value contains one of the TV brands from the list. The property Key for which a Value is most often found in the TV list, is then denoted as the brand key. All brands which are not present in the list, but belong to the brand key are added to the brand list. When this is done all the products that do not have the same brand are no longer compared by the algorithm.

3.2.2   Finding the Screen Size for All Products

The third heuristic assumes product duplicates can not have different screen sizes. Almost all products comprise the screen size in their product title, where the screen sizes that are in the titles are represented by some digits followed by the "(inch) sign and can easily be extracted from the title. The products that do not contain their screen size in the product title are included in the comparison. All of the products with different screen sizes are no longer compared, since it is impossible that two products with different screen sizes are duplicates. This heuristics is only used to decrease the number of unnecessary comparisons of products and henceforth decrease the time it takes to run the algorithm.

3.2.3   Scoring of the Products on Their Titles

The similarity score of compared products based on their titles is found using the lexical similarity between these titles. First both titles are cleaned by removing the brand and two common words: 'refurbished' and 'open box:'. We consider these words to be meaningless in comparing products and we found these by analyzing the available data. Next, the similarity score is calculated using the same lexical similarity measure as for the property alignment framework, i.e., k-shingles Jaccard Similarity. This final score represents the title score for the two products that are being compared.

3.2.4   Scoring of the Products on the Values of Their Matching Keys

For the two products that are being compared a score is calculated for all their properties that either have the same Key or have been matched by our property matching algorithm. Each pair of matched properties adds value to the total property score. Note, each score of the separate pairs of matched properties is weighted with a learned weight dependent on the type of the properties.

If two properties are both of type Qualitative, Boolean, or are non-matching types, the similarity is calculated using the lexical similarity measure. For two properties that are both Measure properties, the score is simply 1, if the absolute value of the difference of the values is smaller than 1. The idea is that small

---

[2] https://en.wikipedia.org/wiki/List_of_television_manufacturers.

differences in numerical values do not indicate any dissimilarity, however, we need to take integers into account. For example, the number of HDMI ports can be 0,1,2,... If one product has 1 port and the compared product has 2, then this indicates dissimilarity between the two products. This is why we chose for an absolute difference smaller than 1. If either the type of measurement differs or the absolute value of the difference between the Values is larger than 1, the score is −1. For two properties that are both of type Quantitative the score is again 1 if the absolute value of the difference of the two Values is smaller than 1 and the score is −1, otherwise. However, if the Values do not solely consist of one numerical Value, the remainder is also taken into account by using the lexical similarity measure. The score of the two properties of type Quantitative is then the average of the score of the numerical part and the score of the non-numerical part. For Compound properties the score is simply the average of the score of each individual part. Each individual part is treated as a Quantitative type comparison and is simply scored 1 or −1 dependent on the absolute value of the difference of the two Values.

At last, the weighted scores are aggregated and the total score is normalized by dividing by the number of scores added up, such that the total property score ranges from 0 to 1. This can be seen in Eq. 4:

$$Property\ score = \frac{\sum_{p_i = p_j} score(p_i, p_j) \times w_i}{Number\ of\ times\ p_i\ matches\ p_j}, \tag{4}$$

where p_i is property i and w_i is the weight dependent on the type of the two properties. This score on the similarity of the matching properties is used later on to indicate duplicate products.

3.2.5   Final Score Based on Matching Keys and Titles

The final score between two products is a weighted average of the score calculated between the two titles and the score calculated for all the matched properties. This results in the following equation:

$$Product\ score = (1 - \tau) \times title\ score + \tau \times property\ score, \tag{5}$$

where τ is a trained parameter. If this product score exceeds a certain trained threshold the products being compared are indicated as product duplicates.

4   Evaluation

In Subsect. 4.1 we describe the process of evaluating our algorithms on property matching and product duplicate detection. Subsect. 4.2 provides the results on the framework for property matching and Subsect. 4.3 evaluates the obtained results from the product duplicate detection framework.

4.1 Evaluation Method

The so-called bootstrap method, i.e., random sampling with replacement, is used to assign measures of accuracy to sample estimates. The number of random samples is set to 50. Each random sample consisting of numerous TVs represents the training set, whereas the remainder of the TVs not allocated to the random sample is used as a test set for validation. Additionally, all the parameters used throughout are trained using a genetic algorithm. This includes: the two thresholds, the weights in the Key score, Property Score, and Property Score, and the weights in the Product Score.

When evaluating the algorithm, a 'match' for the property matching framework means two different properties are considered the same and a 'match' for the product duplicate detection framework means two products are considered as duplicates. The F_1-measure, which is the harmonic mean of the precision and recall, is used as the performance measure. Additionally, the precision and recall themselves are used for measuring the performance. The precision is how accurate the algorithm is, i.e., the correct matches divided by the correct matches plus incorrect matches. The recall is an indication of how much is missed by the algorithm, i.e., the correct matches divided by the correct matches plus the missed matches, that our algorithm should have indicated.

The evaluation of the property matching framework and of the product duplicate detection framework are conducted separately. In order to be able to find performance measures for the algorithms the results of the algorithms should be related to a golden standard. For the property matching framework this golden standard is based on the so-called inter annotator agreement (IAA)[3]. Matches between Keys are added to the golden standard in case a sufficient percentage of selections by the annotators overlap. In our case we have used four annotators. If three out of four annotators indicate the same match then it is considered a sufficient match. In our case the four annotators agreed unanimously in 82.2 % of the cases. When two products share the same ModelID, they are considered as duplicates, which acts as our golden standard. Note that the ModelIDs are often missing on the Web, which makes our framework interesting for duplicate detection, but we have used datasets, where the ModelID is present, to evaluate our algorithm and to be able to train the parameters.

4.2 Property Alignment Framework

In Table 3, the reported F_1-measures of the 50 bootstraps for the property matching framework can be seen. These results are with using the semantic similarity and with the Jaccard similarity using different k-shingles as the lexical similarity measure. When using the semantic similarity measure, the results were higher than without, which is why we only reported these. The algorithm was executed for k varying between 2 and 8. The highest results were achieved for $k = 3$. For

---

[3] https://corpuslinguisticmethods.wordpress.com/2014/01/15/what-is-inter-annotator-agreement/.

Table 3. The average F_1-measure for the property alignment framework. The results are for $k = 2$ to $k = 8$ and with the semantic similarity measure

Metric	Mean	Std. Dev
Jaccard 2-shingle	79.69	2.60
Jaccard 3-shingle	81.55	1.96
Jaccard 4-shingle	81.14	1.86
Jaccard 5-shingle	80.26	2.06
Jaccard 6-shingle	78.32	2.48
Jaccard 7-shingle	73.19	2.66
Jaccard 8-shingle	71.13	2.30

both $k = 7$ and $k = 8$, there was a significant drop of around 5% points in performance for the algorithm. This is quite an intuitive results, since a lot of the Keys are quite short. This means that for a high k, the Keys are completely in a single shingle and would have to be identical otherwise the similarity is 0. This is, obviously, not always the case and this is probably why the results start dropping around $k = 7$.

4.3   Product Duplicate Detection Framework

In Table 4, the reported F_1-measures of the 50 bootstraps of the product duplicate detection framework can be seen. These results have been gained using the semantic similarity and with k ranging from 2 to 8. For the product duplicate detection the highest performance was achieved for $k = 4$. For the product duplicate detection framework, the drop in performance is much less severe, when k is increased. This is probably due to the fact that the titles are often quite long for products. As this is the application of the lexical similarity measure with the most impact, it seems that a high k has less negative influence, since the titles are still compared with a lot of shingles instead of a single shingle.

To compare our method to that of MSM, we have executed our algorithm on the same bootstraps as MSM, with 50 bootstraps in total. The results can be seen in Table 5. On average our methods scores 30.21% better than the MSM method. The standard deviations can also be found in Table 5. Since these are paired observations we can use a paired t-test. Our method significantly outperforms MSM, even with a 99.9% significance level.

Regarding the complexity analysis of the developed algorithms, we can state that the Property Alignment framework has complexity $O(p^2)$ where p is the number of (unique) properties in the dataset, and the Product Duplicate Detection Framework has complexity $O(n^2 \bar{p}^2)$ where n is the number of products and $\bar{p}$ is the average number of properties per product in the dataset.

Table 4. The average F_1-measure for the product duplicate detection framework. The results are for $k = 2$ to $k = 8$ and with the semantic similarity measure

Metric	Mean	Std. Dev
Jaccard 2-shingle	76.41	2.84
Jaccard 3-shingle	77.88	2.68
Jaccard 4-shingle	78.13	2.65
Jaccard 5-shingle	77.37	2.75
Jaccard 6-shingle	76.65	2.69
Jaccard 7-shingle	76.32	2.52
Jaccard 8-shingle	76.10	2.63

Table 5. This table shows the F_1-measures for MSM and our proposed algorithm, as well as the difference in performance. Our algorithm was executed using the semantic similarity measure and $k = 4$-shingles

Method	F_1-measure	Std. Dev	95 % Interval
MSM	47.91	3.05	[41.93 - 53.89]
Our method	78.13	2.65	[72.94 - 83.32]
Difference	30.21	3.65	[23.01 - 37.36]

5 Conclusion

In this paper we have proposed and implemented an extensive property alignment framework, which makes use of inferred data types. It has an F_1 of 81.55 % with the best performing similarity measure on 50 bootstraps. Additionally, we have significantly improved upon the MSM algorithm with regard to the product duplicate detection. Our results are on average 30.21 % higher, when using the F_1-measure. The significance of the improvement has been tested with a t-test and our results are significantly better than those of MSM, even with a 99.9 % significance level. Moreover, the framework can easily be adapted to other classes of products.

The first suggestion for future work is to introduce a framework that can handle different measurement units, such that, for example, properties measured in feet can be compared to properties measured in meters. Furthermore, our method searches for the similarity of property pairs based on the similarity of property Keys *and* property Values. Additionally, it could prove useful to exploit the similarity of Keys with Values for finding additional property matches. For example, for the property 'Parental Control' in our dataset the Value was most often 'V-Chip', while there is also a property with Key 'V-Chip', where the Values are 'Yes' or 'No'.

References

1. Bakker, M., Frasincar, F., Vandic, D.: A hybrid model words-driven approach for web product duplicate detection. In: Salinesi, C., Norrie, M.C., Pastor, Ó. (eds.) CAiSE 2013. LNCS, vol. 7908, pp. 149–161. Springer, Heidelberg (2013). doi:10. 1007/978-3-642-38709-8_10

2. van Bezu, R., Borst, S., Rijkse, R., Verhagen, J., Vandic, D., Frasincar, F.: Multi-component similarity method for web product duplicate detection. In: 30th Symposium On Applied Computing (SAC 2015), pp. 761–768. ACM (2015)

3. Bilenko, M., Mooney, R.J.: Adaptive duplicate detection using learnable string similarity measures. In: Proceedings of the Ninth ACM SIGKDD International Conference on Knowledge Discovery and Data Mining (KDD 2003), pp. 39–48. ACM (2003)

4. Elmagarmid, A.K., Ipeirotis, P.G., Verykios, V.S.: Duplicate record detection: a survey. IEEE Trans. Knowl. Data Eng. **19**(1), 1–16 (2007)

5. eMarketer: Retail Sales Worldwide Will Top $22 Trillion This Year. http://www.emarketer.com

6. Lesk, M.: Automatic sense disambiguation using machine readable dictionaries: how to tell a pine cone from an ice cream cone. In: 5th Annual International Conference on Systems Documentation (SIGDOC 1986), pp. 24–26. ACM (1986)

7. Mann, H.B., Whitney, D.R.: On a test of whether one of two random variables is stochastically larger than the other. Ann. Math. Stat. **18**, 50–60 (1947)

8. Miller, G., Beckwith, R., Felbaum, C., Gross, D., Miller, K.: Introduction to Word-Net: an on-line lexical database. Int. J. Lexicography (Special Issue) **3**(4), 235–312 (1990)

9. Nederstigt, L.J., Aanen, S.S., Vandic, D., Frasincar, F.: FLOPPIES: a framework for large-scale ontology population of product information from tabular data in e-commerce stores. Decis. Support Syst. **59**, 296–311 (2014)

10. Rajaraman, A., Ullman, J.D.: Finding similar items. In: Mining of Massive Datasets, vol. 77, pp. 73–80. Cambridge University Press, Cambridge (2012)

11. Salton, G., Fox, E.A., Wu, H.: Extended Boolean information retrieval. Commun. ACM **26**(11), 1022–1036 (1983)

12. Ukkonen, E.: Approximate string-matching with Q-grams and maximal matches. Theoret. Comput. Sci. **92**(1), 191–211 (1992)

13. Vandic, D., van Dam, J.W., Frasincar, F.: A semantic-based approach for searching and browsing tag spaces. Decis. Support Syst. **54**(1), 644–654 (2012)

14. Vandic, D., Van Dam, J.W., Frasincar, F.: Faceted product search powered by the semantic web. Decis. Support Syst. **53**(3), 425–437 (2012)

15. Xiao, C., Wang, W., Lin, X., Yu, J.X., Wang, G.: Efficient similarity joins for near-duplicate detection. ACM Trans. Database Syst. **36**(3), 15:1–15:41 (2011)

A Semantic Data Parallel Query Method Based on Hadoop

Liu Yang[1], Liu Yang[1], Jiangbo Niu[1], Zhigang Hu[1],
Jun Long[2], and Meiguang Zheng[1](✉)

[1] School of Software, Center South University, Changsha 410073, China
{yangliu,zhengmeiguang}@csu.edu.cn
[2] School of Information Science and Engineering,
Center South University, Changsha 410073, China
jlong@csu.edu.cn

Abstract. To achieve efficient large-scale RDF data queries, we designed a parallel two-phase query strategy-PAQS for large-scale RDF data based on MapReduce, which is divided into two stages: the SPARQL pretreatment stage and the distributed query execution stage. In the SPARQL pretreatment stage, a SPARQL query classification algorithm is implemented, which determines the join order of connection variables by calculating the correlation between the variables in a SPARQL query statement; then, the join between SPARQL clauses is divided into the minimum number of MapReduce jobs according to the connection variables. The distributed query execution phase accomplishes large-scale RDF data query concurrently based on MapReduce jobs from the SPARQL pretreatment stage. The experimental results on the LUMB benchmark set indicate that PAQS can query large-scale RDF data with good efficiency, stability, and scalability.

Keywords: Parallel processing · Semantic query · MapReduce · SPARQL · RDF store

1 Introduction

In the era of big data, the volume of global data is growing at an annual rate of over 50 % [1]. Rapid Semantic Web growth has caused RDF (Resource Description Framework) [2] semantic data such as Wikipedia [3], bioinformatics [4], media data [5] and social networks [6] to dramatically increase. The amount of Linked Open Data (LOD) is rapidly growing, increasing from 295 available RDF data sets in 2011 to 1014 RDF data sets in 2014 [7]. The Semantic Web was estimated to contain 4.4 billion triples in 2009 and has now reached over 20 billion triples. Traditional single-nodes have difficulty handling large-scale data due to limitations in the centralized architecture. RDF continues to grow in volume and variety [8,9], so it brought significant challenges in scalable and efficient querying. Distributed paralleling methods are required to meet new demands for processing large amounts of RDF data.

© Springer International Publishing AG 2016
W. Cellary et al. (Eds.): WISE 2016, Part I, LNCS 10041, pp. 396–404, 2016.
DOI: 10.1007/978-3-319-48740-3_29

In recent years, researchers have proposed Semantics-based data storage and query strategies in cloud environments such as Hadoop [10,13]. However, further research and optimization are needed in storage space and query efficiency. Our main work is as follows in this paper:

(1) Three HBase tables (SP_O, PO_S, and OS_P) are designed to store massive amounts of RDF data using the storage method proposed by paper [10] based on the storage strategy of "two element combination" line key PO.
(2) The SPARQL query partitioning algorithm PAQS is designed and implemented. The connection order of the connection variables was determined by calculating the correlation degree of variables in the SPARQL statement and according to the connection variables. This method shortens query time for large-scale RDF data.
(3) RDF data parallel query is efficiently implemented based on the MapReduce distributed application development framework.

2   RDF Data Storage Strategy Based on HBase

RDF with the S(Subject), P(Predicate), O (Object) triple $<S, P, O>$ describe resources on the Web [11]. Entity information on the Web (or concepts) are represented on the subject S with URI, predicate P describes the attributes of the entity, and object O describes the value of the attributes [12]. SPARQL is the standard query language proposed by W3C for RDF data, which is similar to the syntax of SQL and the SELECT query method to find data to meet the conditions.

RDF distributed storage can be divided into HDFS and HBase programs. Various RDF query algorithms have been proposed based on these two storage methods.

(1) Myung et al. [14] store RDF data files in HDFS and create multiple MapReduce task to iteratively process SPARQL clauses of a join operation. However, this method directly stored RDF data into HDFS and lacked an efficient index structure. Husain et al. [15] proved that reducing MapReduce tasks can reduce RDF query times. They also stored RDF data in HDFS and lacked an efficient indexing structure.
(2) Sun et al. [16] designed six tables (S_PO, P_SO, O_SP, SP_O, PO_S and SO_P) to store RDF triples in HBase. However, this storage method occupies large storage space and creates redundant data because the RDF triples have 6 replicated copies. Franke et al. [17] designed two HBase tables of Tsp and Top, based on "Fixed-Column" SPO storage. However, a full table scan is necessary when S and O are unknown, which significantly increases query time.

This paper presents an SPO triple storage method, and it considers the parallelism, space cost, and time cost of each method. Three HBase tables (O_SP, PO_S, OS_P) are designed to store data. This method reduced overhead space compared with the *SPO* storage strategy based on "binary" row keys; row keys indicate the *SP*, *PO*, and *OS*, which can meet all of the combinations of the SPARQL triple pattern query matching conditions. It also reduces overhead time compared to the "column-fixed" row keys of the *SPO* storage strategy.

3   Two-stage Query Strategies for RDF Data

This paper uses two-stage of the RDF data query strategy based on SPARQL search to achieve massive RDF data parallel queries. To facilitate the description, we first set the following SPARQL query as an example and then define several concepts.

Definition 1: *TP(U)* represents the triple pattern in the SPARQL query statement. The U is a variable set of three triples, $\{X, Y, Z, XY, XZ, YZ\} \in U$.

There is at least one variable in the items of *TP(U)*, otherwise the SPARQL query statement will be meaningless. Above shows that the instance query in the triple model is represented as *TP1(X)*, *TP2(Y)*, *TP3(Z)*, *TP4(XZ)*, and *TP5(XY)*.

Definition 2: Connection variables are variables that appear in two or more $<S, P, O>$ triple patterns for multiple query clauses.

Definition 3: The correlation degree refers to the number of other connected variables directly related to a connected variable, expressed as $R(V)$, $\{X, Y, Z\} \in V$.

The two linking variables Y and Z directly relate to X in the query example shown in above, and only the linking variable X directly related to Y and Z, that is, $R(X) = 2$, $R(Y) = 1$, $R(Z) = 1$.

Definition 4: *IRS(U)* is the variable U of the intermediate results set, which is generated in the MapReduce task of the query process, $\{X, Y, Z, XY, XZ, YZ\} \in U$.

Definition 5: The query time refers to the total amount of time spent on the execution of all MapReduce tasks in query Q, expressed in Cost (Q). The time spent on each MapReduce task is expressed in Cost (job), so the sum of time spent is expressed as: $Cost(Q) = \sum_{i=1}^{n} Cost(job_i), 1 \leq n \leq m$.

Q represents the current SPARQL query, job_i is the current MapReduce task, n is the number of MapReduce tasks, and m is the number of connection variables.

The two-stage query strategy of the RDF data contains two stages of SPARQL preprocessing and distributed query execution. As shown in Fig. 1.

(1) SPARQL preprocessing is proposed based on SPARQL variables correlation query partitioning algorithm PAQS (join on Variable Relation). First, PAQS selects the linking variables in order from the SPARQL query triples, and then assigns the connection operation of the SPARQL sub-clause into the minimum MapReduce task.

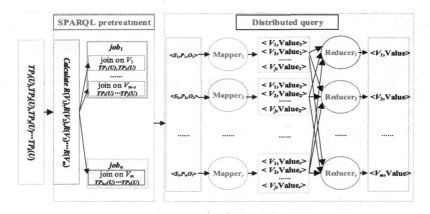

Fig. 1. The two-stage query strategy for RDF data

(2) The data should be read out from the corresponding HBase table when the query clauses are connected in the distributed query execution stage, and then the data are filtered and assembled in the Map stage. Finally, the connection task is completed in the Reduce stage.

3.1 SPARQL Preprocess PAQS Algorithm

The PAQS algorithm determines the connection order of the connection variables by calculating the SPARQL variable correlation degree according to the connection variable greedy partition SPARQL query statement. This achieves the minimum number of distributed query phase jobs (MapReduce task).

In the above PAQS algorithm, the time complexity is $O(mlogm)$ to quick sort of m connection variables. The outer loop (*while* loop) can execute n times at most, and the inner loop (for loop) can execute m times at the most, so the maximum time complexity of the algorithm is $(m(n+logm))$. m is the number of query statement connection variables, n is the number of MapReduce jobs, and $1 \leq n \leq m$.

In the example of the SPARQL query statement, the correlation degree of variables X, Y and Z are $R(X) = 2$, $R(Y) = 1$ and $R(Z) = 1$ respectively, according to the definitions in Sect. 3.1. The PAQS algorithm will choose the connection variable Y, Z, X by correlation degree in a non-decreasing order, because the query corresponds to 2 MapReduce jobs. Figure 2 shows the query partitioning process of the PAQS.

The query partitioning process of the PAQS algorithm can be analyzed by $Cost(query)$. $Cost(query) = Cost(job_1) + Cost(job_2)$.

Researchers have been divided over the SPARQL query based on the JOVF algorithm [15]. The JOVF algorithm chooses the connection variable that appeared most in the SPARQL query statement by greedy choice, and then assigns MapReduce jobs according to the connection variable in order. In the example of the SPARQL query statement shown in Fig. 3, the JOVF algorithm

---

Algorithm 1. PAQS Algorithm

Input: Q (SPARQL query)

Output: *job(MapReduce task) set*

1: $n \leftarrow 1$

2: $U \leftarrow$ sortOnRel($\{u_1, u_2, \ldots u_m\}$); //Sort of connected variables by non-decreasing degree of correlation

3: **while** $Q \neq$Empty **do**

4:    $job_n \leftarrow$Empty; //Current job

5:    $tmp \leftarrow$Empty; //Store intermediate connection results

6:    **for** $i \leftarrow 1$ to m **do**

7:       **if** canJoin(Q, u_i) = true //The Q neutron set can be connected to UI **then**

8:          $tmp \leftarrow tmp \cup$joinResult(Q, u_i); //Save connection results

9:          $Q \leftarrow Q - TP(Q, u_i)$; //Remove connected subset //from Q

10:          $job_n \leftarrow job_n \cup$(join($Q, u_i$); //Join the current job

11:       **end if**

12:    **end for**

13:    **if** tmp = Empty //Connection operation of the three triple is not exist **then**

14:       break;

15:    **end if**

16:    $Q \leftarrow Q \cup tmp$; //New variable set generated by adding intermediate links in Q

17:    $n \leftarrow n + 1$;

18: **end while**

19: return $\{ job_1, job_2, \ldots job_n \}$;

---

computes the appearance times of the connection variable X, Y, Z, which are 3, 3, and 2, respectively. It chooses the connection variable X, Y, Z sequentially, and then assigns 3 MapReduce jobs. Figure 3 shows the query partitioning process of the JOVF algorithm.

The query partitioning process of the JOVF algorithm can be analyzed by $Cost(query)$. $Cost(query) = Cost(job_1) + Cost(job_2) + Cost(job_3)$.

Comparing the different query partitioning processes of the same query statement shown in Fig. 3 with those of Fig. 2, the PAQS algorithm had a lower query time than the JOVF algorithm, because the number of assigned jobs in the PAQS algorithm was lower than that of the JOVF algorithm.

3.2   Distributed Query Execution

The SPARQL preprocessing stage is divided into a lot of job. The distributed query execution stage is based on RDF to realize the parallel query of the MapReduce data, as shown in Fig. 4.

(1) The HBase data read stage: When the triple in the query clause is involved in the connection operation, the data should be read from the corresponding HBase table.

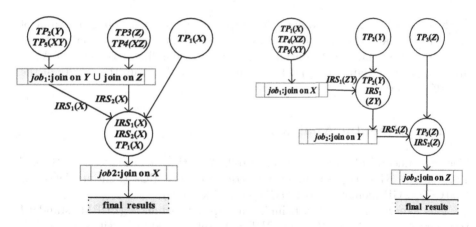

Fig. 2. Query partitioning process in PAQS

Fig. 3. Query partitioning process in JOVF

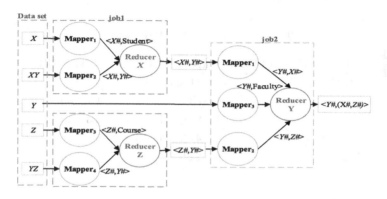

Fig. 4. Instances of the SPARQL query execution process

(2) The Map stage: The query clause in the triple of the corresponding data is set to $<key, value>$, which serves as the key for connecting the corresponding value of the variable. The $value$ is divided into two cases: (1) The $value$ of the triple with one variable is set to the constant value, and the $value$ of the data set Y in job_1 shown in Fig. 4 is set to be the constant $University$ as the object; (2) The data value of the triple with two variables is the non-connected variable, and the key of data set XY is X#, and the value of variable Y is Y#.

(3) The Reduce stage: complete the connection of multiple query clauses that correspond to the same connection variable. In Fig. 4, after the clause is connected to the Y value of the key, the key of the data is still Y, and the value is the combination of all values that participated in the connection operations, so the value is $University\#$, which is output by the self-defined Reduce() function.

In the case of multiple job values, the output of the previous job is the input of the latter job, and the input of job2 are derived from the read of X data set and the job1 output data set to output the final corresponding data value of X, Y, and Z after the Map stage and the Reduce stages. That are the results of the SPARQL query.

4   Experimental Analysis

The experimental platform used Hadoop-2.5.2 as the operation system, HBase-1.0.0 as the RDF triple storage database, and four PCs (Pentium IV CPU 3.00 GHz, 2 GB memory, and 160 GB hard disk space).

This experiment uses the Lehigh University benchmark (LUBM) standard test data set. The 1, 50, 100, 300 RDF data sets of 500 universities are tested. The data set corresponded to the number of triples, and the file size is shown in Table 1.

Table 1. The size of LUBM test set

School number in LUBM	Three tuple number	File size (GB)
10	1, 311, 409	0.2
50	6, 863, 227	1.1
100	13, 824, 437	2.3
300	41, 474, 311	6.8
500	69, 222, 185	10.6

To ensure the accuracy of the experimental results, we have tested the data of each query in different data sets 5 times and obtained the average value. Table 2 shows the average time in each query.

Table 2. The average query time of each query in LUBM

	LUBM(10)		LUBM(50)		LUBM(100)		LUBM(300)		LUBM(500)	
	JOVF	PAQS	JOVF	PAQS	JOVF	PAQS	JOVF	PAQS	JOVF	PAQS
Q1	36.1	35.4	44.3	41.1	45.7	45.6	63.5	62.9	70.6	68.6
Q2	107.8	74.3	124.4	85.6	136.4	96.1	159.4	109.1	184.6	113.4
Q4	37.7	41.4	43.0	43.1	47.3	46.5	64.5	65.6	73.8	72.6
Q8	61.3	44.2	67.3	45.1	80.3	47.1	95.2	69.1	109.4	76.5
Q9	121.8	74.5	136.7	79.1	150.6	90.2	176.2	102.3	192.7	111.4

(1) **Average query time.** From Table 2, the average time of JOVF and PAQS are almost the same for the two query statements $Q1$ and $Q4$, because $Q1$ and $Q4$ both correspond to one job in these two algorithms. However, for $Q2$, $Q8$, and $Q9$, the time spent on the JOVF algorithm was about 1.5 times that of PAQS algorithm.

(2) **Average query time from different scale of data set.** The average query time of JOVF and PAQS increased with the increase of the data set, as shown in Figs. 5 and 6. With the continuous expansion of the scale of the test data set, the average query time of the two algorithms did not show an exponential growth trend, but rather a gradual rise. The average query time of PAQS was significantly smaller, which indicates that PAQS performed better than JOVF did in query stability.

(3) **Scalability.** From Table 2, the average query time of JOVF and PAQS only expanded by about 1.8 and 1.7 times, respectively, when the test data set was expanded by 50 times. These results indicate that JOVF and PAQS both have good expansibility.

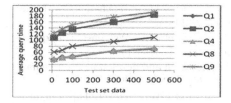

Fig. 5. Average query time of JOVF

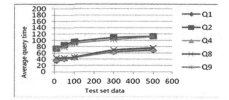

Fig. 6. Average query time of PAQS

5 Conclusion

This paper presents a massive RDF data query strategy in two stages, and proposes a query partitioning algorithm of PAQS based on SPARQL variable correlation degree. This approach achieves the minimum number of query tasks in the distributed query process. Experiments in the LUBM standard data set show that the PAQS algorithm performed better than the existing JOVF algorithm in terms of query efficiency and stability, and can better support the query of massive RDF data. In this paper, the PAQS algorithm mainly focuses on the simple SPARQL query model. We will study the distribute query method of the RDF graph in a future work.

Acknowledgement. The National Natural Science Foundation of China under Grant No. 61301136, No. 61272148 and No. 61602525.

References

1. Big data white paper in 2014. Ministry of Industry and Information Technology Telecommunications Research Institute (2014)
2. Manola, F., Miller, E.: RDF Primer [EB/OL]. W3C Recommendation (2004). http://www.w3.org/TR/rdf-syntax/
3. Hoffart, J., Suchanek, F.M., Berberich, K., et al.: YAGO2: a spatially and temporally enhanced knowledge base from Wikipedia. Artif. Intell. **194**, 28–61 (2013)
4. Belleau, F., Nolin, M.A., Tourigny, N., et al.: Bio2RDF: towards a mashup to build bioinformatics knowledge systems. J. Biomed. Inf. **41**(5), 706–716 (2008)
5. Kobilarov, G., et al.: Media meets semantic web – how the BBC uses DBpedia and linked data to make connections. In: Aroyo, L., et al. (eds.) ESWC 2009. LNCS, vol. 5554, pp. 723–737. Springer, Heidelberg (2009). doi:10.1007/978-3-642-02121-3_53
6. Mika P.: Social networks and the semantic web. In: Proceedings of the IEEE/WIC/ACM International Conference on Web Intelligence, Beijing, 20–24 September 2004, pp. 285–291. IEEE, New Jersey (2004)
7. The Linked Open Data Project (LOD), 06 August 2015. http://www.w3.org/wiki/SweoIG/TaskForces/CommunityProjects/LinkingOpenData
8. Xiao-feng, M.E.N.G., Xiang, C.I.: Big data management: concepts, techniques and challenges. J. Comput. Res. Dev. **50**(1), 146–169 (2013)
9. Wang, S., Wang, H.-J., Tan, X.-P., et al.: Architecting big data: challenges, studies and forecasts. Chin. J. Comput. **34**, 1741–1752 (2011)
10. Li, R.: Research on key technologies of large-scaled Semantic Web ontologies querying and reasoning based on Hadoop. Chongqing University (2013)
11. Xiao-yong, D.U., Yan, W.A.N.G., Bin, L.U.: Research and development on Semantic Web data management. J. Softw. **20**(11), 2950–2964 (2009)
12. Bechhofer, S., Harmelen, F.V., Hendler, J., et al.: OWL web ontology language reference. W3C Recommendation **40**(8), 25–39 (2004). http://www.w3.org/2004/OWL
13. Shi, H.-J.: Research of massive semantic information parallel inference method based on cloud computing.Shanghai Jiaotong University (2012)
14. Myung, J., Yeon, J., Lee, S.G.: SPARQL basic graph pattern processing with iterative MapReduce. In: Proceedings of the 2010 Workshop on Massive Data Analytics on the Cloud, 26 April 2010, pp. 1–6. ACM, New York (2010)
15. Husain, M., Mcglothlin, J., Masud, M.M., et al.: Heuristics-based query processing for large RDF graphs using cloud computing. IEEE Trans. Knowl. Data Eng. **23**(9), 1312–1327 (2011)
16. Cure O, Naacke H, Randriamalala T, et al.: LiteMat: a scalable, cost-efficient inference encoding scheme for large RDF graphs. In: IEEE International Conference on Big Data, pp. 1823–1830. IEEE (2015)
17. Liu, B., Huang, K., Li, J., et al.: An incremental and distributed inference method for large-scale ontologies based on MapReduce paradigm. IEEE Trans. Cybern. **45**(1), 53–64 (2015)

A Strategy to Improve Accuracy
of Multi-dimensional Feature Forecasting
in Big Data Stream Computing Environments

Dawei Sun[1(✉)], Hao Tang[1], Shang Gao[2], and Fengyun Li[3]

[1] School of Information Engineering, China University of Geosciences,
Beijing 100083, People's Republic of China
sundaweicn@cugb.edu.cn, cugb_TH10@163.com
[2] School of Information Technology, Deakin University,
Geelong, Victoria 3216, Australia
shang.gao@deakin.edu.au
[3] School of Computer Science and Engineering, Northeastern University,
Shenyang 110819, People's Republic of China
lifengyun@mail.neu.edu.cn

Abstract. High accuracy of multi-dimensional feature forecasting is very important for online scheduling in big data stream computing environments. Currently, most stream computing systems only consider historical features, with future features ignored. In this paper, a strategy to improve accuracy of multi-dimensional feature forecasting for online data stream is proposed. It includes the following contributions. (1) Profiling principles of accurate future feature forecasting objectives from multi-dimensional big data streams. (2) Extracting future features from multi-dimensional historical features of data stream via an improved hybrid IGA-BP (Immune Genetic Algorithm and Back Propagation) algorithm. (3) Evaluating accuracy of future feature forecasting and acceptable latency objectives in big data stream computing environments. Experimental results conclusively demonstrate the efficiency and effectiveness of the proposed strategy.

Keywords: Big data · Data stream · Feature forecasting · Multi-dimensional features · Hybrid IGA-BP algorithm

1 Introduction

1.1 Background and Motivation

The era of big data has led to the emergence of big data stream computing platform for online, real-time, and distributed data stream computing. Many organizations rely heavily on real time streaming. Big data stream computing helps organizations spot opportunities and risks across real time data stream, which can be employed in many different scenarios, such as trading, emergency response, fraud detection, system monitoring, to name just a few. Compared with batch data, big data stream is usually difficult to process in real time with traditional data computing infrastructure, as it has

© Springer International Publishing AG 2016
W. Cellary et al. (Eds.): WISE 2016, Part I, LNCS 10041, pp. 405–413, 2016.
DOI: 10.1007/978-3-319-48740-3_30

the following distinctive characteristics [1–3]. (1) The data are not all available at once. (2) The order of each data tuple's arrival cannot be controlled. (3) The input data stream rate is often at a high speed level, and might fluctuate with time.

High accuracy of multi-dimensional feature forecasting plays an important role in online scheduling in big data stream computing environments. Future feature forecasting is even more important than a complete historical feature aware for online scheduling. A scheduling scheme is not one-size-fit-all situation. Moreover, many scheduling strategies provide an efficient scheduling in static stream computing environments, but fail in online dynamic environments, as the historical information and future features are always needed in latter. Currently, most stream computing systems only consider historical features, with future features ignored. It is beneficial to consider accuracy of feature forecasting for online scheduling [4, 5].

Our strategy is partially inspired by the following observations.

(1) Future features of data stream are more important than that of historical for online scheduling.
    Data stream is continually changing, which makes historical information has less reference value for online scheduling. If we can forecast the changing pattern of data stream, and get future features of data stream from historical information, appropriate online scheduling strategy can be derived from both future and historical features, meeting future needs in a dynamic data stream environment.
(2) Accuracy of future features of data stream is expected.
    Future features can be forecasted from historical information of data stream. However, if those future features are not reasonably accurate, it might not help us get an ideal online scheduling strategy, or is possible to make the online scheduling strategy even worse. The accuracy of future features is determined by two factors, one is the forecast approach, and the other is the number of dimensions and the size of time window of historical information.

1.2 Contributions

Motivated by the above discussion, we propose a strategy to improve accuracy of multi-dimensional feature forecasting for online data stream (accuracy-MFF). In this paper, we cover all three aspects of accuracy-MFF, summarized as follows:

(1) Profiling principles of accurate future feature forecasting objectives from multi-dimensional big data streams.
(2) Extracting future features from multi-dimensional historical features via an improved hybrid IGA-BP algorithm.
(3) Evaluating accuracy of future feature forecasting and acceptable latency objectives in big data stream computing environments.

Experimental results conclusively demonstrate the efficiency and effectiveness of the proposed strategy.

1.3   Paper Organization

The remainder of this paper is organized as follows. In Sect. 2, the quantitating description of each utility for each dimension is presented. Section 3 focuses on the detailed discussion of the accuracy-MFF. Section 4 addresses the simulation environment, parameter setup and performance evaluation of the accuracy-MFF. Conclusions and future directions are given in Sect. 5.

2   Problem Statement

This section focuses on the quantitating definition of each utility for each dimension in big data stream computing environments [6–8].

Data stream Ds_i is described by d dimension, which is $Ds_i = (ds_i(1), ds_i(2), \cdots, ds_i(d))$. The quality of utility for each dimension is the basis for further forecasting future feature of each dimension. In this paper, three types of utility function are emphasized.

(1) Utility function with rigid time required

In this dimension, the time required by user is rigid, such as running time. If data stream is processed before the deadline of running time d, the utility is 1. Otherwise, the utility is 0, the utility function is described as (1).

$$Uti_{ik}(t) = 1 \cdot \left[Uti'_{ik}(t) - Uti'_{ik}(t - d) \right] = \begin{cases} 1, & t \leq d, \\ 0, & t > d. \end{cases} \tag{1}$$

(2) Utility function with rigid-elastic time required

In this dimension, the time required by user is rigid with in a time range, and will be elastic with other time ranges, such as communication time. Assume rt_1 and rt_2 are two different levels of communication time, if the communication time is in $[0, rt_1)$, the utility is 1; if the communication time is in $[rt_1, rt_2)$, with communication time increasing, the utility is decreasing in linear pattern; if the communication time is more than rt_2, the utility is 0, the utility function is described as (2).

$$
\begin{aligned}
Uti_{ik}(t) = &1 \cdot \left[Uti'_{ik}(t) - Uti'_{ik}(t - t_1) \right] \\
&+ 1 \cdot \left(1 - \frac{t - t_1}{t_2 - t_1} \right) \cdot \left[Uti'_{ik}(t - t_1) - Uti'_{ik}(t - t_2) \right] \\
= &\begin{cases} 1, & t \leq t_1, \\ 1 \cdot \left(1 - \dfrac{t - t_1}{t_2 - t_1} \right), & t \in (t_1, t_2), \\ 0, & t \geq t_2. \end{cases}
\end{aligned}
\tag{2}
$$

(3) Utility function with elastic time required

In this dimension the time required by user is elastic with time ranges, such as waiting time. Assume w is the longest waiting time, if the waiting time is within

$[0, w]$, with waiting time increasing, the utility is decreasing in linear pattern; if the waiting time is greater than w, the utility is 0, the utility function is described as (3).

$$Uti_{ik}(t) = 1 \cdot \left(1 - \frac{t}{w}\right) \cdot \left[Uti'_{ik}(t) - Uti'_{ik}(t - w)\right] = \begin{cases} 1 \cdot \left(1 - \frac{t}{w}\right), & t \leq w, \\ 0, & t > w. \end{cases} \quad (3)$$

3   Accuracy-MFF Overview

To provide a bird's-eye view of the accuracy-MFF, this section focuses on the detailed discussion of accuracy-MFF, including BP model, IGA model, and IGA-BP algorithm [9–12].

3.1   BP Model

BP is a supervised learning approach. It is used to deal with approximation of nonlinear maps. It can also be employed to forecast future features. The structure of BP neural network is shown in Fig. 1.

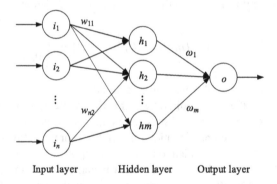

Input layer   Hidden layer   Output layer

Fig. 1. Structure of BP neural network

The input i_k to the kth neuron is described by (4).

$$i_k = \sum_{j=1}^{n} w_{j,k} \cdot o_j, \quad (4)$$

where $w_{j,k}$ is the weight from the jth neuron in the previous layer to the kth neuron. The output o_k of the kth neuron is described by (5).

$$o_k = f(i_k + \delta_k), \quad (5)$$

where $f(x)$ is the activation function of the neurons. We employ bipolar sigmoid activation function, which is described by (6). δ_k is the biases input to the neuron.

$$f(x) = \frac{2}{1+e^{-2x}} - 1. \tag{6}$$

The error of BP is evaluated by mean square error, which is described by (7).

$$er = \frac{1}{n} \cdot \sum_{i=1}^{n} \left(d_i - d_i'\right)^2, \tag{7}$$

where n is the number of training patterns, d_i is the observation data, and d_i' is the forecasted data.

The training continues until the er falls below a threshold or tolerance level.

3.2 IGA Model

IGA operation includes clone, cloning recombinant and cloning variation. The initial solution to the first generation in the ith antibody CA_i is described by $CA_i(1) = (ca_i^1, ca_i^2, \cdots, ca_i^{n_u})$, $ca_i^k \in \{1, 2, \cdots, n\}$.

The fitness $AgAb_i$ between antibody CA_i and antigen is described by (8). The fitness $AgAb_i$ reflects the fitness of antibody CA_i to the antigen. The larger the fitness $AgAb_i$, the higher the utility.

$$AgAb_i = er. \tag{8}$$

The affinity Ab_iAb_j between antibody CA_i and CA_j is described by (9). The affinity Ab_iAb_j reflects the difference between antibody CA_i and CA_j. The less the affinity Ab_iAb_j, the stronger the inhibition between antibody CA_i and CA_j.

$$Ab_iAb_j = \left(1 + sim\left(Ab_iAb_j\right)\right)^{-1}, \tag{9}$$

where $sim\left(Ab_iAb_j\right)$ is the proximity degree between antibody CA_i and CA_j, which is described by (10).

$$sim\left(Ab_iAb_j\right) = \frac{1}{n} \cdot \sum_{k=1}^{n} \sqrt{\left(ca_i^k - ca_j^k\right)^2}, \quad i \neq j. \tag{10}$$

The selection probability q_i of antibody CA_i is described by (11).

$$q_i = AgAb_i \cdot e^{-\alpha \cdot Ab_iAb_j} \bigg/ \sum_{i=1}^{n} AgAb_i \cdot e^{-\alpha \cdot Ab_iAb_j}, \tag{11}$$

where α is the regulation parameter between fitness $AgAb_i$ and affinity Ab_iAb_j. n is the number of antibody in a population.

SBX (Simulated Binary Crossover) helps exchange information between antibodies. PM (Polynomial Mutation) adaptively adjusts antibody, is described by (12).

$$ca_i'^k = ca_i^k + \delta_k \cdot \left(ca_u^k - ca_l^k\right), \quad i \in \left\{1, 2, \cdots, N_{Ab}'\right\}, \tag{12}$$

where ca_i^k and $ca_i'^k$ denote the value of before mutation and after mutation of the kth bit in antibody, respectively, δ_k is the adjustment parameter of PM.

3.3 IGA-BP

When only BP is employed, some issues are more likely to arise, such as local optimal solution, over-training or under-training, and/or difficulty in the number of hidden layers and neurons. To overcome the inherent shortcoming of BP, IGA is used to optimize weights and thresholds of BP, increasing the accuracy of forecasting future features of high dimensional data stream.

Weight or threshold of BP is one to one mapped to one component of a particle' position vector in a particle swarm. For a particle, each component of the current particle is mapped to a weight or threshold of BP sequentially, such as, weight connecting input layer with hidden layer, weight connecting hidden layer with output layer, threshold of hidden layer, or threshold of output layer.

4 Simulation and Performance Evaluation

Experiments are conducted to evaluate the performance of the proposed accuracy-MFF. Simulation environment and parameter settings are firstly discussed in this section, followed by performance evaluation results.

4.1 Simulation Environment and Parameter Setup

Storm platform, used as the experimental environment [13]. The proposed accuracy-MFF is implemented based on Storm 0.8.1, and installed on top of Linux Ubuntu Server 13.04. The cluster consists of 10 machines.

4.2 Performance Evaluation

The experimental setup contains forecasting result and devious of relative error.

(1) Forecasting result

With 5-dimension features, compared with actual data, the forecasted data produced by accuracy-MFF is much closer to the actual data. As shown in Fig. 2, 100 data are forecasted, and result shows nearly all the forecasted data overlap with the actual

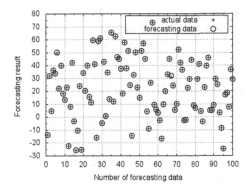

Fig. 2. Forecasting result of 5-dimensions

data. If we use the forecasted data for the following task rescheduling, the rescheduling result will be identical to the result produced by actual data. But with accuracy-MFF, all the work is done before hand.

(2) Devious of relative error

The devious of relative error D_{re} reflects the stability of accuracy of forecasted result. Specifically, devious of relative error D_{re} is obtained by (13).

$$D_{re} = \sqrt{\frac{1}{n} \cdot \sum_{i=1}^{n} \left(RE_i - \overline{RE_i}\right)^2},$$ (13)

where RE_i is the ith relative error in all n forecasted results.

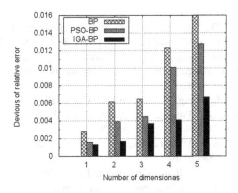

Fig. 3. Average devious of relative error.

Compared with BP and PSO-BP algorithms, the accuracy-MFF (also named IGA-BP) algorithm has a better deviousness of relative error by combining IGA and BP. As shown in Fig. 3, in each dimension of data stream, the deviousness of relative error of each dimension employing accuracy-MFF is smaller than that of employing the BP and PSO-BP. The higher dimension of data stream, the more significant performance improved between accuracy-MFF and BP.

5 Conclusions and Future Directions

In this paper, accuracy-MFF is proposed. Our work and contributions are summarized as follows.

(1) Profiling principles of accurate future feature forecasting objectives from multi-dimensional big data streams.
(2) Extracting future features from multi-dimensional historical features of data stream via an improved hybrid IGA-BP algorithm.
(3) Evaluating accuracy of future features forecasting and acceptable latency objectives in big data stream computing environments.

Future works will focus on the following.

(1) Improving latency of forecasting future features of each dimension in data stream with a fast response time guarantee.
(2) Deploying the accuracy-MFF on real big data stream computing environments.

Acknowledgment. This work is supported by the National Natural Science Foundation of China under Grant No. 61602428; the Fundamental Research Funds for the Central Universities under Grant No. 2652015338 and No. N130316001.

References

1. Demirkan, H., Delen, D.: Leveraging the capabilities of service-oriented decision support systems: putting analytics and big data in cloud. Decis. Support Syst. **55**(1), 412–421 (2013)
2. Tien, J.M.: Big data: unleashing information. J. Syst. Sci. Syst. Eng. **22**(2), 127–151 (2013)
3. Zeng, X.Q., Li, G.Z.: Incremental partial least squares analysis of big streaming data. Pattern Recogn. **47**(11), 3726–3735 (2014)
4. Peng, B., Hosseini, M., Hong, Z., Farivar, R., Campbell, R.: R-Storm: resource-aware scheduling in Storm. In: Proceedings of the 16th Annual Middleware Conference, Middleware 2015, pp. 149–161. ACM Press (2015)
5. Zeng, J., Liu, Z.Q., Cao, X.Q.: Fast online EM for big topic modeling. IEEE Trans. Knowl. Data Eng. **28**(3), 675–688 (2016)
6. Sheikhalishahi, M., Wallace, R.M., Grandinetti, L., Vazquez-Poletti, J.L., Guerriero, F.: A multi-dimensional job scheduling. Future Gener. Comput. Syst. **54**, 123–131 (2016)
7. Sun, D.W., Chang, G.R., Li, F.Y., Wang, C., Wang, X.W.: Optimizing multi-dimensional QoS cloud resource scheduling by immune clonal with preference. Acta Electronica Sinica **20**(8), 1824–1831 (2011)
8. Yang, S.X., He, Z., Chen, Y.P.P.: Workload-based ordering of multi-dimensional data. IEEE Trans. Knowl. Data Eng. **28**(3), 831–844 (2016)

9. Wang, L., Zeng, Y., Chen, T.: Back propagation neural network with adaptive differential evolution algorithm for time series forecasting. Expert Syst. Appl. **42**(2), 855–863 (2015)
10. Kim, J.S., Jung, S.: Implementation of the RBF neural chip with the back-propagation algorithm for on-line learning. Appl. Soft Comput. **29**, 233–244 (2015)
11. Lee, S., Choi, W.S.: A multi-industry bankruptcy prediction model using back-propagation neural network and multivariate discriminant analysis. Expert Syst. Appl. **40**(8), 2941–2946 (2013)
12. Kuo, R.J., Lee, Y.H., Zulvia, F.E., Tien, F.C.: Solving bi-level linear programming problem through hybrid of immune genetic algorithm and particle swarm optimization algorithm. Appl. Math. Comput. **266**, 1013–1026 (2015)
13. Toshniwal, A., Taneja, S., Shukla, A., Ramasamy, K., Patel, J.M., Kulkarni, S., Jackson, J., Gade, K., Fu, M., Donham, J., Bhagat, N., Mittal, S., Ryaboy, D.: Storm@twitter. In: Proceedings of the 2014 ACM SIGMOD International Conference on Management of Data, pp. 147–156. ACM Press (2014)

Cloud Computing

Automatic Creation and Analysis of a Linked Data Cloud Diagram

Alexander Arturo Mera Caraballo[1(✉)], Bernardo Pereira Nunes[1,4],
Giseli Rabello Lopes[2], Luiz André Portes Paes Leme[3],
and Marco Antonio Casanova[1]

[1] Department of Informatics, Pontifical Catholic University of Rio de Janeiro,
Rio de Janeiro, RJ, Brazil
{acaraballo,bnunes,casanova}@inf.puc-rio.br
[2] Federal University of Rio de Janeiro, Rio de Janeiro, RJ, Brazil
giseli@dcc.ufrj.br
[3] Fluminense Federal University, Niterói, RJ, Brazil
lapaesleme@ic.uff.br
[4] Federal University of the State of Rio de Janeiro, Rio de Janeiro, RJ, Brazil
bernardo.nunes@uniriotec.br

Abstract. Datasets published on the Web and following the Linked Open Data (LOD) practices have the potential to enrich other LOD datasets in multiple domains. However, the lack of descriptive information, combined with the large number of available LOD datasets, inhibits their interlinking and consumption. Aiming at facilitating such tasks, this paper proposes an automated clustering process for the LOD datasets that, thereby, provide an up-to-date description of the LOD cloud. The process combines metadata inspection and extraction strategies, community detection methods and dataset profiling techniques. The clustering process is evaluated using the LOD diagram as ground truth. The results show the ability of the proposed process to replicate the LOD diagram and to identify new LOD dataset clusters. Finally, experiments conducted by LOD experts indicate that the clustering process generates dataset clusters that tend to be more descriptive than those manually defined in the LOD diagram.

Keywords: Linked data cloud analysis · Automatic clustering · Domain identification · Community detection algorithms

1 Introduction

The Linked Data principles established a strong basis for creating a rich space of structured data on the Web. The potentiality of such principles encouraged the government, scientific and industrial communities to transform their data to the Linked Data format, creating the so-called Linked Open Data (LOD) cloud. An essential step of the publishing process is to interlink new datasets with those in the LOD cloud to facilitate the exploration and consumption of existing data. Although frameworks to help create links are available, such as LIMES [1] and SILK [2], the selection of datasets to interlink with a new dataset is still a manual and non-trivial task. One possible direction to

© Springer International Publishing AG 2016
W. Cellary et al. (Eds.): WISE 2016, Part I, LNCS 10041, pp. 417–432, 2016.
DOI: 10.1007/978-3-319-48740-3_31

facilitate the selection of datasets to interlink with would be to classify the datasets in the LOD cloud by domain similarity and to create expressive descriptions of each class. Thus, the publisher of a new dataset would select the class closest to his dataset and try to interlink his dataset with those in the class.

The *LOD diagram* [3, 4], perhaps the best-known classification of the datasets in the LOD cloud, adopted the following categories: "Media", "Government", "Publications", "Life Sciences", "Geographic", "Cross-domain", "User-generated Content" and "Social Networking".

However, the fast growth of the LOD cloud makes it difficult to manually maintain the LOD diagram. To address this problem, we propose a community analysis of the LOD cloud that leads to an automatic clustering of the datasets into communities and to a meaningful description of the communities. The process has three steps. The first step creates a graph to describe the LOD cloud, using metadata extracted from dataset catalogs. The second step uses community detection algorithms to partition the LOD graph into *communities* (also called *clusters*) of related datasets. The last step generates descriptions for the dataset communities by applying dataset profiling techniques. As some of the datasets may contain a large number of resources, only a random sample of each dataset is considered. For each dataset community, this step generates a *profile*, expressed as a vector, whose dimensions correspond to relevance scores for the 23 top-level categories of Wikipedia.

The resulting partition of the LOD graph into communities, with the descriptions obtained, may help data publishers search for datasets to interlink their data as follows. Consider a new dataset d to be published as Linked Data; the same profiling technique used in the process we propose may be used to generate a profile for d, expressed as a vector, as in Step 3. Then, a similarity measure (e.g., cosine-based) may be used to compute the similarity between the profile of d and the profile of each dataset community. Finally, the data publisher may receive recommendations for the community with the highest similarity value. This suggested recommendation process is not the focus of this paper, but it is one of the major motivations of this work.

To summarize, the main contributions of this paper are: (i) an automatic clustering of the LOD datasets which is consistent with the traditional LOD diagram, taken as ground truth; and (ii) an automatic process that generates descriptions of dataset communities. The remainder of this paper is structured as follows. Section 2 presents background concepts. Section 3 presents the proposed process. Sections 4, 5 and 6 describe the experimental setup, the results and an extensive analysis of the generated communities and their descriptions, respectively. Section 7 reviews the literature. Finally, Sect. 8 summarizes the contributions and outlines further work.

2   Background

2.1   LOD Concepts

A *dataset* is simply a set t of RDF triples. A resource, identified by an RDF URI reference s, is *defined in t* iff s occurs as the subject of a triple in t. Given two datasets t and u, a *link* from t to u is a triple of the form (s, p, o), where s is an RDF URI reference identifying a resource defined in t and o is an RDF URI reference identifying a resource defined in u. A *linkset* from t to u is a set of links from t to u.

The set of RDF datasets publicly available is usually referred to as the *LOD cloud*.

The *LOD graph (or the LOD network)* is an undirected graph $G = (S, E)$, where S denotes a set of datasets in the LOD cloud and E contains an edge (t, u) iff there is at least one linkset from t to u, or from u to t.

A *LOD catalog* describes the datasets available in the LOD cloud. Datahub[1] and the Mannheim Catalog[2] are two popular catalogs. LODStats[3] collects statistics about datasets to describe their internal structure (e.g. vocabulary/class/property usage, number of triples, linksets). The LOD Laundromat[4] generates a clean version of the LOD cloud along with a metadata graph with structural data.

A *LOD diagram* is a visual representation of the structure of the LOD cloud. At least three versions of the structure of the LOD cloud are currently available [3]. Schmachtenberg et al. [4] provides the most comprehensive statistics about the structure and content of the LOD cloud (as of April 2014). This version of the LOD cloud comprises 1,014 datasets, of which only 570 have linksets. In total, 2,755 linksets (both in- and outlinks) express a relationship between the datasets contained in this version of the LOD cloud. The datasets are divided into eight topical domains, namely, "Media", "Government", "Publications", "Life Sciences", "Geographic", "Cross-domain", "User-generated Content" and "Social Networking". The datasets are not uniformly distributed per topical domain: "Government" and "Publication" are the largest domains, with 23.85 % and 23.33 % of all datasets, respectively; "Media" is the smallest domain, containing only 3.68 % of all datasets. Table 1 presents the number of datasets in each topical domain, for which linksets are defined. We highlight that the wide variation of the size among the domains represents an additional challenge to community detection/clustering algorithms [5].

Table 1. Number of datasets and linksets per topical domain.

Topical domain	#Datasets	#Inlinks	#Outlinks
Media	21	55	39
Government	136	271	330
Publications	133	772	862
Geographic	24	171	56
Cross-domain	40	345	180
Life sciences	63	144	161
Social networking	90	912	986
User-generated content	42	85	141

---

[1] http://datahub.io/.

[2] http://linkeddatacatalog.dws.informatik.uni-mannheim.de/.

[3] http://stats.lod2.eu/.

[4] http://lodlaundromat.org.

2.2   Communities and Community Detection Algorithms

Let $G = (S, E)$ be an undirected graph and $G_C = (S_c, E_c)$ be a subgraph of G (that is, $S_c \subseteq S$ and $E_c \subseteq E$). Let $|s|$ denote the cardinality of a set s.

The *intra-cluster density* of G_C, denoted $\delta_{int}(G_C)$, is the ratio between the number of edges of G_C and the number of all possible edges of G_C and is defined as follows:

$$\delta_{int}(G_C) = \frac{|E_C|}{|S_C|.(|S_C| - 1)/2}$$

Let $\gamma(G_C)$ denote the set of all edges of G that have exactly one node is in S_C. The *inter-cluster density* of G_C, denoted $\delta_{ext}(G_C)$, measures the ratio between the cardinality of $\gamma(G_C)$ and the number of all possible edges of G that have exactly one node is in S_C and is defined as follows:

$$\delta_{ext}(G_C) = \frac{|\gamma(G_C)|}{|S_C|.(|S| - |S_C|)}$$

The average link density of $G = (S, E)$, denoted $\delta(G)$, is the ratio between the number of edges of G and the maximum number of possible edges of G:

$$\delta(G) = |E|/((|S|-1)/2)$$

For the subgraph G_C to be a community, $\delta_{int}(G_C)$ has to be considerably larger than $\delta(G)$ and $\delta_{ext}(G_C)$ has to be much smaller than $\delta(G)$.

The *edge betweenness* [6] of an edge (t, u) in E is the number of pairs (w, v) of nodes in S for which (t, u) belongs to the shortest path between w and v.

Community detection algorithms search, implicitly or explicitly, for the best trade-off between a large $\delta_{int}(G_C)$ and a small $\delta_{ext}(G_C)$. They are usually classified as *non-overlapping* and *overlapping*. In *non-overlapping* algorithms, each node belongs to a single community. An example is the *Edge Betweenness Method* (EBM) [6], which finds communities by successively deleting edges with high edge betweenness. In *overlapping* algorithms, a node may belong to multiple communities. An example is the *Greedy Clique Expansion* algorithm (GCE) [7], which first discovers maximum cliques to be used as seeds of communities and then greedily expands these seeds by optimizing a fitness function. Another example is the *Community Overlap Propagation Algorithm* (COPRA) [8], which follows a label propagation strategy (where the labels represent the communities).

2.3   Clustering Validation Measures

Clustering validation measures are used to validate a clustering (or community detection) strategy against a ground truth.

Let U be the *universe*, that is, the set of all elements. Let $C = \{C_1, C_2, ..., C_m\}$ and $T = \{T_1, T_2, ..., T_n\}$ be two sets of subsets of U.

The definitions that follow are generic, but the reader may intuitively consider U as the set of all datasets in the LOD cloud, C as a set of dataset clusters, obtained by one of the clustering algorithms, and T be a set of sets of LOD datasets taken as the ground truth (i.e., the topical domains of the LOD diagram).

Purity [9] is a straightforward measure of cluster quality that is determined by simply dividing the number of elements of the most frequent domain contained in each cluster by the total number of elements. Purity ranges from 0 to 1, where higher values indicate better clusters with respect to the ground truth, and is defined as follows:

$$purity(C, T) = \frac{1}{|U|} \sum_{i=1,\dots,m} \max_j (|C_i \cap T_j|)$$

Unlike purity, the *normalized mutual information* (NMI) [9] offers a trade-off between the number of clusters and their quality. Intuitively, NMI is the fraction of mutual information that is contained in the current clustering representation. NMI ranges from 0 to 1, where higher values indicate better clusters with respect to the ground truth, and is defined as follows:

$$NMI(C, T) = \frac{I(C, T)}{(H(C) + H(T))/2}$$

where $I(C, T)$ represents the *mutual information* between C and T and is defined as:

$$I(C, T) = \sum_{i=1,\dots,m} \sum_{j=1,\dots,n} \frac{|C_i \cap T_j|}{|U|} \log \left(\frac{|U| \cdot |C_i \cap T_j|}{|C_i| \cdot |T_j|} \right)$$

and $H(C)$ is the *entropy* of C and is defined as:

$$H(C) = - \sum_{i=1,\dots,m} \frac{|C_i|}{|U|} \log \left(\frac{|C_i|}{|U|} \right)$$

and likewise for $H(T)$, the entropy of T.

The *Estimated Mutual Information* (EMI) [10] measures the dependence between C and T (intuitively, the identified clusters and the topical domains in the LOD diagram). EMI is an $m \times n$ matrix, where each element is defined as follows:

$$EMI_{i,j} = \frac{m_{i,j}}{M} . log \left(M . \frac{m_{i,j}}{\sum_{a=1}^{n} m_{i,a} . \sum_{b=1}^{m} m_{b,j}} \right)$$

where

- $[m_{i,j}]$ is the *co-occurrence matrix* of C and T, with $m_{i,j} = |C_i \cap T_j|$, for $i \in [1, m]$ and $j \in [1, n]$
- $M = \sum_{i=1}^{m} \sum_{j=1}^{n} m_{i,j}$

2.4   Dataset Profiling Techniques

Profiling techniques address the problem of generating dataset descriptions. We will use in this paper the profiling technique described in [11], that generates *profiles* or *fingerprints* for textual resources. The method has five steps:

1. Extract entities from a given textual resource.
2. Link the extracted entities to English Wikipedia articles.
3. Extract English Wikipedia categories for the articles.
4. Follow the path from each extracted category to its top-level category and compute a vector with scores for the top-level categories thus obtained.
5. Perform a linear aggregation in all dimensions of the vectors to generate the final profile, represented as a histogram for the 23 top-level categories of the English Wikipedia.

3   The Dataset Clusterization and Dataset Community Description Process

The proposed process has three main steps (see Fig. 1):

1. Construction of the LOD graph.
2. Dataset clusterization.
3. Dataset community description.

The first step of the process creates a graph that describes the LOD cloud, using metadata extracted from metadata catalogs (c.f. Sect. 2.1).

The second step clusters the datasets represented as nodes of the LOD graph. It applies community detection algorithms to partition the LOD graph into *communities* (also called *clusters* or *groups*) of related datasets. Intuitively, a set of datasets forms a

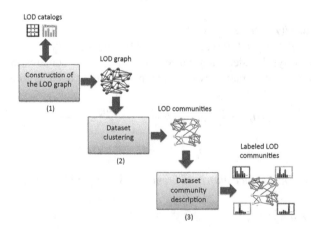

Fig. 1. Community analysis process of the LOD.

community if there are more linksets between datasets within the community than linksets interlinking datasets of the community with datasets in rest of the LOD cloud (c.f. Sect. 2.2).

The last step generates descriptions for the dataset communities by applying a dataset profiling technique to the datasets in each community C_i identified in the previous step. As some of the datasets may contain a large number of resources, only a random sample of each dataset is considered. Furthermore, to generate the labels that describe C_i, the profiling technique considers the literals of the datatype properties rdfs:Label, skos:subject, skos:prefLabel and skos:altLabel of the sampled resources. We recall that this step adopts the profiling technique described in Sect. 2.4 to generate community descriptions.

4 Evaluation Setup

This section details the evaluation setup of the proposed process. Section 4.1 covers the construction of the LOD graph and describes the ground truth. Section 4.2 introduces the community detection algorithms used and discusses how the resulting communities are evaluated by taking into account the clustering validation measures described in Sect. 2.3. Finally, Sect. 4.3 analyses the labels assigned to the resulting communities, considering the expressiveness and the ability to represent the content of the datasets belonging to each community.

4.1 Construction of the LOD Graph and Description of the Ground Truth

To construct a LOD graph, we extracted all datasets from the Mannheim Catalog, along with their content metadata: title, description, tags and linksets. For the sake of simplicity and comparison between the ground truth and the proposed approach, we refer to the topical domains also as communities.

As ground truth, we adopted the LOD diagram described in [4] (see Sect. 2.1).

4.2 Setup of the Dataset Clusterization Step

Three algorithms traditionally used in community detection and clustering problems were considered as an attempt to reproduce the LOD diagram: Greedy Click Expansion (GCE), Community Overlap PRopagation Algorithm (COPRA) and the Betweenness Method (EBM) (see Sect. 2.2). The choice of these three algorithms was based on their previously reported performance in real world scenarios [12, 13].

We used Purity, Normalized Mutual Information (NMI) and Estimated Mutual Information (EMI) (see Sect. 2.3) as clustering validation measures. Again, these measures are estimated by comparing the results obtained with the community detection algorithms and the ground truth.

Table 2. Top 10 best configurations for EBM by decreasing order of NMI.

Number of removed edges	Purity	NMI
600	0.60291	0.49287
550	0.60109	0.48619
300	0.56831	0.47381
650	0.54645	0.46870
700	0.51730	0.45848
500	0.60474	0.45061
750	0.49545	0.44958
450	0.58106	0.44949
800	0.46812	0.44551
850	0.39891	0.42707

A brief description of parameterization of the three algorithms goes as follows:

- EBM: Table 2 shows the top 10 best configurations for EBM in order to reproduce the results found in the ground truth. Very briefly, the number of edges with the highest betweenness that must be removed from the LOD graph in order to detect the communities was used as stopping criterion.
- GCE: Table 3 shows the top 10 best configurations for GCE in order to reproduce the results found in the ground truth. Very briefly, the *Alpha* and *Phi* parameters were used to control the greedy expansion and to avoid duplicate cliques/communities, respectively.
- COPRA: Table 4 shows the best configuration for COPRA. As COPRA is non-deterministic, the tuning of its parameters was obtained by the average of 5-cycle runs.

Unlike EBM, GCE and COPRA are capable of finding overlapping communities. However, as the ground truth defines non-overlapping communities, these algorithms obtained the best results when the overlapping rate/parameter was set to 0 (no overlap between datasets) and 1 (one label per dataset), respectively.

Table 3. Top 10 best configurations for GCE by decreasing order of NMI.

Clique size	Overlapping rate	Alpha	Phi	Purity	NMI
3	0.0	0.8	0.2	0.42076	0.57263
3	0.0	1.0	0.8	0.36430	0.55509
3	0.0	1.0	0.2	0.38251	0.54227
3	0.1	0.8	0.6	0.49362	0.51040
3	0.0	1.2	0.2	0.46630	0.51022
3	0.1	1.2	0.2	0.48816	0.50926
3	0.0	0.8	0.4	0.34426	0.50534
3	0.0	1.2	0.8	0.50820	0.50148
3	0.2	1.0	0.2	0.56648	0.49747
3	0.3	0.8	0.2	0.48452	0.49542

4.3 Setup of the Dataset Community Description Step

Although the Mannheim Catalog lists 1,014 datasets, only a fraction of the listed datasets has SPARQL endpoints available. At the time of this evaluation, approximately 56 % of the SPARQL endpoints were up and running. For each available dataset, a sample of 10 % of its resources were extracted and used as input to the *fingerprints* algorithm (see Sects. 2.4 and 3), which assigned labels to the communities automatically generated by the best performing parameterization of the GCE algorithm.

Table 4. Best quality results for the community detection/clustering algorithms.

Algorithm	#Clusters	Purity	NMI
GCE	6	0.42	0.57
COPRA	4	0.30	0.32
EBM	18	0.60	0.49

5 Results

The first part of the discussion addresses the performance of the dataset clusterization step. The second part presents the results for the dataset community description step.

5.1 Performance of the Dataset Clusterization Step

Quality of the Generated Communities. As shown in Table 4, GCE obtained the highest NMI value, 0.57, and EBM the highest purity value, 0.60. The high NMI value achieved by GCE indicates a mutual dependence between the communities found by the algorithm and those described in the ground truth. Despite the highest purity value obtained by EBM, this technique was not consistent with the communities in the ground truth. COPRA obtained low values for both purity and NMI, indicating that the resulting communities and those induced by the ground truth do not match.

Communities Detected. Table 5 shows the co-occurrence and estimated mutual information matrices for the best performing parameterization of the GCE algorithm. The first column shows the communities (domains) of the ground truth, whereas columns labeled 0-5 represent the communities found by GCE. The light gray cells mark the highest dependencies between the topical domains extracted from the ground truth and the communities generated by GCE. Note that, due to the low level of dependency between the ground truth categories "Cross-Domain" and "User-Generated Content" (UGC) and the clusters found by GCE, datasets in these ground truth categories communities are possibly split over several clusters.

5.2   Performance of the Dataset Community Description Step

Table 6 shows the labels generated by the dataset community description method adopted (see Sect. 2.4). These labels were assigned to the communities found by the best performing parameterization of the GCE algorithm. The first column shows the 23 top-level categories of Wikipedia, whereas columns labeled 0–5 represent the communities found by GCE. To facilitate a comparison between the labels in different communities, we normalized the scores assigning 1.0 to the category with the highest score. The light gray cells mark the strongest relations between the categories from the generated labels and the communities generated by GCE. We recall that Table 1 shows the labels assigned to the communities in the ground truth.

Table 5. Co-occurrence and EMI matrices of the GCE result.

Domain/Community	0	1	2	3	4	5	0	1	2	3	4	5
Social Networking	0	88	0	0	0	0	0	0.262	0	0	0	0
UGC	0	4	0	0	0	0	0	−0.005	0	0	0	0
Geographic	0	2	4	0	0	0	0	−0.003	0.013	0	0	0
Publications	37	4	1	1	0	0	0.092	−0.013	−0.00	−0.002	0	0
Cross-Domain	1	2	0	0	0	0	−0.002	−0.005	2	0	0	0
Life Sciences	0	2	0	13	24	0	0	−0.007	0	0.046	0.095	0
Government	1	1	10	1	0	59	−0.004	−0.006	0	−0.003	0	0.150
Media	0	2	0	1	0	0	0	−0.003	0.018	0.001	0	0
							0					

Table 6. Histograms of top-level categories for each community structure.

Category / Community	0	1	2	3	4	5
Agriculture	0	0	0.39	0.03	0.02	0.03
Applied Science	0.80	0.34	0.37	0.06	0.11	0.03
Arts	0.03	0.11	0	0	0.01	0.03
Belief	0.03	0.04	0	0	0	0.02
Business	0.59	0.53	0.11	0.03	0.03	0.27
Chronology	0.04	0.15	0.02	0.01	0	0.06
Culture	0.13	0.19	0.27	0	0.03	0.11
Education	0.20	0.06	0.06	0.04	0.12	0.08
Environment	0.01	0.03	0.40	0.02	0.02	0.10
Geography	0.05	0.11	1.00	0.13	0.03	0.70
Health	0.05	0.06	0.41	0.18	0.65	0.03
History	0.06	0.03	0.11	0.06	0.02	0.13
Humanities	0.04	0.08	0	0	0.2	0
Language	0.20	0.10	0.01	0	0.02	0.03
Law	0.04	0.45	0.10	0.01	0.02	0.24
Life	0.08	0.03	0.96	1.00	1.00	0.02
Mathematics	0.60	0.03	0.03	0.03	0.03	0.02
Nature	0.29	0.08	0.24	0.03	0.06	0.03
People	0.02	0.52	0.02	0.01	0.03	1.00
Politics	0.05	0.35	0.12	0.03	0.01	0.65
Science	1.00	0.16	0.26	0.03	0.10	0.03
Society	0.32	1.00	0.14	0.06	0.05	0.32
Technology	0.61	0.37	0.11	0.01	0.02	0.08

6 Discussion and Analysis

6.1 An Analysis of the Dataset Clusterization Results

This section analyses the dataset clusterization results. The analysis compares the dataset communities found in the clustering step – referred to as *Community 0* to *Community 5* – with the dataset topical domains defined in the LOD diagram [4] – "Media", "Government", "Publications", "Geographic", "Cross-Domain", "Life Sciences", "Social Networking" and "User-generated content" – taken as ground truth.

As shown in Sect. 5, the GCE algorithm did not recognize as communities the datasets classified in the "Cross-domain" and "Media" domains. A possible reason for the lack of a cross-domain community lies in its own nature, that is, cross-domain datasets tend to be linked to datasets from multiple domains, acting as hubs for different communities. Another (interesting) reason is that cross-domain datasets do not contain a large number of links between themselves. The lack of links between cross-domain datasets results in a subgraph with low density, which GCE does not consider a new community. Nevertheless, if overlapping rates are considered, datasets that belong to several communities may generate a cross-domain community. Likewise, the "Media" community presented a low density due to its low number of linksets.

Community 0 presents a high concentration of datasets from the "Publications" domain, including datasets from the ReSIST project[5], such as `rkb-explorer-acm`, `rkb-explorer-newcastle`, `rkb-explorer-pisa` and `rkb-explorer-budapest`. This led us to assume that this community is equivalent to the "Publications" domain.

Community 1 is the largest community among those recognized and contains mostly datasets from the "Social Networking" domain. This community includes datasets such as `statusnet-postblue-info`, `statusnet-fragdev-com`, `statusnet-bka-li` and `statusnet-skilledtestes-com`.

Contrasting with the previous communities, *Community 2* includes datasets from two different domains, "Government" and "Geographic". Note that datasets in these two domains share a considerable number of linksets, which led GCE to consider them in the same community. Government datasets often provide statistical data about places, which may justify such a large number of linksets between them. *Community 2* includes datasets from the "Government" domain, such as `eurovoc-in-skos`, `gemet`, `umthes`, `eea`, `eea-rod`, `eurostat-rdf` and `fu-berlin-eurostat`. It also includes datasets from the "Geographic" domain, such as `environmental-applications-reference-thesaurus` and `gadm-geovocab`.

Communities 3 and *4* are equivalent to only one domain, "Life Sciences". Intuitively, the original "Life Sciences" domain was split into *Community 3*, containing datasets such as `uniprot`, `bio2rdf-biomodels`, `bio2rdf-chembl` and `bio2rdf-reactome`, and into *Community 4*, containing datasets such as `pub-med-central`, `bio2rdf-omim` and `bio2rdf-mesh`. A distinction

---

[5] http://www.rkbexplorer.com/

between these two communities becomes apparent by inspecting the datasets content: *Community 3* is better related to Human Biology data (about molecular and cellular biology), whereas *Community 4* is better related to Medicine data (about diagnosis and treatment of diseases).

Finally, *Community 5* groups datasets from the "Government" domain. Examples of datasets in this community are `statictics-data-gov-uk`, `reference-data-gov-uk`, `opendatacommunities-imd-rank-2010` and `opendata-scotland-simd-education-rank`.

6.2   An Analysis of the Dataset Community Description Results

This section analyses the dataset community description results (see Table 6). For each dataset community, the analysis compares the 23-dimension vector description automatically assigned by the fingerprint approach with the labels manually assigned by the ground truth. In what follows, we say that a vector v has a peak for dimension i iff $v_i \geq 0.50$.

Community 0, which is equivalent to the "Publications" domain, is described by a vector with peaks for "Applied Science", "Business", "Mathematics", "Science" and "Technology". The presence of five categories shows the diversity of the data in this community. We consider that the label "Publications" assigned by the ground truth classification is better related to the tasks developed in this community than the semantics of the data itself. The rationale behind this argument is that the data come from scholarly articles published in journals and conferences.

Community 1, which is equivalent to the "Social Networking" domain, is described by a vector with peaks for "Business", "People" and "Society". Clearly, the vector was able to capture the essence of social data, covering topics related to the society in general.

Community 2, which has datasets from two different domains, "Government" and "Geographic", is described by a vector with peaks for "Geography" and "Life". Geographic data are available in various domains and, for this reason, the data cannot be described by a single category.

Community 3, which is partially equivalent to the "Life Sciences" domain, is described by a vector with a single peak for "Life", which is similar to the manually assigned domain. *Community 3* is complemented by *Community 4*, whose vector has peaks for "Health" and "Life". Taking into account these two vectors, we may identify datasets in this community with two different content profiles.

Community 5, which is equivalent to the "Government" domain, is described by a vector with peaks for "Geographic", "People" and "Politics". The vector also has significant values for "Business", "Law" and "Society". In general, datasets in this community are related to government transparency. For this reason, the vector for *Community 5* shows an interesting presence of "People", "Society" and "Politics".

7 Related Work

Analyses of the LOD cloud structure followed a wide variety of strategies, ranging from the use of community detection algorithms [12, 13], statistical techniques [4, 14] to dataset profiling techniques [11, 15–17]. Similarly to previous approaches, we combined and applied several techniques from different fields to analyze and generate an automatic version of the manually created LOD diagram. As already know by the LOD community, every couple of years a manual analysis of the state of the LOD cloud is performed and a new LOD diagram is published (see [3, 4]). At the time of the experiments described in this paper, the most recent report was conducted by Schmachtenberg et al. [4] in late 2014 showing the increasing adoption of the LOD principles, the most used vocabularies by data publishers, the degree distribution of the datasets, an interesting manual classification of datasets by topical domain (media, government, publications, geographic, life sciences, cross-domain, user generated content and social networking), among others.

Although such sequence of analyses was widely accepted and adopted by the LOD community, other works presented similar analyses under different perspectives, as those presented by Rodriguez [14]. Based on community detection algorithms, he identified more clusters/communities (Biology, Business, Computer Science, General, Government, Images, Library, Location, Media, Medicine, Movie, Music, Reference and Social) in the LOD cloud. We remind that the main purpose of this work is not only to assign labels to clusters of LOD cloud but to automatically identify and generate a more up to date version of the LOD diagram alleviating the arduous task of data publishers to link their data to others and finding popular vocabularies and others relevant statistics of the actual state of the LOD cloud.

Community detection algorithms are crucial towards an automatic method to generate LOD diagrams. A number of techniques for identifying communities in graph structures were studied by Fortunato [12]. Basically, a community is represented by a set of nodes that are highly linked within a community and that have a few or no links to other communities. Fortunato also presented techniques to validate the clusters found, which we also adopted (see Sect. 4). Xie et al. [13] also explored community detection algorithms. Unlike Fortunato's work, they also considered in their analysis the overlapping structure of communities, i.e., when a community (of datasets) belongs to more than one category. From the 14 algorithms examined by Fortunato, we used the top two best performing overlapping algorithms, GCE and COPRA, in our experiments, as well as a non-overlapping algorithm, which we called the Edge Betweenness Method [6].

As community detection algorithms essentially analyze graph structures to find communities, profiling techniques also play an important role in the identification, at a content level, of the relatedness between datasets. For instance, Emaldi et al. [17], based on a frequent subgraph mining (FSM) technique, extracted structural characteristics of datasets to find similarities among them. Lalithsena et al. [16] relied on a sample of extracted instances from datasets to identify the datasets topical domains. Topics were extracted from reference datasets (such as Freebase) and then ranked and assigned to each dataset profile. Analogously, Fetahu et al. [15] proposed an automated technique to create structured topic profiles for arbitrary datasets through a combination of sampling, named entity recognition, topic extraction and ranking techniques. A more generic approach to create profiles on the Web was presented by Kawase et al. [11]. Kawase's approach generates a histogram (called *fingerprints*) for any text-based resource on the Web based on the 23 top-level categories of the Wikipedia ontology. In this paper, we evaluated Kawase's technique, which demonstrated to be suitable to determine the topical domain of dataset communities. The drawback of Fetahu's approach in our scenario is the large number of categories assigned to a given dataset, which hinders the identification and selection of the most representative topics of a dataset and, consequently, of a community.

8   Conclusions

This paper presented a novel, automatic analysis of the Linked Open Data cloud through community detection algorithms and profiling techniques. The results indicate that the best performing community detection algorithm is the GCE algorithm, with NMI and purity values of 0.57 and 0.42, respectively. Although the EBM algorithm obtained the highest purity value, the high number of communities led to a low NMI value. The mutual dependence between the communities generated using GCE and those from the ground truth is also not high, but, as discussed in Sect. 6, the lack of linksets between datasets in some domains, such as "Cross-Domain", implies a need for the re-organization of datasets as well as the merging and splitting of communities.

The next part of the evaluation focused on comparing the labels manually assigned by the ground truth with the description automatically generated by the profiling technique. The manual labeling process considered as its classification criterion the nature of the data, whereas the automatic process relied on the contents of the datasets to generate the community labels. The experimental results showed that the proposed process automatically creates a clusterization of the LOD datasets which is consistent with the traditional LOD diagram and that it generates meaningful descriptions of the dataset communities. Moreover, the process may be applied to automatically update the LOD diagram to include new datasets. For additional information, including graphical visualizations and detailed results, we refer the reader to the Web site available at http://drx.inf.puc-rio.br:8181/Approach/communities.jsp.

As for future work, we plan to define a recommendation approach based on previous works [18–21], which includes the proposed process, to help data publishers search for datasets to interlink their data.

Acknowledgments. This work was partly funded by CNPq under grants 153908/2015-7, 557128/2009-9, 444976/2014-0, 303332/2013-1, 442338/2014-7 and 248743/2013-9 and by FAPERJ under grants e E-26-170028/2008 and E-26/201.337/2014. The authors would also like to thank the Microsoft Azure Research Program by the cloud resources awarded for the project entitled "Assessing Recommendation Approaches for Dataset Interlinking".

References

1. Ngomo, A.-C.N., Auer, S.: LIMES - a time-efficient approach for large-scale link discovery on the web of data. In: Presented at the 22nd International Joint Conference on Artificial Intelligence (2011)
2. Volz, J., Bizer, C., Gaedke, M., Kobilarov, G.: SILK - a link discovery framework for the web of data. In: Presented at the Workshop on Linked Data on the Web Colocated with the 18th International World Wide Web Conference (2009)
3. Jentzsch, A., Cyganiak, R., Bizer, C.: State of the LOD Cloud. http://lod-cloud.net/state/
4. Schmachtenberg, M., Bizer, C., Paulheim, H.: adoption of the linked data best practices in different topical domains. In: Mika, P., Tudorache, T., Bernstein, A., Welty, C., Knoblock, C., Vrandečić, D., Groth, P., Noy, N., Janowicz, K., Goble, C. (eds.) ISWC 2014. LNCS, vol. 8796, pp. 245–260. Springer, Heidelberg (2014). doi:10.1007/978-3-319-11964-9_16
5. Ertöz, L., Steinbach, M., Kumar, V.: Finding clusters of different sizes, shapes, and densities in noisy, high dimensional data. In: Presented at the SIAM International Conference on Data Mining, San Francisco, CA (2003)
6. Girvan, M., Newman, M.E.J.: Community structure in social and biological networks. PNAS **99**, 7821–7826 (2002)
7. Lee, C., Reid, F., McDaid, A., Hurley, N.: Detecting highly overlapping community structure by greedy clique expansion. In: Presented at the 4th International Workshop on Social Network Mining and Analysis Colocated with the 16th ACM SIGKDD International Conference on Knowledge Discovery and Data Mining (2010)
8. Gregory, S.: Finding overlapping communities in networks by label propagation. New J. Phys. **12**, 103018 (2010)
9. Manning, C.D., Raghavan, P., Schütze, H.: Introduction to Information Retrieval. Cambridge University Press, Cambridge (2008)
10. Pereira Nunes, B., Mera, A., Casanova, M.A., Fetahu, B., Paes Leme, L.A.P., Dietze, S.: Complex matching of RDF datatype properties. In: Decker, H., Lhotská, L., Link, S., Basl, J., Tjoa, A.M. (eds.) DEXA 2013, Part I. LNCS, vol. 8055, pp. 195–208. Springer, Heidelberg (2013)
11. Kawase, R., Siehndel, P., Nunes, B.P., Herder, E., Nejdl, W.: Exploiting the wisdom of the crowds for characterizing and connecting heterogeneous resources. In: Presented at the 25th ACM Conference on Hypertext and Social Media, New York, New York, USA (2014)
12. Fortunato, S.: Community detection in graphs. Physics Reports, vol. 486 (2010)
13. Xie, J., Kelley, S., Szymanski, B.K.: Overlapping community detection in networks: the state-of-the-art and comparative study. In: CSUR, vol. 45 (2013)
14. Rodriguez, M.A.: A Graph Analysis of the Linked Data Cloud. ArXiv e-prints (2009)
15. Fetahu, B., Dietze, S., Pereira Nunes, B., Antonio Casanova, M., Taibi, D., Nejdl, W.: A scalable approach for efficiently generating structured dataset topic profiles. In: Presutti, V., d'Amato, C., Gandon, F., d'Aquin, M., Staab, S., Tordai, A. (eds.) ESWC 2014. LNCS, vol. 8465, pp. 519–534. Springer, Heidelberg (2014)
16. Lalithsena, S., Hitzler, P., Sheth, A.P., Jain, P.: Automatic domain identification for linked open data. In: Presented at the International Conference on Web Intelligence and Conference on Intelligent Agent Technology (2013)
17. Emaldi, M., Corcho, O., López-de-Ipiña, D.: Detection of related semantic datasets based on frequent subgraph mining. In: Presented at the Workshop on Intelligent Exploration of Semantic Data Colocated with the 14th International Semantic Web Conference (2015)

18. Rabello Lopes, G., Paes Leme, L.A.P., Pereira Nunes, B., Casanova, M.A., Dietze, S.: Two approaches to the dataset interlinking recommendation problem. In: Benatallah, B., Bestavros, A., Manolopoulos, Y., Vakali, A., Zhang, Y. (eds.) WISE 2014, Part I. LNCS, vol. 8786, pp. 324–339. Springer, Heidelberg (2014)
19. Caraballo, A.A.M., Nunes, B.P., Lopes, G.R., Paes Leme, L.A.P., Casanova, M.A., Dietze, S.: TRT - a tripleset recommendation tool. In: Presented at the 12th International Semantic Web Conference (2013)
20. Leme, L.A.P., Lopes, G.R., Nunes, B.P., Casanova, M.A., Dietze, S.: Identifying candidate datasets for data interlinking. In: Daniel, F., Dolog, P., Li, Q. (eds.) ICWE 2013. LNCS, vol. 7977, pp. 354–366. Springer, Heidelberg (2013)
21. Lopes, Giseli Rabello, Leme, Luiz André PPaes, Nunes, Bernardo Pereira, Casanova, Marco Antonio, Dietze, Stefan: Recommending tripleset interlinking through a social network approach. In: Lin, Xuemin, Manolopoulos, Yannis, Srivastava, Divesh, Huang, Guangyan (eds.) WISE 2013, Part I. LNCS, vol. 8180, pp. 149–161. Springer, Heidelberg (2013)

Fast Multi-keywords Search over Encrypted Cloud Data

Cheng Hong[1]($\boxtimes$), Yifu Li[2], Min Zhang[1], and Dengguo Feng[1]

[1] Trusted Computing and Information Assurance Laboratory,
Institute of Software, Chinese Academy of Sciences, Beijing, China
hongcheng@tca.iscas.ac.cn
[2] National Computer Emergency Response Team
and Coordination Center of China, Beijing, China

Abstract. Searchable encryption (SE) allows a client to store his data in the Cloud in a way that it is encrypted but still searchable by the server. However, when doing multi-keywords search (MKS), most SE schemes seems to be rather inefficient. In this paper, we design and implement an SE scheme that allows the MKS indices to be pre-built secretly at the server side, making fast MKS on very large databases possible. Detailed experiments and analysis are given, showing that our scheme is efficient, while the possible information leakages during the construction are proved to be minimized.

Keywords: Searchable encryption · Multi-keywords search · Cloud storage

1 Introduction

With the rapid development of the Internet, more and more new-generation businesses began to outsource their data to the Cloud Storage, to achieve better scalability and cheaper price. However, the Cloud users are often concerned with security and privacy issues. The data outsourced, such as medical treatments, credit card bills, or personal photographs, may be quite sensitive, thus the data owner may not want to expose them to untrusted parties, even the storage server itself. In fact, accidents have already taken place in many famous Cloud Storages, such as iCloud (the Hollywood celebrities), Google Docs [1] and Dropbox [2]. Encryption would be an intuitive solution, but only encryption is not enough. Many Cloud services claim to support encryption, but most of their encryption are server-transparent, which means the secret keys are available to the server administrator, thus it still does not solve the problem. The proper solution is to keep the secret key available only for the data owner himself, but it would be hard to search on such data. e.g. a user wants to find all the documents that contains keyword "family" from the Cloud Storage, but the documents are encrypted and the server does not know the keys, thus it is unable to respond to such queries. Plenty of researches have been made in order to achieve searchable

© Springer International Publishing AG 2016
W. Cellary et al. (Eds.): WISE 2016, Part I, LNCS 10041, pp. 433–446, 2016.
DOI: 10.1007/978-3-319-48740-3_32

encryption [3–5], and some of them are already good enough for Cloud usage [6,7]. But an important problem is, they either pay few attentions to MKS, or don't yield a good MKS performance [8,9,11,12]. Since MKS is a common requirement in practical searches, this problem must be solved.

In this paper, we propose a secure searchable encryption called FMS-SE that allows **F**ast and secure **M**ulti-keywords **S**earch. Our main contributions could be described as follows:

1. Compared with the state-of-art MKS schemes, FMS-SE is much faster.
2. We are the first to propose the idea of pre-building the MKS index at the Cloud server, while still maintaining the data privacy.

The remainder of this paper is organized as follows: Sects. 2 and 3 describe the preliminary knowledge for our construction. In Sect. 4 we present the detailed construction and security analysis. Experiments and further discussions are in Sect. 5, and the conclusion is in Sect. 6.

2   Searchable Encryption Backgrounds

2.1   Honest-but-Curious Server

We assume the Cloud servers to be "honest-but-curious". That is to say, they are curious about the plaintext of the data contents or the queried keywords, and will try to use all the resources they have to figure it out, but they still obey the search protocol and execute the proposed tasks correctly as the client requires. Under this assumption, we can mainly focus on the data confidentiality, rather than the data availability.

2.2   Symmetric Searchable Encryption

A typical scenario of searchable encryption on Cloud Storage is illustrated in Fig. 1. In order to securely store and search documents $\mathcal{D} = \{D_1, \ldots, D_m\}$, the

Fig. 1. A typical search over encrypted cloud data

client encrypts $\mathcal{D}$ and uploads them to the Cloud, along with a secure index $\mathcal{I}^1$ and encrypted meta data $\mathcal{M}$.[2]

When making a query, the client sends a trapdoor $\mathcal{T}$ to the honest-but-curious Cloud server, who searches in $\mathcal{I}$ and returns the corresponding $\mathcal{M}$ to the client. Then the client could decrypt the meta data, and download the documents accordingly.

2.3   The Challenges of Multi-keywords Search

For a practical Cloud Storage, a common demand is to submit an MKS query (e.g. given several keywords, find the documents that contain all of the keywords). Currently there are three kinds of MKS solutions described as follows.

(a) Scan the document indices $\mathcal{I} = \{I_1, \ldots, I_m\}$ one by one to find if each document contains all the keywords [10,12]. This method is straightforward, and won't generate extra information leakage, but its time complexity is linear with the number of documents m. This method could be useful for small data sets, but it turns out to be too slow for bigger ones.

(b) Search each individual keyword in the invert indices, and then do an intersection between their result sets [5,6]. Thanks to the invert indices, the time complexity is sublinear with the number of documents, but the intersection operation may be very expensive, and would generate extra information leakage. For example, searching documents containing keyword A and B will reveal the result set matching keyword A and B respectively, thus the server could respond to searches of keyword A (or B) even without legal trapdoors.

(c) Search for a single keyword in the invert indices, and filter its result set to find if they contains the rest keywords [9]. By choosing a keyword whose result set is relatively small (We call such keyword **S-term**), the complexity could be significantly reduced. Compared to approach b, this method won't expose the individual result sets except that of **S-term**, and there's no need of intersections, thus it's the best MKS scheme so far, and it's what FMS-SE based on, but there still exists two major drawbacks in it:

- The time complexity grows with the number of keywords queried. e.g. Let the query be "$A, B_1, B_2 \ldots, B_i$", and the **S-term** be A, it requires a full check on the result set of A, to find whether the documents contains $B_1, B_2, \ldots$ and B_i. The price grows linearly as i increases. According to an investigation of online search habits, about 25 % of the queries are composed of three or more keywords, so this drawback could not be neglected.
- The performance depends heavily on the **S-term** chosen. Actually it won't always be that lucky to find a proper **S-term**. e.g. the client makes a three-keyword query "men's, sports, XXL" on a database of clothes.

---

[1] Here $\mathcal{I}$ could either be in the form of document index: $\mathcal{I} = \{I_1, \ldots, I_m\}$, which build independent indices for each document, or in the form of inverted index: $\mathcal{I} = \{I_1, \ldots, I_n\}$, which build independent indices for each keyword. n is the total number of keywords.

[2] e.g. $\mathcal{M}$ could be the URL or disk address of the document.

In this case, each keyword has a large result set, so whichever keyword chosen as **S-term** would introduce a high search latency.

As it shows, all of the current MKS schemes are inefficient for big data sets. The main reason is, all of their secure indices are built in a single-keyword way, thus they had to do heavy computations on the single-keyword indices to respond to an MKS query. What if the indices are already pre-built in a multi-keyword way (e.g. The server could find the documents containing both A and B directly via an MKS index $\mathcal{I}_{AB}$)? **How to allow a server that is not fully trusted to construct the MKS indices securely?** This is our main motivation to propose FMS-SE.

3   Notations and Preliminaries

This section describes the notations and preliminaries that are necessary to understand FMS-SE.

Documents and keywords: Let $\triangle = \{w_1, ..., w_n\}$ be a dictionary of n words in alphabetical order, and D be a collection of m documents $D = \{d_1, ..., d_m\}$, each containing several keywords. We denote by $D(w)$ the list consisting of all documents in D that contain the word w, and $W = \{W_1, ..., W_q\}$ a list of MKS queries that each W_i contains multiple keywords.

Symmetric Encryption: A symmetric encryption scheme is a set of polynomial-time algorithms $SKE = \{Enc, Dec\}$. Enc takes a key k and a message m and returns a ciphertext c; Dec takes a key k and a ciphertext c and returns m if k was the key that producing c. We call SKE a secure symmetric encryption scheme, if for any m and K, the ciphertexts c are indistinguishable from random texts. Under the current computing abilities, it's quite safe to consider the commonly used encryption schemes (e.g. AES128) secure.

Pseudo-random functions: A pseudo-random function (PRF) is a polynomial-time computable function that its output cannot be distinguished from random functions by any probabilistic polynomial-time adversary. Its difference from encryption is that PRF does not require a decrypt function.

Bilinear Mapping: Let G_1 and G_2 be two multiplicative cyclic groups of prime order p. Let g be a generator of G_1. A bilinear map is an injective function $e : G_1 \times G_1 \rightarrow G_2$ with the following properties:

- Bilinearity: for all $u, v \in G_1$ and $a, b \in Z_p$, $e(u^a, v^b) = e(u, v)^{ab}$.
- Non-degeneracy: $e(g, g) \neq 1$.
- Computability: for all $u, v \in G_1$, $e(u, v)$ is efficiently computable.

4   FMS-SE Construction

This section gives the full construction of FMS-SE, and make analysis on its privacy.

4.1   Basic Algorithms

We begin by proposing the following algorithms for two-keyword searches. The client initializes the scheme by **GenParms**, and runs **UploadFiles** when each document is uploaded. Then he runs **Build2KSIndex** for the chosen keyword pairs, which could be searched later using **IndexedSearch**. Since we are not able to run **Build2KSIndex** for all the possible keyword combinations because of efficiency, the other keyword pairs could be searched using **NonIndexedSearch**.

Algorithm 1. GenParms

1: Choose a bilinear group G_1 with order p, generator g, and bilinear map $e : G_1 \times G_1 \to G_2$, a unified keyword length l, a symmetric encryption algorithm SKE, two pseudo-random functions $F, H : \{0,1\}^* \times Z_p \to Z_p$, then send F and H to the server;

//We choose l that is big enough (e.g. 1024) to hide the length differences. Each keyword and meta data was padded with specified characters (e.g. blank) so that their ciphertexts have the same length l.

2: Generates keys $\{k_1, k_2, k_3, k_5\} \xleftarrow{R} Z_p$, $k_4, k_6 \xleftarrow{R} \{0,1\}^{256}$. k_6 is sent to the server, the else are kept secretly;

3: Create three empty tables $S_1(Z_p, Z_p, \{0,1\}^l)$, $S_2(Z_p)$ and $S_3(\{0,1\}^l, \{0,1\}^l)$ at the server side;

Algorithm 2. UploadFiles

1: Parse the document collection $\mathcal{D}$ into the reverted index form: $\{w_i, D(w_i)\}_{i=1}^n \leftarrow \mathcal{D}$;

2: **for** each w_i **do**

3:     Compute $a_i = F(w_i, k_2)$, $b_i = F(w_i, k_3)$;

4:     init a counter $c_i = 1$;

5:     **for** each $d_j \in D(w_i)$ **do**

6:         Compute $X_j = F(M_j, k_1), Y_j = SKE.Enc(M_j, k_4 \oplus w_i)$;

         // Let M_j be the meta data of d_j

7:         Insert $(F(c_i, H(w_i, k_5)), g^{X_j * a_i}, Y_j)$ into S_1;

8:         Insert $e(g,g)^{X_j * b_i}$ into S_2;

9:         $c_i + +$;

10:     **end for**

11:     Insert $(SKE.Enc(w_i, k_4), SKE.Enc(c_i, k_4 \oplus w_i))$ into S_3;

12: **end for**

Here are the explains why our algorithms are correct: If a document containing w_1 and w_2 is uploaded, **UploadFiles** will insert tuples $(s[1] = F(c_1, H(w_1, k_5)), s[2] = g^{X * a_1}, s[3] = Y)$ into S_1, and $(e(g,g)^{X * b_2})$ into S_2; **Build2KSIndex**(w_1, w_2) will find out $e(h, s[2]) = e(g,g)^{X * b_2}$ exists in S_2(because $h = g^{a_1^{-1} * b_2}$), and inserts $(F(c_{12}, H(w_{12}, k_5)), s[2]^f, s[3])$ into S_1; Then **IndexedSearch**$(\{w_1, w_2\})$ could directly fetch the search result $s[3]$

Algorithm 3. Build2KSIndex

Require: w_i, w_j

1: The client sends $T = \{SKE.Enc(w_i, k_4), SKE.Enc(w_j, k_4)\}$ to the server;

2: The server searches T in S_3 and returns $SKE.Enc(c_i, k_4 \oplus w_i)$ and $SKE.Enc(c_j, k_4 \oplus w_j)$ to the client if they exist;

3: The client decrypts to get c_i and c_j (Without loss of generality, we assume $c_i \leq c_j$);

4: The client computes $w_{ij} = concat(\{w_i, w_j\}), h = g^{a_i^{-1} * b_j}, f = a_i^{-1} * F(w_{ij}, k_2)$ and sends $T' = (H(w_i, k_5), H(w_{ij}, k_5), h, f, SKE.Enc(w_{ij}, k_4))$ to the server;

   // Let $concat(W) = w_1 \| w_2 \| .. w_n$ ($w_1, w_2, ... w_n \in W$, sorted in alphabetical order);

5: The server initialize counters $c_1 = 1$ and $c_2 = 0$;

6: **for** $c_1 = 1$ *to* ∞ **do**

7:     Find s in S_1 that $s[1] = F(c_1, H(w_i, k_5))$;

8:     **if** $e(h, s[2])$ exists in S_2 **then**

9:         Insert $(F(c_2 + 1, H(w_{ij}, k_5)), s[2]^f, s[3])$ into S_1;

10:        $c_2 + +$;

11:     **end if**

12:     **if** No such s found **then**

13:         Insert $(SKE.Enc(w_{ij}, k_4), SKE.Enc(c_2, k_6))$ into S_3;

14:        break;

15:     **end if**

16: **end for**

Algorithm 4. NonIndexedSearch

Require: w_i, w_j

1: Do the same steps as line 1-4 of **Build2KSIndex**;

2: The server initialize counters $c_1 = 1$;

3: **for** $c_1 = 1$ *to* ∞ **do**

4:     Find s in S_1 that $s[1] = F(c_1, H(w_i, k_5))$;

5:     **if** $e(h, s[2])$ exists in S_2 **then**

6:        Return $s[3]$;

7:     **end if**

8:     **if** No such s found **then**

9:        break;

10:     **end if**

11: **end for**

12: The client decrypts the result, read the meta data and download the documents needed;

Algorithm 5. IndexedSearch

Require: keyword set W

1: The client sends $T'' = H(concat(W), k_5)$ to the server;

2: **for** $c_1 = 1$ *to* ∞ **do**

3:     The server finds s in S_1 that $s[1] = F(c_1, T'')$, and returns its $s[3]$ as search result;

4: **end for**

5: The client decrypts the result, read the meta data and download the documents needed;

from S_1, given $H(w_{12}, k_5)$. On the other hand, if we choose not to run **Build2KSIndex** on (w_1, w_2), **NonIndexedSearch**(w_1, w_2) will yield the correct result for the same reason.

4.2 Supporting 3+ Keywords

The algorithm **Build2KSIndex** presented in Sect. 4.1 seems to support two keywords only, but it can be easily extended to a keyword set W with size above two. The condition is that there must exist a keyword set W' satisfying:

(1) $W' \subset W$
(2) $|W'| = |W| - 1$
(3) The MKS index for W' is already built.
(4) $|concat(W)| \leq l$

It means that an N-keyword index cannot be directly built, but must be built based on the N−1 ones recursively, starting from the 2-keyword ones. Since a keyword set always appears in more documents than its superset, and has a higher search priority, this is reasonable.

As long as the above conditions hold, the MKS index for W could be built by the following algorithm:

The algorithm **NonIndexedSearch**(w_i, w_j) could be modified to **NonIndexedSearch**(W', w_j) similarly.

4.3 Picking Optimal Keywords

In real usage, we cannot run **BuildMKSIndex** for all the possible keywords combinations because of efficiency, so a careful evaluation must be done to decide which keywords are picked. In our practices, we pick keywords whose occurrences are all above a certain threshold.

E.g. In the following table, if we set the threshold at 10 %, **Build2KSIndex** has to be run on {"Alice", "computer"}, {"Bob", "computer"} and {"Alice", "Bob"}. Note that the occurrence of {"Bob", "computer"} is still above 10 %, so we also have to run **BuildMKSIndex({"Bob", "computer"},"Alice")**.

After choosing a proper threshold and building the indices, any MKS query W will belong to one of the following cases:

(a) The MKS index for W is already built (e.g. W={"Alice", "Bob"}). In this case, just run **IndexedSearch**(W) directly.
(b) There exist a max subset $W' \subset W$ that W''s MKS index is already built, and its occurrence is below threshold (e.g. W={ "Alice", "Bob", "family"}, W'={"Alice", "Bob"}), in this case, run **NonIndexedSearch**(W', w) for each $w \in W - W'$, and return their intersection.
(c) There exist a single keyword $w \in W$ whose occurrence is below threshold (e.g. W={"Alice", "family"}, w="family"). In this case, run **NonIndexedSearch**(w, w') for each $w' \in W - \{w\}$, and return their intersection.

Either (a), (b) or (c) are more efficient than the approaches without MKS indices.

Algorithm 6. BuildMKSIndex

Require: W', w_j
$\quad$ // Let $W - W' = \{w_j\}$;
1: The client sends $\mathcal{T} = \{SKE.Enc(concat(W'), k_4), SKE.Enc(w_j, k_4)\}$ to the server;
2: The server searches $\mathcal{T}$ in S_3 and returns $SKE.Enc(c_{W'}, k_6)$ and $SKE.Enc(c_j, k_4 \oplus w_j)$ to the client if they exist;
3: The client decrypts to get $c_{W'}$ and c_j(Without loss of generality, we assume $c_{W'} \le c_j$);
4: The client computes $w = concat(W), h = g^{a_i^{-1} * b_j}, f = a_i^{-1} * F(w, k_2)$ and sends $\mathcal{T}' = (H(concat(W'), k_5), H(w, k_5), h, f, SKE.Enc(w, k_4))$ to the server;
5: The server initialize counters $c_1 = 1$ and $c_2 = 0$;
6: **for** $c_1 = 1$ to ∞ **do**
7: $\quad$ Find s in S_1 that $s[1] = F(c_1, H(concat(W'), k_5))$;
8: $\quad$ **if** $e(h, s[2])$ exists in S_2 **then**
9: $\quad\quad$ Insert $(F(c_2 + 1, H(w, k_5)), s[2]^f, s[3])$ into S_1;
10: $\quad\quad$ $c_2 + +$;
11: $\quad$ **end if**
12: $\quad$ **if** No such s found **then**
13: $\quad\quad$ Insert $(SKE.Enc(w, k_4), SKE.Enc(c_2, k_6))$ into S_3;
14: $\quad\quad$ break;
15: $\quad$ **end if**
16: **end for**

Table 1. Example of keywords distribution

Keyword	Appears in ?% of documents
Alice	12%
Alice, computer	4%
Alice, Bob	3%
Bob	15%
Bob, computer	11%
Computer	20%
Family	5%

4.4 Supporting Updates

FMS-SE supports a basic form of file deletion and upload (update is considered as a deletion plus an upload). We divide the system into two parts: the static storage SS, with the MKS indices already built, and the dynamic storage DS without MKS indices. When a file is to be uploaded, the owner runs **Upload-Files** on it, inserts the generated single-keyword index $\mathcal{I}$ into DS. When a file is to be deleted, the owner also generates the corresponding $\mathcal{I}$, and inserts it into DS, marked as "deleted". When a query comes, the server searches SS and DS respectively, and combines the search results (Since we didn't run **BuildMKSIndex** on DS, only **NonIndexedSearch** could be run on DS). If the newly updated files reaches a certain amount, the user can decide whether to run **BuildMKSIndex** on DS, and allocate a new dynamic storage.

We note that supporting efficient updates is an open problem for all SE schemes, and our effort only solved the problem for a certain extent, but it is still a proper solution, especially for storages that won't change significantly.

4.5   Security Analysis

Before stating the security analysis for FMS-SE, we make the following auxiliary definition:

Definition 1 (Keyword Pattern KP). *We use $K(D) = \{w_{D1}, ..., w_{Dn}\}$ to denote all the keywords appear in D. Given two document sets $D' = \{d'_1, ..., d'_m\}, D'' = \{d''_1, ..., d''_m\}$ of equal size, we say D' and D'' have the same keyword pattern: $KP(D') = KP(D'')$ if the following stands:*

(a) $|K(D')| = |K(D'')|$;

(b) There exists a one-to-one mapping $P : K(D') \to K(D'')$ that:
For any $w \in K(D') \cap K(D''), P(w) = w$;
For any $w \in K(D'), |D'(w)| = |D''(P(w))|$;
For any $w_1, w_2 \in K(D'), |D'(w_1) \cap D'(w_2)| = |D''(P(w_1)) \cap D''(P(w_2))|$;

Then we state the "game-based" security definition for FMS-SE, like [9,13] does. The game is between a challenger C and an adversary A as follows:

Setup: C creates a set of documents D and gives it to A. A chooses a non-empty subset D' from D and sends D' to C. C runs **GenParms** and **Upload-Files** on D', and sends the index $\mathcal{I} = \{S_1, S_2, S_3\}$ to A.

BuildIndex: A runs either of the following:

(a) A chooses two keywords w_1, w_2 and sends them to C. C computes the corresponding $\mathcal{T}$ and $\mathcal{T}'$ and returns it to A. A runs **Build2KSIndex**(w_1, w_2) on $\mathcal{I}$.

(b) A chooses a keyword set $W(|W| > 1)$, a keyword w and sends them to C. C computes the corresponding $\mathcal{T}$ and $\mathcal{T}'$ and returns it to A. A runs **BuildMKSIndex**(W, w) on $\mathcal{I}$.

Query: A runs either of the following:

(a) A chooses a keyword set W and sends W to C. C computes the corresponding $\mathcal{T}''$ and returns it to A. A runs **IndexedSearch**(W) on $\mathcal{I}$.

(b) A chooses a keyword set $W(|W| \geq 1)$, a keyword w and sends them to C. C computes the corresponding $\mathcal{T}$ and $\mathcal{T}'$ and returns it to A. A runs **NonIndexedSearch**(W, w) on $\mathcal{I}$.

(c) A chooses two keywords w_1, w_2 and sends them to C. C computes the corresponding $\mathcal{T}$ and $\mathcal{T}'$ and returns it to A. A runs **NonIndexedSearch**(w_1, w_2) on $\mathcal{I}$.

Challenge: After making some BuildIndex and Query, A picks two nonempty subset D_0, D_1 from D satisfying $|D_0| = |D_1|$, and $KP(D_0) = KP(D_1)$, and A must not have run BuildIndex and Query on any keyword in $D_0 \cup D_1$.

Next, A send D_0, D_1 to C who chooses b randomly from $0, 1$, invokes **Upload-Files** on D_b, and sends the generated index $\mathcal{I}_b$ to A.

Response: A output b' as a guess of b.

Definition 2 (KP-Secure). *The advantage of A in winning the game above is $Adv_A = |PR[b = b'] - 1/2|$. If Adv_A is neglectable, we define that FMS-SE is **KP-Secure**, which means running **BuildIndex** and **Query** doesn't give the adversaries any advantages in breaking the cipher indices.*

Then we have:

Lemma 1. *FMS-SE is KP-Secure.*

Here's the proof by comparing $\mathcal{I}_0$ with $\mathcal{I}_1$ from the view of A.

Proof. As algorithm **UploadFiles** shows, each index is made up of three data tables, let's mark them as $\mathcal{I}_0 = \{S_{10}, S_{20}, S_{30}\}$, $\mathcal{I}_1 = \{S_{11}, S_{21}, S_{31}\}$. Note that $KP(D_0) = KP(D_1)$ ensures $|S_{10}| = |S_{11}|, |S_{20}| = |S_{21}|, |S_{30}| = |S_{31}|$.

For S_{10}: Each tuple $s_{10} \in S_{10}$ is made up of three components: $s_{10}[1] = F(c_i, H(w_i, k_5))$, $s_{10}[2] = g^{X_j * a_i}$, $s_{10}[3] = SKE.Enc(M_j, k_4 \oplus w_i)$. Since F, H are pseudo random functions, SKE is a secure symmetric algorithm, and A doesn't know k_4 and k_5, $s_{10}[1], s_{10}[3]$ looks like random data for A. $s_{10}[2]$ does not look like random, because if $d_1, d_2 \in D_0(w_1) \cap D_0(w_2)$, there will exist four tuples $g^{X_1 * a_1}, g^{X_1 * a_2}, g^{X_2 * a_1}, g^{X_2 * a_2} \in \{s_{10}[2]\}$ satisfying $e(g^{X_1 * a_1}, g^{X_2 * a_2}) = e(g^{X_2 * a_1}, g^{X_1 * a_2})$.

For S_{11}: Thanks to $KP(D_0) = KP(D_1)$, any document in $D_0(w_1) \cap D_0(w_2)$ has its couterpart in $D_1(P(w_1)) \cap D_1(P(w_2))$, thus $S_{11}[2]$ has the same features with $S_{10}[2]$. A cannot distinguish S_{10} from S_{11}.

For S_{20}: Tuple $s_{20} \in S_{20}$ has the similar feature with $s_{10}[2]$: If $d_1, d_2 \in D_0(w_1) \cap D_0(w_2)$, there will exist four tuples $g^{X_1 * b_1}, g^{X_1 * b_2}, g^{X_2 * b_1}, g^{X_2 * b_2} \in S_{20}$ satisfying $e(g^{X_1 * b_1}, g^{X_2 * b_2}) = e(g^{X_2 * b_1}, g^{X_1 * b_2})$.

For S_{21}: For the similar reason with S_{11}, S_{21} has the same features with S_{20}. A cannot distinguish S_{20} from S_{21}.

For S_{30} and S_{31}: Since SKE is a secure symmetric algorithm, and A doesn't have access to k_4, S_{30} and S_{31} looks like random data for A.

Table 2. Number of operations in FMS-SE

	UploadFiles (1 file)	BuildMKSIndex (W', w)	NonIndexedSearch (W', w)	IndexedSearch (W)
Client	$i * (4 * t_f + t_s + t_e + t_p)$	Constant	$c_r * t_S$	$c_W * t_S$
Server	Constant	$c_{W'}(t_p + t_f) + c_r * (t_f + t_e)$	$c_{W'}(t_p + t_f)$	$c_W * t_f$

In conclusion, as long as $KP(D_0) = KP(D_1)$, the adversary A cannot tell whether the index $\mathcal{I}_b$ is built upon D_0 or D_1, FMS-SE is KP-Secure.

4.6 Complexity Analysis

The number of operations required in each algorithm is shown in Table 2. Briefly speaking, the cost of **BuildMKSIndex** mainly lies in the server side, and grows linearly with $c_{W'}$, **NonIndexedSearch** cost more than **IndexedSearch**.

$*i$ is the number of keywords in each file, $c_{W'}$ is the number of documents containing W', c_r is the number of documents containing both W' and w, and c_W is the number of documents containing W. t_S is the time cost per SKE encryption/decryption, t_p is the time cost per bilinear pairing, t_e is the time cost per exponentiation, and t_f is the time cost per PRF function.

4.7 Summary

The advantages of FMS-SE are threefold:

(a) The **BuildMKSIndex** procedure is the most time-consuming part of FMS-SE, but luckily it's mainly done at the server side. All the client has to do is to send the proxy information to the server, go offline, and all the rest computation could be done by the server. The security analysis show that the outsourced computation does not sacrifice privacy.

(b) FMS-SE could respond to MKS queries faster than any other approaches. As Sect. 4.3 shows, we build indices for those frequent keyword sets via **BuildMKSIndex**, even if an MKS query does not have its corresponding MKS indices built, it must have a keyword subset which is relatively rare, and could be used to run **NonIndexedSearch** efficiently.

(c) We have neglectable false-positive rate. The only possibility for FMS-SE to return false positive comes from: (1) the string concat function. (2) the PRF collision. As long as the string concat function is carefully designed (e.g. add a proper separator before concatenation), and a big enough prime p is chosen, such probability would be low enough to be neglected.

5 Experiments

In this section, we will show the performance of FMS-SE. The database we used is Apache Hbase 0.96, running on three CentOS 6.2 servers, each with a Intel Xeon 2.4 GHz CPU and 4 GB memory. The client is Windows 7 with Intel I5-2500 CPU and 4 GB memory. We use the Enron E-mail dataset[3] and Sogou news dataset[4].

---

[3] http://www.cs.cmu.edu/~enron/, since the dataset only contains up to 1 million documents, we made several copies of the datasets and merge them together to simulate bigger ones.

[4] http://www.sogou.com/labs/.

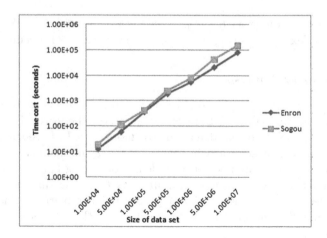

Fig. 2. The time cost of FMS-SE index construction. We set the threshold at 1 %.

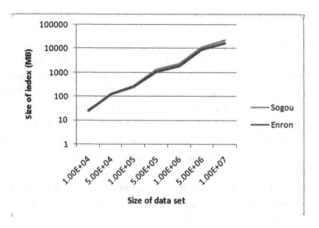

Fig. 3. The space cost of FMS-SE index construction.

The time cost of FMS-SE is shown in Fig. 2. Generally speaking, the time cost increases linearly with the number of documents, but it is tolerable because: (a) This phase could be done at the server side as the data owner is offline. (b) The algorithm could easily be modified to run in parallel. (c) It only need to be run once, for a huge dataset, spending a one-time effort of several hours to make all of the related subsequent queries faster is worth trying.

The space cost of FMS-SE construction is shown in Fig. 3. Generally speaking, the cost increase linearly with the number of documents, but it is relatively small compared to the volume of documents, especially when big documents are concerned. E.g. The enron e-mail data set with $5 * 10^5$ documents takes up to 2.5 GB of disk space, and it requires an extra 1 GB to store the FMS-SE index.

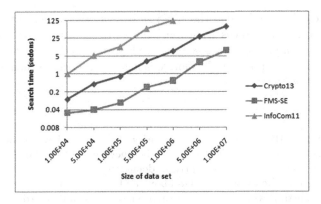

Fig. 4. The performance of a two-keywords search on Enron E-mail dataset. Each keyword appears in 10 % of the total data set.

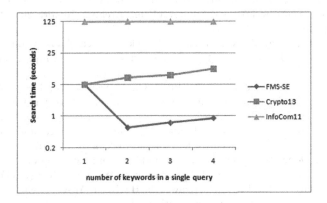

Fig. 5. The performance of multi-keywords search on Enron E-mail dataset. The size of the data set (number of documents) is 1 million.

The sogou news data set with $5 * 10^5$ documents takes up to 10 GB of disk space, and it requires an extra 1.2 GB to store the FMS-SE index.

The time cost of MKS searches is shown in Figs. 4 and 5. We pick two state-of-the-art MKS approaches, Crypto13 [9] and InfoCom11 [12] as comparisons[5]. It is shown that, thanks to the MKS indices, the MKS query cost of FMS-SE is much lower than the others, especially when the number of keywords increases.

6   Conclusion

We discussed how to support multi-keywords search (MKS) in searchable encryption. On top of the discussion, we designed FMS-SE, which is the first searchable

---

[5] We omitted InfoCom11 experiments on datasets above 1 million, because it's too slow.

encryption scheme that builds the MKS indices at the server side. The main advantage is that the indices could be built at the semi-trusted Cloud server without sacrifice of privacy. We prove FMS-SE secure, and make experiments to show that FMS-SE support MKS more efficiently than other schemes.

References

1. Fulton, S.M., III.: Google Docs security hole may have exposed private documents.betanews.com (2009). http://betanews.com/2009/03/09/google-docs-security-hole-may-have-exposed-private-documents
2. Leonhard, W.: Re-examining Dropbox and its alternatives.windowssecrets. com (2011). http://windowssecrets.com/top-story/re-examining-dropbox-and-its-alternatives
3. Song, D.X., Wagner, D., Perrig, A.: Practical techniques for searches on encrypted data. In: IEEE Symposium on Security and Privacy, pp. 44–55 (2000)
4. Goh, E.J.: Secure indexes.Cryptology ePrint Archive, Report 2003/216 (2003). http://eprint.iacr.org/2003/216/
5. Curtmola, R., Garay, J., Kamara, S., Ostrovsky, R.: Searchable symmetric encryption: improved definitions and efficient constructions. J. Comput. Secur. **19**(5), 79–88 (2011)
6. Cash, D., Jaeger, J., Jarecki, S., Jutla, C., Krawczyk, H., Roşu, M.C. et al.: Dynamic searchable encryption in very-large databases: data structures and implementation. In: Network and Distributed System Security Symposium (2014)
7. Pappas, V., Krell, F., Vo, B., Kolesnikov, V.: Blind seer: a scalable private dbms. In: IEEE Symposium on Security and Privacy, pp. 359–374 (2014)
8. Jin, W.B., Dong, H.L.: On a security model of conjunctive keyword search over encrypted relational database. J. Syst. Softw. **84**(8), 1364–1372 (2011)
9. Cash, D., Jarecki, S., Jutla, C., Krawczyk, H., Roşu, M.-C., Steiner, M.: Highly-scalable searchable symmetric encryption with support for boolean queries. In: Canetti, R., Garay, J.A. (eds.) CRYPTO 2013. LNCS, vol. 8042, pp. 353–373. Springer, Heidelberg (2013). doi:10.1007/978-3-642-40041-4_20
10. Golle, P., Staddon, J., Waters, B.: Secure conjunctive keyword search over encrypted data. In: Jakobsson, M., Yung, M., Zhou, J. (eds.) ACNS 2004. LNCS, vol. 3089, pp. 31–45. Springer, Heidelberg (2004). doi:10.1007/978-3-540-24852-1_3
11. Wang, P., Wang, H., Pieprzyk, J.: An efficient scheme of common secure indices for conjunctive keyword-based retrieval on encrypted data. In: Chung, K.-I., Sohn, K., Yung, M. (eds.) WISA 2008. LNCS, vol. 5379, pp. 145–159. Springer, Heidelberg (2009). doi:10.1007/978-3-642-00306-6_11
12. Cao, N., Wang, C., Li, M., Ren, K., Lou, W.: Privacy-preserving multi-keyword ranked search over encrypted cloud data. IEEE Trans. Parallel Distrib. Syst. **25**(1), 222–233 (2014)
13. Li, R., Liu, A.X., Wang, A.L., Bruhadeshwar, B.: Fast range query processing with strong privacy protection for cloud computing. Proc. VLDB Endowment **7**(14), 1953–1964 (2014)

A Scalable Parallel Semantic Reasoning Algorithm-Based on RDFS Rules on Hadoop

Liu Yang[1], Xiao Wen[1], Zhigang Hu[1], Chang Liu[1],
Jun Long[2], and Meiguang Zheng[1(✉)]

[1] School of Software, Center South University, Changsha 410073, China
{yangliu,zhengmeiguang}@csu.edu.cn
[2] School of Information Science and Engineering,
Center South University, Changsha 410073, China
jlong@csu.edu.cn

Abstract. The rapid growth of semantic web utilization in the cloud has resulted in massive amounts of RDF data, which is challenging large-scale RDF semantic reasoning. The traditional semantic reasoning process is very time-consuming and lacks scalability. In this paper, we present a scalable method for RDFS rule-based semantic reasoning using a distributed framework of Hadoop MapReduce, and propose an optimized semantic reasoning algorithm based on RDFS rules. The reasoning algorithm first classifies RDFS entailment rules to build different reasoning rule models, and then orders the rule execution sequences according to the relation of RDFS entailment rules to reduce reasoning time. During algorithm execution in MapReduce, the reasoning work handles RDFS rules in the Map process phase, and data duplication elimination is handled in the Reduce process phase. The experiment results on the LUBM benchmark show that our optimized reasoning algorithm outperforms Urbani's reasoning method in efficiency, stability, and scalability. The average reasoning time of our algorithm is only 1/3 that of Urbani's algorithm with different RDF dataset scales.

Keywords: Ontology reasoning · RDF · Semantic web · MapReduce · Big data

1 Introduction

In the era of big data, large volumes of Semantic Web data have dramatically increased the amount of semantic RDF data. The RDF (Resource Description Framework) [1] is a primary representation of ontology used to describe semantic data. Diverse applications have developed in domains such as life science [2], media data [3], web services composition [4], and social networks. The Linked Open Data project [5] was estimated to contain 1040 RDFS data sets in 2014. Since then, the growth rate of this data has continued to increase. Traditional semantic reasoning methods include those of Pellet [6], Racer [7], etc. However, these reasoning systems run on single-node environment, and their computing

© Springer International Publishing AG 2016
W. Cellary et al. (Eds.): WISE 2016, Part I, LNCS 10041, pp. 447–456, 2016.
DOI: 10.1007/978-3-319-48740-3_33

performance and scalability are not ideal, so algorithms capable of searching large amounts of data have become increasingly important.

2 Related Works

RDF is a data model recommended by the W3C as a standard for data interchange for the Semantic Web, and uses triples in the form of $<s, p, o>$ to describe web resources. Entity information on the Web (or concepts) are represented on the subject s with URI (Uniform Resource Identifier), predicate p describes the attributes of the entity, and object o describes the value of the attributes [8,9]. RDFS (RDF schema) [10] is an extension of RDF that provides basic elements for the ontology description. It defines a set of rules respectively with the ability to represent implicit information. Table 1 shows all 13 RDFS rules about class and sub-class, property of the class.

Table 1. RDFS rules

No.	Antecedent	Consequence
1	*s p o(if o is a literal)*	*_:n rdf:type rdfs:Literal*
2	*p rdfs:domain x&s p o*	*s rdf:type x*
3	*p rdfs:range x &s p o*	*o rdf:type x*
4a	*spo*	*s rdf:type rdfs:Resource*
4b	*spo*	*o rdf:type rdfs:Resource*
5	*p rdfs:subPropertyOf q&q rdfs:subPropertyOf r*	*p rdfs:subPropertyOf r*
6	*p rdf:type rdf:Property*	*p rdfs:subPropertyOf p*
7	*s p o &p rdfs:subPropertyOf q*	*sqo*
8	*s rdf:type rdfs:Class*	*s rdfs:subClassOf rdfs:Resource*
9	*s rdf:type x&x rdfs:subClassOf y*	*s rdfs:subClassOf y*
10	*s rdf:type rdfs:Class*	*s rdfs:subClassOf s*
11	*x rdfs:subClassOf y &y rdfs:subClassOf z*	*x rdfs:subClassOf z*
12	*p rdf:type rdfs:ContainerMembershipProperty*	*p rdfs:subProperyOfrdfs:member*
13	*o rdf:type rdfs:Datatype*	*o rdfs:subClassOf rdfs:Literal*

Among the rule set of Rules 1, 4, 6, 8 and 10, each have only one antecedent, which means it does not affect the reasoning results during parallelization reasoning, and Rule 12 and Rule 13 rarely appear in the RDF data and are not represented, so we will only discuss the rest rules related reasoning shown in the Table 1.

In recent years, many researchers have developed different kinds of semantic data reasoning systems. The RDF reasoning methods can be divided into the following five categories based on RDFS entailment rules. (1) Reasoning based on relational database, such as Minerva [11]. (2) Reasoning based on Hash Table, such as paper [12] proposed a parallel reasoning method based on the distributed Hash Table. (3) Parallel reasoning method based on P2P network. Paper [13] proposed a parallel reasoning method based on the P2P self-organization network. (4) Parallel reasoning based on data partitioning. Soma and Prasanna [14] proposed a parallel reasoning method through data partitioning. Weaver and Hendler [15] proposed a data partitioning model based on MPI. (5) Distributed reasoning methods over MapReduce. Mika and Tummarello [16] first proposed a method to achieve reasoning with the MapReduce framework. Gu et al. [17] designed a distributed reasoning engine named Cichlid on Spark for the RDFS and OWL Horst rule sets. Urbani et al. [18] analyzed the dependent relationship among the RDFS rules and built dependence relationships to ensure the execution order of RDFS rules. However, the reasoning process presented by Urbani was divided into four iteration processes, and during each iteration process the data transmission accounted for much reasoning, so their computing performance and scalability were not ideal.

This paper presents a RDFS reasoning algorithm with MapReduce that was optimized based on Urbani's [19] method. Firstly, our reasoning algorithm classifies RDFS rules to build three reasoning types, including rules of class, rules of property, and the rules of domain and range. Secondly, it orders the execution sequence of the reasoning rules according to the dependency relationship among them. In the process of MapReduce, the algorithm executes the RDFS rules sequentially during the Map phase. Finally, in the Combine and Reduce phases, it eliminates duplicate data generated during the reasoning process. The proposed RDFS reasoning algorithm reduces reasoning to a single job to improve reasoning efficiency, and achieves excellent stability and fault tolerance.

3 Design of Parallel Reasoning Algorithm-Based RDFS Rules

3.1 Construction of Reasoning Rule Models

To elaborate the reasoning method based on RDFS rules, we define the following concepts of the schema triples, instance triples, rule models, and reachability concepts. Schema triples is RDFS triples with *rdfs:domain, rdfs:range, rdfs:subPropertyOf* or *rdfs:subClassOf* as the predicate. Instance triple refers to RDFS triples except for schema triples. *subP* expresses the schema triple reasoning model has the predicate *rdfs:subPropertyOf*. Each node in the model stands for the subject or object, and the directed link between them shows their *subProperty* relation. *subC* expresses the schema triple reasoning model that has the predicate *rdfs:subClassOf*, in which the directed link shows the subclass relation between the node pairs. *domR* expresses the schema triple reasoning model that have the predicate *rdfs:domain* or *rdfs:range*, in which each node in the model stands for subject or object, and the directed link between them shows their domain or range relation. Reachability shows the directed link between the node pairs in the reasoning model.

The rules in Table 1 are divided into three categories. (1) reasoning based on *rdfs:domain* or *rdfs:range*, and corresponding Rule 2 and Rule 3; (2) reasoning based on *rdfs:subPropertyOf* and corresponding Rule 5 and Rule 7; (3) reasoning based on *rdfs:subClassOf* and corresponding Rule 9 and Rule 11. Accordingly, we classify schema triples into three types, and constructs *domR*, *subP*, *subC* model respectively.

Reasoning Principles of *domR* model. If an instance triples $<s_0, p_0, o_0>$ have predicates *rdf:domain* or *rdf:range*, two new triples can be generated from Rule 2 and Rule 3 according to the *domR* model. From the Fig. 1, the two new triples are $<s_0, rdf:type, x_1>$ and $<o_0, rdf: type, x_2>$, in which the predicates *rdf:type* replaces the original predicates *rdf:domain* and *rdf:range*, x_1 replaces the object o_0 of $<s_0, p_0, o_0>$ in the first new triple, and the object o_0 as the new subject and x_2 as the new object of new rule in the second new triple. The same is to $<s_1, p_1, o_1>$, and the two referred new rules are $<s_1, rdf:type, x_1>$ and $<o_0, rdf: type, x_3>$ by *domR* model.

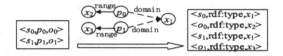

Fig. 1. Reasoning based on the *domR* model

Reasoning Principles of *subP* model. If an instance triples $<s_0, p_0, o_0>$ have predicates *rdfs:subPropertyOf*, new triples can be generated from Rule 5 and Rule 7 according to the *subP* model by transitivity. From the Fig. 2, the new triples are $<s_0, p_1, o_0>$, $<s_0, p_2, o_0>$, $<s_0, p_3, o_0>$ by *subP* model.

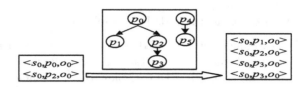

Fig. 2. Reasoning based on *subP* model

Reasoning Principles of *subC* model. If an instance triples $<s_0, rdf:type, o_0>$ have predicates *rdf:type and have object subClass*, new triples can be generated from Rule 9 and Rule 11 according to the *subP* model by inheritance. From the Fig. 3, the new triples are $<s_0, rdf:type, o_1>$, $<s_0, rdf:type, o_2>$, $<s_0, rdf:type, o_3>$, $<s_0, rdf:type, o_4>$ by *subC* model.

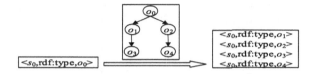

Fig. 3. Reasoning based on *subC* rule model

3.2 Execution Sequence of the Reasoning Rule Models

We analyze the output data of reasoning model. As shown in Figs. 1 and 3, the *domR* model and *subC* model output the triples which have predicate of *rdf:type*, and the subject and the object are the same as the original triples. According to Fig. 2, the *subP* model output the normal instance triples. By analyzing the input triples and the output triples of the above reasoning models, the dependency relations among the *domR*, *subP*, and *subC* models can be determined, as shown in Fig. 4.

Fig. 4. Dependency relations among reasoning models

On the basis of dependency relations among the *domR*, *subP*, and *subC* model, it is possible to determine that the application order of the reasoning model is *subP->domR->subC*. After reasoning by applying the *subP* model, the output triples of instance triple $<s, p, o>$ is still $<s, p, o>$, and it is taken as the input triple of the *domR* model. The new triples are referred with the predicate of *rdf:type* as the triple $<s,rdf:type,o>$, and the reasoning result of the triple $<s,rdf:type,o>$ is still $<s,rdf:type,o>$ by *subC* model.

3.3 Reasoning with MapReduce

In this section, we optimize the RDFS reasoning algorithm and complete RDFS reasoning by using one MapReduce Job. Figure 5 shows the reasoning process based on the MapReduce framework.

Mapping Stage. During the map phase, we use our proposed reasoning model to implement RDFS reasoning. After building the *subP*, *domR*, and *subC* model based on schema triples, and set them as global variables, the map function loads instance triples from HDFS. As shown in Algorithm 1, we execute the reasoning rules based on the following sequence: *subP->domR->subC*. When the reasoning finishes, the map function will emit those triples by using the $<s, p, o>$ triple as the key to eliminate duplicate triples generated during reasoning.

We focus on algorithm complexity. In the outer loop, the number of intermediate results is a constant. In the inner loop, the nodes in the rule models

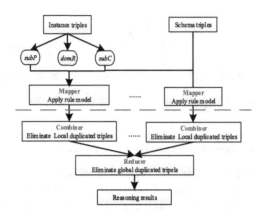

Fig. 5. Reasoning process with MapReduce

Algorithm 1. RDFS Reasoning Rules
___
Input: instance triples
Output: original triples and reasoning results
 1: map($key,value$); //instance triples as key,$value$ is null
 2: $S \leftarrow \{key\}$; //reasoning results stored in S
 3: **for** $s{:}S$ //Traversing S triples, apply the reasoning rule model **do**
 4:    **for** $model$:RDFS inference $models$ //order of application is $sub\text{-}{>}domR\text{-}{>}subC$
       do
 5:       **if** s.$predicate$ or s.$object$ in $model$ //original triples can apply reasoning model
          then
 6:          $Q \leftarrow$ canReach(s.$predicate$ or s.$object$); //get reachable values of this node
 7:          **for** $(q:Q)$ **do**
 8:             add new triple to S //get new triples
 9:          **end for**
10:       **end if**
11:    **end for**
12: **end for**
13: delete key from S; //Removed the original triples from S
14: **for** $(s:S)$ **do**
15:    emit(s,null); //set triples as key
16: **end for**
___

are stored in a Hash Map. The average search time complexity is $O(1)$. Overall, the time complexity of the algorithm is $O(1)$ for each input triple. During the reasoning process in Urbani's algorithm, the instance triples will join the schema triples, and the time complexity is $O(N)$, where N is the number of schema triples. In terms of time complexity, our proposed algorithm outperforms Urbani's algorithm.

Eliminate Duplicated Data Strategy. Duplicated RDF triples may be generated when each reasoning rule is processed. This duplicate triples must be detected and eliminated. There are two types of duplicated RDF triples: (1) new inferred triples that are the same as the original instance triple; and (2) duplicates between the new inferred triples.

We set the *subP* reasoning model shown in Fig. 2 as an example. Based on the reasoning model of *subP*, the reasoning results of $<s_0, p_0, o_0>$ is $<s_0, p_1, o_0>$, $<s_0, p_2, o_0>$, and $<s_0, p_3, o_0>$, the new inferred triple $<s_0, p_2, o_0>$ repeats with the origin instance triple. The instance triple $<s_0, p_0, o_0>$ refer the new triple $< s_0, p_3, o_0>$ by *subP* model, which repeat with the above instance triple $<s_0, p_0, o_0>$ reason result.

We set instance triples as $<key, value>$ pairs in the Map phase based on Hadoop, and all duplicate triples are eliminated which are generated during reasoning, after running the merging operation of local Combiner phase and the global Reduce phase.

The process of local deduplication is as following: The output triples of Map phase are merged by Combiner to achieve local triples aggregation in accordance with the key value, and this process also reduces data transmission between the Map and Reduce phases.

The process of global deduplication is as following: The output triples of the Combiner phase are as the input triples for the Reduce phase, and the triples of different nodes will be merged according to the value of key by the function of *reduce( )*. Finally, all the duplicated triples produced in the reasoning process are eliminated.

4 Evaluation

This experiment is based on a local cluster with 4 nodes running Hadoop-2.5.2. In this cluster, one node acts as the master and the other 3 nodes are used for computation. Each node uses the same configuration: Intel(R) Core(TM) i5-3470 3.20 GHz CPU, 4 GB memory, and a 400 GB 7200 RPM SATA hard drive. The nodes run on the Ubuntu 14.04 Operating System.

We used artificial datasets LUBM [19], and generate 5 different data sets, which contain ontology data from 10, 50, 100, 300, and 500 universities under the same experimental environment, we evaluate the performance of our optimized RDFS reasoning algorithm and Urbani's [19]. The original triples of 10, 50, 100, 300, and 500 universities are 1,311,409, 6,863,227, 13,824,437, 41,474,311 and 69,222,185 respectively, and after executing the RDFS reasoning algorithm, we get the triples of 1,611,159, 8,471,795, 16,989,539, 50,914,016 and 84,937,776, including the inferred triples after reasoning. Figure 6 shows the number of triples before and after reasoning and the average reasoning time.

To achieve better accuracy, tests on each dataset are executed 5 times. Table 2 lists the average execution time.

As shown in Table 2, our proposed method is faster than Urbani's under the same test data set. When the test data expands 50 times, the reasoning time of the two reasoning method increases 18 times and 25 times, respectively, indicating that our method has more advanced scalability.

Table 2. Reasoning time comparison

Number of universities	Reasoning time(s)	
	Optimized reasoning	Urbani's reasoning
10	5.8	18.9
50	16.4	64.3
100	30.6	130.6
300	83.1	324.2
500	106.8	490.8

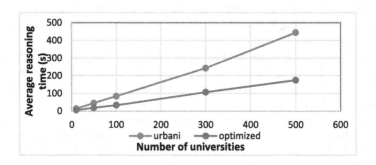

Fig. 6. Average reasoning time on different data set

As shown in Fig. 6, the execution time of both methods increases as the variable increases. The average reasoning time of both methods does not exponentially grow as the data set size grows; rather, they smoothly increase. Our optimized method's reasoning time increases more slowly than Urbani's method.

5   Conclusion

In the era of big data, scalable semantic reasoning has become an important issue with respect to the development of knowledge management systems. This paper presents a RDFS reasoning algorithm with MapReduce optimized based on Urbani's method. The reasoning algorithm classifies RDFS rules to build different types of reasoning models and deeply analyses the three kinds of reasoning models. It analyzes dependency relations among the reasoning rule model to ensure the application order of each kind of reasoning model. The implementation of MapReduce jobs in the Map phase sets RDF instance triples as the key value, and then applies *subP*, *domR*, and *subC* reasoning rule models. Finally, it eliminates the duplicate data generated in the process of reasoning during the Combine and Reduce phase. Our method is implemented based on the MapReduce and Hadoop by using a cluster of up to four nodes. We have evaluated our system on LUBM. In the future, we will validate our methods on more datasets, such as DBpedia and WordNet datasets, and extend the reasoning algorithm with OWL rules or other ontology reasoning rules.

Acknowledgement. The National Natural Science Foundation of China under Grant No. 61301136, No. 61572525 and No. 61602525.

References

1. Manola, F., Miller, E.: RDF Primer [EB/OL]. In: W3C Recommendation (2004). http://www.w3.org/TR/RDFSyntax/
2. Marshall, M.S., et al.: Emerging practices for mapping and linking life sciences data using RDF–a case series. J. Web Semant. **14**, 2–13 (2012)
3. Kobilarov, G., Scott, T., Raimond, Y., Oliver, S., Sizemore, C., Smethurst, M., Bizer, C., Lee, R.: Media meets semantic web – how the BBC uses DBpedia and linked data to make connections. In: Aroyo, L., et al. (eds.) ESWC 2009. LNCS, vol. 5554, pp. 723–737. Springer, Heidelberg (2009). doi:10.1007/978-3-642-02121-3_53
4. Cheng, J., Liu, C., Zhou, M.C., Zeng, Q., Ylä-Jääski, A.: Automatic composition of semantic web services based on fuzzy predicate petrinets. IEEE Trans. Autom. Sci. Eng. (2013, to be published)
5. The Linked Open Data Project (LOD) (2015). http://www.w3.org/wiki/SweoIG/TaskForces/CommunityProjects/LinkingOpenData
6. Cure, O., Naacke, H., Randriamalala, T., et al.: LiteMat: a scalable, cost-efficient inference encoding scheme for large RDF graphs. IEEE International Conference on Big Data, pp. 1823–1830. IEEE (2015)
7. Hermit [EB/OL]. http://hermit-reasoner.com/
8. Xiao-yong, D.U., Yan, W.A.N.G., Bin, L.U.: Research and development on semantic web data management. J. Softw. **20**(11), 2950–2964 (2009)
9. Bechhofer, S., Harmelen, F.V., Hendler, J., et al.: OWL web ontology language reference. In: W3C Recommendation (2004)
10. Hayes, P., (Ed.) RDF Semantics, W3C Recommendation (2004)
11. Zhou, J., Ma, L., Liu, Q., Zhang, L., Yu, Y., Pan, Y.: Minerva: a scalable OWL ontology storage and inference system. In: Mizoguchi, R., Shi, Z., Giunchiglia, F. (eds.) ASWC 2006. LNCS, vol. 4185, pp. 429–443. Springer, Heidelberg (2006). doi:10.1007/11836025_42
12. Kaoudi, Z., Miliaraki, I., Koubarakis, M.: RDFS reasoning and query answering on top of DHTs. In: Sheth, A., Staab, S., Dean, M., Paolucci, M., Maynard, D., Finin, T., Thirunarayan, K. (eds.) ISWC 2008. LNCS, vol. 5318, pp. 499–516. Springer, Heidelberg (2008). doi:10.1007/978-3-540-88564-1_32
13. Muhleisen, H., Dentler, K.: Large-scale storage and reasoning for semantic data using swarms. IEEE Comput. Intell. Mag. **7**(2), 32–44 (2012)
14. Soma, R., Prasanna, V.: Parallel inferencing for OWL knowledge bases. In: Proceedings of the 37th International Conference on Parallel Processing, pp. 75–82 (2008)
15. Weaver, J., Hendler, J.A.: Parallel materialization of the finite RDFS closure for hundreds of millions of triples. In: Bernstein, A., Karger, D.R., Heath, T., Feigenbaum, L., Maynard, D., Motta, E., Thirunarayan, K. (eds.) ISWC 2009. LNCS, vol. 5823, pp. 682–697. Springer, Heidelberg (2009). doi:10.1007/978-3-642-04930-9_43
16. Mika, P., Tummarello, G.: Web semantics in the clouds. IEEE Intell. Syst. **23**(5), 82–87 (2008)

17. Gu, R., Wang, S., Wang, F., Yuan, C., Huang, Y.: Cichlid: efficeinet large scale RDF/OWL reasong with spark. In: 2015 IEEE 29th International Parallel and Distributed Processing Symposium, pp. 700–709 (2015)
18. Urbani, J., Kotoulas, S., Maassen, J., et al.: WebPIE: a web-scale parallel inference engine using mapreduce. J. Web Semant. **17**(2), 59–75 (2012)
19. Guo, Y., Pan, Z., Heflin, J.: LUBM: a benchmark for OWL knowledge base systems. Web Semant. Sci. Serv. Agents World Wide Web **3**(2–3), 158–182 (2005)

Cloud Resource Allocation from the User Perspective: A Bare-Bones Reinforcement Learning Approach

Alexandros Kontarinis[1]([✉]), Verena Kantere[2], and Nectarios Koziris[1]

[1] School of Electrical and Computer Engineering,
National Technical University of Athens, Athens, Greece
`alexandroskontarinis@gmail.com, nkoziris@cslab.ntua.gr`
[2] Centre Universitaire d' Informatique, University of Geneva, Geneva, Switzerland
`verena.kantere@unige.ch`

Abstract. Cloud computing enables effortless access to a seemingly infinite shared pool of resources, on a pay-per-use basis. As a result, a new challenge has emerged: designing control mechanisms to precisely meet the actual workload requirements of cloud applications in an online manner. To this end, a variety of complex resource management issues have to be addressed, because workloads in the cloud are of a dynamic and heterogeneous nature, and traditional algorithms do not cope well within this context. In this work, we adopt the point of view of the user of a cloud infrastructure and focus on the task of controlling leased resources. We formulate this task as a Reinforcement Learning problem and we simulate the decision-making process of a controller implementing the Q-learning algorithm. We conduct an experimental study, the outcomes of which offer valuable insight into the advantages and shortcomings of using Reinforcement Learning to implement such adaptive cloud resource controllers.

Keywords: Cloud computing · Resource allocation · Reinforcement learning · Q-learning

1 Introduction

A basic feature of the cloud computing paradigm is "pay-on-demand" service provisioning. Cloud clients pay a fee corresponding to the number of transactions realized or to the duration of resource utilization Cloud providers try to offer competitive pricing, but at the same time maximize their return on investment by not keeping their resources idle. At the same time, the granularity of rental time in IaaS clouds is expected to become finer and finer, and in some cases be at the scale of seconds [1]. These trends result in great flexibility for both cloud providers and clients, including end users [14]. Unfortunately, human intuition and manual implementation are neither rapid nor precise enough to take advantage of the elasticity of cloud infrastructure resources for control purposes.

W. Cellary et al. (Eds.): WISE 2016, Part I, LNCS 10041, pp. 457–469, 2016.
DOI: 10.1007/978-3-319-48740-3_34

Instead, software is needed to automatically manage cloud resources, a very broad task studied through different perspectives and methods [4, 16].

This work focuses on how a user of an IaaS provider chooses the number of leased homogeneous Virtual Machines (VMs) in real-time. The goal is to serve the incoming workload but at the same time not over-consume resources, and thus avoid unnecessary expenses. Such scaling decisions are difficult to make when the applications are running on third-party infrastructure. Especially in the case of interactive services, cloud workloads are not uniform in time and unpredictable demand spikes are frequent. Therefore, well-defined scaling strategies are needed to dictate intelligent resource adjustments. To this end, researchers have studied numerous methods: threshold-based rules [20], wavelet-based resource demand prediction and proactive VM cloning [21], finite state machines and cloud application profiling [28], queuing theory for multi-tier applications [12], neural networks and linear regression coupled with sliding window techniques [14], linear programming for cloud-based shared Hadoop clusters [18], stochastic gradient-descent learning [25], closed queuing networks and hill-climbing [2], etc.

We propose to obtain such scaling strategies using Reinforcement Learning (RL), a scientific area considered as one of the major Machine Learning approaches, along with Supervised Learning and Unsupervised Learning. RL sets itself apart from both, because learning comes from acting in the environment and receiving reward signals (Fig. 1), without ever being explicitly told the optimal actions. RL is particularly attractive for the complex cloud environment, because a RL agent is not required to explicitly learn a model of the world (e.g. an application performance model). Moreover, RL does not require teachers, experts, or any domain-specific knowledge, while it can harness such assets.

We first describe Markov Decision Processes (MDPs), the mathematical foundation over which RL algorithms run, and define the RL problem setting (Sect. 2). We formulate the aforementioned resource allocation problem as a RL problem, inspired by an existing management framework [26] (Sect. 3). We run simulations implementing the most iconic RL algorithm, namely Q-Learning, and interpret the results (Sect. 4). Our experiments reveal how RL can be used to improve resource allocation in the cloud and shed light into the good practices of designing such solutions. Finally, we overview how RL has been used in similar research work (Sect. 5) and present our conclusions (Sect. 6).

2   Mathematical Preliminaries

The environment over which RL is implemented is typically a MDP [5]. Optimization problems formulated with the help of a MDP are called Markov Decision Problems and can be solved using Dynamic Programming (DP) algorithms [13]. RL algorithms are based on DP techniques.

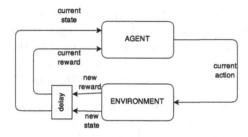

Fig. 1. The process of learning through interaction: the agent takes an action, the environment responds, and the agent receives a (possibly delayed) feedback.

2.1 Markov Decision Problem

A MDP is a mathematical framework for modeling sequential decision-making, widely used in Control Theory, Operations Research, Robotics, and other disciplines [6]. It is a stochastic control process that makes transitions between members of a countable set of states. What differentiates a MDP from a Markov chain is the addition of actions and rewards. In some or all of the system states, the decision-maker chooses one among a set of available actions. The probability distribution of the system's next state only depends on the current state and the action selected, not on past states or actions. This lack of memory is known as the Markov Property [19]. Each state transition produces an immediate reward for the system to be controlled, either positive, negative (a penalty), or zero.

Definition 1. *A Markov Decision Process (MDP) is a tuple (S, A, T, R) of:*

1. the set of states (or state-space): S
2. the set of actions (or action-space): $A \equiv \bigcup_{s \in S} A(s)$
3. the transition probabilities (or transition function):
   $T(s, a, s') = \Pr(S_{t+1} = s' \mid S_t = s, A_t = a)$
4. the transition rewards (or reward function): $R(s, a, s')$,

$s, s' \in S$, $a \in A(s)$, S_t *is a random variable denoting the state at discrete time step t, A_t is a random variable denoting the action taken at discrete time step t.*

For the problem in hand, a straightforward MDP modeling approach would be to map the allocated resource configurations to states and the resource adjustments to actions. For example, if RAM size was the only resource type, then a "4GB RAM" state could lead to a "6GB RAM" state through an "add 2GB" action. The choice of a reward function $R(s, a, s') = \psi(metrics)$ could be left to the user, so that each may differently take into account system metrics (throughput, response time, monetary cost, etc.) monitored in real-time.

A deterministic stationary MDP policy is a function that prescribes exactly which (scaling) action to choose in every decision-making state. Formally:

Definition 2. *A deterministic stationary MDP policy is a mapping from the set of states to the set of actions: $\pi : S \rightarrow A$, $\pi(s) \in A(s), \forall s \in S$.*

To deal with the multiple (often conflicting) criteria of policy optimality, the notion of a utility function[1] is used. This incorporates the decision maker's preferences and goals and is usually a mathematical function of the immediate rewards earned over a pre-determined time-horizon. The most often used type of utility function is the discounted utility [7], commonly expressed as:

Definition 3. *The expected sum of discounted rewards:*

$$U_\pi(s) = \lim_{l \to \infty} E[\sum_{k=1}^{l} \gamma^{k-1} r(x_k, \pi(x_k), x_{k+1}) \,|_{x_1=s}],$$

where π is the policy being followed over an infinite time-horizon, s is the starting state, x_k is the state occupied before the kth transition, E is the expectation operator, $0 < \gamma < 1$ is the discount factor.

Finally, the Markov Decision Problem is the problem of, given a MDP, finding the policy which optimizes a predetermined utility function:

Definition 4. *The Markov Decision Problem for discounted rewards is the problem of finding a policy π^* such that $U_{\pi^*}(s) \geq U_\pi(s), \forall s \in S, \forall \pi$.*

2.2 Dynamic Programming

To find the optimal policy (Definition 4) a value function is used. The Q-value function in particular, indicates how good each action is considered to be in each state. For example, if in the "4GB RAM" state the "add 2GB" action has a Q-value higher than the "add 1GB" action, then whenever there are 4GB of RAM allocated, adding 2GB is considered to be better than adding only 1GB. Although Q-values are defined with respect to the utility function, the two are not the same.

Definition 5. *The Q-value function $Q^*(s, a)$ is the optimal value of the utility function written as a function of the Q-state (the state-action pair), in other words, it is the expected utility for starting in a state $s \in S$, taking an action $a \in A(s)$ available in that state, and acting optimally ever after.*

Given Definitions 3 and 5, a recursive formula for the Q-value function follows:

$$Q^*(s, a) = \sum_{s' \in S} T(s, a, s')[R(s, a, s') + \gamma \max_{a' \in A(s')} Q^*(s', a')] \qquad (1)$$

The discount factor γ denotes the factor by which a future reward must be multiplied, in order to obtain its present significance, and is simply one of the many alternative ways to model temporal discounting [9].

DP [10] is an algorithmic process of solving decision-making problems in the form of mathematical optimization problems. A classical DP algorithm for the

---

[1] Also referred to as an objective function or return function.

Markov Decision Problem is Value Iteration (VI) [6]. Its Q-value counterpart (Q-VI) successively implements (1) as an update rule:

$$Q_{k+1}(s,a) = \sum_{s' \in S} T(s,a,s')[R(s,a,s') + \gamma \max_{a' \in A(s')} Q_k(s',a')], \forall s \in S, \forall a \in A(s)$$

(2)

If the update is run infinitely long, then the estimate will converge to the true Q-value function: $\lim_{k \to \infty} Q_k(s,a) = Q^*(s,a)$ [22].

The solution to the Markov Decision Problem can then simply be extracted:

$$\pi^*(s) = \arg \max_{a \in A(s)} Q^*(s,a), \forall s \in S \qquad (3)$$

2.3   Reinforcement Learning

Inspired from Experimental Psychology and the Optimality Argument[2], Watkins introduced RL in [27] as a model of learning from rewards, in the form of incrementally optimizing control of a MDP. Thus, the RL problem setting is mathematically the same as the Markov Decision Problem setting [24]. However, the decision-maker (or learning agent) is located in the real world and is unaware of the transition probabilities, $T(s,a,s')$, and of the transition rewards, $R(s,a,s')$. The agent can only observe the immediate reward of each action, r, and the environmental state, s, that it is currently found in. This is nonetheless enough to eventually learn the optimal policy. Q-Learning [27] (Algorithm 1) is essentially a sample-based version of Q-VI which implements the following update rule:

$$Q_{t+1}(s,a) \leftarrow (1-\alpha)Q_t(s,a) + (\alpha)sample_t \Leftrightarrow \qquad (4)$$

$$Q_{t+1}(s,a) = (1-\alpha)^t Q_1(s,a) + \sum_{i=1}^{t} [\alpha(1-\alpha)^{t-i} sample_i], \qquad (5)$$

$$sample_t = R(s,a,s') + \gamma \max_{a' \in A(s')} Q_t(s',a') \qquad (6)$$

If the update is run infinitely long, then: $\lim_{k \to \infty} Q_k(s,a) = Q^*(s,a)$. In the cloud context however, there is little practical value in asymptotic convergence. Obtaining good sub-optimal policies fast-enough is more important. Thanks to the learning rate α, the old Q-value estimates are updated based on the newest sample. More specifically, (5) is an exponentially weighted average, where newer samples are more important than older samples, in a one-to-one comparison.

3   Cloud Resource Allocation Modeling

We model a common version of the cloud resource allocation problem, in which a cloud client (not necessarily the same as the end-user, e.g. a NoSQL cluster

---

[2] Optimality arguments explain empirical regularities through objective maximization.

Algorithm 1. Q-Learning

Require: S, A, TH (time-horizon, can be ∞)
1: Initialize $Q(s, a)$, $\forall s \in S$, $\forall a \in A(s)$.
2: Choose s_{init} (perhaps arbitrarily).
3: $s = s_{init}$
4: $time\_step = 1$
5: **while** $time\_step < TH$ **do**
6:    Choose an action $a \in A(s)$ based on the policy derived from the Q-values.
7:    Implement a and observe s' and $R(s, a, s')$.
8:    $Q(s, a) \leftarrow (1 - \alpha)Q(s, a) + \alpha[R(s, a, s') + \gamma \max\limits_{a' \in A(s')} Q(s', a')]$

9:    $s = s'$
10:    $time\_step = time\_step + 1$
11: **end while**

---

administrator [17]) controls the number of VMs allocated from a cloud provider. The cloud-based application is using between 1 and N VMs. An incoming workload enters the system with a mean rate of λ measured in (tens of thousands of) requests per second. Each VM can serve up to a fixed number μ of requests per second, and has an abstract running cost vm_{cost} (e.g. a monetary cost). Adding more VMs increases both the throughput and the running cost of the application. The goal is to deploy the minimal number of VMs that is needed to serve all of the requests, so that the total running cost of these VMs is also minimal.

Each state s_n represents the respective number n of running VMs, ranging from 1 to N. An increase or decrease by 1 VM are the only resizing actions allowed. Immediate rewards are non-negative, produced by a (fixed) function of throughput and cost, where increases in the former and decreases in the latter give higher rewards: $\max(\min(\mu * N, load) - vm\_cost * n, 0)$. Each VM can deal with μ K req/sec. Throughput is thus equal to, either the number of VMs times μ, or to the current incoming workload, whichever is smaller. Ideally however, the reward function should be user-specific and even non-linear and complex.

1. The set of states (or state-space): $S = \{s_1, s_2, ..., s_n, ..., s_N\}$.
2. The set of actions (or action-space): $A = \{no\_op, add\_1, remove\_1\}$.
3. The transition probabilities (or transition function):
   $T(s_n, remove\_1, s_{n-1}) = T(s_n, no\_op, s_n) = T(s_n, add\_1, s_{n+1}) = 1$, $\forall s_n \in S \setminus \{s_1, s_N\}$,
   $T(s_1, remove\_1, s_1) = T(s_1, no\_op, s_1) = T(s_1, add\_1, s_2) = 1$,
   $T(s_N, remove\_1, s_{N-1}) = T(s_N, no\_op, s_N) = T(s_N, add\_1, s_N) = 1$,
   $T(s, a, s') = 0$, otherwise.
4. The transition rewards (or reward function):
   $R(s_n, a, s') = \max(\min(\mu * n, load) - vm\_cost * n, 0)$, $\forall n \in [1, N]$,
   $R(s_1, remove\_1, s_1) = R(s_N, add\_1, s_N) = 0$.

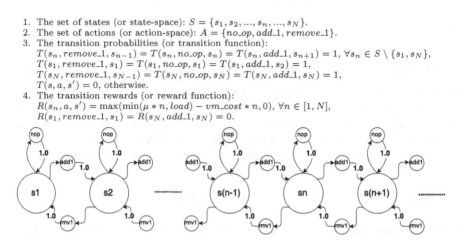

Fig. 2. The formal definition of the (1-dimensional state) simulation MDP model

States should be rich enough to represent environmental characteristics that considerably affect performance. This is why in latter experiments (Fig. 5), the state-space of the MDP is extended, with each state $s_{n,m}$ representing both the number n of running VMs and the level m of workload currently observed: $S = \{s_{1,1}, s_{1,2}, ..., s_{1,m}, ..., s_{1,M}, ..., s_{N,1}, s_{N,2}, ...s_{N,m}, ..., s_{N,M}\}$

4   Experimental Study

For each experiment, mean values over 20 Matlab simulation runs are given.

4.1   Setting

Tabular Q-Learning (Algorithm 1) is implemented over the MDP of Fig. 2. Table 1 contains the values of all the features of the model and of the algorithm.

4.2   Results

Figure 3 confirms that the longer Q-Learning runs the better its proposed policy gets. Actually, thousands of decisions need to be made before the agent starts to follow a decent policy. For example, in Fig. 3a, after 1000 time-steps the Q-values in states s_2, s_3, s_4 all wrongly suggest a decrease in the number of VMs to be the best action. In Fig. 3c, after 10000 time-steps the agent has managed to identify the optimal state, as indicated by the great number of times action *no_op* in state s_5 was taken. This slow convergence to the optimal policy is due to the agent not possessing any prior knowledge. However, a real cloud system would not afford to make that many mistakes, which is why initial Q-values should be set according to knowledge obtained offline. In Fig. 3d, it is apparent that the agent is reluctant at exploring the state-space. Due to the Q-values being initialized to 0, any action tried out will have a higher Q-value than any action not yet taken. As a result, a new action will only be taken as part of an exploration step, the chances of it being $\epsilon * \frac{1}{3} = 0.05$.

In Fig. 3e, Q-values are initially set equal to 5, a value unreachable within the selected time-horizon (but still not arbitrarily large). This way, whenever an action gets chosen, its estimated Q-value will decrease, getting closer to its true Q-value. This technique implicitly favors exploration of new actions which have overly optimistic Q-values. "Optimism in the face of uncertainty" substantially outperforms the pessimistic approach of Fig. 3d. Initial Q-values should not be set exceptionally high either, because then they would take too long to drop to their true levels. In Fig. 3f, initial Q-values are zero, but the learning rate α is now maximally set equal to 1. This allows for bold Q-value updates. Here the workload is static and disregarding past observations in favor of the newest one is harmless. In general, however, relying too much on a single experience is an error-prone way of learning, especially in the volatile cloud environment.

In Figs. 4 and 5, the workload is modeled after a sinusoidal signal, to test if the system can, scale out to more VMs during the highs of demand, and scale

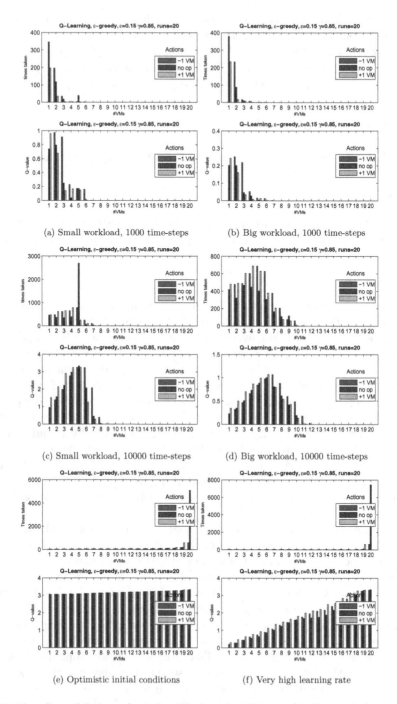

(a) Small workload, 1000 time-steps

(b) Big workload, 1000 time-steps

(c) Small workload, 10000 time-steps

(d) Big workload, 10000 time-steps

(e) Optimistic initial conditions

(f) Very high learning rate

Fig. 3. The effect of the time-horizon and of two learning acceleration techniques under a static workload.

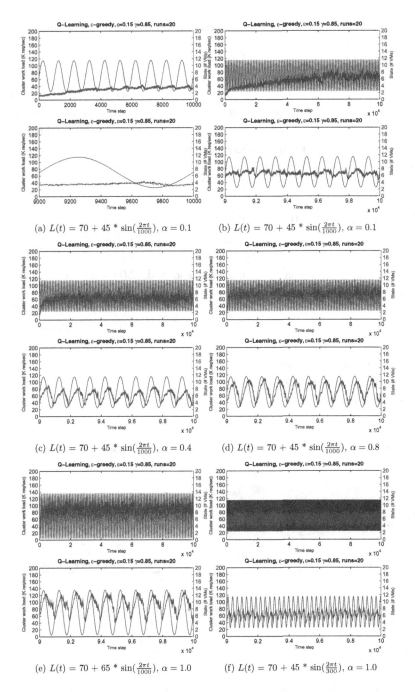

(a) $L(t) = 70 + 45 * \sin(\frac{2\pi t}{1000})$, $\alpha = 0.1$

(b) $L(t) = 70 + 45 * \sin(\frac{2\pi t}{1000})$, $\alpha = 0.1$

(c) $L(t) = 70 + 45 * \sin(\frac{2\pi t}{1000})$, $\alpha = 0.4$

(d) $L(t) = 70 + 45 * \sin(\frac{2\pi t}{1000})$, $\alpha = 0.8$

(e) $L(t) = 70 + 65 * \sin(\frac{2\pi t}{1000})$, $\alpha = 1.0$

(f) $L(t) = 70 + 45 * \sin(\frac{2\pi t}{300})$, $\alpha = 1.0$

Fig. 4. The workload (blue) and the number of running VMs (red) under a dynamic workload, for 100000 time-steps (except (a) spanning 10000 time-steps). The second graph in each case is a zoom in on the latter 10 % of time-steps.

Table 1. Various parameter values change throughout the experiments.

Fixed simulation features		Changing simulation features	
Algorithm	Q-Learning	state-space	1-D (#VMs)
			2-D (#VMs, workload level)
Initial state	$s_{init} = s_1$	initial Q-values	pessimistic (equal to 0)
	(1 VM)		optimistic (equal to 5)
Exploration	ϵ-greedy	learning rate	$\alpha = 0.1$
strategy	with $\epsilon = 0.15$		$\alpha = 0.4$
			$\alpha = 0.8$
			$\alpha = 1.0$
Discount factor	$\gamma = 0.85$	time-horizon	$TH = 1000$
		(time steps)	$TH = 10000$
			$TH = 100000$
Maximum	$N = 20$	incoming workload	$L(t) = \lambda = 50$
number of VMs		(K requests/second)	$L(t) = \lambda = 200$
			$L(t) = 70 + 45 * \sin(\frac{2\pi t}{1000})$
			$L(t) = 70 + 65 * \sin(\frac{2\pi t}{1000})$
			$L(t) = 70 + 45 * \sin(\frac{2\pi t}{300})$
VM running cost	$vm\_cost = 5$		
VM capacity (K requests/second)	$\mu = 10$		

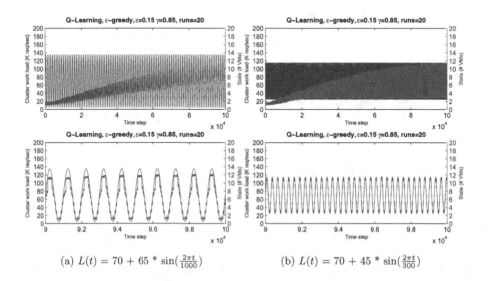

(a) $L(t) = 70 + 65 * \sin(\frac{2\pi t}{1000})$ (b) $L(t) = 70 + 45 * \sin(\frac{2\pi t}{300})$

Fig. 5. Adding a state dimension helps the agent adapt to workload changes.

in to fewer VMs during the lows of demand. In Figs. 4a and b, the agent does not react fast enough, irrespective of the time-horizon. The 1-dimensional state-space only has $20 * 3$ states, and therefore the agent can only keep 60 Q-values[3]. The dynamic workload makes rewarding actions quickly turn into bad choices, and vice versa, but the corresponding Q-values do not change fast enough to reflect this. In Figs. 4c and d, by raising the learning rate, the agent better adapts to the workload's periodic changes. There is still a lot of wasteful VM provisioning, as well as inability to serve a portion of the workload. More importantly, a very high learning rate is not a trustworthy solution, because it places too much emphasis on a few experiences. In addition, increasing the workload amplitude (Fig. 4e) or frequency (Fig. 4f) will still completely break down learning.

A better approach is to generalize less by treating different workloads separately. This is achieved by adding a second dimension to the MDP state description, namely the (quantized) measured level of workload. The agent now keeps more Q-values and no longer has to change the same ones extremely fast. It evaluates the three possible actions differently depending on the workload level last observed. Figure 5 indicates that this helps the agent take the right actions.

5 Related Work

In [15] the authors propose a hybrid method combining Fuzzy Logic and RL. The controller incorporates domain expert knowledge in the form of continuously tuned (through Q-Learning) fuzzy rules that dictate auto-scaling the number of VMs. Self-tuning fuzzy control is also used in [23]. In [3], the authors use Q-Learning and its eligibility traces-based on-policy counterpart algorithm, SARSA(λ), to maximize the net profit of a Cloud Management Broker[4]. In [8], the authors use a hybrid approach combining the simplex method with RL for the coordinated automatic tuning of VM allocation and resident application parameter setting. The simplex algorithm reduces the state-space of the learning process, while CPU and memory utilization monitoring guides the exploration.

Our work is closer to [11, 26]. In [11], the authors use RL to obtain IaaS and PaaS management layer VM allocation policies, which minimize the costs for the service providers while respecting SLAs. The monitored average request response time is part of the MDP state, whereas in our proposed model it would be part of the user-specified reward function. SLAs are taken into consideration through the reward function. The authors stress the importance of initial Q-values and obtain learning speedups using a model-based version of Q-Learning. In [26], the authors use a MDP similar to Fig. 2, but each monitoring experience is stored instead of used to update the Q-value function and then discarded. In [17], they also consider different workload types through the reward function.

---

[3] -2 for the two unavailable actions.

[4] A Broker allocates resources from multiple cloud providers.

6   Conclusions

We introduce RL modeling with respect to cloud resource management, we develop a discrete-time MDP for controlling the number of VMs according to the incoming workload, and we implement the Q-Learning algorithm over that model. Our simulation experiments indicate that RL is promising for improving real-time scaling decision-making. More importantly, they reveal how different methods, techniques, and parameter values can have a huge effect on the accuracy of the decisions. In particular, defining proper MDP states was found to be of paramount importance. Apart from the unacceptably slow convergence to good policies without the use of an initial knowledge base, our work also verifies some of Q-Learning's theoretical advantages. Q-Learning is easy to implement, especially in its tabular form. In addition, each decision only takes constant time $\mathcal{O}(1)$, thanks to the incremental way of learning from each experience. This is imperative for cloud systems that need to make informed decisions in a fine-grained temporal resolution. It is also a very generic algorithm. This is attractive in the cloud domain where there exist multiple stakeholders and objectives. Finally, it provides a wealth of techniques (e.g. optimistic/pessimistic initial ignorance, exploration-exploitation balancing strategy) and parameters (e.g. learning rate, discount factor, epsilon) for fine-tuning the learning process.

References

1. Agmon Ben-Yehuda, O., Ben-Yehuda, M., Schuster, A., Tsafrir, D.: The rise of raas: the resource-as-a-service cloud. CACM **57**(7), 76–84 (2014)
2. Aldhalaan, A., Menasc, D.A.: Near-optimal allocation of vms from iaas providers by saas providers. In: ICCAC, pp. 228–231 (2015)
3. Antonio, P., Stefano, B., Francisco, F., Alessandro, G., Guido, O., Martina, P., Vincenzo, S.: Resource management in multi-cloud scenarios via reinforcement learning. In: Control Conference (CCC), pp. 9084–9089 (2015)
4. Ardagna, D., Casale, G., Ciavotta, M., Pérez, J.F., Wang, W.: Quality-of-service in cloud computing: modeling techniques and their applications. J. Internet Serv. Appl. **5**(1), 1–17 (2014)
5. Bellman, R.: A markovian decision process. Ind. Univ. Math. J. **6**, 679–684 (1957)
6. Bertsekas, D.P.: Dynamic Programming and Optimal Control, vol. II, 4th edn. Athena Scientific, Belmont (2012)
7. Bertsekas, D.P.: Abstract Dynamic Programming, 1st edn. Athena Scientific, Belmont (2013)
8. Bu, X., Rao, J., Xu, C.Z.: Coordinated self-configuration of virtual machines and appliances using a model-free learning approach. IEEE Trans. Parallel Distrib. Syst. **24**(4), 681–690 (2013)
9. Doyle, J.R.: Survey of time preference, delay discounting models. Judgm. Decis. Making **8**(2), 116–135 (2013)
10. Dreyfus, S.: Richard bellman on the birth of dynamic programming. Oper. Res. **50**(1), 48–51 (2002)
11. Dutreilh, X., Kirgizov, S., Melekhova, O., Malenfant, J., Rivierre, N., Truck, I.: Using reinforcement learning for autonomic resource allocation in clouds: towards a fully automated workflow. In: ICAS, pp. 67–74, May 2011

12. Han, R., Ghanem, M.M., Guo, L., Guo, Y., Osmond, M.: Enabling cost-aware and adaptive elasticity of multi-tier cloud applications. Fut. Gen. Comp. Syst. **32**, 82–98 (2014)
13. Howard, R.A.: Dynamic programming and markov processes (1960)
14. Islam, S., Keung, J., Lee, K., Liu, A.: Empirical prediction models for adaptive resource provisioning in the cloud. Fut. Gen. Comp. Syst. **28**(1), 155–162 (2012)
15. Jamshidi, P., Sharifloo, A.M., Pahl, C., Metzger, A., Estrada, G.: Self-learning cloud controllers: fuzzy q-learning for knowledge evolution. In: ICCAC, pp. 208–211 (2015)
16. Jennings, B., Stadler, R.: Resource management in clouds: survey and research challenges. J. Netw. Syst. Manag. **23**(3), 567–619 (2014)
17. Kassela, E., Boumpouka, C., Konstantinou, I., Koziris, N.: Automated workload-aware elasticity of nosql clusters in the cloud, pp. 195–200 (2014)
18. Malekimajd, M., Ardagna, D., Ciavotta, M., Rizzi, A.M., Passacantando, M.: Optimal map reduce job capacity allocation in cloud systems. SIGMETRICS Perform. Eval. Rev. **42**(4), 51–61 (2015)
19. Markov, A.A., Schorr-Kon, J.J.: The theory of algorithms (1954)
20. Marshall, P., Keahey, K., Freeman, T.: Elastic site: using clouds to elastically extend site resources. In: CCGRID, pp. 43–52. IEEE Computer Society (2010)
21. Nguyen, H., Shen, Z., Gu, X., Subbiah, S., Wilkes, J.: AGILE: elastic distributed resource scaling for infrastructure-as-a-service. In: ICAC, pp. 69–82 (2013)
22. Puterman, M.L.: Markov Decision Processes: Discrete Stochastic Dynamic Programming, 1st edn. Wiley, New York (1994)
23. Rao, J., Wei, Y., Gong, J., Xu, C.Z.: Dynaqos: model-free self-tuning fuzzy control of virtualized resources for qos provisioning. In: IEEE International Workshop on Quality of Service (IWQoS), pp. 1–9 (2011)
24. Sutton, R.S., Barto, A.G.: Introduction to Reinforcement Learning, 1st edn. MIT Press, Cambridge (1998)
25. Tan, Y., Xia, C.H.: An adaptive learning approach for efficient resource provisioning in cloud services. SIGMETRICS Perform. Eval. Rev. **42**(4), 3–11 (2015)
26. Tsoumakos, D., Konstantinou, I., Boumpouka, C., Sioutas, S., Koziris, N.: Automated, elastic resource provisioning for nosql clusters using TIRAMOLA. In: CCGrid, pp. 34–41 (2013)
27. Watkins, C.: Learning from delayed rewards. Ph.D. thesis, King's College (1989)
28. Yang, J., Qiu, J., Li, Y.: A profile-based approach to just-in-time scalability for cloud applications. In: IEEE CLOUD, pp. 9–16 (2009)

Event Detection

Event Phase Extraction and Summarization

Chengyu Wang[1], Rong Zhang[1], Xiaofeng He[1(✉)], Guomin Zhou[2],
and Aoying Zhou[1]

[1] Institute for Data Science and Engineering, East China Normal University,
Shanghai, China
chywang2013@gmail.com, {rzhang,xfhe,ayzhou}@sei.ecnu.edu.cn
[2] Zhejiang Police College, Hangzhou, Zhejiang, China
zhouguomin@zjjcxy.cn

Abstract. Text summarization aims to generate a single, concise representation for documents. For Web applications, documents related to an event retrieved by search engines usually describe several event phases implicitly, making it difficult for existing approaches to identify, extract and summarize these phases. In this paper, we aim to mine and summarize event phases automatically from a stream of news data on the Web. We model the semantic relations of news via a graph model called *Temporal Content Coherence Graph*. A structural clustering algorithm *EPCluster* is designed to separate news articles corresponding to event phases. After that, we calculate the relevance of news articles based on a vertex-reinforced random walk algorithm and generate event phase summaries in a relevance maximum optimization framework. Experiments on news datasets illustrate the effectiveness of our approach.

Keywords: Event phase summarization · Structural clustering · Vertex-reinforced random walk

1 Introduction

The information overload on the Web motivates the automatic generation of event summaries from documents [1–4] which aims to generate a single, concrete representation of the event. The accuracy and conciseness of summaries are essential for Web applications, such as Web search, news recommendations, etc.

It can be noticed that, existing approaches model an event as one unit and generate a single summary, paying little attention to the fact that there exist several *phases* in long-span, complicated events. Take the case *Egypt Revolution* as an example. Major phases include *Protests against Hosni Mubarak*, *Egypt under the Supreme Council of the Armed Forces*, *Egypt under President Mohamed Morsi*, *June 2013 Protests against President Morsi*, etc[1]. More recently, the task of timeline generation produces a series of correlated component summaries,

---

[1] See background info at: https://en.wikipedia.org/wiki/Egyptian_Revolution_of_2011.

© Springer International Publishing AG 2016
W. Cellary et al. (Eds.): WISE 2016, Part I, LNCS 10041, pp. 473–488, 2016.
DOI: 10.1007/978-3-319-48740-3_35

ordered by time [5,6]. However, the entries in a timeline are simply arranged in a sequence, lacking a more structured representation of event phases.

To facilitate deeper analysis on these events, the task we aim to solve in this paper is: *how to automatically extract event phases from a news collection and generate event phase summaries*. It is interesting for several reasons: (i) it groups news articles describing each phase together, instead of considering content similarity only; (ii) it helps readers achieve a better understanding of complicated events by event phase summaries; and (iii) it potentially improves the performance of other tasks such as timeline generation.

To solve the problem, we employ a "divide-and-conquer" method to generate summaries individually after identifying event phases. Because these phases are implicitly expressed in the form of natural language text, we first define two semantic relations (i.e., content coherence and temporal influence) in a news collection via a graph model called *Temporal Content Coherence Graph* (TCCG). A structural clustering algorithm *EPCluster* is designed to extract event phases based on TCCG, in which each phase is represented by a subset of news articles.

For new articles related to a single phase, we design a ranking algorithm based on vertex-reinforced random walk process to calculate the relevance scores of news articles. Based on previous research [6], we model an event phase summary as top-k news headlines and their publication time, and employ a greedy, approximate optimization algorithm to select the corresponding news articles.

In summary, this paper makes the following major contributions:

- We propose and formalize the event phase extraction and summarization problem. A graphical structure TCCG is proposed to model the content coherence and temporal influence relations among news articles.
- A structural clustering algorithm *EPCluster* is designed to group news articles related to the same event phase. We introduce a relevance optimization framework to select top-k news articles to generate event phase summaries.
- We conduct extensive experiments and a case study on news datasets to illustrate the effectiveness of our approach.

The rest of this paper is organized as follows. Section 2 summarizes the related work. We define the event phase extraction and summarization problem formally in Sect. 3. Details of the proposed algorithms are described in Sects. 4 and 5. Experiments are presented in Sect. 6. We conclude our paper and discuss the future work in Sect. 7.

2   Related Work

Given a collection of news articles regarding the same event, various approaches have been proposed to provide users a more concrete representation of the event. Most of the approaches can be classified into two categories: *Multi-Document Summarization* (MDS) and *Timeline Generation* (TG). In this section, we provide an overview of research on these fields.

MDS is a technique of extracting the most salient information from a document collection and transferring it into a brief and informative sentence collection. This problem has been addressed using various paradigms, categorized into two types: extraction-based and abstraction-based. Extraction-based methods assign importance scores to sentences or paragraphs and extract ones with highest scores. Score assignment can be determined by using heuristic and NLP rules, and considering semantic relationships between textual units [1]. There are also some machine learning models for this task. Conroy and O'Leary [2] employ an HMM model to tag important textual units as summaries. More recently, He et al. [3] introduce a sparse coding approach to model each sentence in documents as a linear combination of summary sentence. Additionally, graph-based methods are efficient to rank sentences in documents, such as LexRank [7], cluster-based link analysis [8], etc. Abstraction-based methods utilize the natural language generation technique to create a summary that is closest to the corresponding human-generated summary. In Qian and Liu's work [4], smaller units such as words and phrases are used in the generation process, resulting in more informative summaries.

TG is another research effort to summarize evolutionary news articles by generating component summaries along the timeline. Timelines can also be generated by applying MDS on news articles on each individual date. However, the constraints among temporal components are not modeled in the above approaches [5]. For example, Yan et al. [5] model the trans-temporal characteristics among these component summaries by temporal projection. The headlines of news articles are more informative than the contents, which are exploited by Tran et al. [6] to generate timelines directly via influence-based random walk. Ng et al. [9] construct timelines and incorporate them into an MDS system. It shows that the usage of timelines can improve the performance of MDS.

In summary, both MDS and TG provide concrete information for readers. However, for long-span events, it is necessary to decompose the events into more fine-grained event phases. The identification and summarization of event phases can provide a research foundation for deeper analysis and better understanding of complicated events in the future.

3   Problem Formulation

In this section, we introduce some key concepts used in this paper, formally defining the problem of event phase extraction and summarization.

A news article d_i is a triple, represented as $d_i = (h_i, t_i, s_i)$ where h_i, t_i and s_i denote the headline, the publication time and the sentence collection of text contents. A news collection is a set of news articles $D = \{d_i\}$ where d_i is a news article. In classical aging theory, the life cycle of an event is modeled as four stages: birth, growth, decay and death [10]. However, in a real-life, complicated event, it is difficult to capture the characteristics of the event using only four stages. To overcome this problem, we regard an event as a collection of several event phases. We first introduce the concept of event phase summary as follows:

Definition 1. Event Phase Summary. *An event phase summary P is a collection of k news headline and publication time pairs, denoted as $P = \{(h_i, t_i)\}_{i=1}^k$.*

Event phases, however, are unknown before the summarization process, and thus need to be identified beforehand. The task of event phase extraction and summarization is defined as follows:

Definition 2. Event Phase Extraction and Summarization. *Given a news collection D and a positive integer k, the goal is to generate a collection of N event phases $\mathbf{P} = \{P_j\}_{j=1}^N$ where P_j is an event phase summary, i.e., $P_j = \{(h_i, t_i)\}_{i=1}^k$.*

Based on the definition, the number of phases N is not pre-defined for an event. Therefore, given a news collection regarding any event, we can produce multiple summaries as a more fine-grained event representation.[2]

4   Event Phase Extraction

In this section, we present our approach for event phase extraction in detail. The high-level framework is illustrated in Fig. 1.

The major challenge is to determine how to measure the degree that two news articles report the same event phase so that they can be grouped into the same cluster. Here, we consider two key factors in terms of content space and time by defining two semantic relations between news articles. Next, the collection of news articles is mapped into a graph representation TCCG which captures the *local* semantic relations among these articles. A structural clustering algorithm *EPCluster* separates news articles into candidate event phases by partitioning TCCG into several subgraphs after noise removal. To achieve higher accuracy, we add an additional postprocessing step to filter out clusters that are not related to event phases via a logistic regression classifier. In the following, we will present details of the proposed approach.

4.1   Semantic Relations Between News Articles

Relations have been extensively employed to model the semantic connections between entities. However, little has been done to define relations between news articles. In this paper, we study the characteristics of news articles, and introduce two relations, namely content coherence and temporal influence.

Content Coherence. If two news articles are related to the same event phase, they are not necessarily similar in content due to difference in reported aspects and writing styles. Different from traditional measures such as VSM with TF-IDF weights (which suffers from curse of dimensionality), we define the content

---

[2] One issue that needs to be discussed here is that because our dataset is relatively large and there are over k news articles in each cluster regarding an event phase, we set a uniform parameter k for all the event phases. We can also modify the definition such that k varies for different event phases without changing our algorithm.

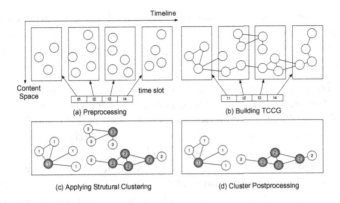

Fig. 1. General framework of event phase extraction.

coherence relation considering both *topic level* and *entity level* similarity. We calculate the strength of the relation by a content coherence score, denoted as $w_c(d_i, d_j) \in [0, 1]$.

Based on the previous research, it is found that in a stream of news articles, there is a change in distribution of topics over time called *topic drift* [11]. We regard it as a signal for identifying the change in event phases. To learn the topics, we employ Latent Dirichlet Allocation (LDA), a well-established topic model for documents [12]. For each news article $d_i \in D$, LDA associates it with a topic distribution vector $\boldsymbol{\theta}_i$. For two articles d_i and d_j, the difference between topic distributions are captured by Jansen-Shannon divergence, defined as:

$$D_{JS}(\boldsymbol{\theta}_i \| \boldsymbol{\theta}_j) = \frac{D_{KL}(\boldsymbol{\theta}_i \| \overline{\boldsymbol{\theta}}) + D_{KL}(\boldsymbol{\theta}_j \| \overline{\boldsymbol{\theta}})}{2} \tag{1}$$

where $\overline{\boldsymbol{\theta}} = \frac{\boldsymbol{\theta}_i + \boldsymbol{\theta}_j}{2}$ is the average topic distribution of d_i and d_j, and $D_{KL}(\boldsymbol{\theta}_i \| \boldsymbol{\theta}_j)$ is the KL divergence between $\boldsymbol{\theta}_i$ and $\boldsymbol{\theta}_j$. We set $n = 2$ in the base of logarithm for KL divergence to ensure $D_{JS}(\boldsymbol{\theta}_i \| \boldsymbol{\theta}_j) \in [0, 1]$.

Another observation is that, entities (e.g. people, locations and organizations) play a vital role in news reports. If an event goes through different phases, the statistics about these entities are likely to change. Due to the unstructured nature of texts, noisy, incorrect or unnormalized entities will be extracted if we directly apply NER techniques. Instead, we utilize our *NERank* algorithm [13] to extract key entities in the news collection D, denoted as E_D. Let $\boldsymbol{c}_i$ be an $|E_D|$-dimensional count vector of entity collection E_D in news article d_i. The entity level similarity between d_i and d_j is calculated by Tanimoto coefficient:

$$TC(\boldsymbol{c}_i, \boldsymbol{c}_j) = \frac{\boldsymbol{c}_i^T \cdot \boldsymbol{c}_j}{\|\boldsymbol{c}_i\|^2 + \|\boldsymbol{c}_j\|^2 - \boldsymbol{c}_i^T \cdot \boldsymbol{c}_j} \tag{2}$$

Therefore, the content coherence score between d_i and d_j is defined as follows:

$$w_c(d_i, d_j) = \alpha \cdot (1 - D_{JS}(\boldsymbol{\theta}_i \| \boldsymbol{\theta}_j)) + \beta \cdot TC(\boldsymbol{c}_i, \boldsymbol{c}_j) \tag{3}$$

where α and β are tuning parameters that control the strength of entity level and topic level similarity measures. We require $\alpha, \beta \in [0,1]$ and $\alpha + \beta = 1$. For simplicity, we set $\alpha = \beta = \frac{1}{2}$ in this paper and leave automatic learning for future research.

Temporal Influence. Content coherence alone is not sufficient because it does not capture the temporal dynamics of news. Consider the previous example of *Egypt Revolution*. There were news articles published in 2011 and 2012 regarding the street protests in Tahrir Square, Cairo. However, although similar in topics and entities, they were in fact related to different event phases, i.e., protests against Hosni Mubarak and the military government, respectively.

The temporal influence relation models to the phenomenon that if publication time of d_i and d_j are close, they are likely to report the same event phase and vice versa. Here, we define the temporal influence score $w_t(d_i, d_j)$ to reflect the strength of the relation by mapping the publication time gap between d_i and d_j into a different space using kernel density estimation. Given d_i and d_j, the publication time gap is calculated by $\Delta t_{i,j} = |t_i - t_j|$. We employ the Hamming (cosine) kernel $\Gamma(\cdot)$ [14] to map $\Delta t_{i,j}$ to a real number in $[0,1]$:

$$\Gamma(\Delta t_{i,j}) = \begin{cases} \frac{1}{2}(1 + \cos(\frac{\Delta t_{i,j} \cdot \pi}{\sigma})) & (\Delta t_{i,j} \leq \sigma) \\ 0 & (\Delta t_{i,j} > \sigma) \end{cases} \tag{4}$$

where σ is a parameter that controls the spread of kernel curves. If $\Delta t_{i,j} > \sigma$, it assumes that there is no direct temporal influence between d_i and d_j. Therefore, the temporal influence score is $w_t(d_i, d_j) = \Gamma(\Delta t_{i,j})$.[3]

4.2   *EPCluster*: A Structural News Clustering Algorithm

With the semantic relations between two news articles properly defined, we now present the *EPCluster* in detail, which is a structural algorithm based on TCCG.

TCCG. A first issue to be considered is that given two relation strength scores $w_c(d_i, d_j)$ and $w_t(d_i, d_j)$, how we can determine there is a strong semantic relation between d_i and d_j. In this paper, we introduce two parameters μ_1 and μ_2 where $\mu_1, \mu_2 \in (0,1)$. We say d_i and d_j are *directly semantic related* iff $w_c(d_i, d_j) > \mu_1$ and $w_t(d_i, d_j) > \mu_2$. In this way, news articles in D can be interconnected and form an undirected graph. See the example in Fig. 1(b). Here, we formally define the concept of TCCG as follows:

Definition 3. Temporal Content Coherence Graph. *A Temporal Content Coherence Graph w.r.t. parameters μ_1 and μ_2 and news collection D is an undirected graph $G_D = (V, E)$ such that:*

---

[3] In the implementation, we set one day as a time slot and compute $w_t(\cdot)$ based on publication date difference. See Fig. 1(a) and (b).

– V is the set of nodes where each node $v_i \in V$ represents a news article $d_i \in D$;
– E is the set of undirected edges where $(v_i, v_j) \in E$ iff $w_c(d_i, d_j) > \mu_1$ and $w_t(d_i, d_j) > \mu_2$.[4]

EPCluster Algorithm. Structural clustering has been extensively exploited to summarize and analyze various types of networks [15]. Based on the definition of TCCG, we can extend structural clustering techniques for news clustering. The high-level procedure of *EPCluster* is illustrated in Algorithm 1.

While traditional structural clustering algorithm *SCAN* [15] requires two parameters, *EPCluster* takes three parameters as input, namely μ_1, μ_2 and *MinPts*, where μ_1 and μ_2 are similarity thresholds, which are employed to construct the TCCG given the news article collection D. *MinPts* is the minimum number of objects within μ_1 and μ_2 similarity of an object. Here, we first define the concept of (μ_1, μ_2)-neighborhood:

Definition 4. (μ_1, μ_2)-Neighborhood. *The (μ_1, μ_2)-neighborhood w.r.t. d_i is a node collection $N(d_i) = \{d_j | (d_i, d_j) \in E\}$.*

We can see that $d_j \in N(d_i)$ is equivalent of $w_c(d_i, d_j) > \mu_1$ and $w_t(d_i, d_j) > \mu_2$. In *EPCluster*, the algorithm categorizes news articles into three types: core, border and noise objects based on (μ_1, μ_2)-neighborhood, defined as follows:

Definition 5. Core Object. *A core object is a news article $d_i \in D$ that satisfies $|N(d_i)| \geq MinPts$.*

Definition 6. Border Object. *A border object is a news article $d_i \in D$ that is not a core point and satisfies $d_i \in N(d_j)$ where $d_j \in D$ is a core object.*

Definition 7. Noise Object. *A noise object is a news article $d_i \in D$ that is neither a core object nor a border object.*

In the algorithm, with the TCCG constructed, it starts with an object $d_i \in D$ and retrieves all the neighbors in $N(d_i)$ (Line 4). If d_i is a core object, a cluster C (i.e., a news article subset) is created. After that, the cluster is expanded by adding the objects in d_i's neighborhood to the cluster C. For each $d_j \in N(d_i)$, if it is a core object, the cluster should be expanded by adding d_j's neighbors to the cluster (Line 6); otherwise, it is a border object. This process continues until a complete cluster C is formed. Thus the algorithm repeats to search for new clusters until all of the objects have been processed. Objects that are not in any cluster are treated as noise objects and discarded.

Complexity Analysis. In *EPCluster*, there is a neighborhood query for each $v_i \in V$, of which the complexity is linearly proportional to $deg(v_i)$ (the degree of v_i) with an adjacent list implementation. The entire runtime complexity is $O(\sum_{v_i \in V} deg(v_i))$, which is equivalent of $O(|E|)$. Therefore, *EPCluster* is an algorithm of which the complexity is linear in terms of edges.

---

[4] Based on the definition, we can see that each news article d_i and node v_i has a one-to-one correspondence relationship. In the following, without ambiguity, we will use d_i to represent a node and a news article interchangeably.

Algorithm 1. *EPCluster* Algorithm

Input: News collection D, parameters $\mu_1, \mu_2, MinPts$.
Output: Cluster collection **C**.
1: $\mathbf{C} = \emptyset$, $clusterID = 1$;
2: **for each** $d_i \in D$ **do**
3:    **if** d_i is not visited **then**
4:       $N(d_i)$ =SearchNeighbors(d_i, μ_1, μ_2);
5:       **if** $|N(d_i)| \geq MinPts$ **then**
6:          $C_{clusterID}$=ExpandCluster$(d_i, \mu_1, \mu_2, MinPts)$;
7:          $\mathbf{C} = \mathbf{C} \cup \{C_{clusterID}\}$;
8:          $clusterID = clusterID + 1$;
9:       **end if**
10:    **end if**
11: **end for**
12: **return C**;

4.3   Cluster Postprocessing

We notice that a few clusters generated by *EPCluster* do not necessarily represent event phases. Instead, they are "small" clusters with similar articles. To improve the accuracy of event phase extraction, we design a quality assessment function to filter such clusters. We consider the following four quality metrics:

Article Quantity. For cluster $C_i \in \mathbf{C}$, denote $N(C_i) = \frac{|C_i|}{|D|} \times 100\%$ as the percentage of articles in that cluster.

Time Interval. For cluster $C_i \in \mathbf{C}$, denote $(t_1^i, t_2^i, \cdots, t_{|C_i|}^i)$ as the sequence of publication dates sorted chronologically. Let t_{Q1}^i and t_{Q3}^i be the first and third quantiles of the empirical temporal distribution. Based on the statistics theory, we estimate the time interval of C_i as $T(C_i) = t_{max}^i - t_0^i$ where

$$t_0^i = \max\{t_1^i, t_{Q1}^i - 1.5 \cdot |t_{Q3}^i - t_{Q1}^i|\} \tag{5}$$

$$t_{max}^i = \min\{t_{|C_i|}^i, t_{Q3}^i + 1.5 \cdot |t_{Q3}^i - t_{Q1}^i|\} \tag{6}$$

Pairwise Topic Similarity. Articles reported the same phase should be similar in topic distributions. We define the average pairwise topic similarity as a quality metric:

$$ATS(C_i) = 1 - \frac{2\sum_{d_m, d_n \in C_i (m < n)} D_{JS}(\boldsymbol{\theta}_m \| \boldsymbol{\theta}_n)}{|C_i| \cdot (|C_i| - 1)} \cdot \tag{7}$$

Pairwise Entity Similarity. Similarly, we define the average pairwise entity similarity as follows:

$$AES(C_i) = \frac{2\sum_{d_m, d_n \in C_i (m < n)} TC(\boldsymbol{c}_m, \boldsymbol{c}_n)}{|C_i| \cdot (|C_i| - 1)} \tag{8}$$

For each cluster C_i, we generate a feature vector consisting of four quality metrics: $F(C_i) = < N(C_i), T(C_i), ATS(C_i), AES(C_i) >$. A weight vector $\boldsymbol{w}$
gives different weights for each feature in $F(C_i)$. Therefore, for each cluster C_i,
we define a score function $Score(C_i) = \boldsymbol{w} \cdot F(C_i)$ to indicate the degree that it is
related to an event phase. To classify the clusters based on the score function, we
construct a logistic regression classifier, with the prediction function as follows:

$$f(C_i) = \frac{1}{1 + e^{-\boldsymbol{w} \cdot F(C_i)}} \tag{9}$$

We learn the weight vector $\boldsymbol{w}$ via gradient ascent on a labeled dataset. After
the model f is trained, we can filter out a news cluster C_i if $f(C_i) < 0.5$. The
rest of the clusters (denoted as $\mathbf{C}^*$) are corresponding to event phases.

5   Event Phase Summarization

In this section, we introduce our steps to generate event phase summaries based
on the previous extraction results. While the relevance between a news article
and an event (represented as keywords e.g. *Egypt Revolution*) can be easily
estimated by IR techniques, it is challenging to determine which articles are more
relevant to an event phase. In this paper, we design a vertex-reinforced random
walk based approach to calculate the relevance scores. Event phase summaries
can be generated by relevance maximum optimization with constraints.

5.1   News Article Ranking

For each $C_i \in \mathbf{C}^*$, we construct a *subgraph* $G_{C_i} = (V_{C_i}, E_{C_i})$ out of the TCCG
G_D where $d_j \in V_{C_i}$ iff $d_j \in C_i$ and $(d_j, d_k) \in E_{C_i}$ iff $d_j \in C_i$, $d_k \in C_i$ and
$(d_j, d_k) \in E$. Refer to a simle example in Fig. 1(d).

While the standard PageRank algorithm [16] employs a time-homogeneous
random walk process on a graph, it tends to assign high scores to closely connected communities, which is capable of selecting nodes with high *centrality*. To
generate representative articles that better summarize the event phase, we need
to pay attention to *diversity* as well. We adopt the *vertex-reinforced random walk
process* framework [17, 18] to balance *centrality* and *diversity* in ranking.

In vertex-reinforced random walk process, denote $M_{m,n}^{(0)}$ as the prior transition probability from d_m to d_n. $N_k(n)$ is the number of visits of random walker
up to the kth iteration. The transition probability from d_m to d_n in the $(k+1)$th
iteration is $M_{m,n}^{(k+1)} \propto M_{m,n}^{(0)} N_k(n)$. Therefore, $M_{m,n}^{(k+1)}$ is reinforced by $N_k(n)$.
This results in a "rich-gets-richer" effect on ranking scores in a community.

In our paper, we calculate the relevance scores of news articles by extending
the vertex-reinforced random walk to the subgraph of TCCG. The implementa-
tion is shown in Algorithm 2. Denote $\boldsymbol{R_0}$ as a $|C_i| \times 1$ prior ranking vector for
articles in C_i. Without prior knowledge, $\boldsymbol{R_0}$ is set uniformly, i.e., $\boldsymbol{R_0} = \frac{1}{|C_i|} \boldsymbol{e}$
where $\boldsymbol{e}$ is a $|C_i| \times 1$ vector with all elements assigned to 1. $M_{m,n}^{(0)}$ (the element

in the mth row and nth column of the prior transition matrix $\mathbf{M_0}$) is defined using the fusion of the two relation strength scores:

$$M_{m,n}^{(0)} = \begin{cases} \frac{1}{Z} \cdot w_c(d_m, d_n) \cdot w_t(d_m, d_n) & (d_m, d_n) \in E_{C_i} \\ 0 & otherwise \end{cases} \tag{10}$$

where Z is a normalization factor and λ is a damping factor, typically set to 0.85. Let $\mathbf{M}_{n+1}$ be the transition probability matrix in the $(n+1)$th iteration, which is updated according to the ranking values and transition probability matrix in the previous iteration:

$$\mathbf{M}_{n+1} = \lambda \mathbf{T}_n \cdot \mathbf{M}_n + (1 - \lambda)\mathbf{M}_0 \tag{11}$$

where $\mathbf{T}_n = [\mathbf{R}_n \mathbf{R}_n \cdots \mathbf{R}_n]$ is a $|C_i| \times |C_i|$ matrix which is utilized to update the transition matrix based on the ranking values in the previous iteration. The update rule for ranking values is defined as:

$$\mathbf{R}_{n+1} = \lambda \mathbf{M}_{n+1} \cdot \mathbf{R}_n + (1 - \lambda)\mathbf{R}_0 \tag{12}$$

The above iterative formula defines an ergodic random walk process in a Markov chain. As shown in [18], it also converges to a stationary distribution. After sufficient large times of iteration N^*, we obtain $r(d_j) = \sum_{d_k \in C_i} M_{j,k}^{(n)} \cdot r(d_k)$ as the relevance score of d_j when $n > N^*$.

Algorithm 2. News Article Ranking Algorithm

Input: News cluster C_i, parameter λ.
Output: Ranking vector $\mathbf{R}$.
1: Compute $\mathbf{M}$ based on C_i;
2: $\mathbf{R}_0 = \frac{1}{|C_i|}e$, $\mathbf{M}_0 = \mathbf{M}$, $n = 0$;
3: **while** not converge **do**
4:    $\mathbf{T}_n = [\mathbf{R}_n \mathbf{R}_n \cdots \mathbf{R}_n]$;
5:    $\mathbf{M}_{n+1} = \lambda \mathbf{T}_n \cdot \mathbf{M}_n + (1 - \lambda)\mathbf{M}_0$;
6:    $\mathbf{R}_{n+1} = \lambda \mathbf{M}_{n+1} \cdot \mathbf{R}_n + (1 - \lambda)\mathbf{R}_0$;
7:    $n = n + 1$;
8: **end while**
9: **return** $\mathbf{R} = \mathbf{R}_n$;

5.2 Event Phase Summary Generation

The final step of our method is to generate an event summary P_i by extracting headlines and publication time of k selected news articles (denoted as S_i). We formulate the news article selection task as an optimization problem that can be solved by a greedy, approximate algorithm.

Ideally, the selected news articles must be relevant to the event phase. However, we notice that the generated summary must not contain redundant information. Therefore, we add an additional constraint such that for any two select news articles d_m and d_n, we require $w_c(d_m, d_n) \leq \mu_1$ and $w_t(d_m, d_n) \leq \mu_2$. Here, we present our *News Selection* optimization problem:

$$\max_{S_i \subset C_i} \quad R(S_i) = \sum_{d_j \in S_i} r(d_j)$$

$$\text{subject to} \quad |S_i| = k \tag{13}$$

$$w_c(d_m, d_n) \leq \mu_1, w_t(d_m, d_n) \leq \mu_2, \forall d_m, d_n \in S_i$$

The proposed optimization problem can be seen as a special case of the *budgeted maximum coverage problem* [19], which is proved to be NP-hard. Because the optimization objective is submodular and monotone, we can employ a greedy algorithm to solve the problem approximately. Here, we present our approximate algorithm for *News Selection* in Algorithm 3. The worst-case approximation ratio is proved to be $1 - \frac{1}{e}$, as shown by Khuller et al. [19]. It selects a news article from S_i that maximizes that objective function without violating any constraints at each iteration. When it stops with k news articles selected, we extract the publication time and headlines in S_i as the event phase summary P_i.

Algorithm 3. News Article Selection Algorithm

Input: News cluster C_i, parameter k.
Output: Selected news collection S_i.
1: $S_i = \emptyset$;
2: **while** $C_i \neq \emptyset$ and $|S_i| < k$ **do**
3:　　Select $d_n = \text{argmax}_{d_n \in C_i} R(S_i \cup \{d_n\}) - R(S_i)$
　　　　subject to $w_c(d_m, d_n) \leq \mu_1, w_t(d_m, d_n) \leq \mu_2, \forall d_m \in S_i$;
4:　　$S_i = S_i \cup \{d_n\}$;
5:　　$C_i = C_i \setminus \{d_n\}$;
6: **end while**
7: **return** S_i;

---

6   Experimental Results

In this section, we conduct experiments on news datasets to evaluate the performance of our approaches and compare it with baselines. All the codes are written in JAVA, and run on a PC with an Intel CPU 2.9 GHz and 16 GB memory.

6.1   Datasets

The news datasets we used in this paper are publicly available from [6], which contain four English news datasets regarding long-span recent armed conflicts. The news articles are collected from 24 news agencies (e.g. Associated Press, Reuters, Guardian, etc.), obtained using the Google search engine. The detailed statistics are illustrated in Table 1.

Table 1. Summary of datasets.

Dataset	Event	#Article	Time range
D_1	Egypt Revolution	3,869	2011.1.11 - 2013.7.24
D_2	Libya War	3,994	2011.2.16 - 2013.7.18
D_3	Syria War	4,071	2011.11.17 - 2013.7.26
D_4	Yemen Crisis	3,600	2011.1.15 - 2013.7.25

6.2 Evaluation on Event Phase Extraction

Experimental Settings. To our knowledge, there is no prior work regrading event phase extraction. However, the proposed approach can be seen as an application of document clustering. To obtain the ground truth, we employ a pairwise judgment method introduced in [20]. For each dataset D_i, we randomly generate news article pairs, denoted as $T_i = \{(d_m, d_n)\}$. We ask human annotators to label whether d_m and d_n are related to the same event phase. Denote $v_{m,n} \in \{1, 0\}$ as the human judgment result and $v'_{m,n}$ as the clustering result, where 1 and 0 represent the same and different phases, respectively. We use precision, recall and F1 score as the evaluation metrics, defined as:

$$Precision(T_i) = \frac{|\{(d_m, d_n) \in T_i | v_{m,n} = 1 \wedge v'_{m,n} = 1\}|}{|\{(d_m, d_n) \in T_i | v'_{m,n} = 1\}|} \tag{14}$$

$$Recall(T_i) = \frac{|\{(d_m, d_n) \in T_i | v_{m,n} = 1 \wedge v'_{m,n} = 1\}|}{|\{(d_m, d_n) \in T_i | v_{m,n} = 1\}|} \tag{15}$$

$$F1\ Score(T_i) = \frac{2 \cdot Precision(T_i) \cdot Recall(T_i)}{Precision(T_i) + Recall(T_i)} \tag{16}$$

In total, we have 300 labeled new article pairs for each dataset. We report macro-average precision, recall and F1 score in the following experiments.

Parameter Tuning. We tune three parameters in *EPCluster*, namely μ_1, μ_2 and *MinPts*. We fix two parameters and vary the remaining one at each time. The results are illustrated in Fig. 2. It can be seen that when $\mu_1 = 0.4$, $\mu_2 = 0.5$ and $MinPts = 10$, *EPCluster* achieve the best results.

Method Comparison. While document clustering is a well-studied problem, we compare our method with classical approaches and the variant of our method, introduced as follows:

- **VSMCluster** - KMeans using word features of TF-IDF weights.
- **TopicCluster** - KMeans using topic distributions based on LDA [12].
- **SCAN** [15] - structural clustering algorithm for network partitioning.
- **EPCluster-C** - our *EPCluster* algorithm without postprocessing.

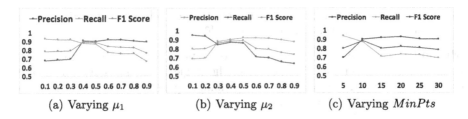

Fig. 2. Clustering results of *EPCluster* under different parameter settings.

In the implementation, because we consider publication time in *EPCluster*, we add it as a feature in *VSMCluster* and *TopicCluster* to make them comparable with ours. To compare our method with the state-of-the-art structural clustering algorithm *SCAN* [15], we first construct a TCCG and then apply *SCAN* on the graph. The results are presented in Table 2.

Table 2. Experimental results of event phase extraction.

Method	VSMCluster	TopicCluster	SCAN	EPCluster-C	Our method
Precision	0.35	0.52	0.78	0.81	**0.89**
Recall	0.74	0.67	0.72	**0.79**	0.78
F1 score	0.48	0.59	0.75	0.80	**0.83**

Based on the experimental results, our method outperforms *VSMCluster* and *TopicCluster* because these classical methods rely on distance computation of high-dimensional vectors. Since these news articles are related to the same event and thus are similar in content, these methods are not suitable for clustering-based event phase extraction method. *SCAN* algorithm has a relative good performance based on TCCG, which indicates that although structural clustering is originally designed for networks, it can be employed for text analysis as well. The comparison between *EPCluster-C* and our method shows that the postprocessing step is effective to improve the performance of event phase extraction.

6.3   Evaluation on Event Phase Summarization

Ground Truth. The ROUGE framework [21] has been extensively used to evaluate the effectiveness of document summarization. However, the summaries we generate are headlines, rather than documents. Tran et al. [6] previously propose a headline summary evaluation framework based on the relevance of machine-generated timelines compared with ground truth timelines. In this paper, we obtain the timeline summaries manually created by professional journalists from Tran et al. These timeline summaries are served as ground truth to be provided to human annotators for the evaluation of our method. The detailed statistics of ground truth summaries can be found in [6].

Method Comparison. Although there is no prior work addressing the event phase summarization issue, if we consider the single summary of an event phase, our task can be regarded as a headline summary generation task. We compare our method with the following baselines:[5]

- **Tran et al.** [6] - the timeline generation method especially for headlines.
- **Chieu et al.** - [22] a timeline generation method based sentence popularity.
- **Our Method (PageRank)** - the variant of our approach which adopts simple PageRank method for relevance calculation.

We also consider the following two benchmark methods:

- **Random** - selects k news articles randomly.
- **Longest** - selects top-k longest headlines due to the informativeness.

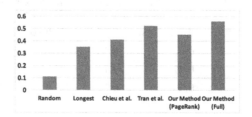

Fig. 3. Relevance evaluation of event phase summarization.

Experiments and Results. To evaluate these methods, we extract generated summaries from 106 dates that are appeared in the ground truth summaries. We present the ground truth and machine-generated summaries to human annotators and ask them to label each headline as relevant or not. We take the average relevance scores for each method as the evolution metrics. The results are presented in Fig. 3. It can be seen that the results of benchmark approaches are not as good as others because they lack textual analysis on news articles. Our method outperforms Chieu et al. and the variant of our method because we pay more attention to the centrality and diversity nature of summaries. The performance of Tran et al. is relatively high because they investigate the characteristics of news headlines and select more informative ones. Our method performs slightly better than Tran et al. in terms of relevance. The unique advantage of ours is that we generate multiple summaries for event phases such that it is easier for readers to track the development phases of long, complicated events.

---

[5] Many other methods focus on timeline generation. However, the summaries we generates are headlines and dates, making it difficult to compare our method with them. We will investigate how to modify these algorithms for our task in the future.

Case Study. We present the event phase summaries of *Egypt Revolution* produced by our approach. Due to space limitation, we only present the publication dates and headlines of two news articles in each event phase. We also manually add a brief description for each phase, shown in Table 3. It shows that our approach can identify and summarize fine-grained event phases effectively.

Table 3. Event phase summaries of *Egypt Revolution*.

Event Phase #1 *Protest against Hosni Mubarak*	
2011.2.2	Egypt protests: Hosni Mubarak to stand down at next election
2011.2.11	Hosni Mubarak resigns and Egypt celebrates a new dawn
Event Phase #2 *Egypt under the Rule of Military Power*	
2011.4.9	Egyptian soldiers attack Tahrir Square protesters
2011.7.10	Protests spread in Egypt as discontent with military rule grows
Event Phase #3 *Mohammed Morsi Won Presidential Election*	
2012.5.23	First round of presidential election
2012.6.24	Election officials declare Morsi the winner
Event Phase #4 *Protest against Morsi and Muslim Brotherhood*	
2013.1.27	Egypt's Mohammed Morsi declares state of emergency, imposes curfew
2013.1.30	Egypt's military chief says clashes threaten the state
Event Phase #5 *Morsi's Ousting*	
2013.7.4	After Morsi's Ousting, Egypt Swears in New Presiden
2013.7.6	Morsi's ouster in Egypt sends chill through political Islam

7   Conclusion and Future Work

In this paper, we formalize the problem of event phase extraction and summarization. We propose a structural clustering algorithm *EPCluster* based on TCCG to group news articles into event phases. For each event phase, we extract top-k news articles by a vertex reinforced random walk based ranking algorithm and generate summaries by relevance maximum optimization. Experiments show that our method can solve the problem effectively. In the future, we will focus on improving the performance of MDS and TG when event phases are considered.

Acknowledgements. This work is partially supported by the National Key Research and Development Program of China under Grant No. 2016YFB1000904, Shanghai Agriculture Science Program (2015) Number 3-2 and NSFC-Zhejiang Joint Fund for the Integration of Industrialization and Informatization under Grant No. U1509219.

References

1. Gong, Y., Liu, X.: Generic text summarization using relevance measure and latent semantic analysis. In: SIGIR, pp. 19–25 (2001)
2. Conroy, J.M., O'Leary, D.P.: Text summarization via hidden markov models. In: SIGIR, pp. 406–407 (2001)
3. He, Z., Chen, C., Bu, J., Wang, C., Zhang, L., Cai, D., He, X.: Document summarization based on data reconstruction. In: AAAI (2012)
4. Qian, X., Liu, Y.: Fast joint compression and summarization via graph cuts. In: EMNLP, pp. 1492–1502 (2013)
5. Yan, R., Kong, L., Huang, C., Wan, X., Li, X., Zhang, Y.: Timeline generation through evolutionary trans-temporal summarization. In: EMNLP, pp. 433–443 (2011)
6. Tran, G., Alrifai, M., Herder, E.: Timeline summarization from relevant headlines. In: Hanbury, A., Kazai, G., Rauber, A., Fuhr, N. (eds.) ECIR 2015. LNCS, vol. 9022, pp. 245–256. Springer, Heidelberg (2015). doi:10.1007/978-3-319-16354-3_26
7. Erkan, G., Radev, D.R.: Lexrank: graph-based lexical centrality as salience in text summarization. J. Qiqihar Junior Teachers Coll. **22**, 2004 (2011)
8. Wan, X., Yang, J.: Multi-document summarization using cluster-based link analysis. In: SIGIR, pp. 299–306 (2008)
9. Ng, J., Chen, Y., Kan, M., Li, Z.: Exploiting timelines to enhance multi-document summarization. In: ACL, pp. 923–933 (2014)
10. Chen, C.C., Chen, Y.-T., Sun, Y., Chen, M.C.: Life cycle modeling of news events using aging theory. In: Lavrač, N., Gamberger, D., Blockeel, H., Todorovski, L. (eds.) ECML 2003. LNCS (LNAI), vol. 2837, pp. 47–59. Springer, Heidelberg (2003). doi:10.1007/978-3-540-39857-8_7
11. Knights, D., Mozer, M.C., Nicolov, N.: Detecting topic drift with compound topic models. In: ICWSM (2009)
12. Blei, D.M., Ng, A.Y., Jordan, M.I.: Latent dirichlet allocation. J. Mach. Learn. Res. **3**, 993–1022 (2003)
13. Wang, C., Zhang, R., He, X., Zhou, A.: Nerank: ranking named entities in document collections. In: WWW, pp. 123–124 (2016)
14. De Kretser, O., Moffat, A.: Effective document presentation with a locality-based similarity heuristic. In: SIGIR, pp. 113–120 (1999)
15. Xu, X., Yuruk, N., Feng, Z., Schweiger, T.A.J.: SCAN: a structural clustering algorithm for networks. In: KDD, pp. 824–833 (2007)
16. Brin, S., Page, L.: The anatomy of a large-scale hypertextual web search engine. Comput. Netw. **30**(1–7), 107–117 (1998)
17. Pemantle, R.: Vertex-reinforced random walk. Probab. Theory Relat. Fields **92**(1), 117–136 (1992)
18. Mei, Q., Guo, J., Radev, D.R.: Divrank: the interplay of prestige and diversity in information networks. In: KDD, pp. 1009–1018 (2010)
19. Khuller, S., Moss, A., Naor, J.S.: The budgeted maximum coverage problem. Inf. Process. Lett. **70**(1), 39–45 (1999)
20. Chen, J., Niu, Z., Fu, H.: A multi-news timeline summarization algorithm based on aging theory. In: Cheng, R., Cui, B., Zhang, Z., Cai, R., Xu, J. (eds.) APWeb 2015. LNCS, vol. 9313, pp. 449–460. Springer, Heidelberg (2015). doi:10.1007/978-3-319-25255-1_37
21. Lin, C., Hovy, E.H.: Automatic evaluation of summaries using n-gram co-occurrence statistics. In: HLT-NAACL (2003)
22. Chieu, H.L., Lee, Y.K.: Query based event extraction along a timeline. In: SIGIR, pp. 425–432 (2004)

ESAP: A Novel Approach for Cross-Platform Event Dissemination Trend Analysis Between Social Network and Search Engine

Yan Tang[2(✉)], Pengju Ma[2], Boyuan Kong[1], Wenqian Ji[2], Xiaofeng Gao[1], and Xuezheng Peng[3]

[1] Shanghai Key Laboratory of Data Science,
Department of Computer Science and Engineering,
Shanghai Jiao Tong University, Shanghai, China
{sjtukong,gaoxiaofeng}@sjtu.edu.cn
[2] College of Computer and Information, Hohai University, Nanjing, China
{tangyan,mpj,innerpeace}@hhu.edu.cn
[3] Baidu, Inc., Beijing, China
pengxuezheng@baidu.com

Abstract. With the rapid development of Internet, new media, such as blogs, wikis, and social media, become a major platform for information dissemination. Numerous studies focus on event dissemination trend analysis for individual media platform, while very few works are conducted to study the dissemination characteristics in a cross-platform manner. In this work, we propose ESAP, a novel cross-platform approach to analyze the event dissemination trend between social network and search engine simultaneously. ESAP includes three models: an event popularity model based on hot word dynamic weight; a trend similarity model to measure the similarity of event popularity across different platforms over time; and an attention degree model to measure event public attention through time. Experimental results based on four real-world event dissemination datasets (two from Baidu and two from Weibo) produce several interesting findings and validate the effectiveness of ESAP in modeling and analyzing event dissemination trend between social network and search engine from different perspectives.

Keywords: Social network · Search engine · Event dissemination

1 Introduction

With the emergence and rapid development of Internet, the primary platform for information dissemination is gradually changed from the traditional media, like newspapers, magazines, radios, into the new media, such as blogs, wikis, and social network. Over the last decade, a vast number of studies focus on analyzing the event dissemination trend for individual new media platform separately [1,7,9,11–13], while very few works are conducted to study the dissemination characteristics in a cross-platform manner [3,15].

© Springer International Publishing AG 2016
W. Cellary et al. (Eds.): WISE 2016, Part I, LNCS 10041, pp. 489–504, 2016.
DOI: 10.1007/978-3-319-48740-3_36

Obviously, the event dissemination will not depend solely on one platform. If an event draws a massive amount of attention, there will be several media platforms that track and report it collectively through a period of time. Thus, the study on cross-platform event dissemination and the interplay among distinct platforms plays a crucial role in understanding the event spreading process, and thus can help track the public concerns, control event dissemination trend, and protect public security, etc. Effective trend analysis could also give third-party companies or business owners a deeper insights for achieving their social values through media platforms. All in all, researches on the cross-platform analysis are evidently of great practical significance and broad application prospect.

In this paper, we aim to conduct a thorough cross-platform event dissemination trend analysis between social network and search engine simultaneously. The reasons for choosing social network and search engine as our study platforms is due to their unique features. Social network, as the source of various public information, has a substantial number of users and has become one of the most influential new media platforms. Instead of receiving information passively, search engine provides active query services, which have become the main entrance for information retrieval. Very little literature studies the relationship between social trends from social network and web trends from search engine [6,8]. Additionally, these works gave few consideration about the event dissemination analysis. This gap is the reason for conducting this research which aims to find appropriate answers for the following two important questions:

1. For social network and search engine, how to describe event/topic trends using data for their dissemination processes, and what are the characteristics of their event dissemination processes?
2. How to describe the interplay between social network and search engine during an event dissemination process?

There exist several challenges for this topic. Firstly, the data from social network and search engine are vastly different, lacking a unified event trending model to compare them side by side. Secondly, social network data has limited content length and may contain large percentage of duplication due to retweeting, whereas search engine query is even more succinct, bringing extra difficulty for modeling and analysis. Lastly, quantitative modeling for cross-platform event dissemination similarity and public attention degree is a new problem, for which finding reasonable and effective solutions is not a trivial task.

To deal with these challenges, we choose Weibo and Baidu as two representatives of social network and search engine. Weibo is one of the most popular online social media and microblogging platform, while Baidu is a major online search platform in China. They both profoundly affect online event dissemination in China and reflect the current public spotlights. In addition, we select two typical events from different fields as case studies to illustrate and validate our models. One is the stock market crash event, the other one is the Eastern Star sinking disaster event. Both of them attracted intensive international attention and lasted for a long period of time with multiple ups and downs, thus being highly worthy for an in-depth study.

Correspondingly, we propose a new cross-platform analytic approach called ESAP, which deals with *Event Similarity, Attention degree and Popularity.* ESAP is a unified approach based on hot words' dynamic weights applicable for both social network and search engine. It provides new perspectives for analyzing event popularity, trend similarity and public attention degree at both macroscopic and microscopic level over a time period.

ESAP contains two phases. First phase consists of data extraction and hot word identification. The second phase contains three models for analyzing event dissemination trend cross-platform from different perspectives, including:

1. An event popularity model in which hot words' dynamic weights are calculated using a TF-IDF based method to obtain the platform event heat degree vector for event popularity analysis.
2. A trend similarity model which uses cosine similarity over event heat degree vector within different time period to quantitatively analyze cross-platform event dissemination similarity.
3. An attention degree model to measure event public attention through time by calculating the area of radar chart generated by the hot words' dynamic weights.

Experimental results based on four real-world event dissemination datasets (two from Baidu and two from Weibo) produce several interesting findings and validate the effectiveness of ESAP in modeling and analyzing event dissemination trend between social network and search engine from different perspectives.

To summarize, the main contributions of this paper are as follows:

- This paper proposes a unified model based on hot words' dynamic weight for event popularity analysis. This model is applicable for social networks and search engines. The advantage of this model is that it does not require any prior knowledge like network topology or user's social relations.
- This paper provides new quantitative models for measuring cross-platform event dissemination similarity and public attention degree.
- We further evaluate ESAP on two real-world events across Weibo and Baidu, the experiment results validates the effectiveness of our approach.

The remainder of this paper is organized as follows. In the next section, we introduce related works. We introduce our approach in Sect. 3 for event popularity, similarity and public attention degree analysis. Section 4 shows experiment results and discussions the findings on four real world datasets from Weibo and Baidu. Finally, the paper is concluded in Sect. 5.

2   Related Work

Individual platform analysis. As early as 2010, Lerman and Hogg [9] used DIGG[1] website data to predict the popularity of news based on social dynamical model. A new technique named Transaction-Based Rule Change Mining is

---

[1] http://digg.com/.

proposed by [1] to detect and track hot events in Twitter dynamically. Zhang et al. [13] presented a novel model to describe the information spread pattern in Weibo. Vaughan et al. [10] collected web search volume data from Google Trends to predict academic fame. In summary, the work mentioned above focuses on either social network or search engine as the single platform for event dissemination. However, exclusively focusing on one platform may not lead to an overall depth depiction of the event dissemination process.

Cross-platform analysis. [5] took into account the relationship between Twitter and YouTube, and studied the role of social cascading in YouTube video diffusion. [3] selects Twitter, New York Times and Flickr as three distinct platforms on microblog, news portal and imaging sharing to study the emerging topic detection. An algorithm based on multiple social networks like Twitter, Weibo, and Facebook to identify anonymous identical users is proposed in [15].

For social network vs. search engine specifically, Kwak et al. [8] made a comparison of duration of hot topics between Twitter and Google and found that the hot topics in Twitter survive longer. Giummolè et al. [6] found that 72 % of the similar trends appear first on Twitter rather than Google. They assessed the relation between similar Twitter and Google trends by testing and comparing three classes of time series regression models.

Aside from the work mentioned above, to the best of our knowledge, no research work has been conducted to study event popularity trend, dissemination similarity and events' user attention degree for both social network and search engine. Our research aims to fill this gap by offering cross-platform trend analysis from different perspectives at both macroscopic and microscopic level over a time period. Unlike some traditional approaches like the work in [6] that requires a graph structure or the work in [14] which requires the restoration of related users' social relations, our model simplifies the modeling process without requiring a topology of the event dissemination network nor any extra effort for restoring users' social relations.

3   The Proposed Approach

3.1   Overview of the ESAP Approach

Figure 1 is the overview of ESAP approach with two phases and three models.

The input into ESPA is raw data extracted from Weibo and Baidu respectively. The raw data is stored in the database together with domain word dictionary and domain specific knowledge as the knowledge base for subsequent processing. Then, for hot words identification, ESAP applies text segmentation (using ansj[2]) on the knowledge base to obtain the bag of words from which the hot words of each event are identified using statistical analysis. After data processing, ESAP has three analytic models. The first model is a unified event popularity model for both social network and search engine, which is the kernel

---

[2] http://github.com/NLPchina/ansjseg/.

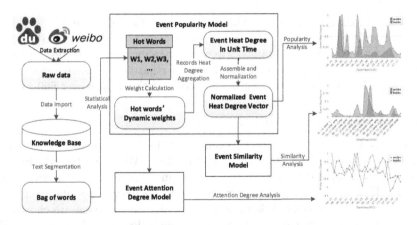

Fig. 1. An overview of ESAP approach

of ESAP. In this model, hot words' dynamic weights are calculated using our proposed method and then used to obtain the normalized platform event heat degree vector for event popularity analysis. The second model deals with cross-platform similarity. ESAP applies cosine similarity calculation on event heat degree vectors to quantitatively analyze cross-platform event dissemination similarity. Lastly, the third model uses hot words' dynamic weights to produce the daily radar chart and use the area size under each radar chart to model and analyze cross-platform event attention degree.

3.2 Data Extraction

In this study, we focus on studying two primary events occurred in 2015 that brought international attention. The first event is the Eastern Star ship sinking disaster occurred in June 2015 and the second event is the crash of Chinese stock market from June to July in 2015. The raw datasets are extracted by Baidu from their search engine records and Weibo data APIs. First, we carefully select signature words for each event. For example, for the Eastern Star ship sinking disaster, the signature words are {'East Star', 'ship sinking'}. Then, Baidu uses these signature words to filter in related Weibo records and search engine query

Table 1. Experimental datasets

Event name	Platform	# Rows (Thousand)	Size(MB)
Eastern star sinking disaster	Weibo	308	336.2
	Baidu	1477	421
Chinese stock matket crash	Weibo	701	606.9
	Baidu	420	77.7

records. Since the record set is too large to process, random sampling is adopted following uniform distribution to draw records from the filtered records.

Four raw datasets are obtained after the data extraction process (Table 1). The largest one is the related Weibo messages regarding Chinese stock market crash in 2015, containing 701 thousands records with the size over 600 MB.

3.3   Hot Words Identification

We use hot words to represent an event. For the first step in the hot words identification process, we store the raw data from Weibo and Baidu into the database, then segment the text (i.e. Weibo message or Baidu query) in the database into words by a word segmentation tool. For a specific event in a domain, a few closely related domain words related to the event are chosen and put into the data dictionary of the word segmentation tool. Also, we establish a stop words lexicon, which includes void words such as "is", "and" and "a". These stop words should be removed from the segmentation results. In brief, the database, the common domain words and stop words lexicon form a simple knowledge base for the word segmentation task.

The word segmentation task produces a bag of words. We then calculate the each word's frequency in order to identify hot words. The word whose frequency is over 10 percent of the whole records number or at top 50 of the word's frequency list is considered a candidate hot word. Finally, we invite experts in relevant fields for hot words filtering. By combining their choices with the word's frequency, we can obtain the final list of hot words for each event.

3.4   Event Popularity Model

The center of our proposed approach is the event popularity model. Firstly, all the notions and the corresponding descriptions in our event popularity model are shown in Table 2.

The popularity of an event in a time unit t is measured by its event heat degree EHD_t. To model the development of an event, we focus on the trend in event heat degree during a time period T. The key challenge here is defining the value for EHD_t, which is influenced by various factors. In this section, we introduce a method based on the hot words' dynamic weights to calculate EHD_t.

Any time period T can be divided into various equal units t, viz. $T = [t_1, t_2, \ldots, t_n]$. Then, within each time unit t, EHD_t is determined by two factors: the degree of public attention and the degree of public involvement, viz. the degree of the correlation between the related records published or queried by the users on the platform about the event. The degree of the correlation for one record is determined by how many hot words are in this record. Therefore, we build an event popularity model based on hot words' dynamic weight through time, as defined below.

Definition 1. *(Hot Word's Dynamic Weight) Given a hot word hw_i of an event E and a time unit t, the Hot Word's Dynamic Weight denoted as $W^t_{hw_i}$ is product of the hot word's dynamic frequency and its inverse records frequency.*

Table 2. Symbol Description for event popularity model

Symbol	Description
T, t	A period of time, and a time unit
E	An event
R, R_i	The set of all the records related to E, and its ith record
$ts(R_i)$	The time stamp of the record R_i
HW	The set of all the hot words
hw_i	The ith hot word in HW
nhw_i^t	the number of occurrences of hw_i within time unit t
$DF_{hw_i}^t$	Dynamic frequency of hw_i within time unit t
IRF_{hw_i}	Inverse record frequency of hw_i
$W_{hw_i}^t$	Dynamic weight of hw_i within time unit t
HDR_i^t	Heat degree of the record R_i within time unit t
EHD_t	Heat degree of event E within time unit t
$\mathbf{EHD}_T^p$	Event heat degree vector in platform p within the time period T
$\mathbf{NEHD}_T^p$	Normalized event heat degree vector in platform p within T

Following Definition 1, the dynamic frequency of the hot word hw_i within time unit t, denoted as $DF_{hw_i}^t$ is defined in Eq. (1):

$$DF_{hw_i}^t = \frac{nhw_i^t}{\sum_k nw_k^t},\tag{1}$$

where nhw_i^t is the number of occurrences of hot word hw_i within a time unit t. $\sum_k nw_k^t$ is the sum of word counts of all the words excluded stop words[3] within t.

The inverse records frequency of hw_i is denoted as IRF_{hw_i}, defined in Eq. (2):

$$IRF_{hw_i} = \log \frac{|R|}{|\{R : hw_i \in R\}| + 1},\tag{2}$$

where $|R|$ is the number of the total records. $|\{R : hw_i \in R\}|$ is the number of records that contain the hw_i. Notice that the IRF_{hw_i} is a static global variable for each hot word. It is introduced to punish the words that have a high frequency but of low importance to an event. Consequently, duplicate records problem (ex. retweeting) can be properly handled.

Based on the TF-IDF numerical statistic, using Eqs. (1) and (2), the dynamic weight of the hot word hw_i within time unit t is defined in Eq. (3):

$$W_{hw_i}^t = DF_{hw_i}^t \cdot IRF_{hw_i}.\tag{3}$$

---

[3] https://en.wikipedia.org/wiki/Stop_words.

After obtaining hot words' dynamic weight, the heat degree of the ith record R_i containing hw_i within a time unit t is defined by Eq. (4). It is the sum of the dynamic weights of all the k hot words hw_i in R_i

$$HDR_i^t = \sum_{i=1,\ldots,N} W_{hw_i}^t | hw_i \in R_i . \tag{4}$$

It should be noted that with the introduction of the inverse records frequency IRF_{hw_i}, the same hot word in any record R_i is calculated only once.

Following Eq. (4), for an event E, its popularity/heat degree within the time unit $t \in [t_i, \ t_i + 1]$ given N related records is defined in Eq. (5):

$$EHD_t = \sum_{i=1,\ldots,N} HDR_i | ts(R_i) \in [t_i, \ t_i + 1] . \tag{5}$$

Based on Eq. (5), within a time period $T = [t_1, \ t_n]$, putting every $EHD_{t_i}^p$ of each platform p (ex. Baidu, Weibo) within a time unit t_i into a vector returns the event heat degree vector defined in Eq. (6):

$$\mathbf{EHD}_T^p = \left[EHD_{t_1}^p, \ EHD_{t_2}^p, \ldots, EHD_{t_n}^p \right] . \tag{6}$$

In order to compare the different platforms side by side, we need to normalize the event popularity of each platform and obtain the normalized event heat degree vector defined in Eq. (7):

$$\mathbf{NEHD}_T^p = \left[EHD_{t_1}^p, \ EHD_{t_2}^p, \ldots, EHD_{t_n}^p \right] / \max(EHD_{t_i}^p) . \tag{7}$$

After the normalization, each value in $\mathbf{NEHD}_T^p$ is between zero and one, thus enabling the analysis and comparison of cross-platform event popularity.

3.5   Event Similarity Model

Definition 2. *(Event Dissemination Trend Similarity) Given an event E, a time period $T = [t_1, \ldots, t_n]$, the interaction between one platform p_a (ex. social network) and another platform p_b (ex. search engine) is modeled by the event dissemination trend similarity defined as the cosine similarity between $\mathbf{NEHD}_T^{p_a}$ and $\mathbf{NEHD}_T^{p_b}$ denoted as $Sim(p_a, p_b)$.*

Following Definition 2, event trend similarity is defined as Eq. (8)

$$Sim(p_a, p_b) = \frac{\sum_{i=1}^{n} \left(NEHD_{t_i}^{p_a} \cdot NEHD_{t_i}^{p_b} \right)}{\sqrt{\sum_{i=1}^{n} \left(NEHD_{t_i}^{p_a} \cdot NEHD_{t_i}^{p_a} \right)} \sqrt{\sum_{i=1}^{n} \left(NEHD_{t_i}^{p_b} \cdot NEHD_{t_i}^{p_b} \right)}} . \tag{8}$$

where $NEHD_{t_i}^{p_k}$ is the normalized event heat degree for a platform p_k calculated using Eq. (7). This is a general model for different platforms.

3.6   Event Attention Degree Model

Event Attention Degree (EAD) Modeling is based on the assumption that the variation of hot words' dynamic weights through time reflects the change of public attention degree on an event. In our EAD model, the public attention degree within the time unit t denoted as EAD^t is represented as a multi-dimensional vector containing the hot words' dynamic weights. Then, the key is to find a reasonable way to calculate EAD based on hot words' dynamic weights vector.

As a useful way to display multivariate observations with an arbitrary number of variables, radar chart has been successfully applied in business and social science studies [2,4]. It is very suitable for modeling public attention degree based on the multi-dimensional hot words' weighs.

Definition 3. *(Event Attention Degree) Within the unit time t, given an event E and its hot words set HW, use the number of words in HW denoted as n as the dimension to establish a radar chart for hot words' dynamic weight $W^t_{hw_i}$. Event attention degree denoted as EAD^t is the area of this radar chart.*

Following Definition 3, event attention degree is calculated by Eq. (9)

$$EAD^t = \sum_{i=0,\ldots,n-1} \sin(\theta) \cdot W^t_{hw_i} \cdot W^t_{hw_{i+1}}/2 + \sin(\theta) \cdot W^t_{hw_n} \cdot W^t_{hw_0}/2 , \qquad (9)$$

where $\sin(\theta)$ is the angle between two dimensions of radar chart.

4   Experiment and Discussion

4.1   Experimental Setup

To verify the effectiveness of our proposed method and model, Sina and Baidu data related to two international events were analyzed using ESAP. To address the two question raised in Sect. 1, the cross-platform event trends analysis consists of three parts: Event Popularity, Event Trend Similarity and Event Attention Degree. Table 1 lists the datasets used in our experiments provided by Baidu Inc. and are extracted using the method described in Sect. 3.2.

The experiment environment is as follows: (1) For the hardware: the experiments are conducted on MacBook Air with 1.6 GHz Intel Core i5 and Memory 8 GB (4 × 2 GB), 1600 MHz, DDR3, running on the operating system of OS X Yosemite; (2) For the software environment: The experiments are carried out in the Java (version 1.8.0) and MySQL (version 5.6.23), Ansj (version 1.4.1) is used to for text segmentation.

4.2   Event Popularity

First, we analyze the stock market crash event, the time period T of the associated datasets is from 2015-06-16 to 2015-07-14. According to the hot words selection method, the following words are chosen as the hot words of

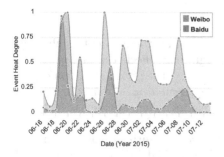

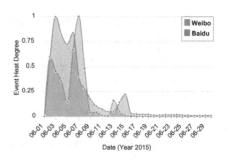

Fig. 2. Heat degree of the stock market crash event

Fig. 3. Heat degree of the Eastern Star sinking disaster event

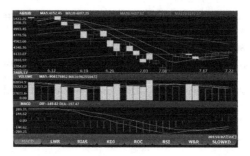

Fig. 4. K line charts from 2015-06-12 to 2015-07-22

Weibo: "stock market", "slump", "Shanghai Index", "steep fall", "stock market crash", "stock", "fall percentage", "stocks slump", "stockholder", "drop", "Growth Enterprise Market (GEM)", "China Securities Regulatory Commission (CSRC)".

And "crash", "Shanghai Index", "stock market crash", "4500 points", "stocks slump", "reason", "Type A stocks", "stock market" as the hot words of Baidu. The cross-platform event popularity results are shown in Figs. 2 and 3.

This event occurred on 06/15/2015 and the stock index dropped drastically until July 8th. Standing at 5423 points, the index of Chinese stock market crashed to 3704 in three weeks, resulting in the vaporization of trillions of RMB and wide spread public fear (Fig. 4).

From the charts we know that the event heat degrees of both Weibo and Baidu on June 19th are very high. On June 19th, Weibo reaches a peak in popularity while Baidu reaches the peak the next day on June 20th. It's suspected that the event dissemination on Baidu has a delay over the actual development of event. June 20th to 22th is the Dragon Boat Festival holiday so that the stock market is closed. Thus, the event heat degree of both two platforms on June 21th decreases substantially. However, on June 22th, the event heat degree of Baidu is high. According to the search query analysis, we find that most users search information on Baidu for the next day's opening index and buying/selling strategy. On June 26th, Weibo reaches another Climax when the stock index declines to 4300 points and over a thousand stocks are crashed flat on the floor. However, it is noticed that Baidu reaches its third peak on June 27th. Hence, the event dissemination delay of Baidu is confirmed again.

From June 28th to July 12th, the event popularity trend of Baidu is becoming relatively stable. In contrast, people are still discussing about the stock market crash event actively on Weibo. We observe that the event popularity trend of Weibo coincides in real-time with the actual development of the stock market. The market tumbles on June 29th, July 3th and reaches the bottom on July 8th, from Fig. 2, it is clear that Weibo's event popularity trend has three peaks that perfectly align with the three days mentioned above. Through the above analysis, our event heat degree model is shown to be reasonable and effective with event popularity trends, aligning very well with the actual event development.

For the Eastern Star ship sinking disaster event. The time period T of the associated datasets is from 06/16/2015 to 07/14/2015. For bettering analyzing the event's dissemination process, we list some important sub-events as follows: On June 1th, the Eastern Star sinking disaster occurred. On June 7th, more than 500 rescue workers and government officials bowed in mourning towards the sunken cruise ship. This was seventh day memorial according to oriental culture tradition. On June 13th, 442 deaths were confirmed with only 12 people rescued, this sub event marks a closure of rescue work for this big tragedy.

From Fig. 3, we see that along the development of this event, there is a big difference between the event popularity degrees before and after June 9th. Because the former is mainly about sinking and rescue work while the latter is mainly public mourning over this tragedy.

On June 2th and June 3th, Chinese government is highly concerned with development of the event, especially on the rescue work. With more and more people paying attention to this news, the event heat degrees on both Weibo and Baidu in the chart are very high in these two days. However, the event popularity does not last for a long time, i.e., the level of public concern about this event declines rapidly. The event popularity trend of Weibo has three peaks on the 2th, 7th and 13th of June, these peaks coincide well with the date of three important sub-events mentioned above.

From the above experimental results, we conclude that (1) The results show that Weibo disseminates the event in real-time. The event heat degree on the Weibo is following the sub-events and the event's development itself; (2) An event must experience the process of germination-climax-dying. We think that the more similar the trends are between Weibo and Baidu, the more interactive these two platforms are for the event dissemination process like the trends over the stock crash event; (3) Experimental results on two events show that our proposed event heat degree model is reasonable and effective in measuring the trend of event popularity.

4.3   Event Trend Similarity

Behind a burst event, there must be some sub-events. Different sub-events have
distinct dissemination characteristics. Therefore, for measuring and analyzing
event trend similarity, we focus on event trend similarity during the time period
of these microscopic sub-events for different platforms (Fig. 6).

The first crash of the stock market takes place on June 19th. Based on Fig. 5,
Baidu has a delay of two days compared with Weibo. Weibo is more responsive
and leads the dissemination of the stock market crash event. It may influence or
stimulate the event dissemination on Baidu. For the event heat degree trends of
Weibo and Baidu, higher similarity indicates stronger inter-relationship between
the two platforms. We selected Weibo heat degrees during the time period from
06/18/2015 20:00 to 06/20/2015 12:00 and Baidu heat degrees during the time
from 06/19/2015 12:00 to 06/20/2015 20:00 (considering the delay effect) as two
trends. Then using Eq. (8), the event trend similarity is 0.812, which represents
a very strong interaction between Weibo and Baidu.

For the Eastern Star sinking disaster event, the records are divided into two
parts. The first part is from 06/01/2015 to 06/06/2015 when sinking and salvage
take place. The second part is from 06/07/2015 to 06/09/2015 when the whole
nation is in mourning. The similarity of Weibo and Baidu on first part is 0.766,
that on the second part is 0.625.

From the above experimental results, we conclude the following: (1) When
using our similarity model for event dissemination platforms, we should first analyze sub-events during the whole event time period; (2) Weibo platform seems to
play a leading role in the event dissemination process; (3) Our model can effectively measure similarity and capture interaction of event dissemination cross
different platforms.

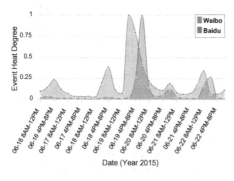
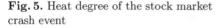
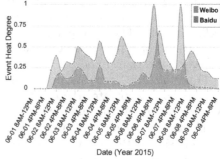

Fig. 5. Heat degree of the stock market crash event

Fig. 6. Heat degree of the Eastern Star sinking disaster event

4.4   Event Attention Degree

In order to make a comparison between the public attention degree and discover more insights about the development of the same event on different platforms, this study applies our event attention degree model to the real-world datasets. Through analyzing records contents from Weibo and Baidu, we observe that Weibo's content is longer and more diverse. Weibo is a platform primarily for people to express and share their feelings and emotions while Baidu is a platform with pre-defined search terms/keywords. Therefore, there are a lot of similar records in Baidu dataset.

Figure 7 visualizes public focus on different hot words every day. We regard every hot word as one dimension to depict this event. The day event attention degree is calculated by Eq. (9) through the aggregation of all the hot words.

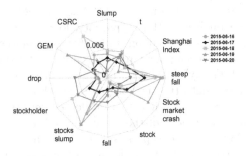

Fig. 7. Public attention degree in the stock market crash event

In order to deeply analyze the fluctuation of public attention, each daily attention degree is subtracted by the average of the daily values. Thus, we have a line indicating the trend in event attention degree over a time period varying up and down around 0 (Figs. 8 and 9):

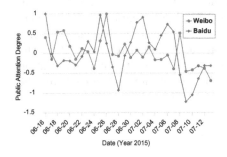

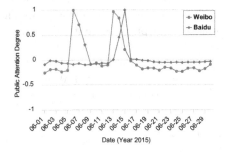

Fig. 8. Public Attention Degree of Weibo and Baidu on stock market crash

Fig. 9. Public Attention Degree of Weibo and Baidu on Eastern Star sinking disaster

Table 3. Public Attention Degree of Weibo and Baidu on stock market crash event

Platform	Maximum	Minimum	Average	Range	Standard deviation
Weibo	0.9999	-0.4668	0.0259	1.4668	0.3532
Baidu	0.9992	-1.2240	0.0119	2.2231	0.5884

Figure 8 shows a distinctive difference in attention degree between Weibo and Baidu over the stock market crash event. To better understand it, we put detailed analysis in Table 3 and observe that the event attention degree varies more significantly for Baidu in both absolute value and Standard Deviation, reflecting a more dramatic change in the public's attention degree. On the contrary, we find that the public attention degree of Weibo corresponds well with the rise and crash of the stock market. When stocks market drops substantially, the popularity reaches two peaks on June 19th and June 26th. Furthermore, the event attention degree on June 19th is apparently lower than that on June 26th. From K line chart, we observe that on June 26th, the market goes through a dramatic down-up-down development pattern and over 2000 stocks drop 10 %. This historical "Black Friday" results in the highest attention degree on Weibo.

Figure 9 shows the event attention degree of the Eastern Star ship sinking disaster. We can see that the public attention degree derived from Baidu is not following the actual development of the event through time. The trend remain steady low most of the time, except a sudden rise on the 14th and 15th of June. This rise is clearly caused by the closure of the rescue work on June 13th. This trend in public's attention degree also confirms the dissemination delay of Baidu. On the contrary, the attention degree trend of Weibo tracks the actual event development accurately. We observe rapid increase in attention degree on Weibo when noticeable sub-events like "the seventh day memorial" and "the closure of rescue work" take place, which can raise intense public attention in a very short period of time and influence other platforms later on.

When putting the trend of Weibo and Baidu attention degrees side by side, we have some interesting findings: Public attention degree on both platforms decreases rapidly after the closure of the event; Weibo is clearly taking a lead in capturing public attention early on. The rise of public attention on Weibo may influence the public attention's increase on Baidu. Overall, we conclude that (1) the event with dramatic ups and downs attracts more attentions. (2) In the attention degree model, with the development of the event, its trend of attention degree fluctuates a lot. (3) Weibo seems to be the platform that captures the public attention in an earlier phase.

5   Conclusion

Platforms like social network and search engine have a tremendous impact on the Internet and contribute to a large fraction of traffic on the web. In this paper, we proposed a unified model called ESAP for cross-platform event trend analysis. Our research fills the current research gap by offering new perspectives for measuring and analyzing event popularity, trend similarity and public attention degree at both macroscopic and microscopic level over a time period. Four real-world datasets on two events from Weibo and Baidu are used to validate the effectiveness of ESAP. We observe that Baidu has a delay effect in event dissemination while Weibo is more timely and more influential. In addition, Weibo is more accurate and steady in capturing public attention. The future work will involve the study of more events.

Acknowledgments. This work has been supported in part by the China 973 project (2014CB340303), the Opening Project of Baidu (No.181515P005267), it is also supported by the Open Project Program of Shanghai Key Laboratory of Data Science (No.201609060001), the Opening Project of Key Lab of Information Network Security of Ministry of Public Security (The Third Research Institute of Ministry of Public Security) Grant number C15602, and National Natural Science Foundation of China (No. 61472252, 61133006), the Natural Science Foundation of Jiangsu Province, China (Grant No. BK20141420 and Grant No. BK20140857).

References

1. Adedoyin-Olowe, M., Gaber, M.M., Dancausa, C.M., Stahl, F., Gomes, J.B.: A rule dynamics approach to event detection in twitter with its application to sports and politics. Expert Syst. Appl. **55**, 351–360 (2016)
2. Angel, M., Benedito, P.: The analysis of behavior of witzerland company by methodology of radar chart. Eur. J. Bus. Soc. Sci. **3**(6), 136–155 (2014)
3. Bao, B.K., Xu, C., Min, W., Hossain, M.S.: Cross-platform emerging topic detection and elaboration from multimedia streams. ACM Trans. Multimedia Comput. Commun. Appl. (TOMM) **11**(4), 54 (2015)
4. Benedito, P., Angel, M.: New trend to evaluate the management of companies: an application of the methodologies of radar chart. Adv. Econ. Bus. **2**(5), 191–199 (2014)
5. Christodoulou, G., Georgiou, C., Pallis, G.: The role of twitter in YouTube videos diffusion. In: Wang, X.S., Cruz, I., Delis, A., Huang, G. (eds.) WISE 2012. LNCS, vol. 7651, pp. 426–439. Springer, Heidelberg (2012). doi:10.1007/978-3-642-35063-4_31
6. Giummolè, F., Orlando, S., Tolomei, G.: A study on microblog and search engine user behaviors: how twitter trending topics help predict google hot queries. HUMAN **2**(3), 195 (2013)
7. Guan, W., Gao, H., Yang, M., Li, Y., Ma, H., Qian, W., Cao, Z., Yang, X.: Analyzing user behavior of the micro-blogging website sina weibo during hot social events. Phys. A Stat. Mech. Appl. **395**, 340–351 (2014)
8. Kwak, H., Lee, C., Park, H., Moon, S.: What is twitter, a social network or a news media? In: Proceedings of the ACM International Conference on World Wide Web (WWW), pp. 591–600 (2010)
9. Lerman, K., Hogg, T.: Using a model of social dynamics to predict popularity of news. In: Proceedings of the ACM International Conference on World Wide Web (WWW), pp. 621–630 (2010)

10. Vaughan, L., Romero-Frías, E.: Web search volume as a predictor of academic fame: an exploration of google trends. J. Assoc. Inf. Sci. Technol. **65**(4), 707–720 (2014)
11. Wu, K., Yang, S., Zhu, K.Q.: False rumors detection on sina weibo by propagation structures. In: Proceedings of the IEEE International Conference on Data Engineering (ICDE), pp. 651–662 (2015)
12. Yu, K.: Large-scale deep learning at baidu. In: Proceedings of the IEEE International Conference on Information and Knowledge Management (CIKM), pp. 2211–2212 (2013)
13. Zhang, H., Zhao, Q., Liu, H., xiao, K., He, J., Du, X., Chen, H.: Predicting retweet behavior in weibo social network. In: Wang, X.S., Cruz, I., Delis, A., Huang, G. (eds.) WISE 2012. LNCS, vol. 7651, pp. 737–743. Springer, Heidelberg (2012). doi:10.1007/978-3-642-35063-4_60
14. Zhang, X., Chen, X., Chen, Y., Wang, S., Li, Z., Xia, J.: Event detection and popularity prediction in microblogging. Neurocomputing **149**, 1469–1480 (2015)
15. Zhou, X., Liang, X., Zhang, H., Ma, Y.: Cross-platform identification of anonymous identical users in multiple social media networks. IEEE Trans. Knowl. Data Eng. (TKDE) **28**(2), 411–424 (2016)

Learning Event Profile for Improving First Story Detection in Twitter Stream

Yongqin Qiu[1,2], Rui Li[1,2($\boxtimes$)], Lihong Wang[1,2], and Bin Wang[1,2]

[1] Institute of Information Engineering, Chinese Academy of Sciences, Beijing, China
{qiuyongqin,lirui,wanglihong,wangbin}@iie.ac.cn
[2] University of Chinese Academy of Sciences, Beijing, China

Abstract. First Story Detection (FSD) in twitter stream is to identify the first report that discusses an event that has not been reported in the posted tweets. FSD offers great assistance for New Event Detection (NED). Traditional methods used online clustering framework as mainstream solutions, but suffering low efficiency and unsatisfied performance and did not consider the event related features. We merge event related features and propose event-profile based FSD method based on online cluster framework. It outperforms traditional methods both in efficiency and effect by replacing tweet-by-tweet comparison with profile-by-profile comparison. In this paper, we take four groups of features into account and propose a learning method for the generation of event profile. Experiments show that the profile produced by our method is more relevant with event, also more robust than the ones produced by rule-based methods, eventually, improves the FSD performance.

Keywords: Twitter · Event profile · First Story Detection

1 Introduction

In recent years, social networks have been developing with amazing pace. Until December 26, 2015, active users in Twitter platform has exceeded 300 million[1], more than 400 million tweets are posted daily [2]. Twitter has become an important information platform for people to broad fresh news [3], share distinctive views and comments. First story detection (FSD) means to identify the first report that discuss an event that has not already been reported in earlier stories.

The mainstream approachs for FSD are online clustering methods, among which the most representative one is SinglePass cluster algorithm. Under the online setting, a new arriving tweet is required to be compared with all the valid tweets that have been identified previously, so as to determine whether the current one is the first story of a certain event. Excessive number of comparisons results in low efficiency, but still unsatisfied performance. Aiming at this problem, local sensitive hash (LSH) [4] and Profile-based method [8] are proposed for the

---

[1] http://www.ebizmba.com/articles/social-networking-websites.

© Springer International Publishing AG 2016
W. Cellary et al. (Eds.): WISE 2016, Part I, LNCS 10041, pp. 505–512, 2016.
DOI: 10.1007/978-3-319-48740-3_37

efficiency. Profile-based method reduces the number of comparison sharply by generating an event profile and replacing tweet-by-tweet comparison with profile-by-profile comparison. It is outstanding for the consideration of event-object and global information.

However, in Qiu's work, the profile was generated by pre-setting rules, and failed to preserve sufficient event-related information. For the improvement of profile, we motivate a learning method incorporating comprehensive features to make a more accurate decision than rule-based method on whether a word is qualified for event profile. Experiments show that the profiles produced by our learning method is more relevant with event, also more robust than the ones produced by rule-based methods.

2   Related Work

To the best of our knowledge, online clustering is the main branch of the solution for FSD.

SinglePass presents a representative incremental clustering framework, which is generally adopted in many scenarios. Sankaranarayanan et al. [9] proposed a system, called TwitterStand, which employed a online clustering algorithm based on weighted term vector according to tf-idf and cosine similarity to form clusters of news. Phuvipadawat [15] presented a method to collect, group, rank, and track breaking news from Twitter. Becker et al. [10] used the classical incremental clustering algorithm, which has been proposed for FSD in news documents [12], to identify of real-world event content and its associated Twitter messages, which continuously clusters similar tweets and then classifies the clusters content into real-world events or nonevents.

A major weakness of SinglePass is the low efficiency caused by massive similarity calculations. Several substantial work was done to improve SinglePass. Petrovic [4] proposed a method based on an adapted variant of the locality sensitive hashing methods [6], which can ensure the effectiveness but cost constant time and space. Then Petrovic [7] integrate paraphrases into locality sensitive hashing algorithm to further improve the effect of first detection in Twitter. The results showed that their method can be very small C_{min} value(0.679), and it is also one of the baseline we take.

Presently, Qiu [8] proposed a Nugget-based method to make further efforts to improve efficiency and effect for FSD task in Twitter Stream, which used of three elements ("@", hashtag and contribute data) containing tweet event information as Nugget of event. Nugget-based FSD method can reduce the C_{min} value to 0.410, and superior to P-LSH method in efficiency, which is recognized as the state-of-the-art method. However, the nugget or profile was generated by pre-setting rules, which is not enough to express the event appropriately. Thus, we propose a learning method to generate event profile.

3   Event Profile-Based First Story Detection

Event Profile-based First Story Detection algorithm (EP-FSD) is proposed to improve the efficiency of SinglePass, the specifics are as follows:

Before detecting a tweet, the system uses machine learning method to train a model to generate tweet profiles. When a new tweet arrives, the algorithm generates a profile for it and compares the tweet profile and every event profile, which is composed of the profiles of tweets in this event. If the tweet profile hits an event profile, it is attributed to the event, else a calculation of similarity between the tweet and all the tweets in every event will be conducted. If the similarity value is below a threshold, which means the tweet is not similar to any currently detected event, and the tweet is also called the first story in this event. Its framework is shown as Fig. 1.

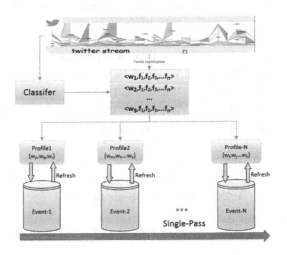

Fig. 1. The framework of Event Profile-based First Story Detection

Next we will describe how to use machine learning algorithm to train a model to generate profiles.

3.1   Learning Classifier for Event Profile

In this section, we build a classifier for each word to check whether it is qualified for the event profile. Formally, given a word w, extracting the corresponding feature vector (detailed in Sect. 3.3) f_w, a decision function G is required to be trained to determine the label y of a certain word according to f_w.

$$y = G(f_w) \tag{1}$$

$f_w = (f_1, f_2, ..., f_n)$ is the feature of word in the tweet. $y = 1$ represents that word w meets the requirements and is deserved to add into profile, while $y = 0$ means that word w is not the desired one.

3.2 Feature Designer

In our study, we have totally considered 17 distinctive features, which can be
further arranged into 4 groups, as detailed in Table 1. These features describe the
word's importance comprehensively, from the self characteristics to the word's
context in the external resources.

Table 1. The list of features

Feature type	No	Feature	Description	Example
Term Feature TF	1	isTitle	Word's first letter capitalized	Nobel Prize
	2	isUp	Words letters are all upper-case	MILLZ
	3	NotAlpha	If it contains characters other than letter	F-16, debate2012
	4	LenMore3	Word length is greater than 3	play , happy
	5	NotWord	The word is in the dictionary	cooooold
	6	Noun	Whether the word is a noun	Quantum
Meta Data (MD)	7	Hashtag	Whether in Hashtag	#HiphopAwards
	8	At	Whether in @	@kendricklamar
Global Feature GF	9	TF	Term Frequency	Term Frequency
	10	IEF	Inverse event frequency	$IEF = \log(N/EF)$
	11	TF*IEF	Product of TF and IEF	$TF*\log(N/IEF)$
	12	In2Events	Word appears in several events	-
Event-Oriented Feature (EOF)	13	Name	The word is a persons name	Mo Yan
	14	Location	The word is a location name	MOSCOW
	15	Time	The word is a time	2012
	16	Action	The word is an action	debate
	17	Concept	The word is Wikipedia entity	HiphopAwards

3.3 Profile Generation

After a classifier is trained, for each newly arrived tweet, every word will be feed
to the classifier to determine whether it can be selected as a profile, thereby
generating a tweet's profile. Event profile consists of these tweets event profile.
Formally: for a given tweet, $t_i = \{w_1, w_2, ..., w_N\}$, its event profile is P_t:

$$P_t = \{w | G(f_w) = 1\} \tag{2}$$

That is says, all event profile words in this tweet make up tweet profile. And event profile P_E consists of all tweets' profile in the event. The definition of P_E is:

$$P_E = \bigcup_t P_t \qquad\qquad (3)$$

4   Experiments

4.1   Dataset and Evaluation

Datasets: We carry out our experiment on two datasets.

1. **Edinburgh Twitter Corpus** [14] (ETC, for short): It is the same corpus as [7], including 2363 tweets and 27 events labeled. This is the corpus Petrovic used to invalidate the efficiency and effect of LSH method. when a new tweet arrives it need to compare to all the history data to decide which event should be attributed to.
2. **Event2012 Corpus** [2]: This corpus includes 81087 tweets and 505 events, which is collected from Twitter API (collected tweets are all in English). The time of corpus lasted from October 10, 2012 to November 7, 2012, a total of 28 days.

Evaluation: Unless stated otherwise, we use the official evaluation metrics and parameters from TDT[2]: *miss probability*, P_{miss} and *false alarm*, P_{FA}. C_{det} is computed as [4], and C_{miss} and C_{FA} are costs of miss and false alarm, P_{target} and $P_{non-target}$ are the prior target and non-target probabilities.

4.2   Experiments Setting

In this section, we will introduce the processing of the experiment.

In our experiment, we organize a number of professionals to label these 3023 words as Profile or not. Finally, we get a 1339 positive cases and 1684 negative cases. We use these 3023 words as a training set to train a classification model of the word. When a new tweet arrives, the classifier will classify every word in tweet to profile or not.

Baseline: We chose four methods as our baseline: Traditional SinglePass, Center vectors method, P-LSH and Nugget-based approach

4.3   Results and Analysis

Experiment 1 Comparison of Efficiency: In this group of experiments, we also compare the comparison number of the 5 methods.

As can be seen from the Fig. 2: (1) Event profile method (EP-FSD) we propose and Nugget-based method is almost the same number of comparisons,

---

[2] ftp://jaguar.ncsl.nist.gov/current_docs/TDT3eval/TDT3fsd.pl.

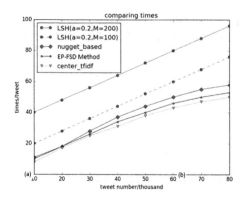

Fig. 2. The comparison number of different methods

because both methods cluster the tweet through comparing the elements like "summary"of events. (2) The comparisons number of our EP-FSD method is significantly better than P-LSH, this is because with the amount of data increases, LSH methods conflict number will also increase, thus the number of comparisons will increase.

Experiment 2 Effect of Different Methods: The second set of experiments, we focus on the effect of five methods:

From Tables 2 and 3 common trends can be seen: (1) EP-FSD is better than P-LSH method. This is because the P-LSH method only used the tweet content. However, EP-FSD method generated profile, contains not only the characteristics of tweet content, but also contains a lot and Event-related features. (2) EP-FSD method is superior to Nugget method. This is because the Nugget method is generated by the rule, will definitely lead to a lot of missed detection rate and the error detection rate, especially when large volumes of data.

<table>
<tr><td colspan="2">Table 2. Effect in ETC</td><td colspan="2">Table 3. Effect in Event2012</td></tr>
<tr><td>System</td><td>C_{min}</td><td>System</td><td>C_{min}</td></tr>
<tr><td>P-LSH-baseline</td><td>0.679</td><td>P-LSH-baseline</td><td>-</td></tr>
<tr><td>Nugget-FSD</td><td>0.410</td><td>Nugget-FSD</td><td>0.595</td></tr>
<tr><td>SinglePass</td><td>0.585</td><td>SinglePass</td><td>0.650</td></tr>
<tr><td>Center-Vector</td><td>0.786</td><td>Center-Vector</td><td>0.866</td></tr>
<tr><td>EP-FSD</td><td>0.355</td><td>EP-FSD</td><td>0.486</td></tr>
</table>

Experiment 3 Effect of Different Features: The third set of experiments we analyzed in our EP-FSD process, how the different signals influence the event

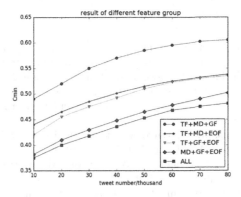

Fig. 3. The effect of different features

detection effect. We cancel a type of features, retaining all the other types to see the experimental results.

From the Fig. 3 we can draw the following conclusions: (1) The event-oriented features are the most important feature: because if you remove the event-oriented features, it has the greatest impact on the results. C_{min} values from the original 0.481 rise to 0.605. The result also reflects the importance of the event-oriented features of the FSD this task. (2) Global features are the second important. This category features from a global perspective to measure the importance of the word as a profile, which considers not only the word itself features such as TF, also takes into account the IEF, which can distinguish different events in globally. The word has higher value of TF * IEF is more representative of an event.

5   Conclusion

This paper proposes a new profile-based event detection method for FSD task. The new method we proposed generate profiles for tweets and events, and then determine which event a tweet attributes to by comparing the similarities of their profiles. This approach greatly reduces the number of comparisons for FSD task, thus greatly improved efficiency. Meanwhile, the effect has been greatly improved, because a lot of event-related features are integrated in the profile generation process. In summary, Profile-based method out performs the current state-of-art methods both on the effectiveness and efficiency. The next job is mainly to optimize profile generation and update, and to extract more effective features both in quality and quantity under the existing framework.

Acknowledgments. We would like to thank the anonymous reviewers for their valuable comments and suggestions. This work is supported by the National Natural Science Foundation of China (grant No. 61572494), the Strategic Priority Research Program of the Chinese Academy of Sciences (grant No. XDA06030200), and the National Key Technology R and D Program (grant No. 2012BAH46B03).

References

1. Allan, J.: Introduction to topic detection and tracking. In: Topic Detection and Tracking. The Information Retrieval Series, vol. 12, pp. 1–16 (2002)
2. McMinn, A.J., Moshfeghi, Y., Jose, J.M.: Building a large-scale corpus for evaluating event detection on twitter. In: CIKM, pp. 409–418 (2013)
3. Petrovic, S., Osborne, M., McCreadie, R., et al.: Can Twitter replace Newswire for breaking news? (2013)
4. Petrovic, S., Osborne, M., Lavrenko, V.: Streaming first story detection with application to twitter. In: NAACL, pp. 181–189 (2010)
5. Yang, Y., Pierce, T., Carbonell, J.: A study of retrospective and on-line event detection. In: SIGIR, pp. 28–36 (1998)
6. Gionis, A., Indyk, P., Motwani, R.: Similarity search in high dimensions via hashing. VLDB **99**(6), 518–529 (1999)
7. Petrovic, S., Osborne, M., Lavrenko, V.: Using paraphrases for improving first story detection in news and Twitter. In: NAACL, pp. 338–346 (2012)
8. Qiu, Y., Li, S., Li, R.: Nugget-based first story detection in twitter stream. In: SMP, pp. 74–82 (2015)
9. Sankaranarayanan, J., Samet, H., Teitler, B.E.: Twitterstand: news in tweets. In: SIGSPATIAL 2009, pp. 42–51 (2009)
10. Becker, H., Naaman, M., Gravano, L.: Beyond trending topics: real-world event identification on twitter (2011)
11. Allan, J., Lavrenko, V., Jin, H.: First story detection in TDT is hard. In: CIKM, pp. 374–381 (2000)
12. Allan, J., Papka, R., Lavrenko, V.: On-line new event detection and tracking. In: SIGIR, pp. 37–45 (1998)
13. Allan, J.: Topic Detection and Tracking: Eventbased Information Organization, pp. 111–122. Kluwer Academic Publishers (2002)
14. Petrovic, S., Osborne, M., Lavrenko, V.: The edinburgh twitter corpus. In: NAACL, pp. 25–26 (2010)
15. Phuvipadawat, S., Murata, T.: Breaking news detection and tracking in twitter. In: WI-IAT, pp. 120–123 (2010)

A Novel Approach of Discovering Local Community Using Node Vector Model

Jinglian Liu[1,2], Daling Wang[1,3(✉)], Shi Feng[1,3], Yifei Zhang[1,3], and Weiji Zhao[2]

[1] School of Computer Science and Engineering, Northeastern University,
Shenyang, People's Republic of China
datamining@163.com,
{wangdaling,fengshi,zhangyifei}@cse.neu.edu.cn
[2] School of Information Engineering, Suihua University, Suihua, People's Republic of China
sdzhaoweiji@163.com
[3] Key Laboratory of Medical Image Computing of Ministry of Education,
Northeastern University, Shenyang, People's Republic of China

Abstract. Local community detection aims at discovering a community from a seed node without global information about the entire network structure, and various local community detection algorithms have been proposed. However, most existing algorithms either are parameter-dependent or have low accuracy. In this paper, we propose a novel approach of discovering local community using node vector model. In detail, we propose node vector model to represent nodes in graphs. Moreover, we define weighted Jaccard similarity coefficient to estimate the similarities between nodes. Based on the model and definition, local community can be detected. Our algorithm gives priority to the node which is most similar to the nodes in the current local community. We compare the proposed algorithm on both synthetic and real-world networks. The experimental results demonstrate that our algorithm is highly effective at local community detection compared to related algorithms.

Keywords: Node vector model · Local community detection · Network graph · Community structure

1 Introduction

A wide variety of complex systems can be modeled as networks, such as social networks [6, 8, 19], collaboration networks [13], the Internet [4], and E-mail networks [20]. A common feature of these networks is community structure that partitions network nodes into groups within which the edges are dense but between which they are sparse [5, 6, 16, 17]. Community detection has a lot of applications in analyzing social networks, collaborative tagging systems and biological networks [22].

Traditional community detection methods discover all communities in a network requiring the global network structure [3, 6, 14, 15, 18, 20]. For huge networks, such

© Springer International Publishing AG 2016
W. Cellary et al. (Eds.): WISE 2016, Part I, LNCS 10041, pp. 513–521, 2016.
DOI: 10.1007/978-3-319-48740-3_38

as Web graph and social graph, it is difficult for us to get their entire network structure nowadays [7]. For solving this problem, local community detection was proposed and has attracted a lot of attention.

Local community detection algorithms discover a community from a start node requiring only the local network structure. Various local community detection algorithms have been proposed [1, 2, 7, 11, 12, 22]. Some local community detection algorithms discover local community by using the nodes' degree, such as l-shell expansion algorithm [1], greedy optimization local modularity metric R [2], greedy optimization sub-graph modularity M [11]. Others discover local community by using similarities between pairs of nodes, such as LTE algorithm [7], GMAC algorithm [12]. However, most existing algorithms either require predefined parameters which are hard to set manually or have low accuracy.

Inspired by LTE and GMAC algorithms, we firstly propose node vector model to represent nodes in graphs and define weighted Jaccard similarity coefficient to estimate the similarities between nodes. Then we propose a novel similarity-based local community detection algorithm. We test our algorithm on both synthetic and real-world networks. The experimental results show that our algorithm is highly effective at local community detection compared to other related algorithms.

The rest of the paper is organized as follows. Section 2 gives problem definition of local community and some related definitions. We describe our approach in Sect. 3 and report experimental results in Sect. 4, followed by conclusions in Sect. 5.

2   Preliminaries

In this section, we define the problem of local community detection in network, and then introduce some necessary definitions.

2.1   Definition of Local Community Detection in Network

Firstly, we give the following definitions, and then describe our problem of local community detection.

Definition 1 (Function *neighbors*). Given an undirected graph $G = (V, E)$, where V is the set of nodes and E is the set of edges. Moreover, $n = |V|$ is the number of nodes, $(u, v) \in E$ indicates an edge between nodes u and v, and $m = |E|$ is the number of edges. The node set adjacent to v is denoted by $\Gamma(v)$, $\Gamma(v) = \{u | u \in V, (u, v) \in E\}$. Function *neighbors*$(G, v)$ returns $\Gamma(v)$. The degree of node v is the number of nodes in $\Gamma(v)$, denoted by k_v.

The problem of local community detection can be described as: For an undirected network $G = (V, E)$, local community detection starts from a seed node s, the work is to discover the local community D that s belongs to.

Note that in the process of detecting local community, the entire network structure of G is unknown. As shown in Fig. 1, D is the local community, the shell node set of D

is denoted by N, $N = \{u|(v,u) \in E, v \in D, u \in V-D\}$. The known partial network is the union of D and N, and the unknown node set in G is denoted by U. Obviously, $U = V-D-N$. Similar definitions of D, N can be found in [2, 10, 21].

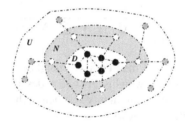

Fig. 1. An Illustration of Division of a Network into Local Community D (block nodes), D's Shell Node Set N (white nodes) and Unknown Node Set U (grey nodes)

When a local community detection algorithm starts, $D = \{s\}$, $N = neighbors(G,s)$. For $v \in N$, let $R = neighbors(G,v)$, to merge node v into D, and update N by adding nodes in $R \cap U$ is the only way to expand the known partial network. The task of local community detection is to discover a local community D from a seed node s by continuously repeating this step until an appropriate stopping criteria is satisfied [21].

2.2 Measurements Based on Node Vector Model

How to represent a node in networks is a key problem in similarity-based local community detection algorithms. Huang et al. [7] represent a node by its adjacent nodes. Ma et al. [12] represent a node by its d-level neighbors, a set of nodes whose shortest path to it is less than or equal to d. Both of them take adjacent nodes equally, neglect the information of their different weights. To overcome this limitation, we propose node vector model to represent a node in networks, and define weighted Jaccard similarity coefficient to estimate similarities between nodes.

Definition 2 (Node Vector Model). Given an undirected graph $G = (V, E)$, for any node V_i in G, we represent V_i by a n-dimensions row vector $NV(V_i)$, $NV(V_i) = (w_1, w_2, ..., w_n)$, where w_j is calculated as follows.

$$w_j = \begin{cases} \dfrac{|\Gamma(V_i) \cap \Gamma(V_j)|}{|\Gamma(V_i) \cup \Gamma(V_j)|} & (V_i, V_j) \in E \\ 0 & (V_i, V_j) \notin E \end{cases} \tag{1}$$

If V_j is adjacent to V_i, then w_j is the Jaccard Similarity Coefficient between them, otherwise, $w_j = 0$.

Based on the assumption that more common neighbor nodes with higher weights two nodes have, more similar they are, we define weighted Jaccard similarity coefficient to estimate the similarities between nodes.

Definition 3 (Weighted Jaccard Similarity Coefficient). Given a graph $G = (V, E)$, the weighted Jaccard similarity coefficient between nodes V_i and V_j is defined as:

$$\varphi(V_i, V_j) = \frac{\sum\limits_{x \in \Gamma(V_i) \cap \Gamma(V_j)} (NV(V_i)[x] + NV(V_j)[x])}{\sum\limits_{y \in \Gamma(V_i)} NV(V_i)[y] + \sum\limits_{z \in \Gamma(V_j)} NV(V_j)[z]} \tag{2}$$

where the numerator is the sum of their common neighbors' weights, and the denominator is the sum of their neighbors' weights. For two nodes, if they are directly connected, the probability of being the same community is much higher. So we calculate the final similarity between two nodes as follows.

$$similarity(V_i, V_j) = \varphi(V_i, V_j) + \alpha \tag{3}$$

If V_j is adjacent to V_i, then $\alpha = \dfrac{k_{v_i} \times k_{v_j}}{2m}$, otherwise $\alpha = 0$.

We adopt Compactness-Isolation (CI for short) [12] to measure the quality of local community.

Definition 4 (Compactness-Isolation Metric). For a local community D with shell node set N, the Compactness-Isolation Metric of D, denoted by $CI(D)$, is defined as

$$CI(D) = \frac{\sum\limits_{V_i, V_j \in D, (V_i, V_j) \in E} similarity(V_i, V_j)}{1 + \sum\limits_{V_i \in D, V_j \in N, (V_i, V_j) \in E} similarity(V_i, V_j)} \tag{4}$$

We use CI to measure a local community's modularity. For a node in N, if agglomerating it into D will cause an increase in CI, then add it to D.

3   Discovering Local Community Using Node Vector Model

In this section, we introduce our local community detection algorithm in detail. Firstly, initialize $D = \{s\}$ and $N = neighbors(G, s)$. At each step, our algorithm keeps choosing node with maximum sum of similarities with nodes in D as candidate node. If agglomerating candidate node into D will cause an increase in CI, add it to D and update N, otherwise, remove it from N, repeat this step until N is empty. Finally, return D as the local community of v. The pseudo code is described in Algorithm 1 as follows.

Algorithm 1: Local Community Detection

Input: a given seed node *s*, network $G=(V,E)$;
Output: local community *D*;
Describe:
  1) initialize $D=\{s\}$, $N=neighbors(G,s)$;
  2) while *N* empty do
  3)   create a new dictionary variable *sim* to store the similarities of nodes
      belonging to *N* with *D*;
  4)   for each *n* *N* do
  5) $sim[n]=\sum_{x\in D}similarity(n,x)$; //see Formula (3);
  6)   end for
  7)   find *a* such that *sim*[*a*] is maximum;
  8)   if $CI(D)>0$ then
  9)   add *a* to *D* and update *N*;
10)   else
11)   remove *a* from *N*;
12)   end if
13) end while
14) return *D*;

---

4   Experiments

In this section, we evaluate the effectiveness of our proposed algorithm on synthetic as well as real-world networks with ground-truth community structure.

4.1   Related Methods and Evaluation Criteria

We compare our algorithm with three representative local community detection algorithms. The first is Clauset's algorithm [2], note that the same as [12, 21], we improve its stopping criteria by detecting changes in local modularity *R*. The second is Luo et al.'s algorithm [11] (LWP for short). The third is GMAC algorithm [12], for parameter *d*, we set $d = 3$ as suggested by authors.

We measure the performance of these algorithms by three evaluation measures: *precision*, *recall*, and *F-score*, which are also adopted in [10, 12]. *Precision* is the fraction of discovered nodes that are relevant, while *recall* is the fraction of relevant nodes that are discovered. *F-score* is the harmonic mean of *precision* and *recall*. In our experiments, we use each node in these networks as a seed node and report algorithms' average *precision*, *recall*, and *F-Score* on each of these networks.

4.2   Evaluation on Synthetic Networks

LFR benchmark networks, introduced by Lancichinetti et al. [9], are widely adopted by many community detection algorithms [7, 12]. The specified parameters of this network generating model are presented as follows: the number of nodes n, the average degree of the nodes k, the maximum degree k_{max}, and mixing parameter μ. The parameters are set as follows: $n = 5000$, $k = 10$, $k_{max} = 50$, others except μ use default values. The higher the mixing parameter μ of a network, the weaker community structure it has. So we generate 10 networks with different mixing parameter μ ranging from 0.05 to 0.5 with a span of 0.05.

The comparison results of *precision*, *recall*, and *F-score* for four algorithms on these networks are shown in Fig. 2. With μ increasing, all the four algorithms suffer varying degree of performance degradation. Because the increasing of μ strongly affects the performances of degree-based methods, the performances of degree-based algorithms, such as LWP and Clauset, drop rapidly. Compared with degree-based algorithms, the increasing of μ less affects similarity-based algorithms. With μ increasing, the performances of similarity-based algorithms, such as GMAC and our algorithm, drop slowly.

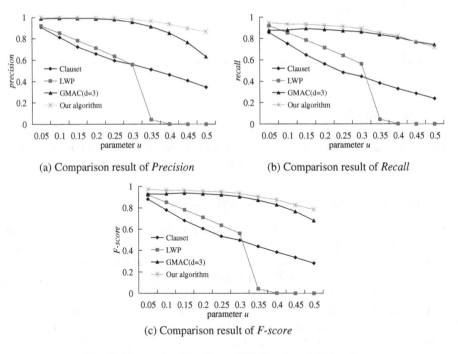

(a) Comparison result of *Precision*   (b) Comparison result of *Recall*

(c) Comparison result of *F-score*

Fig. 2.   Comparison Results on LFR Benchmark Networks

To be more precisely, Clauset's algorithm has lower *precision*, *recall*, *F-score* than other three algorithms when $\mu \leq 0.3$. The *precision*, *recall*, and *F-score* of the LWP algorithm is zero or nearly zero when $\mu \geq 0.35$. The same conclusion was drawn in [7]. GMAC algorithm has better performance compared with LWP and Clauset algorithms.

Because node vector model is a more efficient way to represent node compared with *d*-NS which is used by GMAC algorithm, our algorithm has a good *precision* and *recall*, and gets the highest *F-score* on each of these networks compared with other algorithms.

In general, our algorithm achieves better performance to discovery local community against the other algorithms on LFR benchmark networks.

4.3   Evaluation on Real Networks

In this subsection, we use two real-world networks to evaluate the performance of our algorithm. The first one is Zachary Karate Club Network (Karate for short) [23], in which $n = 34$ and $m = 78$. The second one is NCAA football network (Football for short) [6], in which $n = 115$ and $m = 613$.

Figure 3 shows comparison results of the four algorithms on two real-world networks. The *precision* and *F-score* of our algorithm are usually the highest of four algorithms on the two networks, the *recall* is a little less than Clauset which has highest *recall* on Football dataset. The experimental results show that our algorithm is highly effective at discovering local community structure compared with the other algorithms on real-world networks.

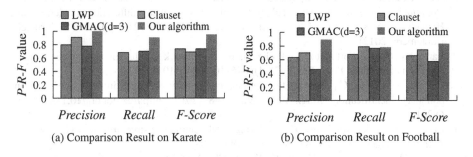

(a) Comparison Result on Karate   (b) Comparison Result on Football

Fig. 3. Comparison Results on Real-world Datasets

5   Conclusion and Future Work

In this paper, we firstly propose node vector model to represent nodes in graphs. In this model, different adjacent nodes have different weights. Next, a weighted Jaccard similarity coefficient is given to estimate the similarity between nodes. Then we propose our similarity-based local community detection algorithm. Compared with other related algorithms, our algorithm doesn't need any manual parameters, and achieves better performance on both synthetic and real-world networks.

In future, we will further improve our algorithm for more effective and efficient results, and apply our algorithm on real networks for discovering local community. We will also explore the community detection on heterogeneous network.

Acknowledgments. The project is supported by National Natural Science Foundation of China (61370074, 61402091), the Fundamental Research Funds for the Central Universities of China under Grant N140404012.

References

1. Bagrow, J., Bolt, E.: A local method for detecting communities. Phys. Rev. E **72**(4), 046108-1–046108-10 (2005)
2. Clauset, A.: Finding local community structure in networks. Phys. Rev. E **72**(2), 026132 (2005)
3. Clauset, A., Newman, M.E., Moore, C.: Finding community structure in very large networks. Phys. Rev. E: Stat., Nonlin, Soft Matter Phys. **70**(6), 264–277 (2004)
4. Faloutsos, M., Faloutsos, P., Faloutsos, C.: On Power-law relationships of the internet topology. In: SIGCOMM 1999, pp. 251–262 (1999)
5. Fortunato, S.: Community detection in graphs. Phys. Rep. **486**(3/5), 75–174 (2010)
6. Girvan, M., Newman, M.: Community structure in social and biological networks. Proc. Natl. Acad. Sci. U.S.A. **99**(12), 7821–7826 (2002)
7. Huang, J., Sun, H., Liu, Y., Song, Q., Weninger, T.: Towards online multiresolution community detection in large-scale networks. PLoS ONE **6**(8), 492 (2011)
8. Jia, G., Cai, Z., Musolesi, M., Wang, Y., Tennant, D.A., Weber, R.J., Heath, J.K., He, S.: Community detection in social and biological networks using differential evolution. In: Hamadi, Y., Schoenauer, M. (eds.) LION 2012. LNCS, vol. 7219, pp. 71–85. Springer, Heidelberg (2012)
9. Lancichinetti, A., Fortunato, S., Radicchi, F.: Benchmark graphs for testing community detection algorithms. Phys. Rev. E **78**(4), 046110-1–046110-5 (2008)
10. Liu, Y., Ji, X., Liu, C., et al.: Detecting local community structures in networks based on boundary identification. Math. Prob. Eng., 1–8 (2014). http://dx.doi.org/10.1155/2014/682015
11. Luo, F., Wang, J., Promislow, E.: Exploring local community structures in large networks. Web Intell. Agent Syst. (WIAS) **6**(4), 387–400 (2008)
12. Ma, L., Huang, H., He, Q., Chiew, K., Wu, J., Che, Y.: GMAC: a seed-insensitive approach to local community detection. In: Bellatreche, L., Mohania, M.K. (eds.) DaWaK 2013. LNCS, vol. 8057, pp. 297–308. Springer, Heidelberg (2013). doi:10.1007/978-3-642-40131-2_26
13. Newman, M.: The structure of scientific collaboration networks. Working Paper. **98**(2), 404 (2000)
14. Newman, M.: Fast algorithm for detecting community structure in networks. Phys. Rev. E Stat. Nonlinear Soft Matter Phys. **69**(6), 066133-1–066133-5 (2004)
15. Newman, M., Girvan, M.: Finding and evaluating community structure in networks. Phys. Rev. E Stat. Nonlinear Soft Matter Phys. **69**(2), 026113-1–026113-15 (2004)
16. Radicchi, F., Castellano, C., Cecconi, F., et al.: Defining and identifying communities in networks. Proc. Natl. Acad. Sci. U.S.A. **101**(9), 2658–2663 (2004)
17. Schaeffer, S.: Graph clustering. Comput. Sci. Rev. (CSR) **1**(1), 27–64 (2007)
18. Shao, J., Han, Z., Yang, Q., Zhou, T.: Community Detection based on distance dynamics. In: Proceedings of the 21th ACM SIGKDD International Conference on Knowledge Discovery and Data Mining, pp. 1075–1084 (2015)
19. Takaffoli, M.: Community evolution in dynamic social networks - challenges and problems. In: ICDM Workshops 2011, pp. 1211–1214 (2011)

20. Tyler, J.R., Wilkinson, D.M., Huberman, B.A.: Email as spectroscopy: automated discovery of community structure within organizations. Inf. Soc. **21**(2), 143–153 (2005)
21. Wu, Y., Huang, H., Hao, Z., Chen, F.: Local community detection using link similarity. J. Comput. Sci. Technol. (JCST) **27**(6), 1261–1268 (2012)
22. Wu, Y., Jin, R., Li, J., Zhang, X.: Robust local community detection: on free rider effect and its elimination. In: VLDB, pp. 798–809 (2015)
23. Zachary, W.: An information flow model for conflict and fission in small groups. J. Anthropol. Res. **33**(4), 452–473 (1977)

Data Mining

Labeled Phrase Latent Dirichlet Allocation

Yi-Kun Tang, Xian-Ling Mao$^{(\boxtimes)}$, and Heyan Huang

Department of Computer Science and Technology,
Beijing Institute of Technology, Beijing 100081, China
{tangyk,maoxl,hhy63}@bit.edu.cn

Abstract. In recent years, topic modeling, such as Latent Dirichlet Allocation (LDA) and its variations, has been widely used to discover the abstract topics in text corpora. There are two state-of-the-art topic models: Labeled LDA (LLDA) and PhraseLDA. LLDA is a supervised generative model which considers the label information, but it does not take into consideration word order under the *bag-of-words* assumption. On the contrary, PhraseLDA regards each document as a mixture of phrases, which partly considers the word order. However, PhraseLDA cannot model the supervised label information. In this paper, in order to overcome the defects of two models above while combining their merits, we propose a novel topic model, called Labeled Phrase LDA, which synchronously considers the supervised information and word order. Lots of experiments were conducted among the proposed model and two state-of-the-art models, which show the proposed model significantly outperforms baselines in terms of case study, perplexity and scalability.

Keywords: Labeled Phrase LDA · Topic model · Multi-labeled corpus

1 Introduction

At the information era, topic modeling is becoming more and more popular in identifying latent semantic components in text corpora. Lots of topic models have been proposed. The existing topic models can be divided into four categories: *Unsupervised non-hierarchical topic models* [1,18,19], *Unsupervised hierarchical topic models* [20–22], and their corresponding supervised counterparts [2,4,17].

Labeled LDA [2] is a supervised generative topic model, which is designed to deal with multi-labeled corpus. It establishes a one-to-one correspondence between latent topics and document's tags. Compared to LDA [1], topics learned by Labeled LDA can be more easily interpreted, since users can use a concise semantic label to understand corresponding latent topic. Labeled LDA usually does not consider the word order under the *bag-of-words* assumption. However, a single word, especially some proper nouns, prepositions or verbs, sometimes fails to express the meaning completely or exactly. For example, the phrase *white sox* is related to baseball, but neither of the single word *white* or *sox* is related to baseball. Thus, Labeled LDA is not the best solution to capture the latent semantic structure of labeled documents.

© Springer International Publishing AG 2016
W. Cellary et al. (Eds.): WISE 2016, Part I, LNCS 10041, pp. 525–536, 2016.
DOI: 10.1007/978-3-319-48740-3_39

PhraseLDA [3], another state-of-the-art topic model, regards each document as a mixture of phrases, under the *bag-of-phrases* assumption. PhraseLDA partly considers the word order, and takes advantage of phrases in documents, which can convey more semantic information than single words. Unfortunately, PhraseLDA cannot deal with the supervised information, such as labels for documents. In fact, there are lots of labeled collections in the world. For example, webpages and their categories, papers and corresponding keywords, and tweets and their hashtags. Therefore, there is a growing need to use topic models to deal with multi-labeled corpus.

In this paper, we propose a novel topic model to overcome the defects of PhraseLDA and Labeled LDA while combining their merits, called Labeled Phrase LDA. We take each document as *bag-of-phrases* and restrict the target topics of each labeled document to the label set of the document. The rest of the paper is organized as follows. In Sect. 2, we review the related work. We introduce our topic model, Labeled Phrase LDA, in Sect. 3. In Sect. 4, we present our experiments on three datasets: paper dataset, Twitter dataset and Yahoo! Answers dataset. Finally, we conclude the paper and suggest directions for future research in Sect. 5.

2 Related Work

2.1 Topic Models on Labeled Corpora

There are many existing topic models to deal with labeled corpora. Several variations of LDA to incorporate supervision have been proposed in the literature. Two such models, Supervised LDA [4] and DiscLDA [5] are first proposed to model documents associated with only a single label. Supervised LDA adds to LDA a response variable associated with each document and assumes that a label is generated from each document's empirical topic mixture distribution; while DiscLDA associates a single categorical label variable with each document and associates a topic mixture with each label. Another category of models, such as the MM-LDA [6], Author TM [7], Flat-LDA [8], Prior-LDA [8], Dependency-LDA [8], Partially Observed Topic (POT) [9] model and Partially LDA (PLDA) [10] etc., are not constrained to one label per document because they model each document as a bag of words with a bag of labels.

Different to those models, Labeled LDA (LLDA) [2], a state-of-the-art supervised topic model to deal with multi-labeled corpora, can be used to correspond latent topics directly with the labels. It restricts the topic of each document to its labels. Unlike the traditional LDA, Labeled LDA considers the topic distribution of each document generating from the topics which have corresponding labels of the document, rather than all the topics in the datasets. And the dimensionality of the Dirichlet prior parameter is the same as the number of labels of each document, rather than the total number of topics of the datasets.

The models mentioned above are all *Supervised non-hierarchical topic models*. Moreover, there are also some *Supervised hierarchical topic models*, such as hLLDA [15], HSLDA [16] and SSHLDA [17], which focus on the dependency structure of labels.

2.2 Topic Models and Phrase Mining

LDA is under the *bag-of-words* assumption, however, in our social life, the meaning of a text is related to the order of words. There has been many existing topic models dealing with texts in the form of phrase. Topical n-grams [11] discovers topics as well as topical phrases, and uses additional latent variables and word-specific multinomials to model bi-grams. PhraseDiscovering LDA [12] is a hierarchical generative probabilistic model of topical phrases. There are other topic models dealing with documents as a mixture of phrases, such as Topic Similarity Model [13] and Constructing a Topical HierarchY [14] etc. PhraseLDA [3], as a state-of-the-art model, combines phrase mining and LDA, which overcomes the high-complexity of Topical n-grams. The top phrases associated with each topic learned by PhraseLDA have more intact meaning than words learned by tradition LDA.

In a word, lots of prior work has been done to deal with text corpora in form of phrases, or use multi-labeled corpora as input. However, as far as we know, few work has been done to deal with multi-labeled corpora in form of phrases. In this paper, we present a novel model that combines the advantages of PhraseLDA and Labeled LDA, called Labeled Phrase LDA.

3 Labeled Phrase LDA

In this section, we will first review Labeled LDA and PhraseLDA, and then introduce our novel supervised topic model, i.e., the Labeled Phrase Latent Dirichlet Allocation (LPLDA).

3.1 Labeled LDA

Labeled LDA [2] is a generative model for multiple labeled corpora. It is a supervised topic model, in which the topics of each document are restricted to its labels. The topic distribution in each document in Labeled LDA is generated from a Dirichlet distribution, whose dimensionality of the prior parameter is the same as the number of labels of each document, rather than the number of the total topics of the datasets.

In Labeled LDA, for each document d and each topic k, $\Lambda_k^{(d)} \in \{0,1\} \sim Bernoulli(\cdot|\Phi_k)$, where Φ_k is the label prior parameter for topic k. For example, suppose the total number of topics in the datasets is five, denoted as a, b, c, d and e, that is K = 5. If the dth document has labels b, c and e, then there will be $\Lambda^{(d)} = \{0,1,1,0,1\}$. And then, $\theta^{(d)}$ is drawn from a Dirichlet distribution with parameters $\alpha^{(d)} = (\alpha^2, \alpha^3, \alpha^5)^T$, where α is the topic prior. The above example explains that the target topics of a document depend on its labels.

3.2 PhraseLDA

PhraseLDA [3] chooses the *bag-of-phrases* as its input. Before that, the algorithm segment the input documents into phrases, and the phrase mining algorithm contains two major steps:

Step 1. Mining the frequent candidate phrases by collecting aggregate counts for all contiguous words in the datasets.

Step 2. Merge frequent candidate phrases and select applicable phrases.

3.3 Labeled Phrase LDA

Labeled Phrase Latent Dirichlet Allocation (LPLDA) is a supervised topic model dealing with multi-labeled corpora. It is restricted that in LPLDA topics of each document are in the domain of the labels in the document. The topic distribution in each document draws from the Dirichlet distribution with a M dimension prior parameters, where M changes with the number of labels of each document. Different from Labeled LDA, LPLDA segments the documents into phrases, and all words in the gth phrase in the dth document has a latent topic, random in $\{z_{d,g,1}, ..., z_{d,g,W_{d,g}}\}$, where $W_{d,g}$ is the number of tokens in g^{th} phrase of dth document (Fig. 1).

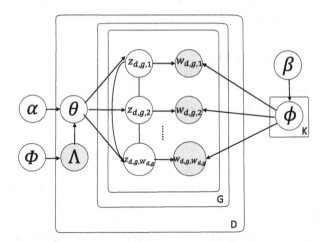

Fig. 1. Graphical model for Labeled Phrase LDA. In Labeled Phrase LDA, we segment each document into phrases, we assume that each word in a phrase shares the same latent topic. And the target topics of a document is chosen from its labels.

The generative process can be divided into three parts. Firstly, drawing the vocabulary distribution from Dirichlet distribution for each topic in the datasets. Secondly, drawing the topic distribution for each document from Dirichlet distribution only in the domain of the labels of each document. Thirdly, generate latent topic for each word in each phrase.

Algorithm 1. Generative process for Labeled Phrase LDA.

1: For each topic $k \in \{1, ..., K\}$:
2:    Generate $\beta_k = (\beta_{k,1}, ..., \beta_{k,V})^T \sim Dir(\cdot|\phi)$
3: For each document d:
4:    For each topic $k \in \{1, ..., K\}$:
5:       Generate $\Lambda_k^{(d)} \in \{0,1\} \sim Bernoulli(\cdot|\Phi_k)$
6:    Generate $\alpha^{(d)} = L^{(d)} \times \alpha$
7:    Generate $\theta^{(d)} = (\theta_{l_1}, ..., \theta_{l_{M_d}})^T \sim Dir(\cdot|\alpha^{(d)})$
8:    For each g in phrases $\{1,...,G_d\}$:
9:       For each i in words $\{1,...,W_{d,g}\}$:
10:          Generate $z_{d,g,i} \in \{\lambda_1^{(d)}, ..., \lambda_{M_d}^{(d)}\} \sim Mult(\cdot|\theta^{(d)})$
11:          Generate $w_{d,g,i} \in \{1, ..., V\} \sim Mult(\cdot|\beta_{z_{d,g,i}})$

The generative process for our LPLDA model can be found in Algorithm 1, where β_k is the parameters of the multinomial distribution of the kth topic. α is the parameters of the Dirichlet distribution of the topic prior, and ϕ is the parameters of the Dirichlet distribution of the word prior. Φ_k is the labeled prior for topic k. $z_{d,g,i}$ is the latent topic for i^{th} token in gth phrase of dth document, that is the latent topic for $w_{d,g,i}$. And $w_{d,g,i}$ is the ith token in g^{th} phrase of document d.

3.4   Learning and Inference

In LPLDA, we define a indicator function $I^{(d)}(k)$ as below:

$$I^{(d)}(k) = \begin{cases} 1 & \text{if the } k^{th} \text{ topic is in the set of labels of the } d^{th} \text{ document.} \\ 0 & \text{otherwise.} \end{cases} \quad (1)$$

Like PhraseLDA, we first segment each document into single and multi-word phrases. We assume that all the words in the same phrase share the same latent topic. Therefore, the Gibbs sampling probability of $C_{d,g}$, phrase g in a document d, for a topic in LPLDA can be:

$$P(C_{d,g} = k|W, Z_{\setminus C_{d,g}}) \propto$$

$$I^{(d)}(k) \cdot \prod_{j=1}^{W_{d,g}} (\alpha_k^{(d)} + N_{d,k \setminus C_{d,g}} + j - 1) \frac{(\beta_{k,w_{d,g,j}} + N_{w_{d,g,j},k \setminus C_{d,g}})}{(\sum_{x=1}^{V} \beta_{k,x} + N_{k \setminus C_{d,g}} + j - 1)} \quad (2)$$

where $N_{d,k \setminus C_{d,g}}$ is the number of tokens assigned to topic k in document d without the gth phrase, $N_{w_{d,g,j},k \setminus C_{d,g}}$ is the number of tokens with value $w_{d,g,j}$ and topic k without the gth phrase in document d, and $I^{(d)}(k)$ is the indicator function.

The above equation is similar to that of PhraseLDA, but the indicator function restricts the target topic of each document to its labels. Moreover, different from PhraseLDA, Labeled Phrase LDA is a supervised topic model.

The parameter estimation for any single Gibbs Sampling is as follows:

$$\hat{\phi}_{w,k} = \frac{N_{w,k} + \beta_{k,w}}{N_{(\cdot),k} + \sum_{x=1}^{V} \beta_{k,x}} \tag{3}$$

$$\hat{\theta}_k^{(d)} = \frac{N_{d,k} + \alpha_k^{(d)}}{N_{d,(\cdot)} + \sum_{i=1}^{K} \alpha_i^{(d)}} \tag{4}$$

where $N_{w,k}$ is the number of tokens with value w and topic k, $\beta_{k,w}$ is the multinomial distribution over words in topic k and word w, and $N_{(\cdot),k}$ is the number of tokens with topic k in Eq. (3). And $N_{d,k}$ is the number of tokens assigned to topic k in document d in Eq. (4). The topic-specific distribution ϕ can be used to obtain topical abstracts for topics; Meanwhile the topic distribution θ for each document can be used to discover the most relevant topics for a document and find documents with similar topics.

4   Experiment

4.1   Experiment Setting

We conducted the experiments on three datasets. One of our datasets, we call it **Conf** in the rest of the paper, and it contains full papers of four conferences (CIKM, SIGIR, SIGKDD and WWW) from the year 2011 to the year 2013. We use the keywords of each paper as labels. And we filtrate the raw datasets by removing the document with only one label. After filtrating, it remains 1169 documents, and all documents in our datasets are multi-labeled.

And the second is a corpus of tweets downloaded from Twitter[1], a website where people can post short messages about their current activities. We call it **Twitter** later. It contains tweets on Twitter from August 2009 to September 2009. And we filtrate the raw datasets by removing labels tagged less than 2000 documents, and then removing tweets that contain only one label, since the LDA family is under the multi-topic assumption. After filtrating, the **Twitter** remains 883799 tweets.

As for the last datasets, we crawled the questions and associated answer pairs (QA pairs) of one of a top category of Yahoo! Answers[2], *Health*. This produced twenty-three subcategories from 2005.11 to 2008.11, and we selected 330000 QA documents from it randomly. We use the category or subcategory of each question or answer as labels of each document. And all documents have more than one label. We refer the Yahoo! Answer data as **Yahoo! Answers**.

The statistics of all datasets are summarized in Table 1. All documents in our three datasets have more than one label. And we conduct all the experiments on a server with an Intel(R) Xeon(R) CPU E5-2683 v3 @ 2.00 GHz and 125 GB memory. In the rest subsections we will compare the proposed model with Labeled LDA and PhraseLDA in terms of case study, perplexity and scalability.

---

[1] http://twitter.com/.

[2] https://answers.yahoo.com/.

Table 1. The statistics of the datasets.

Datasets	Conf	Twitter	Yahoo! Answers
Size of documents	1169	883799	330000
Size of labels	855	209	24
Size of vocabulary	27312	27350	40307

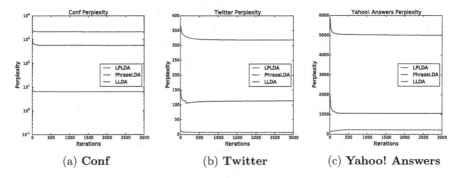

(a) **Conf**   (b) **Twitter**   (c) **Yahoo! Answers**

Fig. 2. A comparison of the perplexity of LPLDA vs PhraseLDA and LLDA during Gibbs sampling inference on three datasets.

4.2   Case Studies

We ran the proposed Labeled Phrased LDA, Labeled LDA and PhraseLDA on the three corpus described above. All models are based on the standard collapsed Gibbs sampling, and set the same initial hyperparameters, where α and β both equal to 0.01.

Tables 3, 4 and 5 show top ten words or phrases from five topics learned on the **Conf, Twitter** and **Yahoo! Answers** respectively. All topics are directly named with the labels of each documents. And the topics learned by unsupervised topic model, PhraseLDA, were matched to a topic in Labeled Phrase LDA by using Kullback-Leibler divergence.

Compared to Labeled LDA, phrases learned by Labeled Phrase LDA show more particular information and contain more distinct meaning than words learned by Labeled LDA. For example, the phrase *breast cancer* learned by

Table 2. A comparison of the run-time of LPLDA vs PhraseLDA.

Method \ Datasets	Conf	Twitter	Yahoo! Answers
LPLDA	0.444h	0.839h	0.175h
PhraseLDA	13.865h	5.464h	0.728h

Table 3. Top ten words from five topics learned on the **Conf.**

Model / Label	LPLDA	LLDA	PhraseLDA
language models	language mode, retrieval models, translation model, negative feedback, difficult queries, ACM SIGIR, general negative language model, statistically significant, information retrieval, query expansion	language, documents, negative, model, document, retrieval, query, feedback, based, models	method outperforms, performance of our approach, significantly outperforms, compare the performance, baseline methods, compare our approach, outperforms the baseline, outperforms significantly the RPF methods, compare our method, compare our algorithm
topic model	topic models, topic distribution, latent topics, topic proportions, multinomial distribution, Dirichlet prior, Gibbs sampling, topic document, generative process, topic assignments	topic, model, topics, words, models, document, distribution, word, lda, number	ACT model, ACTC model, influential authors, author topic, topic models, conferences such as SIGMOD, topical authority, Author Author, topic model for authors, author s most influential
book search	book search, digital libraries, internal and external, Open Library, faceted filtering, advanced search interface, test collections, external search, search behavior, external sessions	internal, book, external, library, filtering, digital, books, interaction, operators, faceted	digital libraries, internal and external, book search, Open Library, Google Books, external search, external and internal, external sessions, searches over books, book recommending
recommendation system	interest patterns, recommender systems, implicit feedback, user interests, exploratory search, recommendation performance, search history, search results, recommends books, book recommending	recommendation, patterns, brek, cluster, book, exploratoriness, long-lastingness, searches, similarity, exploratory	method outperforms, performance of our approach, significantly outperforms, compare the performance, baseline methods, compare our approach, outperforms the baseline, outperforms significantly the RPF methods, compare our method, compare our algorithm
collaborative filtering	collaborative filtering, user item, recommender systems, user ratings binary codes, matrix factorization, data set, rated items, rating of user, web pages	user, users, items, data, ratings, set, recommendation, item, collaborative, filtering	method outperforms, performance of our approach, significantly outperforms, compare the performance, baseline methods, compare our approach, outperforms the baseline, outperforms significantly the RPF methods, compare our method, compare our algorithm

Table 4. Top ten words from five topics learned on the **Twitter**.

Label \ Model	LPLDA	LLDA	PhraseLDA
web	web design, web seo internetmarketing, web domain hosting business seo, tags web jobs hiring, web app, web hosting, hosting web domain seo, design web, web tech, tech web	web, design, seo, article, internetmarketing, hosting, domain, business, marketing, tags	seo jobs, web seo internetmarketing, sem jobs, seo marketing, seo sem, link building, search engine, web site, freelance seo job seo, freelance seo job
media	media noisemachine, japan tech media, fox news, social media news, business media television news markets media, fox media, reuters japan tech media, news media, glenn beck, social media	media, news, fox, japan, noisemachine, social, tech, business, obama, glenn	fox news, rt rt, media noisemachine, rt fox news, mainstream media, news politics, news media, media matters, abc news, rt cnn
un	security council, human rights, iranelection amnesty, tcot gop, amnesty iranelection, general assembly, violence against women un news, global warming, iranelection iran, whn minor	news, iranelection, execution, amnesty, tcot, rights, iran, sentenced, minor, women	rt indico vip, indico vip rt, rt rt indico vip, rt rt, rt rt rt, rt rt rt indico vip, indico vip, rt rt rt rt, rt rt rt rt indico vip, rt rt rt rt rt
showbiz	nieuws showbiz gossip, nieuws showbiz entertainment, reuters nieuws showbiz gossip, ap nieuws showbiz gossip, lady gaga, michael jackson, pictures nieuws showbiz gossip, sex and the city, mad men, chris brown	showbiz, nieuws, gossip, reuters, entertainment, film, video, back, jackson, michael	nieuws sport, nieuws showbiz gossip, nieuws actualiteit, nieuws showbiz entertainment, voor nieuws sport, op nieuws sport, van nieuws sport, nieuws internet ict, bij nieuws sport, met nieuws sport
baseball	mlb baseball, baseball mlb, win mlb baseball, red sox, game mlb baseball, white sox, sports baseball, free agent, sports baseball mlb, season mlb baseball	baseball, mlb, win, giants, dodgers, sox, game, angels, tigers, reds	mlb baseball, yankees angels, yankees mlb, yankees phillies, phillies yankees, angels yankees, support supernatural add twibbon to your avatar, support philadelphia phillies add twibbon to your avatar, world series, dodgers phillies

Table 5. Top ten words from five topics learned on the **Yahoo! Answers**.

Model / Label	LPLDA	LLDA	PhraseLDA
Optical	eye doctor, contact lenses, wear contacts, wear glasses, contact lens, eye drops, dry eyes, eye exam, left eye, colored contacts	eye, eyes, contacts, glasses, lenses, wear, vision, contact, doctor, don't	eye doctor, contact lenses, wear glasses, wear contacts, pink eye, eye drops, close your eyes, contact lens, dry eyes, left eye
Heart Diseases	heart attack, blood pressure, high blood pressure, chest pain, heart disease, heart problems, heart rate, heart beat, heart failure, heart murmur	heart, blood, pressure, doctor, high, attack, normal, chest, pain, rate	heart attack, heart disease, panic attacks, heart problems, heart rate, heart failure, heart beat, kidney stones, gall bladder, bowel movement
Infectious Diseases	chicken pox, sore throat, strep throat, staph infection, wash your hands, immune system, flu shot, HIV AIDS, high fever, bacterial infection	infection, flu, virus, fever, throat, doctor, people, cold, antibiotics, don't	sore throat, yeast infection, cold sores sinus infection, ear infection, strep throat, bacterial infection, staph infection, cold or flu, bladder infection
Respiratory Diseases	sleep apnea, shortness of breath, asthma attack, sore throat, chest pain, sinus infection, cystic fibrosis, carbon monoxide, deep breath, lung disease	asthma, doctor, cough, lungs, breathing, smoking, smoke, sleep, chest, throat	sore throat, yeast infection, cold sores sinus infection, ear infection, strep throat, bacterial infection, staph infection, cold or flu, bladder infection
Cancer	breast cancer, lung cancer, type of cancer, skin cancer, cancer cells, lymph nodes, prostate cancer, cure for cancer, colon cancer, cervical cancer	cancer, breast, doctor, treatment, chemo, years, time, good, people, don't	breast cancer, lymph nodes, cervical cancer, lung cancer, type of cancer, CT scan, cancer cells, prostate cancer, skin cancer, brain tumor

LPLDA obviously contains more clear information than either the word *cancer* or *breast* learned by LLDA for the topic *Cancer* on the **Yahoo! Answers**.

Moreover, comparing Labeled Phrase LDA and PhraseLDA, we can see Labeled Phrase LDA performed better in finding the most appropriate phrases. For example, considering the topic *book search* on the **Conf**, the top one phrase by Labeled Phrase LDA is *book search*, while by PhraseLDA the top one phrase is *digital libraries*.

Consequently, LPLDA performs better than the baseline models in term of case study.

4.3 Perplexity

Besides comparing the case studies of the three models on the above corpus, we also calculated the perplexity, which can measure the quality of different models quantificationally. Fig. 2 shows that LPLDA is much better than the other models on all the three datasets. For example, in the **Yahoo! Answers**, the perplexity of LPLDA stabilize at 227.5374585, however, in PhraseLDA the perplexity stabilize at 1055.42179, and in LLDA it stabilize at 5008.971362.

4.4 Scalability

To evaluate the scalability of our method, we compute our model's runtime on different datasets and compare them to PhraseLDA's. Since LLDA does not contains the phrase ming step, it is not fair to compare the runtime of LPLDA and LLDA. Table 2 shows the result. It is obvious that LPLDA is much more efficient than PhraseLDA, especially when the datasets has a large number of topics. For example, in the **Conf**, the runtime of LPLDA is 0.444 h, however, in PhraseLDA the runtime increases to 13.865 h.

5 Conclusion and Future Work

In this paper, we present a supervised topic model called Labeled Phrase LDA. We simply reviewed two popular topic models, Labeled LDA and PhraseLDA, and expounded the advantages and disadvantages of the two models. Then we introduced our Labeled Phrase LDA model in detail. And then, we conducted three experiments on three different datasets in different domain, using Labeled Phrase LDA and the two original models. The results of the experiments shows that our LPLDA model performs better than the two original models in general.

Further work may focus on methods to manage the following two challenges: (1) Filtrate similar phrases, especially in spoken corpus. (2) Develop the online learning algorithm for LPLDA.

Acknowledgement. This work was supported by 863 Program (2015AA015404), China National Science Foundation (61402036, 60973083, 61273363), Beijing Technology Project (Z151100001615029), Science and Technology Planning Project of Guangdong Province (2014A010103009, 2015A020217002), Guangzhou Science and Technology Planning Project (201604020179).

References

1. Blei, D., Ng, A., Jordan, M.: Latent dirichlet allocation. J. Mach. Learn. Res., 993–1022 (2003)
2. Ramage, D., et al.: Labeled LDA: a supervised topic model for credit attribution in multi-labeled corpora. Empirical Methods in Natural Language Processing (2009)
3. Elkishky, A., et al.: Scalable topical phrase mining from text corpora. In: Proceedings of the Vldb Endowment 8.3, pp. 305–316 (2014)

4. Blei, D., Mcauliffe, J.: Supervised Topic Models. Neural Information Processing Systems (2008)
5. Lacostejulien, S., Sha, F., Ijordan, M.: DiscLDA: discriminative learning for dimensionality reduction and classification. In: Neural Information Processing Systems (2009)
6. Ramage, D., et al.: Clustering the tagged web. In: Web Search and Data Mining (2009)
7. Rosenzvi, M., et al.: The author-topic model for authors and documents. In: Uncertainty in Artificial Intelligence (2004)
8. Nrubin, T., et al.: Statistical topic models for multi-label document classification. Mach. Learn. **88**(1), 157–208 (2012)
9. Xiao, H., Wang, X., Du, C.: Injecting structured data to generative topic model in enterprise settings. In: Asian Conference on Machine Learning (2009)
10. Ramage, D., Dmanning, C., Tdumais, S.: Partially labeled topic models for interpretable text mining. In: Knowledge Discovery and Data Mining (2011)
11. Wang, X., Mccallum, A., Wei, X.: Topical N-Grams: phrase and topic discovery, with an application to information retrieval. In: International Conference on Data Mining (2007)
12. Vlindsey, R., Pheadden, W., Jstipicevic, M.: A phrase-discovering topic model using hierarchical pitman-yor processes. In: Empirical Methods in Natural Language Processing (2012)
13. Xiao, X., et al.: A topic similarity model for hierarchical phrase-based translation (2012)
14. Wang, C., et al.: A phrase mining framework for recursive construction of a topical hierarchy. In: Knowledge Discovery and Data Mining (2013)
15. Petinot, Y., Mckeown, K., Thadani, K.: A hierarchical model of web summaries. In: Meeting of the Association for Computational Linguistics (2011)
16. Perotte, A., et al.: Hierarchically supervised latent Dirichlet allocation. In: Neural Information Processing Systems (2011)
17. Mao, X., et al.: SSHLDA: a semi-supervised hierarchical topic model. In: Empirical Methods in Natural Language Processing (2012)
18. Deerwester, S., et al.: Indexing by latent semantic analysis. J. Am. Soc. Inf. Sci. **41**(6), 391–407 (1990)
19. Hofmann, T.: Probabilistic latent semantic indexing. In: International ACM SIGIR Conference on Research and Development in Information Retrieval (1999)
20. Lgriffiths, T., et al.: Hierarchical topic models and the nested chinese restaurant process. In: Neural Information Processing Systems (2004)
21. Whyeteh, Y., et al.: Hierarchical Dirichlet processes. J. Am. Stat. Assoc., 1566–1581 (2012)
22. Li, W., Mccallum, A.: Pachinko allocation (DAG-structured mixture models of topic correlations). In: Machine Learning (2006)

WTEN: An Advanced Coupled Tensor Factorization Strategy for Learning from Imbalanced Data

Quan Do[1]($^{(\boxtimes)}$), Thanh Pham[2], Wei Liu[1], and Kotagiri Ramamohanarao[2]

[1] Advanced Analytics Institute, University of Technology Sydney, Sydney, Australia
{Quan.Do,Wei.Liu}@uts.edu.au
[2] Department of Computing and Information Systems,
University of Melbourne, Melbourne, Australia
Thanhp1@student.unimelb.edu.au, Kotagiri@unimelb.edu.au

Abstract. Learning from imbalanced and sparse data in multi-mode and high-dimensional tensor formats efficiently is a significant problem in data mining research. On one hand, Coupled Tensor Factorization (CTF) has become one of the most popular methods for joint analysis of heterogeneous sparse data generated from different sources. On the other hand, techniques such as sampling, cost-sensitive learning, etc. have been applied to many supervised learning models to handle imbalanced data. This research focuses on studying the effectiveness of combining advantages of both CTF and imbalanced data learning techniques for missing entry prediction, especially for entries with rare class labels. Importantly, we have also investigated the implication of joint analysis of the main tensor and extra information. One of our major goals is to design a robust weighting strategy for CTF to be able to not only effectively recover missing entries but also perform well when the entries are associated with imbalanced labels. Experiments on both real and synthetic datasets show that our approach outperforms existing CTF algorithms on imbalanced data.

Keywords: Tensor Factorization · Coupled Tensor Factorization · Imbalanced data learning

1 Introduction

Recent innovations on the Internet and social media have made many multi-mode, high dimensional, sparse and imbalanced data available. Together with this explosive dimension growth, Coupled Tensor Factorization (CTF) has become one of the most popular methods for joint analysis of sparse data generated from different sources. It has also been proven to predict missing data entries with high accuracy [2]. In case the actual entries are skewed toward a particular class, generally, we want to achieve a high prediction rate of the class of interest in spite of its rarity. Nevertheless, even if the reconstructed tensor predicts everything to be

© Springer International Publishing AG 2016
W. Cellary et al. (Eds.): WISE 2016, Part I, LNCS 10041, pp. 537–552, 2016.
DOI: 10.1007/978-3-319-48740-3_40

of the majority class, the overall accuracy is still very high. For example, in the event of a binary class (such as high and low ratings of movies) and the actual entries are skewed towards a particular class, for instance, the ratio of negative class (e.g., low ratings of movies) to positive class (e.g., high ratings) is 99 to 1, any CTF would easily achieve 99 percent overall approximation accuracy by just approximating all missing entries to the negative class. This 99 percent precision rate is impressive enough if we ignore the 0 percent accuracy of the positive class. This bias accuracy not only reduces the robustness of the model but also might cause severe consequences, especially in cases that most of the observed samples are normal and just a few rare cases are anomaly ones. For example in disease diagnosing application, even though most of the training data are healthy specimens, predicting an unhealthy sample as a healthy one costs extremely high, in many cases a human life. However, it might be acceptable to classify a healthy person as an unhealthy and perform a few other diagnoses. Achieving a high rate on classifying the rare case without jeopardizing the majority class is, therefore, a crucial requirement in this instance.

Learning from imbalanced data has attracted considerable attention in knowledge discovery community. Sampling [7] and cost-sensitive [8] approaches have been studied to deal with imbalanced datasets. Although they have been proposed to decision trees and neural networks, they have not yet been applied to multi-mode, high dimensional, sparse and imbalanced tensor data (and importantly, its decompositions). This significant theoretical gap motivates us to take the advantages of both CTF and imbalanced data learning techniques to address the problem of recommending missing entries from imbalanced yet sparse heterogeneous datasets. In particular, we adjust CTF's objective function by a weighting strategy that lowers the significance of wrongly recommending the majority class and strengthens the importance of correctly estimating the rare case. Here we introduce a weighting strategy for CTF, called WTEN, as the first CTF approach for imbalanced data learning. Although this paper targets binary missing label estimation problem, the weighting strategy can be straightforwardly applied to multiple labels, such as integer ratings where the frequencies of integers are imbalanced.

In brief, our main contribution in this paper are the following:

(1) **Performance**: we propose a novel weighting strategy, named WTEN (Weighted Tensor Factorization), for missing entry recommendation using CTF. Our model robustly assigns effective weights with respect to different classes' approximation, and consequently performs significantly better on the minority class estimation without jeopardizing the majority one. WTEN is the first method, to our best knowledge, that enables CTF to handle imbalanced missing data entries.

(2) **Foundation**: we study the effectiveness of joint analysis of the main tensor and the additional coupled data in CTF techniques for handling sparsity and imbalance over Tensor Factorization. Our theoretical analysis and experimental results suggest CTF to serve as a foundation for a general purpose latent factor imbalanced data learning (Fig. 1).

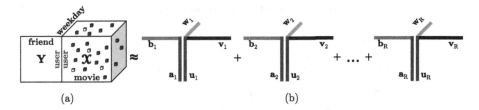

Fig. 1. Factorization of coupled imcomplete data sets for missing entries recovery. (a) Correlation among different aspects of a dataset. $\mathcal{X}$ is a tensor of ratings made by users for movies on weekdays. Dark boxes are observed low ratings (which are majority) and white boxes are known minority high ratings. Matrix $\mathbf{Y}$ represents user information. Movie rating tensor $\mathcal{X}$ is, therefore, coupled with user information matrix $\mathbf{Y}$ in 'user' mode. (b) $\mathcal{X}$ is factorized as a sum of low rank factors that can be used to recover missing majority as well as minority cases.

(3) **Usability and reproducibility**: Our factorization method with weighting scheme can be easily extended to different datasets and applications. Performance of WTEN is validated by both real-world and synthetic datasets. To promote the reusability of our idea, we open our source code with this paper.[1]

The rest of this paper is organized as follows. We introduce the background of tensor factorization in Sect. 2 followed by a review of existing work in Sect. 3. Section 4 explains our proposed idea. Experimental results together with our discussion are included in Sect. 5. We finally conclude our work in Sect. 6.

2 Preliminary

This section provides a brief introduction of core definitions and preliminary concepts of tensor, tensor factorization and coupled tensor factorization.

2.1 Tensor and Our Notations

Tensors are multidimensional arrays which are often specified by their number of modes (a.k.a., orders or ways). In specific, a mode-1 tensor is a vector; a matrix is a mode-2 tensor. A mode-3 or higher-order tensor is often called tensor in short. We denote tensors by boldface Euler script letters, e.g. $\mathcal{X}$. We use boldface capitals, e.g. $\mathbf{A}$, for matrices. A boldface Euler script with indices in its subscript is used for an entry of a tensor while a boldface capital with indices in its subscript is for an entry of a matrix. For example, $\mathbf{A}_{i,j}$ is an entry at row i and column j of matrix $\mathbf{A}$; the $(i, j, k)^{\text{th}}$ entry of $\mathcal{X}$ is $\mathcal{X}_{i,j,k}$. Table 1 lists all the symbols we throughout use in this paper.

---

[1] Our source code is available at https://github.com/quanie/WTEN.

Table 1. Symbols and their description

Symbol	Description
$\mathcal{X}$	A tensor
$\mathbf{X}$	A matrix
$\mathcal{X}_{i,j,k}$	An entry of a tensor
$\mathbf{X}_{i,j}$	An entry of a matrix
$\hat{\mathcal{X}}_{i,j,k}$	A reconstructed missing entry of tensor $\mathcal{X}$
$\|\mathbf{A}\|_F$	Frobenius norm
$\mathbf{U}^{(n)}$	A n-th mode factor
I*J*K	Dimensions of tensor $\mathcal{X}$
I*M	Dimensions of matrix Y
R	Decomposition rank
$\circ$	Khatri-Rao product
$\circledast$	Hadamard (elementwise) product
$\mathcal{L}$	Loss function

2.2 Tensor Factorization (TF) and Coupled Tensor Factorization (CTF)

Tensor factorization, based on PARAFAC decomposition [11], approximates a high-order tensor into a sum of a finite number of low rank factors.

$$\mathcal{X} \approx \sum_{r=1}^{R} \prod_{n=1}^{N} \mathbf{U}_{I_n,r}^{(n)}$$

where $\mathcal{X} \in \mathbb{R}^{I_1 * I_2 * \cdots * I_N}$ is a N-mode tensor and its N rank-R factors are $\mathbf{U}^{(n)} \in \mathbb{R}^{I_n * R}, \forall n \in [1, N]$.

The goal of PARAFAC decomposition is to find the best low-dimensional approximation of $\mathcal{X}$ [14]. In other words, PARAFAC decomposition finds

$$\min_{\hat{x}} \|\mathcal{X} - \hat{\mathcal{X}}\|_F \quad \text{with} \quad \hat{\mathcal{X}} = \sum_{r=1}^{R} \prod_{n=1}^{N} \mathbf{U}_{I_n,r}^{(n)}$$

and $\mathcal{L} = \|\mathcal{X} - \hat{\mathcal{X}}\|_F$ is defined as the loss function of the factorization.

In case $\mathcal{X}$ is a mode-3 tensor, TF decomposes $\mathcal{X}$ into a Khatri-Rao product of its factors, and thus the loss function is defined by:

$$\mathcal{L}(\mathbf{U}, \mathbf{V}, \mathbf{W}) = \frac{1}{2} * \|\mathcal{X} - \mathbf{U} \circ \mathbf{V} \circ \mathbf{W}\|_F^2$$

where $\mathbf{U} \in \mathbb{R}^{I*R}$, $\mathbf{V} \in \mathbb{R}^{J*R}$ and $\mathbf{W} \in \mathbb{R}^{K*R}$.

We often have additional information in a format of a matrix or a tensor which has one or more modes in common with the main tensor. These side information along with the main data can help to deepen our understanding of the underlying patterns in the data, and to improve the accuracy of tensors composition. For example, Acar et al. [2] defined an objective function for joint analysis of a tensor $\mathcal{X}$ coupled with a matrix $\mathbf{Y}$ in its first mode by:

$$\mathcal{L}(\mathbf{U}, \mathbf{V}, \mathbf{W}, \mathbf{A}) = \frac{1}{2} * \|\mathcal{X} - \mathbf{U} \circ \mathbf{V} \circ \mathbf{W}\|_F^2 + \frac{1}{2} * \|\mathbf{Y} - \mathbf{U}\mathbf{A}^T\|_F^2 \tag{1}$$

where $\mathbf{U}$ is the common factor of both $\mathcal{X}$ and $\mathbf{Y}$.

By solving this Eq. (1) with an optimizer, low rank factors $\mathbf{U}$, $\mathbf{V}$, $\mathbf{W}$ and $\mathbf{A}$ can be obtained. These factors can then be used to approximate both tensor $\mathcal{X}$ and matrix $\mathbf{Y}$.

2.3 Missing Data Completion

Latent factors discovered by TF or CTF above can be used to recover missing data from the original input tensor. The most widely used approach [2] is to utilize these latent factors to reconstruct $\hat{\mathcal{X}}$ for missing entries recovery. Suppose a tensor $\mathcal{X}$ coupled with a matrix $\mathbf{Y}$ are factorized by (1). A missing entry (i,j,k) of $\mathcal{X}$ is estimated by:

$$\hat{\mathcal{X}}_{i,j,k} = \sum_{r=1}^{R} \mathbf{U}_{i,r} \mathbf{V}_{j,r} \mathbf{W}_{k,r} \tag{2}$$

In the event of binary entries, a simple method to decide a label of $\hat{\mathcal{X}}_{i,j,k}$ is to use a threshold ϵ. An entry (i,j,k) of $\mathcal{X}$ belongs to negative label if $\hat{\mathcal{X}}_{i,j,k} \leq \epsilon$ or else positive label.

3   Literature Review

Learning from imbalanced data has attracted considerable attention in knowledge discovery community. Imbalanced data learning algorithms proposed in the literature can be categorized into sampling, cost sensitive, kernel based approaches [12]. Kernel based methods that mainly focus on modifying SVM kernel for imbalanced learning [25] or applying sampling to SVM framework [3] are out of scope of this paper. In this section, we provide a brief overview of sampling and cost sensitive methods.

When a training data is skewed toward a particular class, a straightforward strategy [7] is sampling to create a more balanced data distribution for both classes. Two sampling techniques, oversampling and undersampling, are widely proposed for imbalanced learning. Oversampling increases the minority class population by creating more data samples. The extra data samples can be made by replicating minority samples [13] or synthesized by various techniques such as Synthetic Minority Oversampling Technique (SMOTE) [6]. Overfitting is often considered a potential disadvantage of oversampling [7]. Undersampling, on the other hand, reduces the majority class by eliminating some of its samples. This reduction can be done randomly [20] or based on statistical knowledge [19].

Both randomly and statistically undersampling have the possibility of losing important data.

An alternative method to overcome data imbalance is a cost-based approach. Instead of balancing the data distribution by sampling, cost sensitive learning [8] associates different penalties with misclassifying different classes correspondingly. For example, in case a training data is skewed towards negative label, the total cost of misclassifying negative classes as positive ones outweighs that of misclassifying positive labels as negative. Any learning algorithm that minimizes total misclassification cost mostly optimizes the negative class only [23]. By associating a higher cost with misclassifying a positive class as a negative one than with the contrary, the algorithm now balances better for the positive class. As a result, this cost-sensitive approach improves the classification performance with respect to the rare class. Although this technique has been successfully applied to decision tree via subtree pruning [5] or data split [16], and neural networks [26], it has not ever been proposed for high-dimensional decompositions of tensors with imbalanced data entries.

Tensor factorization (TF) has been used for multi-mode, high-dimensional and sparse data analysis with a goal to capture the underlying low rank structures. This analysis has become a new trend since the Netflix Prize competition [15] where it is used to predict movie ratings with high accuracy. Researchers has extended TF to do joint data analysis. Early work by Singh and Gordon [24] introduced Collective Matrix Factorization (CMF) to take an advantage of correlations between different coupled matrices and simultaneously factorized them. CMF techniques have been successfully applied to capture the underlying complex structure of data [15,21]. Acar, Kolda and Dunlavy [2] later expanded CMF to CTF to handle Coupled Matrix and Tensor Factorization by modeling heterogeneous data sets as higher-order tensors and matrices in a coupled loss function. They also proved the possibility of using these low rank factors to recover missing entries. Tensor methods have been studied for factorization with labeled information [17] and also compression with tensor representations [18]. Papalexakis et al. [22] and Beutel et al. [4] scaled CTF up to parallel and distributed environments but with the same loss function as proposed by other authors.

Motivations for this Research

Despite the popularity of CTF on high-dimensional datasets, improvements of CTF on imbalanced data has not been studied. If we apply CTF with its traditional objective function (1) on imbalanced data, it will tend to ignore the minority cases and approximate most missing values to be the majority ones. The objective function is still optimal thanks to the fact that almost all predicted instances are correct. This is because the loss function (1) assumes that errors of factorizing the majority class in $\mathcal{X}$ and that of decomposing the minority one contribute equally to the final loss of the CTF. Apparently, this is not the case for imbalanced data as the majority class extremely out-represents the minority one. Thus, the loss of factorizing the majority class totally outweighs all the loss of decomposing the rare one. The algorithms, hence, focus on optimizing the major class to achieve a lower loss.

Another problem with CTF on imbalanced learning is that both oversampling and undersampling do not work effectively. First of all, oversampling does not balance out the data distribution. Suppose $\mathcal{X}_{i,j,k}$ is an entry of the minority class, oversampling by duplicating $\mathcal{X}_{i,j,k}$ does not add anything to the tensor as an entry at the index $\{i,j,k\}$ is already there in $\mathcal{X}$. Hence, the data distribution does not change. Secondly, even though undersampling creates a more balanced data distribution, it may remove some important observed data. This is especially critical as the observed data is sparse. Losing more data might prevent CTF from achieving its optimization, thus, reducing its accuracy. Last but not least, sampling cannot straightforwardly be done on the additional data in a form of coupled matrices or tensors, and doing so again does not make any change on the data distribution of the main tensor.

Motivated by the above significant theoretical gaps, in this paper we propose a novel cost-sensitive weighting strategy to overcome the imbalanced data problem in high dimensional and heterogeneous datasets. Our algorithm optimizes the factorization of both the majority and minority class in a balanced manner, significantly improving missing entry estimations of the minority class.

4   WTEN: Weighted Tensor Factorization for Imbalanced Data

In this section, we introduce our proposed WTEN to handle imbalanced datasets. Suppose $\mathcal{X} \in \mathbb{R}^{I*J*K}$ is a mode-3 tensor coupled with a matrix $\mathbf{Y} \in \mathbb{R}^{I*M}$ in their first mode, and suppose their data entries are binary. In an event when $\mathcal{X}$ is skewed toward class 0, algorithms [2,4,22] with the objective function (1) show their drawbacks in estimating class 1 as they approximate everything to be of the majority class. Yet the objective function is still considered as optimal because almost all predicted instances are correct. This hence reduces the robustness of the methods in dealing with the imbalanced input.

One of a few possible improvements of the above problem is to properly highlight the impact of errors in approximating the rare cases. This can be done by adjusting the objective function (1) to a weighted version based on observed frequencies of different classes. So the objective function (1) becomes:

$$\mathcal{L}(\mathbf{U}, \mathbf{V}, \mathbf{W}, \mathbf{A}) = w_0 * \|\mathcal{X}_0 - \mathbf{U} \circ \mathbf{V} \circ \mathbf{W}\|_F^2 + w_1 * \|\mathcal{X}_1 - \mathbf{U} \circ \mathbf{V} \circ \mathbf{W}\|_F^2 \\ + \|\mathbf{Y} - \mathbf{U}\mathbf{A}^T\|_F^2 \tag{3}$$

where $\mathcal{X}_0$ and $\mathcal{X}_1$ are tensor entries containing negative and positive labels, respectively; w_0 is a weight of precisely estimating class 0 and w_1 is that of correctly approximating class 1. The first term represents prediction error of class 0 whereas the second term captures that of class 1.

If appropriate weights are used, w_0 and w_1 will have an effect of balancing out the impact of misclassifying different classes with respect to their observation ratio. An effective approach is to assign $w_0 = N_1/size(\mathcal{X})$ and $w_1 = N_0/size(\mathcal{X})$ where N_0 and N_1 are the number of observed 0s and 1s in $\mathcal{X}$, respectively, and $size(\mathcal{X})$ denotes the number of observed elements of $\mathcal{X}$. Doing so lowers importance of approximating the majority class. Thus, this weighting strategy will likely improve the estimation of the minority label.

The objective function (1) can also be adjusted by a weighting tensor as the following:

$$\mathcal{L}(\mathbf{U}, \mathbf{V}, \mathbf{W}, \mathbf{A}, \mathbf{B}) = \|\mathcal{W} \circledast (\mathcal{X} - \mathbf{U} \circ \mathbf{V} \circ \mathbf{W})\|_F^2 + \|\mathbf{Y} - \mathbf{U}\mathbf{A}^T\|_F^2 \qquad (4)$$

where $\mathcal{A} \circledast \mathcal{B}$ is a Hadamard (element-wise) product of $\mathcal{A}$ and $\mathcal{B}$ which yeilds a tensor $\mathcal{C}$ with entries $\mathcal{C}_{i,j,k} = \mathcal{A}_{i,j,k} * \mathcal{B}_{i,j,k}$ and $\mathcal{W}$ is a weighting tensor having the same size of $\mathcal{X}$, but its entries' value is determined by

$$\mathcal{W}_{i,j,k} = \begin{cases} \frac{size(\mathcal{X})}{N_0} & \text{when } \mathcal{X}_{i,j,k} = 0 \\ \frac{size(\mathcal{X})}{N_1} & \text{when } \mathcal{X}_{i,j,k} = 1 \end{cases}$$

This weighting tensor, $\mathcal{W}$ as illustrated in Fig. 2, will have an effect of increasing the impact of errors in approximating the minority class. This weighting scheme produces the same result as the approach suggested in (3). Yet, an implementation of (4) might be simpler as its gradient is likely to be more straightforward to be computed in the optimization processes.

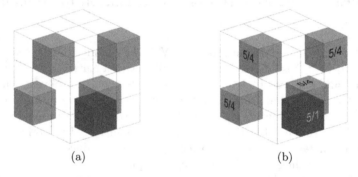

(a)   (b)

Fig. 2. Imbalanced and sparse tensor of size $R^{3 \times 3 \times 3}$, and its weighting tensor. (a) Data tensor $\mathcal{X}$ where minority class, majority class and missing entries are represented by blue (darker), green (lighter) and transparent boxes, respectively. (b) Weighting tensor $\mathcal{W}$ where each observed entries of $\mathcal{X}$ will be assigned a weight (5/4 for the majority and 5/1 for the minority). (Color figure online)

Equation (3) as well as (4) overcomes the problem of the conventional loss function (1) in dealing with imbalanced data. By introducing different weighting parameters, they balance out importances of predicting both the majority class and the minority one. In other words, this non-uniform weighting strategy either emphasizes the impact of errors in estimating the minority class or reduces the significance of losses in approximating the majority case so that WTEN is very well balanced in optimizing the performance of both class labels.

5   Performance Evaluation

Our goals of conducting experiments below are to assess: (1) the contribution of an additional coupled matrix to the tensor factorization accuracy and (2) the effectiveness of our proposed weighting strategy on the estimation of missing entries when it is applied to imbalanced datasets. For better validating our work, both synthetic datasets in which the imbalance rate and coupled relationship are controlled and two real-world data are used.

5.1   Data Used in Our Experiments

We use two synthetic and two real-world datasets for our experiments. The following subsections explain how synthetic data is generated and introduce two real-world datasets.

5.1.1   Synthetic Data

Two datasets are synthesized by the following steps:

- Step1: A symmetric user-by-user matrix Y_{50} with 0s and 1s is randomly created to represent friendship among users. Value 1 means a pair of users is friend, value 0 means otherwise. Each user has a certain percentage of all users as friends. In this experiment, we randomly generated Y_{50} with about 50 % of all users as friends.
- Step 2: For every user, a set of random ratings of 1s (for 5-star ratings) and 0s (other ratings) for movies over twelve months is generated following a rule that ensures any pair of users with 1 in Y_{50} has almost the same rating patterns. This is to capture the fact that users who are friends usually have similar preferences for movies over the year. The generated ratings are in a tensor format of (users, movies, months). Two different sparse tensors are synthesized for this experiment with the ratio between 0 and 1 ratings of 100:1 (X_0) and 1,000:1 (X_1) as summarized in Table 2.
- Final step: just like real-world scenarios where users make friends with those who have similar preferences or unfriend those who do not while their ratings in the past do not change, we analyze each of the two tensors to find pairs of users having the same rating patterns. 1s are then added to these pairs in Y_{50} to create other two matrices Y_{80}^0 and Y_{80}^1 of about 80 % of friendship for each tensors, respectively. The same process is done to form other two matrices Y_{20}^0 and Y_{20}^1 of about 20 % of friendship by removing 1s in Y_{50} for pairs of users having unique rating patterns. These relationships are showed in Table 3.

5.1.2   ABS Data

Australian Bureau of Statistics (ABS) [1] publishes a comprehensive data about people and families for all Australia geographic areas. This ABS dataset has

Table 2. Ground truth distributions of the two synthesized tensors $\mathcal{X}_0$ and $\mathcal{X}_1$ of size $100 \times 100 \times 12$, a real-world ABS tensor $\mathcal{X}_2$ of size $153 \times 88 \times 3$ and a real-world MovieLens tensor $\mathcal{X}_3$ of size $943 \times 1{,}682 \times 7$

Label	$\mathcal{X}_0$ (100:1)		$\mathcal{X}_1$ (1,000:1)		$\mathcal{X}_2$ (9:1)		$\mathcal{X}_3$ (4:1)	
	Training	Testing	Training	Testing	Training	Testing	Training	Testing
1	192	48	19	5	664	167	16,744	4,457
0	19,200	4,800	19,200	4,800	5,799	1450	63,256	15,543

Table 3. Matrices coupled with synthetic tensors $\mathcal{X}_0$ and $\mathcal{X}_1$ for CTF.

Tensor	Matrix		
	Friendship rate		
	20%	50%	80%
$\mathcal{X}_0$	$\mathbf{Y}_{20}^0$	$\mathbf{Y}_{50}$	$\mathbf{Y}_{80}^0$
$\mathcal{X}_1$	$\mathbf{Y}_{20}^1$	$\mathbf{Y}_{50}$	$\mathbf{Y}_{80}^1$

income ranges of different family types within 153 New South Wales' areas, so-called "local government areas", in 2001, 2006 and 2011, forming a tensor $\mathcal{X}_2$ of (area, income range, year) of size 153 by 88 by 3. $\mathcal{X}_2$ has 8080 observations whose values are 1 s for nontrivial income ranges and 0 s for trivial ones. ABS dataset also includes population, number of services provided, and Socio-Economic Indexes for Areas that rank areas with respect to their relative socio-economic advantage and disadvantage. This additional information is compiled into a 153 by 3 matrix $\mathbf{Y}_2$ of (area, profile). In this paper, we train our model with 80% of known $\mathcal{X}_2$'s entries, together with a fully observed $\mathbf{Y}_2$. The rest 20% of known entries of $\mathcal{X}_2$ are for testing. Table 2 summarizes this ABS data distribution.

5.1.3   MovieLens Data

MovieLens dataset [10] includes ratings from 943 users for 1,682 movies. It is compiled into tensor $\mathcal{X}_3$ of (users, movies, weekdays) whose entries are ratings, matrix $\mathbf{Y}_3$ of (users, users' profile) and matrix $\mathbf{Z}_3$ of (movies, genres). Matrix $\mathbf{Y}_3$ has the size of 943 by 83 in which a user is specified by her gender (0 or 1), is grouped in one of 61 age groups, and have one of 21 occupations. Matrix $\mathbf{Z}_3$ categories 1,682 movies into 19 different genres. One movie belongs to one or more genres. Finally, values of $\mathcal{X}_3$'s entries are 1 s for high ratings (e.g. 5-star) and 0 s for observed low ratings (e.g. 1-star to 4-star). In this paper, we train our model with 80,000 known ratings, together with $\mathbf{Y}_3$ and $\mathbf{Z}_3$ of 2,159 and 2,893 observed nonzeros, respectively. 20,000 ratings are for testing. MovieLens data distribution is also shown in Table 2.

5.2   Factorization Accuracy

We investigate effects of an additional matrix $\mathbf{Y}$ to a tensor $\mathcal{X}$'s factorization accuracy by comparing the performance of decomposing only $\mathcal{X}$ (TF in this case) and that of joint factorizing $\mathcal{X}$ and $\mathbf{Y}$ (CTF whose coupled relationships are defined in Table 3). Both mean squared errors (MSE) (5) of the training sets and approximation results of the testing sets are the metrics for our evaluation.

$$\mathrm{MSE} = \frac{\|\mathcal{X} - \mathbf{U} \circ \mathbf{V} \circ \mathbf{W}\|_F^2}{\mathrm{size}(\mathcal{X})} \tag{5}$$

where $\mathrm{size}(\mathcal{X})$ denotes the size of tensor $\mathcal{X}$. In case $\mathcal{X}$ is a sparse tensor, it is the number of observed elements of $\mathcal{X}$.

By having additional data in a form of coupled matrix, CTF improves the factorization accuracy of the main tensor over TF as illustrated in Fig. 3 and Fig. 4. As one may anticipate, having additional information in the event of extremely skewness towards one class is very crucial. Figure 3b shows the additional matrices help improve the MSE of factorizing $\mathcal{X}_1$ 60 times on average compared with factorizing $\mathcal{X}_1$ alone. The more interesting points lie in Fig. 3a where the lower friendship rate, in other words, less informative, a coupled matrix is, the lower training MSE of factorizing $\mathcal{X}_0$ achieves. In particular, as information richness increases from left to right ($\mathbf{Y}_{20}^0$, $\mathbf{Y}_{50}$, then $\mathbf{Y}_{80}^0$), the MSEs of $\mathcal{X}_0$ when joint factorizing $\mathcal{X}_0$ with $\mathbf{Y}_{20}^0$, $\mathbf{Y}_{50}$ and $\mathbf{Y}_{80}^0$ raise correspondingly, even to higher than that of decomposing tensor $\mathcal{X}_0$ alone. This does not mean joint factorizing a tensor $\mathcal{X}_0$ with a stronger constraint and more informative matrix performs worse. Actually, a stronger constraint and more essential user-user information in $\mathbf{Y}_{80}^0$ guides the factorization of $\mathcal{X}_0$ towards a resistant of the conventional trend that approximates everything to be the majority class label to achieve a better estimation of the minority case. This resistance, thus, increases the training MSEs.

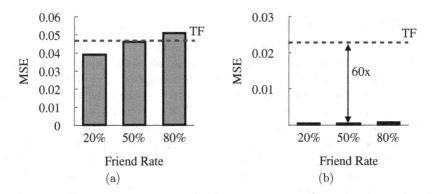

Fig. 3. MSE of a) $\mathcal{X}_0$ and b) $\mathcal{X}_1$ when they are joint factorized with different matrices. Reference lines (blue) in (a) and (b) are MSEs of decomposing $\mathcal{X}_0$ and $\mathcal{X}_1$ alone (by TF). (Color figure online)

CTF with coupled richer information matrices enables more tested minority class (class 1) to be correctly approximated. As illustrated in Fig. 4a, many more tested minority class is correctly recovered in case 80 % friendship matrix $\mathbf{Y}_{80}^{0}$ is coupled factorized with $\mathcal{X}_0$. When $\mathbf{Y}_{50}$ is used, CTF has similar performance with TF since information in $\mathbf{Y}_{50}$ already includes in $\mathcal{X}_0$ which has been done in the second step of generating synthetic data. So $\mathbf{Y}_{50}$ does not really add any extra information to $\mathcal{X}_0$ decomposition. Less informative $\mathbf{Y}_{20}^{0}$ has less meaningful information, compared to the other two matrices, to guide CTF toward correct direction, hence, performs worst. The same trend observed in Fig. 4b with $\mathcal{X}_1$ suggests a dominance of CTF over TF when an extra and meaningful matrix is joint decomposed with a tensor.

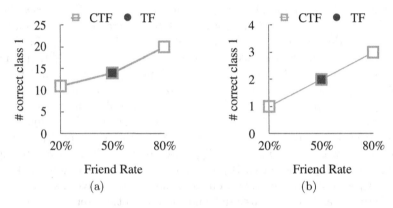

Fig. 4. Number of correctly approximation of missing 1 s in (a) $\mathcal{X}_0$ and (b) $\mathcal{X}_1$. In both cases, the richer additional data is joint decomposition, the higher prediction rate is achieved.

5.3 Missing Entry Recovery

We compare our proposed WTEN with existing CMTF-OPT [2] and Sampling CMTF in which imbalanced data is first randomly under-sampling and then factorized by CMTF-OPT on missing entry recovery. Our target is to assess how well these algorithms approximate missing entries of the imbalanced ABS and MovieLens tensors. A missing entry (i,j,k) of $\mathcal{X}$ is classified as 0 (a majority or a negative label) if the reconstructed $\hat{\mathcal{X}}_{i,j,k} \leq \epsilon$ or 1 (a minority or a positive label) otherwise. Recall, Precision and the area under a ROC curve [9] (AUC) which are widely used metrics in imbalanced data learning are our measurements. CMTF-OPT is optimized by three different optimization methods including Nonlinear Conjugate Gradient (NCG), Limited-memory BFGS (LBFGS) and Truncated Newton (TN) whereas WTEN is optimized by Stochastic Gradient Descent.

Table 4 summarizes the result of CMTF-OPT, Sampling CMTF and our proposed WTEN on missing imbalanced data recovery for both ABS dataset and

Table 4. Performance of missing entries estimation with real-world ABS and Movie-Lens datasets. In both cases, WTEN achieves the highest accuracy on positive labels.

Algorithms	ABS dataset			MovieLens dataset		
	Precision	Recall	AUC	Precision	Recall	AUC
CMTF-OPT (NCG)	0.7658	0.7246	0.8495	**0.5477**	0.3143	0.6199
CMTF-OPT (LBFGS)	0.7602	0.7784	0.8751	0.4512	0.3330	0.6084
CMTF-OPT (TN)	**0.7697**	0.7605	0.8671	0.4372	0.2165	0.5683
Sampling CMTF (NCG)	0.4139	0.8922	0.8733	0.3653	0.7247	0.6818
Sampling CMTF (LBFGS)	0.3495	0.8623	0.8387	0.3249	0.4451	0.5900
Sampling CMTF (TN)	0.3830	0.8623	0.8511	0.3468	0.6679	0.6536
WTEN	0.4825	**0.9102**	**0.8989**	0.4055	**0.7393**	**0.7142**

MovieLens dataset. Boldface numbers highlight the best among the algorithms for each dataset. As shown in Table 4, CMTF-OPT produces the highest Precision for both datasets as it approximates most of the tested entries to the majority labels (e.g. 0 s), leading to a low false positive rate. This is confirmed for both datasets by CMTF-OPT's lowest Recall measurements, which denote the percentage of correctly estimating the minority labels (e.g. 1 s), compared to the others. Sampling CMTFs with different optimization methods improve CMTF-OPT's performance on imbalanced data with higher Recalls. However, their performances are outweighed by WTEN which accurately estimates the positive labels even more (shown by the highest Recall in both cases) without jeopardizing the negative ones (illustrated by just a little lower Precision compared with the best CMTF-OPT). AUC also confirms the dominance of WTEN over existing algorithms on imbalanced data as WTEN achieves the highest AUC for both ABS and MovieLens datasets. All of these results demonstrate the performance of missing entry recovery on imbalanced data does not improve significantly by using a more sophisticated optimizer or applying sampling on the input imbalanced data, but in fact, our proposed strategy enables CTF to achieve a better performance.

To illustrate the advantage of our proposed WTEN, we present in Fig. 5 the convergence of all the algorithms on MovieLens training data. There are two insights we can observe in this figure. Firstly, CMTF-OPTs' least squares errors are generally lower than that of WTEN. This is because WTEN decomposes the input tensor with respect to both majority and minority labels optimization, whereas CMTF-OPTs focuses on minimizing the lost of the majority ones leading to lower least squares errors. Secondly, the convergence speed of WTEN is the same as, if not better than, different optimizers of CMTF-OPT. They almost reach the optimum after about 10 s. This convergence evidence together with WTEN's enhanced performance of missing entries on both real-world datasets confirms its significance on improving the accuracy of minor class estimation, suggesting WTEN as the most appropriate method for CTF to handle imbalanced data learning.

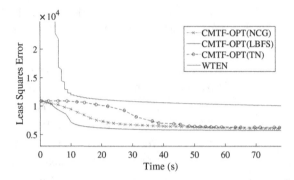

Fig. 5. Convergence of WTEN and CMTF-OPT on MovieLens training data. It is worth to note that the least squares error here is not the error rate that we evaluate in the experiment for final performance comparisons. Since the data is imbalanced, we use AUC as the comparison metric, as shown in Table 4.

6   Conclusion

We proposed a weighting strategy to provide Coupled Tensor Factorization method a capability to handle imbalanced data. Our work suggests three key learning insights. Firstly, our novel weighting strategy enables CTF to perform significantly better on the minority class prediction without jeopardizing the classification of the majority case. Secondly, our experiments demonstrate the impact of the additional matrix on CTF's performance over TF. This finding can serve as a foundation for a general purpose latent factor on imbalanced data learning. Thirdly, our factorization algorithm with weighting scheme can be easily extended to different imbalanced data sets and applications. Although this paper targets binary missing label estimation problem, the weighting strategy can be straightforwardly applied to multiple labels, such as integer ratings where the frequencies of integers are imbalanced. In the future, we are planning to scale up our idea using distributed computing environments.

References

1. Australian bureau of statistics data sets, time series profile of local government areas. http://www.abs.gov.au/websitedbs/censushome.nsf/home/datapacks
2. Acar, E., Kolda, T.G., Dunlavy, D.M.: All-at-once optimization for coupled matrix and tensor factorizations. arXiv preprint arXiv:1105.3422 (2011)
3. Akbani, R., Kwek, S., Japkowicz, N.: Applying support vector machines to imbalanced datasets. In: Boulicaut, J.-F., Esposito, F., Giannotti, F., Pedreschi, D. (eds.) ECML 2004. LNCS (LNAI), vol. 3201, pp. 39–50. Springer, Heidelberg (2004). doi:10.1007/978-3-540-30115-8_7

4. Beutel, A., Talukdar, P.P., Kumar, A., Faloutsos, C., Papalexakis, E.E., Xing, E.P.: Flexifact: Scalable flexible factorization of coupled tensors on hadoop. In: SIAM International Conference on Data Mining (SDM), pp. 109–117 (2014)
5. Bradford, J.P., Kunz, C., Kohavi, R., Brunk, C., Brodley, C.E.: Pruning decision trees with misclassification costs. In: Nédellec, C., Rouveirol, C. (eds.) ECML 1998. LNCS, vol. 1398, pp. 131–136. Springer, Heidelberg (1998). doi:10.1007/BFb0026682
6. Chawla, N.V., Bowyer, K.W., Hall, L.O., Kegelmeyer, W.P.: Smote: synthetic minority over-sampling technique. J. Artif. Intell. Res. **16**, 321–357 (2002)
7. Chawla, N.V., Japkowicz, N., Kotcz, A.: Editorial: special issue on learning from imbalanced data sets. ACM SIGKDD Explor. Newslett. **6**(1), 1–6 (2004)
8. Elkan, C.: The foundations of cost-sensitive learning. In: Proceedings of the 17th International Joint Conference on Artificial Intelligence, vol. 2 (2001)
9. Fawcett, T.: An introduction to ROC analysis. Pattern Recogn. Lett. **27**(8), 861–874 (2006)
10. Harper, F.M., Konstan, J.A.: The movielens datasets: history and context. ACM Trans. Interact. Intell. Syst. **5**(4), 19:1–19:19 (2015)
11. Harshman, R.A.: Foundations of the parafac procedure: models and conditions for an "explanatory" multi-modal factor analysis (1970)
12. He, H., Garcia, E., et al.: Learning from imbalanced data. IEEE Trans. Knowl. Data Eng. **21**(9), 1263–1284 (2009)
13. Jo, T., Japkowicz, N.: Class imbalances versus small disjuncts. ACM SIGKDD Explor. Newslett. **6**(1), 40–49 (2004)
14. Kolda, T.G., Bader, B.W.: Tensor decompositions and applications. SIAM Rev. **51**(3), 455–500 (2009)
15. Koren, Y., Bell, R., Volinsky, C.: Matrix factorization techniques for recommender systems. Computer **8**, 30–37 (2009)
16. Ling, C.X., Yang, Q., Wang, J., Zhang, S.: Decision trees with minimal costs. In: Proceedings of the 21st International Conference on Machine Learning (2004)
17. Liu, W., Chan, J., Bailey, J., Leckie, C., Ramamohanarao, K.: Mining labelled tensors by discovering both their common and discriminative subspaces. In: SIAM International Conference on Data Mining (SDM13), pp. 614–622 (2013)
18. Liu, W., Kan, A., Chan, J., Bailey, J., Leckie, C., Pei, J., Kotagiri, R.: On compressing weighted time-evolving graphs. In: Proceedings of the 21st ACM International Conference on Information and Knowledge Management (CIKM 2012), pp. 2319–2322 (2012)
19. Liu, X.Y., Wu, J., Zhou, Z.H.: Exploratory undersampling for class-imbalance learning. IEEE Trans. Syst. Man Cybern. **39**(2), 539–550 (2009)
20. Mani, I., Zhang, I.: knn approach to unbalanced data distributions: a case study involving information extraction. In: Proceedings of Workshop on Learning from Imbalanced Datasets (2003)
21. Menon, A.K., Elkan, C.: Link prediction via matrix factorization. Mach. Learn. Knowl. Discov. Databases **6912**, 437–452 (2011)
22. Papalexakis, E.E., Faloutsos, C., Mitchell, T.M., Sidiropoulos, N.D.: Turbo-SMT: Accelerating coupled sparse matrix-tensor factorizations by 200x. In: SIAM International Conference on Data Mining (SDM) (2014)

23. Ristanoski, G., Liu, W., Bailey, J.: Discrimination aware classification for imbalanced datasets. In: Proceedings of the 22nd ACM International Conference on Information and Knowledge Management (CIKM 2013), pp. 1529–1532 (2013)
24. Singh, A.P., Gordon, G.J.: Relational learning via collective matrix factorization. In: ACM SIGKDD International Conference on Knowledge Discovery and Data Mining (KDD), pp. 650–658 (2008)
25. Wu, G., Chang, E.Y.: Kba: Kernel boundary alignment considering imbalanced data distribution. IEEE Trans. Knowl. Data Eng. **17**(6), 786–795 (2005)
26. Zhou, Z.H., Liu, X.Y.: Training cost-sensitive neural networks with methods addressing the class imbalance problem. IEEE Trans. Knowl. Data Eng. **18**(1), 63–77 (2006)

Mining Actionable Knowledge Using Reordering Based Diversified Actionable Decision Trees

Sudha Subramani[1(✉)], Hua Wang[1], Sathiyabhama Balasubramaniam[2], Rui Zhou[1],
Jiangang Ma[1], Yanchun Zhang[1], Frank Whittaker[1], Yueai Zhao[3],
and Sarathkumar Rangarajan[1]

[1] Centre for Applied Informatics, College of Engineering and Science, Victoria University,
Melbourne, Australia
{sudha.subramani1,sarathkumar.rangarajan}@live.vu.edu.au,
{hua.wang,Rui.Zhou,jiangang.ma,yanchun.zhang,
Frank.Whittaker}@vu.edu.au
[2] Department of Computer Science and Engineering, Sona College of Technology,
Salem, Tamilnadu, India
sathiyabhama@sonatech.ac.in
[3] Taiyuan Normal University, Taiyuan, Shanxi, China
tysyzya@sina.com

Abstract. Actionable knowledge discovery plays a vital role in industrial problems such as Customer Relationship Management, insurance and banking. Actionable knowledge discovery techniques are not only useful in pointing out customers who are loyal and likely attritors, but it also suggests actions to transform customers from undesirable to desirable. Postprocessing is one of the actionable knowledge discovery techniques which are efficient and effective in strategic decision making and used to unearth hidden patterns and unknown correlations underlying the business data. In this paper, we present a novel technique named Reordering based Diversified Actionable Decision Trees (RDADT), which is an effective actionable knowledge discovery based classification algorithm. RDADT contrasts traditional classification algorithms by constructing committees of decision trees in a reordered fashion and discover actionable rules containing all the attributes. Experimental evaluation on UCI benchmark data shows that the proposed technique has higher classification accuracy than traditional decision tree algorithms.

Keywords: Data mining · Actionable knowledge discovery · Postprocessing · Decision tree

1 Introduction

Data mining is the process of extracting interesting information from large databases [9]. It is also referred as knowledge discovery in databases (KDD). Data mining or knowledge extraction from data may also be viewed as the process of turning the data into information. Actionable knowledge discovery is viewed as turning information into

© Springer International Publishing AG 2016
W. Cellary et al. (Eds.): WISE 2016, Part I, LNCS 10041, pp. 553–560, 2016.
DOI: 10.1007/978-3-319-48740-3_41

action and the action into value or profit. Hence, the term actionable refers to the mined patterns suggesting concrete and profitable actions to the decision maker [7]. The Domain driven Data Mining (D^3M) is an actionable knowledge discovery-based problem-solving system in the data mining process [11].

Extensive research in data mining has been done on the underlying data to discover patterns. An action is useful if it validates the patterns that users are interested and possess profitable value. The current applications of data mining in Customer Relationship Management (CRM) only aim at constructing customer profiles, which predict the customers' characteristics of certain classes. Examples of these classes are: Who are likely loyal and disloyal customers? This interpretation is useful but it does not benefit the enterprise directly [15]. To improve the customer relationship and to increase the profit, the enterprise must know what actions are to be taken to change customers from an undesired status to a desired one.

For example, instilling the business intelligence into the big data warehousing systems is one of the challenges facing system analysts in the past. Limiting human intervention in decision making process and extracting management decisions from data is the need-of-the-hour in real world scenarios. The actionability of a model is the ability to identify a set of changes to the input features that transforms the model prediction of this input to the desired output. This model will be of great help to banking sectors, in ascertaining the credit score of lenders of loans. This will also empower them with suggestions on how to improve a customer's credit score.

2   Related Work

Several studies have been developed to extract knowledge from CRM data. Various KDD techniques such as association rules, Bayesian networks, Support vector machines, decision trees, nearest neighbour, neural networks, and clustering have achieved phenomenal success in CRM of industrial applications, they just stop producing the results at the stage of data-centred knowledge discovery.

Lift chart [3] is the approach to rank the customers by the likelihood of responding to marketing actions and probability estimation. Metacost framework [6] is another methodology to determine accuracy based classification depends on cost matrix. Reinforcement learning is deployed to tackle the issues in sequential decision making [8].

Most of the above techniques have not given consideration to the resource constraints and cost effectiveness, which is the most important part of actionable knowledge discovery. Yang et al. [15] proposed an algorithm to postprocess the decision trees and generate actionable rules. The power of decision trees and postprocessing to make actionable rules in CRM is well illustrated in this paper.

Ensemble decision tree classification methods like Bagging [1], Boosting [4], Random Forests [2], and Cascading and Sharing trees (CS4) [10] have shown promise in achieving higher classification accuracy than the single classifier method like ID3 and C4.5 [12]. Ensemble decision tree classification methods reduce the bias by effectively aggregating decisions from diversified classifiers and it is effective for high-dimensional datasets like CRM, stock exchange and banking.

Bagging [1] and boosting [4] use a bootstrap technique to resample the training data for each iteration in tree construction. As a result, some bagging and boosting rules may not be true when applied to the original training data as they only approximate the true rules. Li et al. [10] presented a method called CS4 to construct committees of decision trees for classification. This method built member trees by considering different top-ranked features as the root nodes. Hence, apart from the top level attribute, other data in the tree are shared and number of trees may use some data repeatedly. Thus, noise from one data may affect most of the other trees.

With the recent advances in actionable knowledge discovery algorithms [13, 14], D^3M becomes more customer oriented method to serve organizations. As many patterns are mined, the organizations are unaware of, what follow-up actions are needed to improve the customer retention rate and decrease customer churning rate. Hence this paper provides the optimized decision support framework named Reordering based Diversified Actionable Decision Trees (RDADT) to promote the paradigm transfer from data centred mining to domain driven knowledge discovery in CRM.

3   Proposed Work

Problem definition: In the existing ensemble decision tree methods, the attributes are overlapped among the multiple trees, which significantly affect the actionable knowledge discovery derived in the latter stage. The noise data of any attribute affects the performance of all trees. Hence, RDADT construct trees in a reordered fashion and guarantees the trees are diversified and unique at both the parent and children node level. Even, if the expression level of one attribute is wrong, it affects that particular tree and doesn't propagate to other trees. This algorithm promises to deliver higher classification accuracy and robustness in actionable knowledge discovery. The overall process is explained in the following four steps.

3.1   Import Customer Data with Data Preprocessing

The data are usually collected from the various databases, data warehouses, and data marts. Data preprocessing usually includes common tasks namely outlier detection, generalization, normalization and scaling [5].

3.2   Construction of Unique and Diversified Decision Trees

The aim is to construct multiple decision trees by reordering of attributes. RDADT uses C4.5, a heuristic method to compute the gain ratio for all the attributes over entire dataset in order to get split thresholds and finally rank them into an ordered list based upon their gain value [12]. In RDADT, however, most are not standard C4.5 trees in the decision committee. This is because instead of using only the top ranked attribute as the splitting node, RDADT employs a wide range of attribute choices for building the trees. Thus reordering approach creates maximum diversity in each tree both globally at root node level and locally at children nodes' level.

For the first decision tree, first few top ranked attributes are considered. Similarly second, third and fourth decision trees are built on subsequent attributes. In each decision committee, unique set of attributes are participated in order to estimate the class labels of tree leaves. As a result, all trees are unique and do not share common attributes. These process repeats until all the attributes are used in the decision committee.

This proposed idea brings the following merits:

- In the tree construction stage, RDADT avoids the overlapping of attributes among alternative trees because the attributes that are used in one tree would not be participated in the successive trees. Thus, RDADT guarantees that constructed trees are truly unique and maximizes the diversity of the final classifiers. Hence, if any attribute contains noise, incomplete, inconsistent or missing values, it only affects a particular tree and not propagates to other trees. Thus, RDADT based classifier model ensures improved robustness than the standard approaches in tolerating noise data and significantly ensures better performance and accuracy.
- RDADT generates multiple trees that are both smaller and more accurate because only limited attributes are participating in each decision committee. Hence, the interpretation of rules will be more precise and accurate and there is no additional pruning required which reduces the size of decision trees.

3.3   Leaf Node Search Algorithm

An action is a change, in which an attribute's value is transformed from x_1 to x_2 where x_1 is in a low status and x_2 is a highly desirable one. After a decision tree is built using RDADT, it is given as input to the leaf-node search algorithm which searches for optimal actions to transfer each leaf node to another leaf node with a higher probability of being in a more desirable class [15].

3.4   Select the Cost Effective Actions

The best action can be selected from the action set by two steps namely attribute type and level. The first step is based on an attribute type. Certain attributes like instalment rate, debtors and month duration are soft attributes and they can be changed with reasonable costs. Remaining attributes like gender and age are core attributes. Hence, changes that made on soft attributes are only considered. The second step is based on the number of levels in attribute changes. The more number of changes in an attribute to attain a desirable class will incur lower gain rather than fewer changes. Thus, the best action is preferred based on the attributes which are having minimal changes.

3.5 RDADT Algorithm

```
1. Build Decision Tree (D,T,Aᵢ)

Input: Customer Dataset D, set of attributes Aᵢ.
Output:  Multiple diversified decision trees T.
Method:
    begin
      let T = Φ;
           for i=1 to n do
               Rank attributes Aᵢ based on gain ratio;
               Build decision trees Tᵢ from D based on
        attribute ranking;
                   T=T U Tᵢ;
               end for
        return T decision trees;
      end.

2. Action set (T,X)

Input: Multiple decision trees Tᵢ where i=1 to n.
Output: Set of K action sets.
Method:
    begin
      for i=1 to n do
          call leaf node search algorithm to calculate
          the net profit for iᵗʰ decision tree;
      end for
    return k action sets correspond to median net profit;
      end.
```

4   Experimental Setup Details

The experiment is conducted on the UCI German dataset [16] to evaluate the performance of algorithms in Java JDK 1.5. This dataset classifies people described by a set of attributes as good or bad credit risks. It has 1000 instances and 20 attributes which includes the status of existing checking account, credit history, duration in month, purpose, credit amount, and so on.

In RDADT, all the 20 attributes are ranked based upon the gain ratio and the different unique trees are constructed until it reaches the pre-set value. The first decision tree is built with the first top ranked attributes and second tree is built with the next ranked attributes. Subsequently remaining trees are built in the same manner.

RDADT is compared with other popular classifiers like c4.5, naïve bayes and also ensemble classifiers such as bagging, boosting and random forest respectively. The accuracy rate is determined by number of correctly classified instances and incorrectly classified instances. The percentage of correctly classified instances of naïve bayes, C4.5, bagging, boosting, random forest, and RDADT are 75.4 %, 70.5 %,

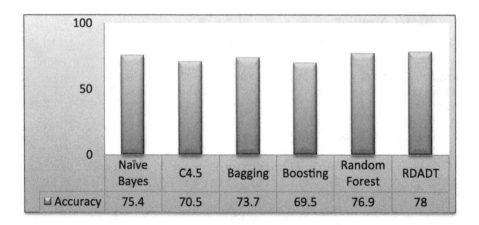

Fig. 1. Classification Accuracy of all models

73.7 %, 69.5 %, 76.9 % and 78 % respectively. The above results show that RDADT attains good accuracy than other classifiers (Fig. 1).

After the decision trees are built, unique set of rules are extracted from each tree. Each of the n trees in the committee has k rules to predict the class labels for the sample T.

Each of $rule_{good}^{i}$ ($1 \leq i \leq k_1$) predicts T to be in the positive class (good risk), while each of $rule_{bad}^{i}$ ($1 \leq i \leq k_2$) predicts T to be in the negative class (bad risk). Here $k_1 + k_2 = k$.

$$Count_G = \sum\nolimits_{i=1}^{k_1} rule_{good} \tag{1}$$

$$Count_B = \sum\nolimits_{i=1}^{k_2} rule_{bad} \tag{2}$$

After the rules are extracted, the best actions are selected using leaf node search algorithm [15]. The possible status changes from negative to positive class are performed in the validation part. Again the $Count_G$ and $Count_B$ are calculated in validation which is denoted by new_$Count_G$ and new_$Count_B$. Thus the status change for each decision tree is calculated as:

$$status\ change = \frac{Count_B - New\_Count_B}{Count_B} * 100\% \tag{3}$$

The first tree is having status change rate of 80 % and second, third, fourth and fifth trees are having change rate of 70 %, 67 %, 85 % and 71 % respectively. Status change rate is calculated based upon the rules generated from a tree and their role in converting customers from bad class to good class.

The goal is to classify the loan applicant into one of two categories either good or bad which is the class attribute and to suggest actions that transfer the loan applicant from bad to good credit risk. When the applicant (customer) is having marginal value in the splitting attribute it is then determined by the appropriate class.

5   Conclusion and Future Work

In this paper, an effective actionable knowledge discovery based classification algorithm namely RDADT is designed to produce potential actionable outcomes rather than passive output. RDADT provides highly intelligible rules that help in actionable knowledge discovery. This method forces all the attributes to participate in the decision committee based on ranking order. By this way, all trees are unique and create maximum diversity both globally at root node level and locally at children nodes' level. To assess the performance of RDADT, German dataset of customers' loan application information is used. RDADT classifies the applicants into appropriate target class either good or bad and provides loan to the desired customers. These actionable outcomes are mainly used in strategic decision making applications such as increasing the profit, improving the quality of products and offering intelligent and effective solutions for banking, finance institutions and insurance companies. Further, Adaptive Ant Colony Optimization technique can be deployed into the RDADT to improve the classification accuracy which may be suitably applied to clinical databases for effective knowledge discovery.

References

1. Breiman, L.: Bagging predictors. In: Machine Learning, pp. 123–140 (1996)
2. Breiman, L.: Random forests–random features. Technical report 567, University of California, Berkley (1999)
3. Huang, J., Ling, C.X.: Using AUC and accuracy in evaluating learning algorithms. IEEE Trans. Knowl. Data Eng. **17**, 299–310 (2005)
4. Freund, Y., Schapire, R.E.: Experiments with a new boosting algorithm. In: International Conference on Machine Learning, pp. 148–156 (1996)
5. Han, J., Kamber, M., Pei, J.: Data Mining Concepts and Techniques, 2nd edn., Morgan Kaufmann (2006)
6. Domingos, P.: Metacost: a general method for making classifiers cost-sensitive. In: Proceedings of the Fifth ACM SIGKDD International Conference on Knowledge Discovery and Data Mining, pp. 155–164. ACM (1999)
7. He, Z., Xu, X.: Datamining for actionable knowledge: a survey. In: CoRR (2005)
8. Pednault, E., Abe, N., Zadrozny, B.: Sequential cost-sensitive decision making with reinforcement learning. In: Proceedings of the Eighth ACM SIGKDD International Conference on Knowledge Discovery and Data Mining, pp. 259–268. ACM (2002)
9. Lakshmi, B.N., Raghunandhan, G.H.: A conceptual overview of data mining. In: Proceedings of the National Conference on Innovations in Emerging Technology, pp. 27–32 (2011)
10. Li, J., Liu, H.: Ensembles of cascading trees. In: Proceedings of the IEEE International Conference on Data Mining (ICDM 2003), pp. 585–588 (2003)
11. Longbing, C.: Introduction to domain driven data mining. In: Data Mining for Business Applications, pp. 3–10. Springer (2009)
12. Quinlan, J.R.: C4.5 Programs for Machine Learning. Morgan Kaufmann (1993)
13. Cui, Z., Chen, W., He, Y., Chen, Y.: Optimal action extraction for random forests and boosted trees. In: Proceedings of the 21th ACM SIGKDD International Conference on Knowledge Discovery and Data Mining, pp. 179–188. ACM (2015)

14. Vadivu, P.S., David, V.K.: Optimized feature extraction and actionable knowledge discovery for Customer Relationship Management (CRM). In: Proceedings of the Second International Conference on Computational Science, Engineering and Information Technology, pp. 275–282. ACM (2012)
15. Yang, Q., Yin, J., Ling, C.X., Pan, R.: Extracting actionable knowledge decision trees. IEEE Trans. Knowl. Data Eng. **18**(12), 43–56 (2007)
16. http://archive.ics.uci.edu/ml/datasets/Statlog+(German+Credit+Data)

Improving Distant Supervision of Relation Extraction with Unsupervised Methods

Min Peng[1], Jimin Huang[1], Zhaoyu Sun[1], Shizhong Wang[2], Hua Wang[3],
Guangping Zhuo[4], and Gang Tian[1(✉)]

[1] School of Computer, Wuhan University, Wuhan, China
{pengm,huangjimin,sunzhaoyu,tiang2008}@whu.edu.cn
[2] School of Economics and Management, Wuhan University, Wuhan, China
szwang@whu.edu.cn
[3] Centre for Applied Informatics, Victoria University, Melbourne, Australia
hua.wang@vu.edu.au
[4] Department of Computer Science, Taiyuan Normal University, Taiyuan, China
zhuoguangping@163.com

Abstract. Distant supervision has been widely adopted in relation extraction task since it does not require any labeled data. It can automatically align knowledge base with corpus to generate training data. However, the intuition base of alignment in this method may fail, resulting in wrong label problem. In this paper, we try to improve the intuition of distant supervision from the perspective of relation mentions, and propose a novel method called *Clustered DS* which employs our improved intuition in an unsupervised manner. By incorporating the information about the distribution of relation mentions, our method achieves a more precise alignment, thus it significantly reduces the number of wrong labels. Experimental results demonstrate the advantage of *Clustered DS* over existing distant supervision methods and show the effectiveness of our improved intuition.

Keywords: Information extraction · Relation extraction · Distant supervision · Unsupervised method

1 Introduction

Distant supervision(DS) [1] is one of the most promising methods in information extraction(IE) to extend traditional supervised methods to web-scale dataset. It automatically generates training data by aligning an external knowledge base with free texts. In this paper, we focus on the task of relation extraction(RE), one of the subproblems of IE, which seeks to extract relations from corpora such as Wikipedia. For example, given an *entity pair* (Barack Obama, United States), the task of RE aims at extracting new relation instance (Barack Obama,

This research is supported by the Natural Science Fundation of China(No. 61472291), and the Natural Science Fundation of Hubei Province, China(No. ZRY2014000901).

W. Cellary et al. (Eds.): WISE 2016, Part I, LNCS 10041, pp. 561–568, 2016.
DOI: 10.1007/978-3-319-48740-3_42

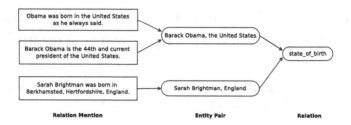

Fig. 1. Training examples generated via DS.

state_of_birth, United States) from the first sentence of Fig. 1. The alignment of DS in this field is based on an intuition, or a hypothesis, which leverages
external knowledge base. Given a knowledge base D and a corpus L, the intuition assumes that for any entity pair (e_1, e_2) belonging to the relation r in D,
all relation mentions of (e_1, e_2) express the same relation r. For instance, we can
generate 3 training examples from sentences in Fig. 1.

Though the intuition can significantly reduce human efforts in generating training data, it introduces two major challenges into the task. First, not
all relation mentions express the same relation as their entity pairs, which
is called the challenge of *false positive*. For instance, as shown in Fig. 1, the
entity pair (Barack Obama, United States) has two relations: state_of_birth
and employed_by. However, relation mentions will always be labeled with
state_of_birth, no matter which relation they actually express. Riedel et al.
[2] reported that there are over 31 % of false positives in the NYT corpus when
they implied the intuition with Freebase [3]. Second, knowledge bases are usually
highly incomplete. There are massive relation mentions explicitly expressing one
of known relations, but their entity pairs are not in the knowledge base. For
example, Min et al. [4] showed that 93.8 % of persons from Freebase have no
state_of_birth. These relation mentions will be generated as *false negatives*
via the intuition of DS.

Different from previous researches, we improve the intuition of DS from the
perspective of relation mentions to address these two challenges. Rather than
focusing on entity pairs, we leverage the similarity between relation mentions
to achieve a more precise alignment. Along this line, we propose a novel two-
stage method, called *Clustered DS*. At the first stage, an unsupervised method
is adopted to divide relation mentions into different bags, using the information
of the relation mentions instead of their entity pairs. At the second stage, we
assign relation mentions from the same bag to the same relation that holds most
of entity pairs in the knowledge base. Experimental results show that *Clustered
DS* yields an average increase of 3 % F1 compared with previous DS methods,
indicating the effectiveness of our method and the improved intuition.

2   Related Work

Since it was first proposed by Craven et al. [1], distant supervision has gained
wide attraction in the relation extraction field as it can automatically generate

training data without labeled data [2,4–14]. To tolerate the wrong labels introduced by the intuition of distant supervision, Riedel et al. [2] adopt Multiple Instance Learning(MIL) with the assumption that at-least-one of the relation mentions in each bag express the relation assigned via the intuition of distant supervision. MultiR [9] and Multi-Instance Multi-label(MIML) learning [10] further support multiple relations expressed by different sentences in a bag.

As these methods bypass the challenge of false negative, Min et al. [4] introduce a new latent variable in MIML, which allows them to learn from only positive and unlabeled data. Takamatsu et al. [11] model the probabilities of a pattern expressing relations and remove mentions that match low-probability patterns. Similar to our method, Min et al. [14] relabel examples to their most likely relations. However, they estimate the probabilities of patterns rather than relation mentions.

Several other researches also attempt to improve distant supervision with supervised learning. Pershina et al. [12] propose *Guided DS* to utilize labeled data to guide MIML during training. Angeli et al. [13] apply active learning to the distant supervision and design a novel criterion to sample examples which are both uncertain and representative. Compared with their methods, our *Clustered DS* gives the same improvement without requiring any labeled data.

3 Intuition Improvement

In this paper, our goal is to reduce the wrong labels introduced by the generation of training data. From the perspective of relation mentions, we observe that relation mentions of the same entity pair may express different relations. At the same time, those expressing the same relation may not share the same entity pair. We turn to explore what is the crux between relation mentions expressing the same relation if their entity pairs differ. Obviously, the relation mentions expressing the same relation are similar to each other in the semantic level. For instance, sentences of (Barack Obama, United States) and of (Sarah Brightman, England) are highly similar and indeed express the same relation, as shown in Fig. 1. Based on our observation, we make an assumption about the distribution of relation mentions in the semantic space. We assume that relation mentions expressing the same relation are more similar than those expressing different relations. We then improve the intuition of DS with our assumption, which also can be deemed as a two-phase procedure:

1. Divide relation mentions into bags according to the similarity between relation mentions.
2. Align the whole bag to the relation that holds most of the entity pairs of the bag in the knowledge base.

The main difference of our improved intuition, motivated by our observation, is that we use the similarity of relation mentions to guide the generation of the division. According to our assumption, it guarantees that relation mentions from the same bag indeed express the same relation. In fact, the intuition of DS

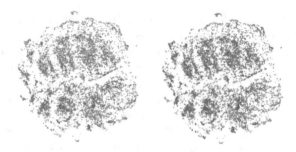

Fig. 2. The division generated by the traditional intuition(Left) and our improved intuition(Right). A point represents a relation mention in the semantic space whose color denote the bag it belongs. Best viewed in color. (Color figure online)

latently makes the same commitment, but the division generated based on entity pairs is not precise enough to support it. In Fig. 2, we present a 2D visualization of the division generated by the traditional intuition and our improved intuition using *t-Distributed Stochastic Neighbor Embedding*(t-SNE)[15].

It is now capable for relation mentions with different entity pairs to be divided to the same bag that is likely to express the same relation. Additionally, relation mentions whose entity pairs never appear in the knowledge base are now possible to be aligned to a known relation, rather than directly labeled with `no_relation`. Therefore, we can address both two issues with our improvement. Based on our improved intuition, we propose a novel two-stage method, called *Clustered DS*.

4   Division Generation

In the first stage of *Clustered DS*, an unsupervised method is adopted to generate the division of relation mentions. To deal with millions of relation mentions, Kmeans clustering, one of unsupervised methods, is a promising choice that is simple and efficient to perform in web-scale. The method iteratively refines centers in two steps with k initial cluster centers:

1. Assign datum to its nearest center.
2. Update cluster centers with the mean of its belonged datum.

The algorithm converges when there is no change in assignment. In our method, we first extract feature vectors from relation mentions by leveraging the feature set developed by Surdeanu et al. [16], which is the finest handcrafted feature set in relation extraction. Kmeans clustering then takes the input as feature vectors and generate k bags. It naturally leverages our improved intuition to reform the division according to the similarity between relation mentions.

The extracted vector is sparse and high-dimensional, which means we are unable to preform methods that automatically decides the cluster number k such as the *Chinese Resturant Process* [17]. Nevertheless, it is reasonble to set k twice the number of known relations or even higher, for the given set of relations obviously is just a small part of all relations. More importantly, even if the cluster number is much higher than the number of relations expressed in the corpus, small clusters can be assigned with the right relation in the following stage if the division is precise enough.

5   Relation Assignment

In the second stage of *Clustered DS*, bags generated in the previous stage are aligned to the known relations or no_relation. Since relation mentions from the same bag no longer share the same entity pair, we align the whole bag to the relation holding most of entity pairs of that bag. Given a bag M, the probability of the relation r is calculated in two steps:

1. For each relation mention m of M, calculate the probability $P(m \in r)$ according to its entity pair.

$$P(m \in r) = \begin{cases} 0 & \text{if entity pair of } m \notin r, \\ 1 & \text{if entity pair of } m \in r. \end{cases} \qquad (1)$$

2. Calculate the probability $P(M \in r)$ by normalizing the sum of $P(m \in r)$.

$$P(M \in r) = \frac{\sum_{m \in M} P(m \in r)}{|M|} \qquad (2)$$

We then assign M the most possible relation r.

6   Experiments

6.1   Dataset

We use the KBP dataset publicly released by Surdeanu et al. [10]. The dataset contains 1.5 million documents and nearly 110 000 entities. We analyze nearly 1 million relation mentions after extracting. The KBP shared task requires to extract all latent candidates from the corpus when the relation and the first entity are known. We use 200 queries from the 2010 and 2011 evaluation during the experiments, where 40 queries are used as development set and the rest are used as testing set. We adopt the adapted KBP scorer as Surdeanu et al. [10].

6.2   Implementation Detail

In the implementation of *Clustered DS*, we apply the *MiniBatchKmeans* [18] clustering from the open *scikit-learn* package [19] rather than Kmeans clustering for its much faster convergence speed. Our method generates a more precise training dataset, which allows us to incorporate previous DS methods as improved classifier. In this paper, we implement our method with several previous DS methods, including: (1) *Mintz++* [5], an improved implementation of DS as a strong baseline. (2) *MultiR* [9](denoted as *Hoffmann*), a multi-instance multi-label algorithm and (3) *MIML* [10], another multi-instance multi-label algorithm but trained with Bayesian framework. We then compare our implementation with the incorporated DS methods in the real world dataset.

6.3 Results

6.3.1 Improve Distant Supervision Performance

In this section, we summarize all of the results in Table 1. It shows that when using the training data generated by *Clustered DS*, both Mintz++, Hoffmann and MIML yield better performance. When employing MIML as the classifier, *Clustered DS* achieves the best performance, which is comparable to the result generated by the MIML model with an external set of labeled data (Angeli) [13]. Though we can not incorporate Angeli in our method for its published model is already trained with DS, it is still reasonable to believe that we can use an external set of labeled data to improve our method as Angeli. Table 1 also shows that the time consumption increased by *Clustered DS* is negligible when compared with the training of classifier.

Table 1. A summary of the results testing over KBP dataset. The bold items denote the best performance among all systems.

System	DS				Clustered DS			
	P	R	F1	Time	P	R	F1	Time
Hoffmann	30.65	19.79	24.05	1 h	25.15(±2.95)	30.15(±2.86)	27.03(±1.09)	1.1 h
Mintz++	28.60	23.78	25.97	3 h	28.04(±2.08)	27.32(±1.57)	27.59(±1.55)	3.1h
MIML	28.00	28.30	28.15	6 h	**30.94(±2.83)**	33.51(±3.63)	31.66(±2.14)	6.1h
Angeli	29.40	**35.07**	**31.99**	6.3 h				

In Fig. 3, we show the P/R curves of our implementations compared with their incorporated DS methods. When we employ Mintz++ as the classifier of our method, the overall performance shows nearly 2 % F1 increase comparing with the original method. Mintz++ is the original DS method in relation extraction, which can be deemed as a basic classifier. It appears that our method can significantly reduce wrong labels in training data introduced by the intuition of DS. Furthermore, for other improved DS methods such as Hoffmann and MIML, which derive a more complex classify model to tolerate noisy inputs, the system

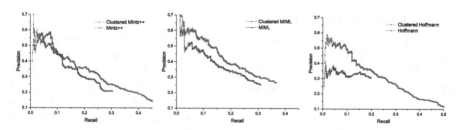

Fig. 3. The result of *Clustered DS* with DS methods incorporated as classifier which is shown in red curve. We also compare our implementations with the incorporated method which is shown in blue curve. (Color figure online)

of *Clustered DS* can yield an improvement of 3.5 % F1. The reason why improved DS methods benefit more is that our method introduces new wrong labels due to an unsuccessful division. Thus those methods are more robust to leverage our training data when they can tolerate wrong labels.

6.3.2   Cluster Number Setting

The Kmeans clustering adopted in our method is highly sensitive to the cluster number k. However, we can set k much higher than the number of known relations. Our method still performs well. Here we implement a series of *Clustered DS* with employing Mintz++ as classifier and set k to {50, 100, 150, 200, 250}, for the number of known relation in the KBP dataset is 42. The context of implementations is guaranteed to be the same except k. The performance is displayed in Fig. 4 with Mintz++ as the baseline. As shown in Fig. 4, we note that when k is set above the number of known relations, our method reaches the best performance around twice of the number of known relations and provides a consistent performance when k keeps growing. There is a slow decline when the cluster number keeps growing, due to the generation of more small clusters with a higher k. Nevertheless, our method still outperforms Mintz++ for we can aggregate small clusters in the stage of relation assignment.

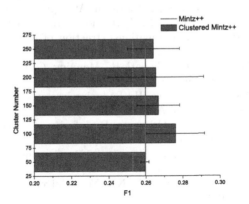

Fig. 4. The performance of a series of Clustered DS with different cluster number.

7   Conclusion

Traditional distant supervision methods in relation extraction face challenges of false positives and false negatives due to the invalid intuition base. In this paper, we proposed a novel two-stage method called *Clustered DS* to tackle with these two challenges. By using our improved intuition of distant supervision, we adopted an unsupervised method to generate a more precise division, resulting in a considerable decrease of wrong labels in training data. We showed that *Clustered DS* can improve the distant supervision methods, with an increment of

3 % in average F-score on the KBP dataset. Empirically, we noted that *Clustered DS* performs well as long as the cluster number is set above the number of known relations. In the future, we plan to combine our method with other approaches to further improve the performance.

References

1. Craven, M., Kumlien, J.: Constructing biological knowledge bases by extracting information from text sources. In: ISMB (1999)
2. Riedel, S., Yao, L., McCallum, A.: Modeling relations and their mentions without labeled text. In: PKDD (2010)
3. Bollacker, K.D., Cook, R.P., Tufts, P.: Freebase: a shared database of structured general human knowledge. In: AAAI (2007)
4. Min, B., Grishman, R., Wan, L., Wang, C., Gondek, D.: Distant supervision for relation extraction with an incomplete knowledge base. In: NAACL (2013)
5. Mintz, M., Bills, S., Snow, R., Jurafsky, D.: Distant supervision for relation extraction without labeled data. In: ACL (2009)
6. Bunescu, R.C., Mooney, R.J.: Learning to extract relations from the web using minimal supervision. In: ACL (2007)
7. Wu, F., Weld, D.S.: Autonomously semantifying wikipedia. In: CIKM (2007)
8. Nguyen, T.V.T., Moschitti, A.: End-to-end relation extraction using distant supervision from external semantic repositories. In: ACL (2011)
9. Hoffmann, R., Zhang, C., Ling, X., Zettlemoyer, L.S., Weld, D.S.: Knowledge-based weak supervision for information extraction of overlapping relations. In: ACL (2011)
10. Surdeanu, M., Tibshirani, J., Nallapati, R., Manning, C.D.: Multi-instance multi-label learning for relation extraction. In: EMNLP (2012)
11. Takamatsu, S., Sato, I., Nakagawa, H.: Reducing wrong labels in distant supervision for relation extraction. In: ACL (2012)
12. Pershina, M., Min, B., Xu, W., Grishman, R.: Infusion of labeled data into distant supervision for relation extraction. In: ACL (2014)
13. Angeli, G., Tibshirani, J., Wu, J., Manning, C.D.: Combining distant and partial supervision for relation extraction. In: EMNLP (2014)
14. Min, B., Li, X., Grishman, R., Sun, A.: New York University 2012 system for KBP slot filling. In: TAC (2011)
15. Van der Maaten, L., Hinton, G.: Visualizing data using t-SNE. J. Mach. Learn. Res. 9(2579–2605), 85 (2008)
16. Surdeanu, M., Gupta, S., Bauer, J., McClosky, D., Chang, A.X., Spitkovsky, V.I., Manning, C.D.: Stanford's distantly-supervised slot-filling system. In: TAC (2011)
17. Aldous, David, J.: Exchangeability and related topics. In: Hennequin, P.L. (ed.) École d'Été de Probabilités de Saint-Flour XIII — 1983. LNM, vol. 1117, pp. 1–198. Springer, Heidelberg (1985). doi:10.1007/BFb0099421
18. Sculley, D.: Web-scale k-means clustering. In: WWW (2010)
19. Michel, V., Pedregosa, F., Passos, A., VanderPlas, J., Weiss, R., Dubourg, V., Duchesnay, E., Grisel, O., Cournapeau, D., Blondel, M., Varoquaux, G., Prettenhofer, P., Thirion, B., Perrot, M., Gramfort, A., Brucher, M.: Scikit-learn: machine learning in python. CoRR abs/1201.0490 (2011)

Author Index

Aksoy, Cem I-199
Allahyari, Mehdi I-263
Anagnostopoulos, Ioannis I-174
Arruda, Narciso II-196

Baas, Frederique II-35
Bai, Quan I-211
Bai, Rufan II-309
Bakaev, Maxim I-252
Balasubramaniam, Sathiyabhama I-553
Balke, Wolf-Tilo I-124
Bartoletti, Massimo II-55
Belghaouti, Fethi II-157
Borland, Ron II-146
Bornea, Mihaela II-298
Bouguettaya, Athman II-223
Bouzeghoub, Amel II-157
Brambilla, Marco I-140
Bu, Ning II-324
Bus, Olivier II-35

Cai, Xiongcai I-61
Cao, Jie I-343
Cao, Jinli I-294
Cao, Jiuxin I-3
Caraballo, Alexander Arturo Mera I-417
Casanova, Marco Antonio I-417, II-196,
    II-238
Ceri, Stefano I-140
Chen, Liang II-223
Chen, Long I-278
Chen, Yunfang II-290
Chiky, Raja II-157
Chowdhury, Nipa I-61
Clifton, David A. II-121
Constantin, Camelia II-275
Cui, Lizhen I-352
Cui, Zhan II-361

Dass, Ananya I-199
Deng, Jiayuan I-19
Deng, Shuiguang I-92
Dimitriou, Aggeliki I-199
Ding, Minjie II-424

Do, Quan I-537
Dolby, Julian II-298
Dong, Dan I-3
Dong, Hai II-223
Dou, Chenxiao II-172
du Mouza, Cedric II-275
Du, Xiaofeng II-361

Feng, Dengguo I-433
Feng, Ling II-121
Feng, Shi I-227, I-513
Fleyeh, Hasan II-371, II-385
Fokoue, Achille II-298
Frasincar, Flavius I-380, II-35
Fu, Jinghua I-352
Fu, Xi II-187

Gaedke, Martin I-252
Gan, Zaobin I-243
Gao, Hong II-259
Gao, Shang I-405
Gao, Xiaofeng I-489
Gao, Yang I-77
Gayoso-Cabada, Joaquín II-43
Guan, Chun II-397
Gulla, Jon Atle II-3
Guo, Guibing I-278
Guo, Lipeng II-309
Gupta, Shivani II-340

Hansson, Karl II-385
He, Xiaofeng I-473
Heil, Sebastian I-252
Hiranandani, Gaurush II-340
Hong, Cheng I-433
Hu, Jun II-397
Hu, Ping II-424
Hu, Zhigang I-396, I-447
Huang, Heyan I-525
Huang, Jimin I-561

Jhamtani, Harsh I-190
Ji, Donghong I-19

Ji, Wenqian   I-489
Jia, Jia   II-121
Jiang, Bo   I-109
Jose, Joemon M.   I-278

Kantere, Verena   I-457
Kazi-Aoul, Zakia   II-157
Kementsietsidis, Anastasios   II-298
Kochut, Krys   I-263
Kong, Boyuan   I-489
Koniaris, Marios   I-174
Kontarinis, Alexandros   I-457
Koziris, Nectarios   I-457
Krishnan, Adit   I-157
Kuzmanovic, Aleksandar   II-105

Lande, Stefano   II-55
Lei, Jiankun   II-309
Leme, Luiz André Portes Paes   I-417, II-238
Leonardi, Chiara   I-140
Li, Dai   II-290
Li, Fengyun   I-405
Li, Jianzhong   II-259
Li, Peng   I-50, II-85, II-187
Li, Qi   II-121
Li, Rui   I-50, I-109, I-505, II-85, II-187
Li, Yaxin   I-35
Li, Yifan   II-275
Li, Yifu   I-433
Liang, Jiguang   I-109
Liang, Shangsong   II-70
Liu, Bo   I-3
Liu, Chang   I-243, I-447
Liu, Fei   I-294
Liu, Jie   I-363
Liu, Jinglian   I-513
Liu, Peng   II-3
Liu, Qing   I-211
Liu, Shijun   I-352
Liu, Wei   I-537
Long, Guoping   II-324
Long, Jun   I-396, I-447
Lopes, Giseli Rabello   I-417, II-238
Lu, Hongwei   I-243
Lu, Jianguo   I-35
Luo, Xucheng   II-105

Ma, Jiangang   I-553
Ma, Jun   II-70

Ma, Pengju   I-489
Ma, Wenjing   II-324
Ma, Yanlin   I-35
Ma, Zhuo   I-3
Maarry, Kinda El   I-124
Maneriker, Pranav   II-340
Mao, Xian-Ling   I-525
Marotta, Adriana   II-436
Massa, Alessandro   II-55
Mauri, Andrea   I-140
Mehta, Sameep   I-157
Menendez, Elisa S.   II-196
Modani, Natwar   I-190, II-340

Nguyen, Thin   II-137, II-146
Niu, Jiangbo   I-396
Niu, Shuzi   II-324
Nunes, Bernardo Pereira   I-417, II-238
Nunes, Nuno   II-349
Nyberg, Roger G.   II-404

Osinga, Alexander   II-35

Padmanabhan, Deepak   I-157
Paes Leme, Luiz A.   II-196
Pang, Jun   II-247
Peng, Min   I-561
Peng, Xuezheng   I-489
Pequeno, Valeria M.   II-196
Pham, Thanh   I-537
Phung, Dinh   II-137, II-146

Qian, Tieyun   II-94
Qin, Yongrui   I-313
Qin, Zhiguang   II-105
Qiu, Qibo   II-223
Qiu, Yongqin   I-505
Qu, Wen   I-227

Ramamohanarao, Kotagiri   I-537
Rangarajan, Sarathkumar   I-553
Ranu, Sayan   I-157
Ren, Yafeng   I-19
Renso, Chiara   II-238
Rodríguez-Cerezo, Daniel   II-43

Schouten, Kim   I-380, II-35
Serra, Flavia   II-436
Sewnarain, Ravi   I-380

Sha, Ying   I-109
Sheng, Quan Z.   I-313, II-19
Shi, Yuliang   I-352
Sierra, José-Luis   II-43
Silva, Thiago   II-349
Sinha, Atanu R.   II-340
Skogholt, Martin   I-380
Song, William Wei   II-411, II-424
Srinivas, Kavitha   II-298
Srinivasan, Balaji Vasan   I-190
Subramani, Sudha   I-553
Subramanian, Vaishnavi   II-340
Sun, Daniel   II-172
Sun, Dawei   I-405
Sun, Zhaoyu   I-561
Szabo, Claudia   II-19

Tang, Hao   I-405
Tang, Yan   I-489
Tang, Yi-Kun   I-525
Tao, Longquan   I-294
Taylor, Kerry   I-313
Theodoratos, Dimitri   I-199
Tian, Gang   I-561

Utpal   II-340

Valente, Pedro   II-349
van der Zaan, Tim   I-380
van Loenhout, Steffie   II-35
van Rooij, Gijs   I-380
van de Ven, Nikki   II-35
Vassiliou, Yannis   I-174
Vaziri, Mandana   II-298
Venkatesh, Svetha   II-137, II-146
Vidal, Vânia M.P.   II-196, II-238
Volonterio, Riccardo   I-140
Vrolijk, Lisanne   II-35

Wan, Yifang   I-50, II-85
Wang, Bin   I-50, I-505, II-85, II-187
Wang, Chengyu   I-473
Wang, Daling   I-227, I-513
Wang, Deyun   I-35
Wang, Dongjing   I-92
Wang, Hao   I-77, II-207
Wang, Hua   I-553, I-561
Wang, Lihong   I-109, I-505
Wang, Ruili   I-77

Wang, Shizhong   I-561
Wang, Tao   I-363
Wang, Xiaopeng   I-343
Wang, Xin   II-411
Wang, Yan   I-35
Wei, Jun   I-363
Wen, Xiao   I-447
Whittaker, Frank   I-553
Winckler, Marco   II-349
Wong, Raymond K.   II-172
Wu, Haijiang   I-363
Wu, Jian   II-223
Wu, Xiaoying   I-199
Wu, Zhiang   I-343

Xiao, Chunjing   II-105
Xie, Min   II-207
Xu, Fanjiang   II-207
Xu, Guandong   I-92, I-343
Xu, Ke   I-328
Xu, Shuai   I-3
Xu, Xiangzhen   I-352
Xue, Yuanyuan   II-121

Yang, Dun   I-343
Yang, Liu   I-396, I-447
Yang, Yi   I-211
Yao, Lina   I-313
Yao, Minghui   II-411
Ye, Dan   I-363
Ye, Feiyue   I-302
Ye, Yongjun   I-50, II-85
Yearwood, John   II-146
Yella, Siril   II-385, II-404
Yin, Hongzhi   II-207
Ying, Haochao   II-223
Ying, Shi   II-94
Yong, Hua-Hie   II-146
You, Zhenni   II-94
Yu, Feng   II-121
Yu, Ge   I-227
Yu, Haitao   I-278
Yu, Lei   II-324
Yu, Yonghong   I-77
Yuan, Fajie   I-278

Zablotskaia, Polina   II-247
Zhang, Anzhen   II-259
Zhang, Baochao   II-94

Zhang, Hao   I-35
Zhang, Haolin   I-302
Zhang, Lemei   II-3
Zhang, Liang   II-309
Zhang, Min   I-433
Zhang, Rong   I-473
Zhang, Shaozhong   II-424
Zhang, Sheng   II-411
Zhang, Wei Emma   I-313
Zhang, Wei   II-290
Zhang, Weinan   I-278
Zhang, Yanchun   I-553
Zhang, Yang   II-247
Zhang, Yifei   I-227, I-513
Zhang, Yihong   II-19
Zhao, Jichang   I-328
Zhao, Liang   II-121

Zhao, Weiji   I-513
Zhao, Yueai   I-553
Zhao, Yukun   II-70
Zheng, Meiguang   I-396, I-447
Zhong, Hua   I-363
Zhou, Aoying   I-473
Zhou, Guomin   I-473
Zhou, Meilin   I-50, II-85
Zhou, Rui   I-553
Zhou, Tao   I-3
Zhou, Tong   I-35
Zhou, Xiangmin   I-227
Zhou, Xiaofang   II-207
Zhou, Zhenkun   I-328
Zhu, Ziqing   I-3
Zhuo, Guangping   I-561
Zou, Zhaonian   II-259

Printed in the United States
By Bookmasters